Building Contract Claims

Building Contract Claims

Fourth Edition

David Chappell
Vincent Powell-Smith
& John Sims

Blackwell
Publishing

Editorial offices:
Blackwell Publishing Ltd, 9600 Garsington Road, Oxford OX4 2DQ, UK
  Tel: +44 (0)1865 776868
Blackwell Publishing Inc., 350 Main Street, Malden, MA 02148-5020, USA
  Tel: +1 781 388 8250
Blackwell Publishing Asia Pty Ltd, 550 Swanston Street, Carlton, Victoria 3053, Australia
  Tel: +61 (0)3 8359 1011

First Edition published by Granada Publishing 1983
Reprinted 1984
Reprinted with minor amendments by Collins Professional and Technical Books 1985
Second Edition published by BSP Professional Books 1988
Reprinted 1989, 1991
Third Edition published by Blackwell Science Ltd, 1998
Fourth Edition 2005

Library of Congress Cataloging-in-Publication Data
Chappell, David.
  Building contract claims / by David Chappell, Vincent Powell-Smith,
John Sims.–4th ed.
      p. cm.
  Rev. ed. of: Powell-Smith & Sims' building contract claims. 3rd ed. /
David Chappell. c1998.
  Includes bibliographical references and index.
  ISBN 1-4051-1763-X (hardback : alk. paper)
1. Construction contracts–Great Britain. 2. Breach of contract–Great Britain. 3. Actions and defenses–Great Britain. 4. Damages–Great Britain. I. Powell-Smith, Vincent. II. Sims, John, 1929-III. Chappell, David. Powell-Smith & Sims' building contract claims. IV. Title.

KD1641.P67 2004
343.41'078624–dc22   2004012941

ISBN 1-4051-1763-X

A catalogue record for this title is available from the British Library

Set in 10/12 pt Palatino
by Kolam Information Services Pvt. Ltd, Pondicherry, India
Printed and bound in Great Britain
by MPG Ltd, Bodmin, Cornwall

For further information on Blackwell Publishing, visit our website:
www.thatconstructionsite.com

Contents

Preface

It was with pleasure, flecked with trepidation, that I accepted the invitation to enlarge the scope and update the third edition of this popular book.

Much of the original text was retained in the third edition, but updated in the light of recent case law; the number of case references was almost doubled and, at the wish of the late Professor Powell-Smith, footnotes were included; a change which has proved very popular. Some restructuring took place within chapters to establish a comprehensible system of sub-headings and a few topics, such as liquidated damages (the subject of many disputes), global claims, causation and concurrency, were given greater importance. The scope of the book was extended to include more contracts.

With the fourth edition, the opportunity has been taken to carry out further fundamental changes to the structure of the book, bringing general principles to the beginning of the book and dealing with their application to specific contracts later. The text has also been substantially revised and almost a hundred additional cases added.

The latest contract editions current at the time of writing replace the previous (mainly 1980) editions and some new contracts have been added: the JCT Construction Management and Major Project contracts, the JCT Standard Form of Domestic Sub-Contract and the Engineering and Construction Contract. Not all contracts have been covered. A deliberate decision had to be made to omit some in favour of others, because of space constraints.

As before, the style has been to use the JCT Standard Form (JCT 98) as the basis and highlight important differences in the other forms. In some cases there are few similarities. The JCT are working on a greatly changed version of JCT 98 which is expected to be published in 2005. For some time to come, however, claims will continue to be made on the basis of the JCT 98 form. For the same reason, extensions of time and money claims under DOM/1 have been kept as Appendix B. JCT 63, which lurked in the appendices of the previous edition, has finally been removed. At the time of writing, the latest official amendments have been taken into account as follows:

Main contract forms:

JCT 98	Amendment 5
IFC 98	Amendment 5
MW 98	Amendment 5
WCD 98	Amendment 5
PCC 98	Amendment 5
MC 98	Amendment 5
TC/C 02	Amendment 1

MPF 03
GC/Works/1 (1998)
ACA 3   2003 Revision
NEC   2000 Amendment

Sub-contract forms:
NSC/C   Amendment 4
NAM/SC   Amendment 4
DSC/C   Amendment 1
DOM/1   Amendment 10
DOM/2   Amendment 10
WC/2   Amendment 5
ACA/SC   2003 Revision

Reference has also been made to the Housing Grants, Construction and Regeneration Act 1996 and the Arbitration Act 1996 where appropriate.

It should be noted that in the reproduction and commentary on the JCT standard forms, for brevity 'Architect' is used to stand for 'Architect/Contract Administrator'. Throughout the book, the masculine pronoun and its derivatives have been used for convenience; they should be taken to mean 'he' or 'she' etc. as appropriate.

Building contract claims are a unique combination of law and practice – the proper preparation and evaluation of building contract claims requires knowledge of building practice as much as of law. This book attempts to set out some of the principles which underlie the whole subject. Readers must not assume that standard contract forms are handbooks of rules sufficient to themselves; they are contracts like any other and they must be read and interpreted against the general background of law.

In my experience, many claims are ill-founded, often because the basic principles are misunderstood. Loss of money or lack of profit is not alone the basis of a claim, although it is usually the trigger. The contractor's entitlement is founded in the terms of the contract and in the general law to which it relates and if a particular contract term confers on a contractor the right to reimbursement of direct loss and/or expense, he is entitled to invoke the machinery laid down in the contract in order to obtain amounts properly due.

The book is addressed to all parties involved in construction. It is not always possible to give a definitive answer to every question, either because the courts have not considered the matter or because there have been apparently conflicting judgments. Where there is doubt, the doubt is expressed and, if practicable, I have taken a view of the situation.

One of the great perks of writing a preface is that it provides the opportunity to thank the people who have assisted me. The late Professor Vincent Powell-Smith LLB(Hons), LLM, DLitt, FCIArb, DSLP, MCL, FSIArb, and John Sims FRSA, FRICS, FCIArb, MAE, were responsible for the first and second editions and their names have been kept as co-authors in recognition of this although little of the original text now remains. Derek Marshall LLB(Hons), MRICS, MCIOB, MCIArb, was responsible for the redraft of

Appendix A for the third edition and it formed a solid foundation for the extensive revision I have undertaken for this edition. Michael Dunn BSc(Hons), LLB, LLM, FRICS, FCIArb, kindly read my first draft of Part 1 of this book and made a great many useful suggestions.

I have endeavoured to state the law from available sources at the end of May 2004.

David Chappell
Wakefield   May 2004

Acknowledgements

Permission to reproduce extracts from the following contracts is gratefully acknowledged:

Extracts from JCT 98, IFC 98, MW 98, WCD 98, PCC 98, MC 98, Works Contract/2, TC/C 02, MPF 03, NSC/C, DSC/C, NAM/SC are reproduced with the permission of RIBA Enterprises Limited, the copyright holders.

Extracts from GC/Works/1 are © Crown copyright material and are reproduced with the permission of the Controller of HMSO and Queen's Printer for Scotland, the copyright holders.

Extracts from the ACA Form of Building Agreement 2003 and the Sub-Contract Form 2003 are reproduced with the permission of the Association of Consultant Architects, the copyright holders.

Extracts from the Engineering and Construction Contract are reproduced with the permission of NEC, a division of Thomas Telford Limited, the copyright holders.

Extracts from DOM/1 are reproduced with the permission of the Construction Confederation, the copyright holders.

Contract abbreviations

ACA 3	Association of Consultant Architects Form of Building Agreement 2003.
CM 02	JCT Construction Management Contract 2002.
DOM/1 and DOM/2	Standard Form of Sub-Contract for Domestic Sub-Contractors.
DSC/C	Standard Form of Domestic Sub-Contract 2002.
FIDIC	Fédération Internationale des Ingénieurs-Conseils
GC/Works/1 (1998)	General Conditions of Government Contracts for Building and Civil Engineering Works 1998.
IFC 98	JCT Intermediate Form of Building Contract 1998.
IN/SC	IFC 98 Domestic Sub-Contract Conditions.
JCT 63	JCT Standard Form of Building contract 1963.
JCT 98	JCT Standard Form of Building Contract 1998.
MC 98	JCT Management Contract 1998.
MPF 03	JCT Major Projects Form of Contract 2003.
MW 98	JCT Agreement for Minor Building Works 1998.
NAM/SC	IFC 98 Named Sub-Contract Conditions 1998.
NEC	New Engineering and Construction Contract 2003 revisions.
NSC/C	JCT Standard Form of Nominated Sub-Contract Conditions 1998.
PCC 98	JCT Standard Form of Prime Cost Contract 1998.
SMM7	Standard Method of Measurement of Building Works, 7th edition.
TC/C 02	Trade Contract for use with the JCT Client and Construction Manager Agreement 2002.
WCD 98	JCT Standard Form of Contract With Contractor's Design 1998.
Works Contract/2	JCT Works Contract Conditions 1998.

PART I

Chapter 1
Introduction

1.1 Structure of the book

The book has been arranged in three parts:

Part I deals with general principles relating to time, liquidated damages and financial claims of various kinds.
Part II looks at the relevant clauses in JCT contracts.
Part III looks at the equivalent clauses in other standard contracts and in some standard sub-contracts.

1.2 Definitions

A dictionary definition of the word 'claim' is 'a demand for something as due'[1]. For the purposes of this book the term may be defined as the assertion of a right, usually by the contractor, to an extension of the contract period and/or to payment arising under the express or implied terms of a building contract. In the construction industry the word 'claim' is commonly used to describe any application by the contractor for payment which arises other than under the ordinary contract payment provisions. In this broad sense, a claim includes an application for an *ex gratia* payment, although this would not fall within the dictionary definition given above as it is not the assertion of a right and it also includes requests for additional payment as a result of variation instructions under the terms of the contract. The word is also used to describe a contractor's application for extension of time under a building contract. In one sense, claims for loss and expense may be considered to be regulated provisions for the payment of damages.

In fact, there are four types of claim that may be made by contractors against employers if the broad definition is accepted. They are: contractual claims, common law claims, *quantum meruit* claims and *ex gratia* claims. It should not be forgotten that an employer may make claims against a contractor for liquidated damages or for payment of a balance owing on the final certificate or after determination of the contractor's employment by the employer.

---

[1] *Concise Oxford Dictionary.*

1.2.1 Contractual claims

These are claims that arise out of the express provisions of the particular contract, e.g. for 'direct loss and/or expense' under certain clauses of the Joint Contracts Tribunal Limited (JCT) Standard Forms. They make use of the machinery in the contract to process the claim and produce a result. The principal reason for having such provisions in the contract is to avoid the necessity for the contractor to have to seek redress at common law and the inevitable expense involved for both parties in doing so. Most standard form contracts in any event preserve the contractor's right to seek damages at common law if he is not satisfied with his reimbursement under the contract.

1.2.2 Common law claims

Common law claims are sometimes and misleadingly called 'ex-contractual' or 'extra-contractual' claims. These terms should not be confused with the Latin term *ex contractu*, which is sometimes found in legal textbooks to refer to claims 'arising from the contract'.

Common law claims are claims for damages for breach of contract under common law and/or legally enforceable claims for breach of some other aspect of the law, e.g. in tort or for breach of copyright or under statute. Entitlement to such claims is expressly preserved to the contractor by the JCT Forms: see JCT 98, clause 26. It is also so preserved by most other standard forms, and a common law claim for breach may avoid some of the restrictions under the contract, regarding the giving of notices and so on, but it may be more restricted in scope than the matters for which a contractual claim can be made, some of which (for example, architect's instructions) are not breaches of contract.

1.2.3 *Quantum meruit* claims

A *quantum meruit* claim ('as much as he has earned') provides a remedy where no price has been agreed. There are four situations:

(1) Where work has been carried out under a contract, but no price has been agreed
(2) Where work has been carried out under a contract believed to be valid, but actually void
(3) Where there is an agreement to pay a reasonable sum
(4) Where work is carried out in response to a request by a party, but without a contract. This is usually termed a claim in quasi-contract or restitution. Work done following a letter of intent is a good example.

The type of claim and the method of valuation are two different things. It is useful to consider the method of valuation under two heads:

(1) Where there is a contract
(2) Where there is no contract.

Where there is a contract

A homely example is where a plumber has been called in by a householder to effect emergency repairs and no price has been agreed in advance. In that case, failing subsequent agreement as to price, the plumber would be entitled in law to 'reasonable' remuneration. A simple reported example of this kind of *quantum meruit* is *Powell* v. *Braun*[2] where an employer wrote to his secretary saying that he was pleased with her past services and that he wished her to undertake added responsibility in the future. Instead of a salary increase, he offered to pay her a yearly bonus on the net trading profits of the business. She accepted this offer, but the manner in which the bonus was to be assessed was never agreed. The Court of Appeal held that the employer was bound to pay her a reasonable sum each year.

An example of a successful *quantum meruit* claim in the construction industry is *Amantilla Ltd* v. *Telefusion PLC*[3], where the claimant carried out building and shop-fitting works for the defendant for an agreed lump sum price of £36,626. The claimant agreed to carry out extensive extra works on a daywork basis, but no price was actually agreed. The defendant made various interim payments amounting to £53,000 but the claimant's request for a further interim payment of £5,000 was not met. Subsequently, correspondence and meetings took place between the parties, during which the claimant submitted a detailed breakdown of the cost of the extra works. At no time did the defendant make any complaint about the claimant's work and, during a lengthy meeting between the parties the defendant's representative expressed satisfaction with it and confirmed that the defendant would shortly submit an offer of 'somewhere between' £10,000 and £132,000 to settle the account. After further correspondence, the defendant offered the sum of £2,000 in full and final settlement, having made a further interim payment. The final offer was rejected and the claimant was held entitled to recover on a *quantum meruit* basis.

An interesting but rather special case in the field of building contracts is *Sir Lindsay Parkinson & Co. Ltd* v. *Commissioners of Works & Public Buildings*[4]. The claimants contracted with the defendants to erect an ordnance factory at a cost of £3.5 million. Under the contract the defendants were entitled to require the claimants to perform additional work and at the date of the contract it was thought that this would cost a further £0.5 million. Later, the parties further agreed that, in order to complete the work by the date originally fixed, uneconomic methods should be employed which would add £1 million to the cost of the work, and that the claimants should be allowed a net profit of at least £150,000 but not more than £300,000 on the

---

[2] [1954] 1 All ER 484.
[3] (1987) 9 Con LR 139.
[4] [1950] 1 All ER 208.

actual cost of the works. It was clear from the evidence that at this time the total cost was estimated to be approximately £5 million. In fact, the total cost of the work exceeded £6.5 million.

The Court of Appeal held that the claimants were entitled to be paid a reasonable profit on the excess cost of £1.5 million in addition to the maximum profit of £300,000 previously agreed. This was because the upper and lower limits of profit were based on an estimated cost of the works made at the time of the further agreement. Because that estimate had been greatly exceeded, at the request of the employer, the contractors were entitled to be paid the additional profit. A reading of the full report of the case illustrates the difficulty of establishing such a claim and the facts were very special.

However, what is in effect a *quantum meruit* approach to the contractor's claim for damages for disruption and delay has been adopted in Canada in the case of *Penvidic Contracting Co Ltd* v. *International Nickel Co of Canada Ltd*[5], where the contractor was held entitled to claim the price he would have charged for the job had he foreseen the problems caused by the employer's breaches of contract at the time he tendered. There was evidence that the figure claimed was a reasonable estimate, and Spence J saw no objection to 'using the method suggested by the claimant ... rather than attempting to reach it by ascertaining items of expense from records which, by the very nature of the contract, had to be fragmentary and probably mere estimations'. In effect, the contractor there performed work which was substantially different from what he tendered for, and effectively the court adopted a 'broad brush' approach to the assessment of damages. The general principle is that damages are purely compensatory and, as will be seen, under JCT and allied forms, loss and expense claims can be equated to common law damages for breach of contract.

Where there is no contract

In *British Steel Corporation* v. *Cleveland Bridge & Engineering Co Ltd*[6], contractors were held entitled to be paid on a *quantum meruit* in respect of work done on a letter of intent, the High Court finding that no contract had come into existence between the parties, although the claimants had done work to the value of some £230,000 on the basis of a letter of intent from the defendants.

The exact meaning of *quantum meruit* in practical terms can be a difficult question. It seems that in the absence of any other indicator, it must be a fair commercial rate[7]. Moreover, it can be valued by reference to any profit on the work made by the other party and to any competitive edge which the provider of the service enjoys – for example, already being on site and,

---

[5] (1975) 53 DLR 748.
[6] (1981) 24 BLR 94.
[7] *Laserbore* v. *Morrison Biggs Wall* (1993) CILL 896.

therefore, avoiding the need for mobilisation costs[8]. Valuable guidance was given on the basis of *quantum meruit* in *Serck Controls Ltd* v. *Drake & Scull Engineering Ltd*[9] where Drake & Scull had given a letter of intent to Serck instructing them to carry out work on a control system for BNFL. Part of the letter said:

> 'In the event that we are unable to agree satisfactory terms and conditions in respect of the overall package, we would undertake to reimburse you with all reasonable costs involved, provided that any failure/default can reasonably be construed as being on our part.'

The way in which the *quantum meruit* was to be calculated was the basis of the trial. Several points of interest were considered.

Judge Hicks had to consider whether, by 'reasonable sum', was meant the value to Drake & Scull of Serck's reasonable costs in carrying out the work. In his view the term *quantum meruit* covered the whole spectrum from one to the other of these positions. Reference to 'reasonable sums incurred' entitled Serck to reasonable remuneration. 'Costs' implied the exclusion of profit and, possibly, overheads, but the judge did not believe that they were excluded in this instance.

What, if any, relevance was to be placed on the tender? Because the tender did not form part of any contract, its use was limited. It could not be the starting point for the calculation of the reasonable sum. Probably its only use was a check on whether the total amount arrived at by other means was surprising.

So far as site conditions were concerned, if the criterion was the value to Drake & Scull, site conditions in carrying out the work would be irrelevant. If the starting point had been an agreed price, the only relevant points would have been any changes to the basis of the price. On the basis of a reasonable remuneration, the conditions under which the work was actually undertaken were relevant: if the work proved to be more difficult than expected, Serck were entitled to be recompensed.

The conduct of the two parties was considered, particularly allegations that Serck had worked inefficiently and what effect that had on the calculation of *quantum meruit*. It was held that if the value was to be worked out on a 'costs plus' basis, deductions should be made for time spent in repairing or repeating defective work, and for inefficient working. If the value was to be worked out by reference to quantities, the claimant gains nothing from such deficiencies and, if attributable to the claimant or his sub-contractors, they are irrelevant to the basic valuation; extra time and expense enter into the picture at this stage only if relied upon by the claimant as arising without fault on his part. Defects remaining at completion should give rise to a deduction, whatever method of valuation was chosen.

---

[8] *Costain Civil Engineering Ltd and Tarmac Construction Ltd* v. *Zanin Dredging & Contracting Company Ltd* (1996) CILL 1220
[9] (2000) CILL 1643.

In another interesting case[10] a letter of intent was sent to the contractor. It was somewhat unusual in nature. A letter of intent is usually an assurance by one party to the other which, if acted upon, will have limited contractual effect such that reasonable expenditure will be reimbursed. Usually, either party is free to stop work at any time. In this instance the letter imposed substantial and detailed obligations on both parties. The contractor had the option whether or not to start work but, once started, he had to continue. The letter envisaged that both parties would continue to negotiate about the form of contract. In the event, no form of contract was finally agreed. The judge referred to the letter of intent as a 'provisional contract' which was intended to be superseded. He concluded that the reasonable remuneration should be that which would be payable under the building contract once entered into, the rates to be derived from the bills of quantities and any extra remuneration to be derived from the terms of the intended JCT contract.

1.2.4 *Ex gratia* claims

An *ex gratia* ('out of kindness') claim is one which the employer is under no legal obligation to meet. It is sometimes called a 'sympathetic' or 'hardship' claim. *Ex gratia* claims are often put forward by contractors but are seldom met unless some benefit may accrue to the employer as a result. For example, an employer might agree to make an *ex gratia* payment to save a contractor from insolvency where the cost of employing another contractor to complete the work would be more than the amount of the *ex gratia* payment.

1.3 *The basis of claims*

'Claim' is often seen as a dirty word in the employer section of the industry. It is easy to understand why this should be so, because it so often results in original budgets being exceeded. There are only two sorts of claim: justified and unjustified. A justified claim is one properly made under the terms of the contract or under common law. An unjustified claim is one which does not comply with the terms of the contract or which does not satisfy the criteria for a common law claim.

There is nothing wrong with a justified claim since most standard form contracts specifically entitle the contractor to apply for reimbursement of direct loss and/or expense which he incurs as a result of certain matters specified in the contract, all of which are within the direct control of the employer or of those for whom he must bear the responsibility in law. On the other hand, unjustified claims, or those that are engineered at the outset of the project or even, on occasion, during the tendering process, can cause a great deal of trouble in the industry. They give rise to the common and

---

[10] *Hall & Tawse South Ltd v. Ivory Gate Ltd* (1997) CILL 1376.

unfortunately not always misconceived view that some contractors embark on a contract with the intention of creating conflict. It is probably not too strong to categorise such claims as fraudulent and the construction industry is perhaps the only one where such practices would be tolerated and treated as the norm. This book is not concerned with that kind of spurious claim.

Undoubtedly there are situations in which the employer will find himself paying substantial sums because of circumstances which are largely if not entirely beyond his control – or, indeed, that of his architect or other advisers; for example, major redesign of foundations resulting from unexpected ground conditions that normal surveys could not have revealed, which are subject to implied terms such as 'that the ground conditions would accord with the hypotheses upon which' the contractors were instructed to work, and in fact the ground conditions were different[11]. Indeed, the JCT and other standard forms of building contract and subcontract proceed on the basis that claims are likely to be made as the contract progresses. The provisions of the contracts, together with any bespoke amendments, will determine the allocation of risk between the parties in such instances.

In fact, contract claims for both time and money are a feature of any construction project. Claims are very simple to generate, but are not always easy to substantiate, and therein lies the employer's protection. He is only bound to meet claims that are based on some express or implied provision of the contract or rule of law and it is for the contractor to prove his claim. Where the claim is brought within the contract procedure, the contractor must also show that he has followed the administrative machinery provided in the contract itself. Above all, contract claims must be founded on facts and these facts must be substantiated by the contractor. Merely because a contractor is losing money on a particular contract does not mean that he is entitled to look to the employer for reimbursement. He must be able to establish that the loss results directly from some act or default of the employer or those for whom he is responsible in law, or else is referable to some express term of the contract entitling him to reimbursement.

Contractual claims for time or for money as defined must originate from a particular clause in the contract. The JCT Standard Forms contain provisions enabling the architect to extend the time for completion of the contract (JCT 98, clause 25; IFC 98, clauses 2.3–2.8). However, an extension of the contract time does not, of itself, give the contractor any right to make a money claim. Conversely, it is not necessary for the contract time to be extended in order for the contractor to have a right to make a money claim. There is considerable confusion in the construction industry on this point.

This confusion no doubt arises because under JCT forms all of the grounds on which the contractor can make a money claim are also grounds that may entitle him to an extension of time. But it is not the granting of an extension of time that gives rise to the entitlement to money but the fact of delay and/or disruption of regular working. There is no necessary link

---

[11] *Bacal Construction (Midlands) Ltd* v. *Northampton Development Corporation* (1976) 8 BLR 88.

between the grant or refusal of an extension of time and the success of a contractor's application for direct loss and/or expense[12].

Generally speaking, unexpected difficulty or expense in completing the contract does not entitle the contractor to refuse to carry out the work or to claim additional payment. This statement is well illustrated by *Davis Contractors Ltd* v. *Fareham Urban District Council*[13].

The claimants agreed with the Fareham Council to build for them 78 council houses within a period of 8 months for a firm price. There was a shortage of skilled labour, and the work of erecting the houses took 22 months to complete as a result. The contractors claimed that by reason of the scarcity of labour the contract had been frustrated and that they were entitled to recover a sum in excess of the contract price on the basis of a *quantum meruit*.

The House of Lords found against the contractors. An unexpected turn of events that renders the performance of a contract more onerous than the parties contemplated does not release the party adversely affected from the contract. Here, the scarcity of labour, which was not due to the fault of either party, was not sufficient to justify a finding that the contract had been brought to an end when the expectations of the parties were not realised. Accordingly, the contractors had not been released from the terms of the contract as regards price and could not maintain a claim for payment on a *quantum meruit*.

In the Court of Appeal – which also ruled against the contractors – Denning LJ, as he then was, put the matter in a nutshell when he said:

> 'We could seriously damage the sanctity of contracts if we allowed a builder to charge more, simply because, without anyone's fault, the work took him much longer than he thought.'[14]

The *Davis* case may be contrasted with the *Parkinson* case quoted earlier in this chapter.

1.4 Architect's and contract administrator's powers and liability to the contractor

Most of the comments in this section will apply whether the contract administrator is an architect or some other construction professional. However, to avoid undue complication, reference is made only to the architect. Under JCT contracts the architect's powers are limited. For example, JCT 98, clause 4.1.1, obliges the contractor to comply only with instructions 'expressly empowered by the Conditions', and a later sub-clause provides a method whereby the contractor may challenge the architect's authority to issue a particular instruction. In a claims situation, the architect can certify

---

[12] *H. Fairweather & Co Ltd* v. *London Borough of Wandsworth* (1987) 39 BLR 106. See in particular p. 120.
[13] [1956] 2 All ER 148 HL.
[14] *Davis Contractors Ltd* v. *Fareham Urban District Council* [1955] 1 All ER 275 at 278 CA.

for payment only sums which the express terms of the contract authorise him to so certify. Under the JCT Standard Forms the architect has no power to certify amounts in respect of common law, *quantum meruit* or *ex gratia* claims. A contract may, of course, endow the architect with authority to do so or the employer may authorise the architect to act for him in respect of such claims.

Architects sometimes assume, incorrectly, that they enjoy inherent powers to act as the employer's agents in all respects. The same mistaken assumption is sometimes made by contractors who consequently are disappointed when the architect correctly refuses to certify payment for common law claims. Under the JCT Forms of Contract the powers of the architect as agent of the employer and to certify sums for payment are closely defined, and the architect may himself be at risk if he exceeds the powers so conferred upon him. The architect's position has been succinctly summed up thus:

> 'The occasions when an architect's discretion comes into play are few, even if they number more than the one which gives him a discretion to include in an interim certificate the value of any materials or goods before delivery on site ... The exercise of that discretion is so circumscribed by the terms of that provision of the contract as to emasculate the element of discretion virtually to the point of extinction.'[15]

It should be remarked that the judge was referring to the JCT 63 form and that the discretion which the architect then had with regard to certification of materials off-site has since been removed.

Under the claims provisions of both JCT 98 and IFC 98 it seems that there must be implied a duty on the architect or quantity surveyor to carry out the ascertainment of direct loss and/or expense within a reasonable time of receiving reasonably sufficient information from the contractor[16]. What is a reasonable time will vary with the circumstances of each case. Indeed, if an architect or quantity surveyor delayed unreasonably in the ascertainment he might be liable personally, either directly or through the employer, to the contractor in damages. This proposition receives some support from an *obiter* observation of Parker J in *F. G. Minter Ltd* v. *Welsh Health Technical Services Organisation* at first instance, where he said: '[If] the period was unreasonable the chain of causation would be completely broken. This might give rise to a claim against the architect...'[17].

Although the first instance judgment of Parker J was reversed in part by the Court of Appeal, it appears that the learned judge's intimation of a possible personal liability of the architect or quantity surveyor is a correct statement of the law. That passage of the judgment of Stephenson LJ

---

[15] *Partington & Son (Builders) Ltd* v. *Tameside Metropolitan Borough Council* (1985) 5 Con LR 99 at 108 per Judge Davies.
[16] *Croudace Ltd* v. *London Borough of Lambeth* (1986) 6 Con LR 70.
[17] (1979) 11 BLR 1 at 13 per Parker J partially reversed by the Court of Appeal, but not on this point (1980) 13 BLR 7.

touching upon this point in the Court of Appeal is far from clear. His lordship disagreed with Parker J as to the breaking of the chain of causation but did not, seemingly, express disagreement with Parker J as to the possible personal liability of architect and quantity surveyor.

In *Michael Salliss & Co Ltd* v. *ECA Calil*[18], the contractor sued Mr and Mrs Calil and the architects, W. F. Newman & Associates. It was claimed that the architects owed a duty of care to the contractor. The claim fell into two categories:

- Failure to provide the contractors with accurate and workable drawings
- Failure to grant an adequate extension of time and undercertification of work done.

The court held that the architect had no duty of care to the contractors in respect of surveys, specifications or ordering of variations, but he did owe a duty of care in certification. It was held to be self-evident that the architect owed a duty to the contractor not to negligently undercertify.

> 'If the architect unfairly promotes the building employer's interest by low certification or merely fails properly to exercise reasonable care and skill in his certification it is reasonable that the contractor should not only have the right as against the owner to have the certificate revised in arbitration but also should have the right to recover damages against the unfair architect.'[19]

In arriving at that conclusion, the court was following the rules laid down by many courts. In *Campbell* v. *Edwards*[20], the Court of Appeal said that the law had been transformed since the decisions of the House of Lords in *Sutcliffe* v. *Thackrah*[21] and *Arenson* v. *Arenson*[22] because contractors now had a cause of action in negligence against certifiers and valuers. Before these cases, certifiers had been protected because the Court of Appeal in *Chambers* v. *Goldthorpe*[23] had held that certifiers were quasi-arbitrators. The House of Lords overruled that in 1974. Until the *Pacific Associates* case, at no time in the history of English law had it been doubted that architects owed a duty to contractors in certifying. After all, there was no need even to invent the doctrine of quasi-arbitrators if there was no liability for negligence. In the *Arenson* case, in reference to the possibility of the architect negligently undercertifying, it was said:

> 'In a trade where cash flow is perceived as important, this might have caused the contractor serious damage for which the architect could have been successfully sued.[24]

---

[18] (1987) 4 Const LJ 125.
[19] (1987) 4 Const LJ 125 at 130 per Judge Fox-Andrews.
[20] [1976] 1 All ER 785.
[21] [1974] 1 All ER 859.
[22] [1975] 3 All ER 901.
[23] [1901] 1 KB 624.
[24] [1975] 3 All ER 901 at 924 per Lord Salmon.

The case of *Pacific Associates* v. *Baxter*[25] appeared to throw doubt on this position. Halcrow International Partnership were the engineers for work in Dubai for which Pacific Associates were in substance the contractors under a FIDIC contract. During the course of the work, the contractors claimed that they had encountered unexpectedly hard materials and that they were entitled to extra payment of some £31 million. Halcrow refused to certify the amount and in due course Pacific Associates sued them for the £31 million plus interest and another item. It was claimed that Halcrow acted negligently in breach of their duty to act fairly and impartially in administering the contract. At first instance, the court struck out the claim, holding that Pacific Associates had no cause of action. The court noted that:

- There was provision for arbitration between employer and contractor; and
- There was a special exclusion of liability clause in the contract (clause 86) to which, of course, the engineers were not a party, whereby the employers were not to hold the engineers personally liable for acts or obligations under the contract, or answerable for any default or omission on the part of the employer.

The question of whether a duty of care exists does not depend on the existence of an exclusion of liability clause, except to the extent that the existence of such a clause suggests acceptance by the engineer that there is a duty of care which, without such a clause, would give rise to such liability. Whether such a clause would be deemed reasonable under the provisions of the Unfair Contract Terms Act 1977 is doubtful[26]. Surprisingly, it was held that the inclusion of an arbitration clause in the contract, General Condition 67, excluded any liability by the engineer to the contractor. Why that should be so is anything but clear. The fact that the employer and the contractor choose to settle any disputes by arbitration rather than litigation does not excuse the engineer from his clear duty to both parties. However, it seems that these two points were decisive in the decision. Moreover, it was upheld by the Court of Appeal. The decision can be criticised on three major points:

(1) In *Lubenham Fidelities* v. *South Pembrokeshire District Council*[27] the Court of Appeal expressly affirmed the principle that the architect owed a duty to the contractor in certifying. The architects in that case were not held liable, because the chain of causation was broken and the contractor's damage was held to be caused by their own breach in wrongfully withdrawing from site. But the court said:

'We have reached this conclusion with some reluctance, because the negligence of Wigley Fox [the architects] was undoubtedly the source from which this unfortunate sequence of events began to flow, but

---

[25] (1988) 44 BLR 33.
[26] *Smith* v. *Eric S. Bush* [1989] 2 All ER 514.
[27] (1986) 6 Con LR 85.

their negligence was overtaken and in our view overwhelmed by the serious breach of contract by Lubbenham.'[28]

It expressly approved the first instance judgment saying:

> 'Since Wigley Fox were the architects appointed under the contracts, *they owed a duty to Lubbenham as well as to the Council to exercise reasonable care in issuing certificates and in administering the contracts correctly.* By issuing defective certificates and in advising the Council as they did, Wigley Fox acted in breach of their duty to Lubenham.'[29] (emphasis added)

(2) The Court of Appeal is bound by its own previous decisions. This decision seemed to be contrary to all the previous cases, including those of the House of Lords by which it was bound, going back for more than a century together with well-established law that had been followed in all common law jurisdictions such as Hong Kong and Australia[30].

(3) It apparently ignored or at any rate failed to consider the fundamental principle that (at that time) parties could not be bound by a term in a contract to which they were not a party and had not consented.

Recent cases[31] provide firm support to the idea that the reliance principle established in *Hedley Byrne & Co Ltd* v. *Heller & Partners Ltd* (1964) is capable of extension to accommodate actions as well as advice given by the architect. In *J. Jarvis & Sons Ltd* v. *Castle Wharf Developments and Others*[32] the Court of Appeal held that a professional who induces a contractor to tender in reliance on the professional's negligent misstatements could become liable to the contractor if it could be demonstrated that the contractor relied on the misstatement[33].

1.5 *Quantity surveyor's powers*

The only major function ascribed solely to the quantity surveyor under the JCT Forms of Contract is the valuation of variations, including any necessary measurement for this purpose (clause 13, JCT 98; clause 3.7, IFC 98)[34]. In JCT 98 the quantity surveyor is also expressly charged with the production of what is called (in clause 30.6.1.2) 'a statement of all adjust-

---

[28] (1986) 6 Con LR 85 at 111 per May LJ.

[29] (1986) 6 Con LR 85 at 101 per May LJ.

[30] See, for example: *Ludbrooke* v. *Barrett* (1877) 46 LJCP 798; *Stevenson* v. *Watson* (1879) 48 LJCP 318; *Demers* v. *Dufresne* [1979] SCR 146; *Trident Construction* v. *Wardrop* (1979) 6WWR 481; *Yuen Kun Yen* v. *Attorney General of Hong Kong* [1988] AC 175; *Edgeworth Construction Ltd* v. *F. Lea & Associates* [1993] 3 SCR 206.

[31] *Henderson* v. *Merritt Syndicates* [1995] 2 AC 145, (1994) 69 BLR 26; *White* v. *Jones* [1995] 1 All ER 691; *Conway* v. *Crow Kelsey & Partner* (1994) CILL 927.

[32] [2001] 1 Lloyds Rep 308.

[33] There is a very perceptive article by John Cartwright (Liability in negligence: new directions or old) in *Construction Law Journal* (1997) vol. 13, p. 157.

[34] See Chapter 13.

ments to be made to the Contract Sum' – in other words the final variation account.

The limited nature of the quantity surveyor's powers under JCT forms has been clearly stated:

> 'His authority and function under the contract are confined to measuring and quantifying. The contract gives him authority, at least in certain instances, to decide quantum. It does not in any instance give him authority to determine any liability, or liability to make any payment or allowance.'[35]

The position appears to be the same under JCT 98 so far as the quantity surveyor's powers are concerned. The terms of the contract, express and implied, give the quantity surveyor no independent authority.

Under the JCT forms the ascertainment of the amount of 'direct loss and/ or expense' incurred by and reimbursable to the contractor is primarily the responsibility of the architect. The quantity surveyor will only carry out that function if expressly so instructed by the architect. In practice, on the vast majority of contracts, although the architect retains the responsibility for deciding whether a claim is valid, it is the quantity surveyor who ascertains the amount reimbursable to the contractor, since his training and experience best fit him for the task. Because any 'claim' put forward by the contractor as the basis of ascertainment will almost invariably have been produced by quantity surveyors in the contractor's own office, any financial discussions will obviously best be conducted from the employer's side by a member of the same profession, speaking the same language[36].

Like that of the architect, the quantity surveyor's position is a strange and sometimes difficult one. He will have been retained directly by the employer (though sometimes through the architect) and, particularly where the employer is a local authority, may even be a member of the employer's own staff. His primary and contractual duty is therefore to the employer. Certainly, in all pre-contract functions, such as the preparation of cost estimates, cost planning, the preparation of the bills of quantities, and the arithmetical and technical checking of the priced bills submitted by the lowest tenderer, etc., his duty is wholly and exclusively to the employer – while, of course, always maintaining proper professional standards of integrity.

Once a contractor is appointed and the contract is let, however, the quantity surveyor, like the architect, assumes a dual function and a dual responsibility. His contractual relationship, whether under a consultancy agreement or under a direct contract of employment, is still solely with the employer; but one of the duties he has under that contract is to carry out the functions ascribed to him under the building contract in accordance with its terms. Therefore, the proper carrying out of those functions in strict

---

[35] *County & District Properties Ltd* v. *John Laing Construction Ltd* (1982) 23 BLR 1 at 14 per Webster J, where this question arose under a contract in JCT 63 form.
[36] See also Chapter 12, section 12.2.4, for a consideration of the relative positions of architect and quantity surveyor where the latter carries out the ascertainment.

compliance with the building contract terms in itself becomes an important part of his contractual duty to the employer. But he will also now have a duty to the employer to act fairly to the parties.

In part that duty arises through the insertion of his name in the appropriate space in the articles of agreement to the building contract where, in effect, employer and contractor together agree that the person so named shall carry out the duties assigned to the quantity surveyor under the contract. If the quantity surveyor fails to carry out those functions in accordance with those terms, the employer may be liable to the contractor for that failure as a breach of a contractual undertaking – he will have failed to procure his quantity surveyor to carry out those functions properly. The quantity surveyor is *not* a party to the contract, no more than is the architect.

In addition to these contractual duties, the quantity surveyor may also owe a duty of care, in the exercise of his professional skills, to others in the building process. Usually this will be where it can be shown that a party relied on the quantity surveyor to exercise reasonable care and skill[37] – that is, not only the main contractor, but also nominated sub-contractors and anyone who may suffer damage as a direct result of the quantity surveyor's breach of duty. For instance, where it is a part of the quantity surveyor's duties to value work executed for the purpose of interim payment, a contractor who suffers damage through negligent undervaluation may be entitled to take legal action against the quantity surveyor for negligent misstatement in a similar way to an employer damaged by negligent overvaluation would be entitled to take action in contract and/or in tort. Action against the quantity surveyor by anyone other than his client is virtually unknown at present, but recent developments in the law point to the possibility of actions of this kind[38]. The remarks in section 1.4 earlier are relevant.

The quantity surveyor's duty is to carry out the functions ascribed to him under the building contract in strict accordance with the terms of that contract. Under clause 26 of JCT 98, for instance, it is the quantity surveyor's duty, if so instructed by the architect, to 'ascertain' the amount of direct loss and/or expense which has been or is being incurred by the contractor as a result of the regular progress of the works or any part of them having been 'materially affected' by one or more of the factors listed in clause 26.2 – the architect having already formed the opinion that regular progress of the works has been or is likely to be so affected. It is therefore his duty to find out the actual amount of loss and/or expense incurred by the contractor as a direct result of the effect upon regular progress. It is certainly his duty to ensure that the employer pays no more than the actual amount of loss and/or expense directly and properly incurred by the contractor; but it is equally his duty to ensure that the contractor recovers no less. It is most emphatically *not* his duty to deprive the contractor of amounts properly recoverable under the contract, nor can it ever be interpreted as part of his contractual or

---

[37] *Hedley Byrne & Co v. Heller & Partners Ltd* [1964] AC 465.
[38] See John Cartwright, *Liability in negligence: new directions or old,* (1997) 13 Const LJ 157.

professional duty to the employer so to do. The quantity surveyor's duty is not to *defend* or to *break* claims, but rather to establish on the architect's instructions and in strict accordance with the contract, the amount payable to the contractor.

In the vast majority of cases, large elements of the ascertainment will in fact involve discussion – the contractor putting forward his view of his entitlement for examination by the quantity surveyor appointed under the contract. In practice, most claims are ultimately settled by agreement. The quantity surveyor, of course, is not normally empowered to 'do a deal' and where some sort of 'broad brush' settlement is clearly to the benefit of the parties, the architect and the quantity surveyor must place the options in front of the employer and obtain instructions.

Chapter 2
Time

2.1 Time of the essence

A term, the breach of which by one party gives the other party a right to treat it as repudiatory, is sometimes said to be of the essence of the contract. At one time the common law took a strict view with the result that a contract had to be performed on the date stated, the only relief being obtained through the Court of Equity. However, for three quarters of a century, time has not been automatically considered as of the essence of a contract unless Equity would have so considered it prior to 1875[39]. There are probably only three instances where time will be of the essence:

- If the contract expressly so stipulates
- If it is a necessary implication of the contract and its surrounding circumstances
- If a party unreasonably delays his performance, time may be made 'of the essence' if the other party serves a notice on the party in breach setting a new and reasonable date for completion.

It must be a term so fundamental that its breach would render the contract valueless, or nearly so, to the other party. It is noteworthy that where a term is not originally of the essence it may be made of the essence by one party giving the other a written notice to that effect[40]. In that case, failure to comply with the notice would be evidence of a repudiatory breach rather than a repudiatory breach itself. This may be of limited use in cases where a contractor consistently fails to meet time targets for reasons which do not entitle him to an extension of time under the contract provisions. However, in the case of most standard form building contracts, the provisions for determination (e.g. for failure to proceed regularly and diligently) adequately cover the situation.

There is authority that time will not normally be of the essence in building contracts unless expressly stated to be so. This is because the contract makes express provision for the situation if the contract period is exceeded in the shape of an extension of time clause and liquidated damages[41]. In that context, making time of the essence would be contradictory and little or no practical benefit to the employer although it was done in *Peak Construction (Liverpool) Ltd* v. *McKinney Foundations Ltd* and apparently gave the

---

[39] Law of Property Act 1925 s. 41.
[40] *Behzadi* v. *Shaftsbury Hotels Ltd* [1992] Ch 1.
[41] *Lamprell* v. *Billericay Union* (1849) 18 LJ Ex 282; *Babacomp Ltd* v. *Rightside* [1974] 1 All ER 142.

employer the right to 'determine the contract at the end of' the period as extended by the architect[42].

2.2 Time at large

In the absence of any agreed contractual mechanism for fixing a new date for completion, no such new date can be fixed and the contractor's duty then will be to complete the works within a reasonable time[43]. Provided a contractor has not acted unreasonably or negligently, he will complete within a reasonable time despite a protracted delay if the delay is due to causes outside his control[44]. In such circumstances time is said to be 'at large'. The question of time being 'at large' and the relationship between the extension of time clause and liquidated damages provisions in JCT contracts has been stated thus:

> '1. The general rule is that the main contractor is bound to complete the work by the date for completion stated in the contract. If he fails to do so, he will be liable for liquidated damages to the employer.
> 2. That is subject to the exception that the employer is not entitled to liquidated damages if by his acts or omissions he has prevented the main contractor from completing his work by the completion date – see for example *Holme* v. *Guppy* (1838) 2 M & W 387, and *Wells* v. *Army & Navy Co-operative Society* (1902) 86 LT 764.
> 3. These general rules may be amended by the express terms of the contract.
> 4. In this case [which involved a contract in terms identical to JCT 63] the express terms of clause 23 of the contract do affect the general rule . . .'[45]

In practice, very few building contracts are without a clause enabling the employer or his agent to fix a new completion date after the employer has caused delay to the contractor's progress. All standard forms have clauses permitting the extension of time although not all of the terms are entirely satisfactory. Even where a building contract contains terms providing for extension of the contract period, time may yet become at large, either because the terms do not properly provide for the delaying event or because the architect has not correctly operated the terms.

The JCT series of contracts (other than MW 98) favour a list of events giving grounds for extension of time. Because the architect's power to give an extension of time is circumscribed by the listed events, there is a danger that the employer may delay the works in a way which does not fall under one of the events. In such a case, time would be at large. For example, the

---

[42] 1 BLR 114 at 120 per Salmon LJ.
[43] *Wells* v. *Army & Navy Co-operative Stores* (1902) 2 HBC 4th Edition (vol. 2) 346.
[44] *Pantland Hick* v. *Raymond & Reid* [1893] AC 22.
[45] *Percy Bilton* v. *Greater London Council* (1982) 20 BLR 1 at 13 per Lord Fraser of Tullybelton, delivering the unanimous decision of the House of Lords.

1980 edition of the JCT Standard Form did not include power for the architect to extend time for the employer's failure to give the contractor possession of the site on the due date. An employer's failure in this respect resulted in time becoming at large and the contractor's obligations were to complete the works within a reasonable time. This was despite it being acknowledged by the court that the contractor had himself subsequently contributed to the delay[46]. By Amendment 4 in 2002, the JCT further improved JCT 98 by adding to the relevant events one which allowed an extension of time for any act or default of the employer – virtually a catch-all category. It has been held that the architect has the power to give an extension of time if the employer causes further delay when the contractor is already in delay through his own fault, i.e. in culpable delay[47].

Where the extension of time clauses are properly drafted, but the architect operates them incorrectly, time may become at large depending on all the circumstances. An example of this would be if the architect was late in delivering necessary drawing information to the contractor, but failed to give any extension of time. This is a clear case of the architect not taking advantage of the available mechanism. Another example is where the contract provision sets out a timetable within which the architect must operate to give an extension of time. If he fails to observe the timetable, his power to give an extension will end and time will become at large. It is often said that such time periods are not mandatory, but simply directory on the authority of the Court of Appeal in *Temloc Ltd* v. *Errill Properties Ltd*[48]. This appears to be an incorrect reading of the decision. The court in *Temloc*, in making that observation, were interpreting the provisions *contra proferentem* the employer who sought to rely upon them. The employer had stipulated '£nil' as the figure for liquidated damages and the Court of Appeal held that this meant that the parties had agreed that if the contractor finished late, no liquidated damages would be recoverable by the employer. The court went on to hold that the employer could not opt to claim unliquidated damages. The contract provided that after practical completion the architect must, within 12 weeks, confirm the existing date for completion or fix a new date. The architect exceeded the 12 weeks and the employer contended that the liquidated damages clause could be triggered only if the architect carried out his duty at the right time. Therefore, the employer could claim unliquidated damages for breach of an implied term. It was in this context that the court, in a view which is probably *obiter* in any event, suggested that the time period was not mandatory. They gave no real reasons, but a clue when it was said:

---

[46] *Rapid Building Group Ltd* v. *Ealing Family Housing Association Ltd* (1984) 1 Con LR 1.
[47] *Balfour Beatty* v. *Chestermount Properties Ltd* (1993) 62 BLR 1, where the judge held that the architect had such power under the slightly amended form of JCT 80 under consideration. The decision has been referred to with approval in *Henry Boot Construction (UK) Ltd* v. *Malmaison Hotel (Manchester) Ltd* (1999) 70 Con LR 32 and *Royal Brompton Hospital NHS Trust* v. *Hammond and Others (No 7)* (2001) 76 Con LR 148.
[48] (1987) 38 BLR 30.

'The whole right of recovery of liquidated damages under clause 24 does not depend on whether the architect, *over whom the contractor has no control*, has given his certificate by the stipulated day.'[49] (emphasis added)

It seems that the court recognised that the architect is the employer's agent. Had the employer's argument succeeded, it would have been contrary to the established principle that a party to a contract cannot take advantage of his own breach[50]. The 12-week review period has been confirmed in a recent case:

'The process of considering and granting extensions of time is to be completed not later than 12 weeks after the date of practical completion and the architect must, within that timescale, either finally fix the completion date or notify the contractor that no further extensions of time are to be granted.'[51]

2.3   *Extension of time clauses in contracts*

If the parties intend that liquidated damages are to be payable if the contractor fails to complete the works, a date for completion must be stipulated in the contract. That is because there must be a definite date from which to calculate liquidated damages[52]. There is an implied term in every contract that the employer will do all that is reasonably necessary to co-operate with the contractor[53] and that he will not prevent him from performing[54]. In the context of a building contract, the employer's co-operation probably extends to little more than that he should ensure that the contractor has all necessary drawings and instructions at the right time and adequate access to the site to enable him to carry out the work. In this respect, the employer also has a duty to ensure that any architect appointed by him carries out his duties properly although the duty does not arise until the employer becomes aware that the architect is not performing properly and that there is a need to remind him of his duties[55].

Alongside the implied term of co-operation, there must be in every contract an implied term that neither party will do anything to hinder or delay performance by the other[56]. Such a term was upheld as generally applicable to building contracts in *London Borough of Merton* v. *Stanley Hugh*

---

[49] (1987) 39 BLR 30 at 39 per Nourse LJ.
[50] *Alghussein Establishment* v. *Eton College* [1988] 1 WLR 587 HL.
[51] *Cantrell and Another* v. *Wright & Fuller Ltd* (2003) 91 Con LR 97 at 147 per Judge Thornton.
[52] *Miller* v. *London County Council* (1934) 50 TLR 479.
[53] *Luxor (Eastbourne) Ltd* v. *Cooper* [1941] 1 All ER 33.
[54] *Cory Ltd* v. *City of London Corporation* [1951] 2 All ER 85.
[55] *Perini Corporation* v. *Commonwealth of Australia* (1969) 12 BLR 82; *Penwith District Council* v. *V. P. Developments Ltd*, 21 May 1999, unreported; *Hong Kong Development Co (Pte) Ltd* v. *Hiap Hong & Co Pte Ltd* (2000) CILL 1787.
[56] *Barque Quilpue Ltd* v. *Brown* [1904] 2 KB 261.

Leach Ltd[57]. If the employer does hinder the contractor, he can no longer insist that the contractor finishes his work by the contractual date for completion. This principle has the weight of judicial authority behind it. In *Holme* v. *Guppy* it was said:

> '...and there are clear authorities, that if the party be prevented, by the refusal of the other contracting party, from completing the contract within the time limited, he is not liable in law for the default.'[58]

That was a case where a builder agreed to construct a brewery in $4\frac{1}{2}$ months subject to liquidated damages of £40 per week. Completion was late due to the default of the employer in failing to give possession of the site on the due date. It was said in a New Zealand judgment:

> '...no person can take advantage of the non-fulfilment of a condition the performance of which has been hindered by himself; that a party is exonerated from the performance of a contract when the performance is rendered impossible by the wrongful act of the other contracting party; or more emotively, that a party cannot take advantage of his own wrong.'[59]

It is clearly not an immutable rule; it will depend on circumstances. For example, where a contractor has undertaken to carry out works including any alterations or additions which the employer might choose to make, he can be bound to his undertaking. In *Jones* v. *St John's College Oxford* it was said:

> '...the plaintiffs undertake not only to do by a given time the works which were specified, and which they had the opportunity therefore of forming their own judgment upon, but they also undertake to do the alterations, that is to say, such alterations as are contemplated by the contract, within the time originally prescribed for the performance of the works.'[60]

It is not clear whether the judge was referred to *Jones* in *Wells* v. *Army & Navy Co-operative Society Ltd* where the contractor was not liable to pay 'penalties' on account of exceeding the contract period[61]. The key facts seem to have been that although the contract provided for the contractor to complete by the due date notwithstanding variations, strikes and weather conditions and subject only to any extension of time which the employer may, but was not obliged to grant, it was not wide enough to cover the employer's own defaults. In general, the courts adopt the approach that 'it is not to be inferred that the one party meant to bind himself so very stringently, unless it is so stated.'[62] In *Wells*, it was said:

---

[57] (1985) 32 BLR 51.
[58] (1838) 3 M & W 387 at 389 per Parke B.
[59] *Canterbury Pipelines Ltd* v. *Christchurch Drainage Board* (1979) 16 BLR 76.
[60] (1870) LR 6 QB 115 at 123 per Mellor J.
[61] (1902) 86 LT 764.
[62] *Roberts* v. *Bury Commissioners* (1870) LR 5 CP 310 at 327 per Kelly CB.

'In the contract one finds time limited within which the builder is to do the work. That means not only that he is to do it within that time but it means also that he is to have that time within which to do it ... in my mind that limitation of time is intended not only as an obligation, but as a benefit to the builder ... In my judgment where you have a time clause and a penalty clause (as I see it) it is always implied in such clauses that penalties are only to apply if the builder has, as far as the builder owner is concerned and his conduct is concerned, that time accorded to him for the execution of the works which the contract contemplates he should have.'[63]

In *Dodd* v. *Churton* it was held that an employer cannot recover liquidated damages if he prevents the contractor completing within the stipulated time[64]. *Jones* was distinguished, because although there was a term which empowered the ordering of additional work and this was done, the contractor had not agreed to complete within the original period despite the ordering of additional work. Very clear words will be needed in order to bind a contractor to a completion date if the employer is the cause of the delay. This principle is now well established[65] and it seems unlikely that a modern court would take so stern a view as the 1870 court in *Jones*.

Extension of time clauses should be drafted so as to include for all delays which may be the responsibility of the employer, for example clause 11.5 of ACA 3. Then, if the employer, either personally or through the agency of his architect, hinders the contractor in a way which would otherwise render the date for completion ineffective, the architect will have the power to fix a new date for completion and thus preserve the employer's right to deduct liquidated damages. The position was set out in *Peak Construction (Liverpool) Ltd* v. *McKinney Foundations Ltd*:

'The liquidated damages and extension of time clauses in printed forms of contract must be construed strictly *contra proferentem*. If the employer wishes to recover liquidated damages for failure by the contractors to complete on time in spite of the fact that some of the delay is due to the employer's own fault or breach of contract, then the extension of time clause should provide, expressly or by necessary inference, for an extension on account of such a fault or breach on the part of the employer. I am unable to spell any such provision out of ... the contract [clause] in the present case.'[66]

In that case, the extension of time clause, after referring to certain events of a neutral character, i.e. they could not be said to be the fault of either contractor or employer, made reference to '...or other unavoidable circumstances...'. This was the phrase on which the employer relied, but 'delay

---

[63] *Wells* v. *Army & Navy Co-operative Stores* (1902) 2 HBC 4th Edition (vol. 2) 346 at 355 per Vaughan Williams LJ.
[64] [1897] 1 QB 562.
[65] *Percy Bilton* v. *Greater London Council* (1982) 20 BLR 1.
[66] (1970) 1 BLR 114 at 121 per Salmon LJ.

due to the employer cannot be said to have been an unavoidable circumstance to anyone save the contractor'[67]. A similar phrase is 'other causes
beyond the control of the contractor'. It has been held that these words
'ought to be construed with reference to the preceding causes of delay, and
ought not to receive such an extension as would make the defendants judges
in respect of their own defaults'[68]. This view was noted with approval in
Perini Pacific v. *Greater Vancouver Sewerage & Drainage District*[69].

2.4 Concurrency

A question which frequently arises in regard to causation[70] is the method of
dealing with loss which may be due to either or both of two causes. It is
important to differentiate between the delaying event or cause and the delay
itself. It is generally recognised that there are times when there are delays
which may be the result of different causes, but that sometimes the causes
will run at the same time or overlap. This makes it difficult to decide how to
treat the delay, particularly if the causes originate from different parties or
the delays are of different kinds. For example, under the Standard Forms of
Contract, some causes of delay may give rise to an extension of the contract
period, some causes may give rise to extension and possibly also loss and
expense, while other causes may not entitle the contractor to any extension
or loss and expense whatsoever. Take, for example, the following situations:

(1) A contractor is just starting to carry out the covering of a large roof when
    he receives an architect's instruction to change the covering to another
    material which will take a few days to arrive on site. Within hours, the
    weather takes a turn for the worse and the contractor has to pull all his
    men off site for several days. If the overall delay is 6 days, is the architect
    responsible or can the contractor only get an extension of time due to
    exceptionally adverse weather conditions – if that?
(2) A contractor is in delay through his own fault after the contract comple-
    tion date and the architect postpones all the works.
(3) The contractor is about to start some complex trench excavation in a
    confined space, but he has not received the architect's detailed draw-
    ings. He decides to make a start where he can, but his machinery breaks
    down. By the time it is in working order, the architect has got his
    drawings to site. Who is responsible for the delay and what can the
    contractor recover?

At first sight, it is difficult to see a clear answer to some of these problems.
'*Keating on Building Contracts*'[71] looks at a number of propositions as follows:

---

[67] (1970) 1 BLR 114 at 126 per Edmund Davies LJ.
[68] *Wells* v. *Army & Navy Co-operative Society Ltd* (1902) 86 LT 764 at 765 per Wright J at first instance.
[69] (1966) 57 DLR (2d) 307 at 321 per Bull JA.
[70] See Chapter 7.
[71] Stephen Furst and Vivian Ramsey, *Keating on Building Contracts*, 7th edition, 2001, Sweet &
Maxwell, p.246.

(1) *The Devlin Approach* which contends that if there are two causes operating together and one is a breach of contract, the party responsible for the breach will be liable for the loss.
(2) *The Dominant Cause Approach* which contends that if there are two causes, the effective, dominant cause is to be the deciding factor.
(3) *The Burden of Proof Approach* which contends that if there are two causes and the claimant is in breach of contract, it is for the claimant to show that loss was caused otherwise than by his breach.

It is sometimes said that the case of *H. Fairweather & Co Ltd* v. *London Borough of Wandsworth*[72] is authority to the effect that the 'dominant cause' approach is incorrect. Fairweather entered into a contract to erect 478 dwellings for Wandsworth on JCT 63 terms. Long delays culminated in the architect giving an extension of time of 81 weeks for strikes. The contractor sought arbitration in an attempt to have the extension allocated under different heads. He mistakenly thought that an extension of time under appropriate heads was necessary before he could become entitled to any loss and/or expense. He wanted at least 18 weeks designated as on account of architect's instructions or late instructions. The arbitrator decided that where it was not possible to allocate the extension among different heads of delay, the extension must be given for the dominant reason. What the judge actually said in that case was:

> ' "Dominant" has a number of meanings: "Ruling prevailing, most influential". On the assumption that condition 23 is not solely concerned with liquidated or ascertained damages but also triggers and conditions a right for a contractor to recover direct loss and expense where applicable under condition 24 then an architect and in his turn an arbitrator has the task of allocating, when the facts require it, the extension of time to the various heads. I do not consider that the dominant test is correct. But I have held earlier in this judgment that assumption is false. I think the proper course here is to order that this part of the interim award should be remitted to Mr Alexander for his reconsideration and that Mr Alexander should within 6 months or such further period as the court may direct make his interim award on his part.'[73]

Besides, being almost certainly *obiter*, this statement is nowhere near the kind of condemnation often suggested. Other cases, indeed, show that the courts have embraced the 'dominant cause approach' quite happily[74].

> 'One has to ask oneself what was the effective and predominant cause of the accident that happened, whatever the nature of the accident may be'.[75]

---

[72] (1987) 39 BLR 106.
[73] *H. Fairweather & Co Ltd* v. *London Borough of Wandsworth* (1987) 39 BLR 106 at 120 per Judge Fox-Andrews.
[74] See, for example, *Fairfield-Mabey Ltd* v. *Shell UK* (1989) 45 BLR 113 and *Yorkshire Dale Steamship* v. *Minister of War Transport* [1942] 2 All ER 6.
[75] *Yorkshire Dale Steamship* v. *Minister of War Transport* [1942] 2 All ER 6 at 10 per Viscount Simon.

In *Fairfield-Mabey Ltd* v. *Shell UK Ltd*, Shell entered into a contract with Fairfield-Mabey (FM) to fabricate parts of a gas platform in the North Sea. Sub-contractors were employed by FM to carry out weld-testing, etc. It was fast track work. Delays occurred and there followed claim and counterclaim. FM sued Shell and joined in Met-Testing (MT) claiming an indemnity against the counterclaim. The settlement reached was £280,000 to FM, but they then claimed £400,000 against MT. MT said that even if they were at fault regarding the testing, there was no damage because another subcontractor had in any case caused the delay. It was held that the absence of approval for certain tests was not a cause of equal efficacy with the subcontractor delays. The test was that of the ordinary bystander who would have said that the cause of delay was due to the sub-contractor.

Another case which is instructive is *Carslogie Steamship Co Ltd* v. *Royal Norwegian Government (The Carslogie)*[76]. In 1941, the Heimgar, belonging to the respondents, collided with the Carslogie, which belonged to the appellants. The Carslogie was at fault. Temporary repairs to the Heimgar were carried out in England and the ship proceeded to the USA for permanent repairs. During her voyage, she suffered heavy weather damage which needed immediate repair. The ship remained in dock for 50 days and repairs to the collision damage and weather damage were carried out concurrently. It was agreed that 10 days should be allocated to the repair of the collision damage and 30 days to repair the weather damage. The respondents claimed damages for loss of charter hire during the ten days attributable to the collision damage. It was held that the appellants were only liable for the loss of profit suffered by the respondents resulting from the appellants' wrongful act. During the time that the Heimgar was detained in dock she was not profit-earning because the heavy weather damage had made her unseaworthy; therefore, the respondents had not suffered any damage, because the vessel was undergoing repairs in respect of the collision damage for 10 days. The case contains reference to further examples which are very instructive.

'It is well established that, if a ship goes into dock for repairs of damage occasioned by a collision brought about by the fault of another vessel, the owners of that other vessel must pay for the resulting loss of time, even although her owners take advantage of her presence in the dock to do some repairs which, though not necessary, are advisable. Thus, in *Ruabon SS Co.* v. *London Assurance* [1900] AC 6, the Ruabon suffered damage on the voyage which made it necessary for her to be put into dry dock. The owners (without causing delay or increase of dock expenses) took advantage of her being in dry dock to have made the survey of the vessel for renewing her classification, though this survey was not then due. It was decided that the expense of getting the vessel into and out of dock, as well as those incurred in the use of the dock, fell on the underwriters alone.'[77]

---

[76] [1952] 1 All ER 20.
[77] *Carslogie Steamship Co Ltd* v. *Royal Norwegian Government (The Carslogie)* [1952] 1 All ER 20 at 24 per Viscount Jowitt.

A case dealing with the question of dominance is *Galoo Ltd and Others* v. *Bright Grahame Murray*[78] where it was held that the 'but for' test of causation was not sufficient and it was clear that if a breach of contract by a defendant was to be held to entitle a claimant to claim damages, it must first be held to be an effective or dominant cause of his loss. In considering whether a breach of duty imposed upon a defendant, whether by contract or in tort in a situation analogous to a breach of contract, was the cause of the loss or merely the occasion for the loss, the court had to arrive at a decision on the basis of the application of common sense. In *Henry Boot (Construction) Ltd* v. *Malmaison Hotel (Manchester) Ltd*, it was said:

> 'Secondly, it is *agreed* that if there are two concurrent causes of delay, one of which is a relevant event, and the other is not, then the contractor is entitled to an extension of time for the period of delay caused by the relevant event notwithstanding the concurrent effect of the other event. Thus, to take a simple example, if no work is possible on site for a week not only because of exceptionally inclement weather (a relevant event), but also because the contractor has a shortage of labour (not a relevant event), and if the failure to work during that week is likely to delay the works beyond the completion date by one week, then if he considers it fair and reasonable to do so, the architect is required to grant an extension of time of one week. He cannot refuse to do so on the grounds that the delay would have occurred in any event by reason of the shortage of labour.'[79] (emphasis added)

This *dicta* was adopted in subsequent cases and noted with approval by some commentators despite the fact that it was clearly not a judicial decision, but rather a note of what the parties had agreed ('...it is agreed...')[80]. Moreover the court qualifies the statement further by the words: '...if he considers it fair and reasonable to do so...'. Later in the judgment, the court appears effectively to contradict itself by accepting that the architect may say that the 'true cause of the delay was other matters, which were not relevant events and for which the contractor was responsible'. A useful view of concurrency and an interpretation of *Henry Boot* was given in a later case:

> 'However, it is, I think, necessary to be clear what one means by events operating concurrently. It does not mean, in my judgment, a situation in which, work already being delayed, let it be supposed, because the contractor has had difficulty in obtaining sufficient labour, an event occurs which is a relevant event and which, had the contractor not been delayed, would have caused him to be delayed, but which in fact, by reason of the existing delay, made no difference. In such a situation although there is a relevant event, *"the completion of the Works is [not] likely to be delayed thereby beyond the Completion Date"*.

---

[78] TLR 14 January 1994.
[79] (1999) 70 Con LR 32 at 37 per Dyson J.
[80] See *Motherwell Bridge Construction Ltd* v. *Micafil Vakuumtechnik and Another* (2002) CILL 1913.

The relevant event simply has no effect upon the completion date. This situation obviously needs to be distinguished from a situation in which, as it were, the works are proceeding in a regular fashion and on programme, when two things happen, either of which, had it happened on its own, would have caused delay, and one is a relevant event, while the other is not. In such circumstances there is a real concurrency of causes of the delay. It was circumstances such as these that Dyson J was concerned with in the passage from his judgment in *Henry Boot Construction (UK) Ltd* v. *Malmaison Hotel (Manchester) Ltd* at paragraph 13 on page 37 of the report which [Counsel] drew to my notice. Dyson J adopted the same approach as that which seems to me to be appropriate to the first type of factual situation which I have postulated when he said, at paragraph 15 on page 38 of the report: *"It seems to me that it is a question of fact in any case whether a relevant event has caused or is likely to cause delay to the works beyond the completion date in the sense described by Coleman J in the Balfour Beatty case".*[81]

This seems to come nearest to the solution, but none of the cases provides a universal solution.

When faced with a problem of concurrent delays, it is always worthwhile pausing and asking whether the delays really are concurrent. Most delays are in fact consecutive. The test is to look at the critical path. Delays must generally be consecutive unless there are two or more critical paths. Several critical paths running in parallel is a common situation, but even in such cases, true concurrency is rare. Usually it can be seen that one delay occurs after the other.

Therefore, before the question of concurrency arises at all, it must be established that there are two competing causes of delay operating at the same time and affecting the critical path or paths of the project. Examining the situations set out at the beginning of this section:

(1) The delay is clearly caused by the architect's instruction which makes it impossible for the contractor to work for the 'several days' it will take to receive the new roof covering. That is the cause of the delay and any extension to the completion date. The bad weather, even if satisfying the criteria for 'exceptionally adverse weather', has no effect on the completion date which is already being delayed. Of course, if the roof covering arrives, but the bad weather continues, the bad weather will take the place of the late roof covering as a cause of delay.

(2) This, in essence, is the *Balfour Beatty* v. *Chestermount* scenario. There is no actual concurrency of either the delaying event or the delay itself, because what happens in this example is that although the contractor is in delay, he is still working on site, trying to finish. When the architect postpones the work, the contractor stops working on site and it is the postponement which is causing the delay until the architect brings the postponement to an end.

---

[81] *Royal Brompton Hospital NHS Trust* v. *Hammond and Others (No.7)* (2001) 76 Con LR 148 at 173 per Judge Seymour.

(3) It can be seen that there is no real delay until the contractor's machinery breaks down. Before it is repaired, the architect's drawings are issued. They are certainly late and it is a delaying event under most contracts, but the late drawings did not delay the completion date.

This is the approach outlined in *Royal Brompton Hospital* and what can be drawn from *Henry Boot* if the *dicta* referred to earlier is read in context with succeeding paragraphs.

Assuming that the criteria for concurrency have been satisfied and assuming further that there are the same two causes in each case (one the fault of the contractor, the other the fault of the employer or the architect), there are four possible situations.

The third situation just discussed is shown in Figure 2.1(a); following the authorities, no extension of time is due to the contractor. Figure 2.1(b) is the converse: work is stopped awaiting the architect's information. During the delay, the contractor's machinery breaks down and is repaired again before the architect's information arrives. In this instance, the machinery breakdown had no effect on the completion date, because it was already being delayed by the late information from the architect and 4 days of extension of time is due.

Figures 2.1(c) and 2.1(d) are not specifically dealt with in either *Royal Brompton Hospital* or *Henry Boot*, but useful conclusions can be drawn from them. In Figure 2.1(c), the late information causes a delay. It continues for 3 days and affects the completion date similarly, because it is on the critical path. On the second day, the contractor's machinery breaks down, but it has no effect on the completion date which is already delayed due to the late information. However, when the information is provided, the machinery remains inoperative for a further day and, during that day, it and not the late information affects the completion date. The total delay is 4 days, of which the appropriate extension of time is 3 days.

The final situation is shown in Figure 2.1(d). In this instance, the machinery breaks down and causes a delay to the completion date lasting 3 days. On the second day, the architect's information should arrive, but it is delayed for 3 days. During the first two of those days, the late information has no effect on the completion date, but when the machinery is repaired, the remaining day of delay is caused by the architect's late information. Therefore, the appropriate extension of time would be 1 day although the total delay is 4 days.

The same principles can be applied if the concurrency involves a cause which would give an entitlement to extension of time and another cause which not only gives an entitlement to extension of time, but also, if the contractor makes application, to loss and/or expense.

2.5 Acceleration

'Acceleration' has been usefully defined as follows:

Time

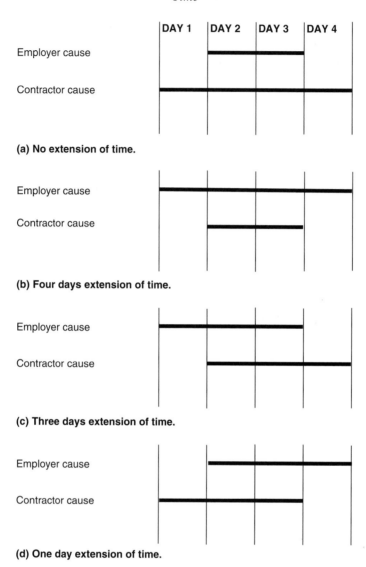

Figure 2.1 Concurrency.

'"Acceleration" tends to be bandied about as if it were a term of art with a precise technical meaning, but I have found nothing to persuade me that that is the case. The root concept behind the metaphor is no doubt that of increasing speed and therefore, in the context of a construction contract, of finishing earlier. On that basis "accelerative measures" are steps taken, it is assumed at increased expense, with a view to achieving that end. If the other party is to be charged with that expense, however, that description gives no reason, so far, for such a charge. At least two further questions are relevant to any such issue. The first, implicit in the description itself, is "earlier than what?". The second asks by whose decision the relevant steps were taken.

The answer to the first question will characteristically be either "earlier than the contractual date" or "earlier than the (delayed) date which will be achieved without the accelerative measures". In the latter category there may be further questions as to responsibility for the delay and as to whether it confers entitlement to an extension of time. The answer to the second question may clearly be decisive, especially in the common case of contractual provisions for additional payment for variations, but it is closely linked with the first; acceleration not required to meet a contractor's existing obligations is likely to be the result of an instruction from the employer for which the latter must pay, whereas pressure from the employer to make good delay caused by the contractor's own fault is unlikely to be so construed.'[82]

The reasons for acceleration usually fall into one the following categories:

(1) By agreement between the parties or, if the contract so provides, on the instruction of the architect.
(2) Unilaterally on the initiative of the contractor, often categorised as 'mitigation' by the contractor or as 'using best endeavours' by the employer.
(3) Constructive acceleration where the contractor argues that he has no real alternative in the circumstances.

By agreement or instruction

Under the general law, the architect has no power to instruct the contractor to accelerate work. The contractor's obligation is to complete the work within the time specified, or where no particular contract period is specified – within a reasonable time. The contractor cannot be compelled to complete earlier than the agreed date unless there is an express contract term authorising the architect to require acceleration.

Few standard form contracts give the architect power to order the contractor to accelerate. There appears to be such a power in the ACA 3 form, clause 11.8, and the MC 98 Management Contract contains a clause of considerable complexity which gives the architect such power if the parties agree (clause 3.6). Clause 13 in MPF 03 is to similar effect. There should be no difficulty in obtaining payment where the architect, in exercise of his powers under a contract, orders acceleration of the work or the employer and the contractor agree acceleration and a claim under the direct loss and/ or expense clause is unnecessary.

---

[82] *Ascon Contracting Ltd* v. *Alfred McAlpine Construction Isle of Man Ltd* (2000) 16 Const LJ 316 at 331 per Judge Hicks.

Unilateral acceleration

This is the situation where a contractor accelerates without any agreement with the employer or instruction from the architect. No pressure has been placed on him by the refusal of an extension of time; indeed in this situation it may be that the contractor is reasonably confident of getting an extension to the contract period. The contractor may nevertheless decide to place more operatives on site. The reason for so doing may be in order to find work for operatives from another site which is drawing to a close. The result may be that some time is recovered and an extension of time is not required.

With the assistance of computer programming, it is possible to indicate the extent to which the completion would have been exceeded had the contractor not accelerated. This can be used to support the cost of acceleration when compared with the alternative prolongation costs. It is by no means clear, however, under what contract provision the contractor could be paid even if the architect was sympathetic.

In most such cases, the contractor will find it difficult to contend that he was doing other than using his best endeavours to reduce delay. However, the contractor may find some degree of comfort in the *Ascon* case below.

Constructive acceleration

An argument sometimes advanced by a contractor is based on the architect's failure to give an extension of time to which the contractor believes he is entitled. A contractor will commonly put more resources into a project than originally envisaged and then attempt to recover the value on the basis that he was obliged to do so in order to complete on time, because the architect failed to make an extension of the contract period. The contractor contends that, as a direct result of the architect's breach, he was obliged to put more resources on the project so as to finish by the date for completion for fear that otherwise he would be charged liquidated damages. This claim is advanced whether or not completion on the due date was actually achieved. A failure in this respect is usually explained as a result of yet more delaying events which the contractor was powerless to control. Duncan Wallace has this to say on the subject:

> 'In the United States, a highly ingenious type of contractor's claim, based on a "constructive acceleration order" theory, has been accepted in the Court of Claims for government contracts in the not uncommon situation where the [Architect/Engineer], in the bona fide belief that the contractor is not entitled to an extension of time and is in default, *presses* a contractor to complete by the original contract completion date, and it is subsequently held that the contractor had been entitled to an extension of time. This is, however, a development of what, in any event, is a largely jurisdictional and fictitious doctrine of "constructive change orders" (CCOs) developed by the Boards of Contract Appeals, and is not founded

on any consensual or quasi-contractual basis which would be acceptable in English or Commonwealth Courts it is submitted.'[83] (emphasis added)

The important question to be asked before this kind of argument can be entertained is the extent to which pressure is put on a contractor. The contractor's problem is one of causation. Where the architect wrongfully fails to make an extension of time, either at all or of sufficient length, the contractor's clear route under the contract is adjudication or arbitration. If, as a matter of fact and law, the contractor is entitled to an extension of time, it may be said that he should confidently continue the work, without increasing resources, secure in the knowledge that he will be able to recover his prolongation loss and/or expense, and any liquidated damages wrongfully deducted, at adjudication or arbitration. If he increases his resources, that is not a direct result of the architect's breach, but of the contractor's decision.

In practice, it must be acknowledged that a contractor in this position may not be entirely confident. The facts may be complex and the liquidated damages may be high. Faith in the wisdom of the adjudicator, arbitrator or even the judge, may not be total. It may be cheaper, even without recovering acceleration costs, for the contractor to accelerate rather than face liquidated damages with no guarantee that an extension of time will ultimately be made. As a matter of plain commercial realism, the contractor may have no sensible choice other than to accelerate and take a chance as to recovery. Unless the contractor can show that the architect has given him no real expectation that the contract period will ever be extended and in those circumstances the amount of liquidated damages would effectively bring about insolvency, this kind of claim has little chance of success[84].

Acceleration has been considered in *Motherwell Bridge Construction Ltd v. Micafil Vakuumtechnik and Another*[85]. In the course of an extremely long judgment, the court concluded that the contractor was entitled to recover the cost of acceleration if an extension of time was justified, but refused, and the liquidated damages were 'significant'. Although the judgment was long, much of it was a recital of facts and the *ratio* for the decision is difficult to decipher although pressure from the employer appears to have been a significant factor. Moreover, the court appeared to agree that the contractor was additionally entitled to loss and/or expense for the prolongation which would, but for the acceleration, have taken place. That seems to give the contractor, not two bites at the cherry but two cherries for the price of one.

It is thought that the better view is the one earlier set out in *Ascon Contracting Ltd* v. *Alfred McAlpine Construction Isle of Man Ltd* where it was said that there could not be

> 'both an extension to the full extent of the employer's culpable delay, with damages on that basis, and also damages in the form of expense

[83] I. N. Duncan Wallace, *Hudson's Building and Engineering Contracts*, 11th edition, 1995, Sweet & Maxwell, p. 909.
[84] *Perini Corporation* v. *Commonwealth of Australia* (1969) 12 BLR 82.
[85] (2002) CILL 1913.

incurred by way of mitigation, unless it is alleged and established that the attempt at mitigation, although reasonable, was wholly ineffective.'[86]

2.6 *Sectional completion*

Employers and contractors often run into difficulties where the employer has chosen to incorporate sectional completion into the contract. It should be noted that, where there is just one date for completion in the contract, sectional completion cannot usually be achieved by simply inserting inter-mediate dates in the specification or bills of quantities[87]. Moreover, although a court may be prepared to imply a date for completion of the works in the absence of agreement, dates for sectional completion will not be implied[88] – certainly not in JCT contracts which have a clause giving priority to the printed form over other contract documents.

In such cases, the single completion date will take precedence despite a multitude of intermediate dates in the subordinate document. Where JCT contracts are not involved and there is no equivalent priority clause, the ordinary rule will apply that 'type prevails over print' and any intermediate or sectional completion dates stated in the bills of quantities or specification would apply. This would immediately give rise to a number of contractual difficulties, for example, most standard form contracts provide for only one certificate of practical completion, one certificate of making good defects and one defects liability period (or their equivalents). Not least, such con-tracts refer to extension of time in relation to the completion *date* or fixing a new completion *date*. Such problems can only be resolved by goodwill on both sides or with the assistance of an adjudicator, arbitrator or judge. The adoption of sectional completion into a standard building contract requires a considerable number of contract amendments. Reference to the multitude of small changes required by the JCT sectional completion supplement illustrates the point.

A common problem occurs where a sectional completion supplement has been properly used, but two or more of the sections are interdependent. For example, a school project may be divided into four sections, but section 3 may be dependent on section 1 in the sense that the contractor cannot be given possession of section 3 until section 1 has reached practical comple-tion. That may be because the occupants of section 3 have to be moved to section 1.

Invariably, the dates for possession and completion of each section are inserted into the contract as a series of dates. The date for completion of section 1 may be 25 March and the date for possession of section 3 may be 30 March to allow the transfer of pupils and staff from one section to another. The problem arises when section 1 is not finished by the completion date.

---

[86] (2000) 16 Const LJ 316 at 332 per Judge Hicks.
[87] *M. J. Gleeson (Contractors) Ltd* v. *Hillingdon Borough Council* (1970) 215 EG 165.
[88] *Bruno Zornow (Builders)* v. *Beechcroft Developments* (1990) 6 Const LJ 132.

This could be because the contractor has been inefficient, or it may be because events have occurred which entitle the contractor to an extension of time. Where a project is split into sections, any extensions of time must be given in respect of the particular section affected by the delaying event. Therefore, if the contractor is entitled to an extension of time, it is in connection with section 1 only. Whether or not he is so entitled is irrelevant, because in any event, when the 30 March arrives, the contractor is still working on section 1 and the occupants of section 3 cannot be transferred. The result is that the employer cannot give the contractor possession of section 3 on the due date. If no extension of time is due for section 1 and the cause of the delay is entirely the fault of the contractor, the architect may say that the contractor has himself to blame and cannot expect possession of section 3 on the due date.

This approach is to misunderstand the situation entirely. The principles of causation must be applied. The cause of the delay to possession of section 3 is not the contractor's delay to section 1, but the fact that the two sections are linked. If they were not linked, the contractor's delay to section 1 would not affect section 3 in the slightest degree. However, where the dates for possession and completion are simply expressed as a series of dates, there is nothing to put the contractor on notice of the likely problem. The contractor is likely to argue that the employer is in breach of contract and he would be correct.

There are two immediate aspects to this problem. First, what can the architect or the employer do to retrieve the situation? Second, what could have been done to avoid it in the first place? If there is provision for the employer to defer possession of any of the sections by the appropriate amount, the employer will be obliged to take that route and the contractor will be entitled to an extension of time and probably whatever amount of loss he has suffered as a result of the deferment of possession. If there is no deferment provision, the correct analysis of the situation appears to be that there is a breach of contract which, dependent upon its likely duration, may become repudiatory in nature. In any event, the contractor would be entitled to recover as damages the amount of loss he has suffered as a result. That is fairly straightforward and the amount payable to the contractor, whether by virtue of a loss and/or expense clause in the contract or as damages for the breach, may not be substantial. The contractor would have to demonstrate his loss and, essentially, the situation is simply that section 3 has been pushed back in time. In the absence of a provision for the delay situation in the contract, the architect would be unable to make any extension of time with the result that the contractor's obligation with regard to section 3 would be to complete within a reasonable time. Therefore, liquidated damages would not be recoverable for this section.

To prevent the situation occurring, the employer should indicate the links in the sections. Therefore, in the example above, section 1 would have a date for possession and a date for completion, but section 3 would not have a date for possession. That section of the contract would simply state 'the date for possession is x days after the date of practical completion of section 1'. In that way, a delay to completion in section 1 would be reflected in the date of possession of section 3, and breach of contract, damages and extension of time to section 3 would not be relevant. Moreover, the contractor would be aware of the situation and he could make some provision for it in his price. The date for completion of section 3 would not be inserted, but rather 'the date for completion is x weeks after the date possession of this section is taken by the contractor'. It seems doubtful that the contractor could make any financial claim on the employer in this situation for delays to section 1 which cause a delay to the possession of section 3.

A rather different situation occurred in *Trollope & Colls Ltd* v. *North-West Metropolitan Regional Hospital Board*[89]. There were three sections or phases. Each had a separate contract sum and set of conditions. Although the start of section 3 was dependent on the completion of section 1, the date for completion of section 3 was given as a specific date. The result was that when completion of section 1 was delayed, the start of section 3 was also delayed, but the date for completion of section 3 remained the same. Effectively, the period for completion was reduced from 30 to 16 months. The House of Lords refused to imply a term into the contract that the completion date for section 3 should be extended accordingly.

2.7 The SCL extension of time protocol

In October 2002, the Society of Construction Law produced its Delay and Disruption Protocol after some months of consultation throughout the industry. The object of the Protocol is stated in the introduction as providing useful guidance on some of the common issues arising in connection with extension of time and claims for compensation for time and resources. The Protocol purports to provide a means for resolving such matters and avoiding disputes.

The first thing that must be said about the Protocol is that it cannot replace the terms of the particular contract in use. Therefore, it will avail the parties nothing to quote the Protocol if the terms of the contract are at variance with it. The Protocol recognises this and suggests that its contents should be considered when contracts are being drafted. The document is divided into sections. In view of the publicity given to the Protocol, the sections will be briefly examined in order and comments given. Where appropriate, references to relevant parts of this book are given in brackets.

2.7.1 Core principles relating to delay and compensation

This section sets out the principal items on which guidance is given later.

---

[89] (1973) 9 BLR 60.

2.7.2 Guidance notes section 1

Extensions of time

> The summary of the extension of time position is generally good and the advice is sound although fairly general in nature.

Float as it relates to extensions of time

> This is a broadly accurate statement, but somewhat confused by references to hypothetical situations which appear to allow the contractor an extension of time although the contract date for completion is not exceeded. This would be an extraordinary outcome which runs counter to the reason for extending time. The contract would have to contain special clauses to permit this result and no standard form contract currently has this type of provision.

Concurrency as it relates to extensions of time

> The Protocol's approach seems to be to take a particular position on the subject of concurrency on the basis that it is a complex topic and a compromise solution is necessary. A basic principle is that no concurrent cause of delay which is the result of any fault of the contractor should reduce the extension of time to which he would otherwise be entitled. For the reasons stated elsewhere in this book, this approach is not considered to represent the true position in law and it is not recommended.

Mitigation of delay

> This is a consideration of the general law duty to mitigate (see Chapter 6, section 6.1.4).

Financial consequences of delay

> This is a clear and accurate statement that entitlement to extension of time does not automatically entitle the contractor to any money.

Valuation of variations

> The Protocol recommends a mechanism similar to the current JCT price statement for dealing with the valuation of variations and associated extension of time and loss and expense.

Compensation for prolongation

It is rightly stressed that ascertainment must be based on actual additional costs incurred by the contractor. However, there appears to be some confusion between a contractor's claims for loss and expense under the contract machinery and claims for damages for breaches of contract. The former are reimbursable under most standard form contracts while the latter, being a claim outside the contract, are not so reimbursable.

There appears to be a half suggestion that payments to the contractor might be simplified by being dealt with by a reverse kind of liquidated damages clause. This approach was suggested many years ago and the appropriate clause to be inserted in the contract was termed 'Brown's clause' after the person who proposed it. Anything which can simplify and cheapen the loss and/or expense process while achieving a result which does not vary too much from the strictly accurate entitlement is to be welcomed, but the Brown's clause had a number of practical and legal difficulties then and there is no reason to suppose that a similar clause now would be any better.

Relevance of tender allowances for prolongation and disruption compensation

It is refreshing to see that the Protocol considers that tender allowances have little or no relevance to the evaluation of the costs of prolongation or disruption.

Concurrency as it relates to compensation for prolongation

This appears to be a straightforward summary of the position.

Time for assessment of prolongation costs

Another straightforward summary.

Float as it relates to compensation

Where a contractor plans to complete before the contract date for completion, the Protocol recommends that he is entitled to compensation, but not an extension of time, if he is prevented from completing to his own planned date, but finishes before the contract date for completion. This is a complicated topic to which the Protocol does not do justice. However, the basic recommendation must be rejected. The position is that in deciding this question, all the circumstances must be taken into account (see Chapter 7, section 7.3).

Mitigation of loss

> A clear exposition of the situation. More could have been said about the contractor's rights, or otherwise, to claim the reasonable costs of mitigation (see Chapter 6, section 6.1.4).

Global claims

> It is good to see that global claims are discouraged (see Chapter 8).

Claims for payment of interest

> Although this survey of the position seems to be broadly correct, it is not made clear that for interest to be claimable, it must be shown to be part of the loss and/or expense and not, as suggested here, a result of it (see Chapter 6, section 6.5.8).

Head office overheads

> This is a clear and concise explanation (see Chapter 6, section 6.5.2).

Profit

> Very brief and to the point (see Chapter 6, section 6.5.3).

Acceleration

> This is a broadly correct interpretation of the position, but the reference to the possibility of accelerating by instructions about hours of working and sequence of working is to be doubted (see section 2.5 of this chapter).

Disruption

> The definition of disruption does not adequately explain that disruption can also refer to a delay to an individual activity not on the critical path where there is no resultant delay to the completion date. It is also stated that most standard forms do not deal expressly with disruption. That, of course, is true. But it is also true that most standard forms do not expressly deal with prolongation. For example, JCT forms refer to regular progress being materially affected. That appears to be quite broad enough to encompass both disruption and prolongation (see Chapter 6, section 6.2.2).

2.7.3 Guidance section 2

This deals with guidance on preparing and maintaining programmes and records. Stress is placed on obtaining an 'Accepted Programme'. That is a programme agreed by all parties. There are several problems with this. Perhaps the foremost is that an architect will be unlikely to have the requisite skills and/or experience or indeed the information required to accept the contractor's programme. He is probably capable of questioning parts of it, but highly unlikely to be possessed of sufficient information to be able to satisfy himself that the programme is workable. The Protocol, rightly, accepts that the contractor is entitled to construct the building in whatever manner and sequence he pleases, subject to any sectional completion or other constraints. The Protocol then states:

> 'Acceptance by the CA merely constitutes an acknowledgement by the CA that the Accepted Programme represents a contractually compliant, realistic and achievable depiction of the Contractor's intended sequence and timing of construction of the works.'

This is placing a responsibility on the architect (or CA as the Protocol prefers) which he is not required to carry. There appears to be no need for a programme to be accepted. It is sufficient if the contractor puts it forward as the programme to which he intends to work. The architect is entitled to question any part which appears to be clearly wrong or unworkable. But, in the light of the contractor's insistence that he can and will carry out the works in accordance with the submitted programme, it is difficult to refuse a programme unless firm objections can be raised. There is not a great deal of guidance on maintaining records generally (see Chapter 9).

2.7.4 Guidance section 3

This section deals with guidelines for dealing with extensions of time during the course of the project. It provides much good practical advice including the importance of calculating extensions of time by means of various programming techniques. Although every architect should be familiar with such techniques, careful consideration should be given to the aptness of any particular technique in a given situation. No doubt it is still possible to work out an extension of time perfectly well using old fashioned inspection techniques applied to the programme if the programme and the delays are not complicated.

2.7.5 Guidance section 4

This deals with disputed extension of time after completion of the project and spends some time examining the different types of analysis that can be employed.

2.7.6 Appendices

There are four appendices. Appendix A is a very useful glossary. Appendix B is a model specification clause intended for use when drafting a specification, although the content appears more suitable for inclusion in the contract itself – if at all. Whether and to what extent one wishes to use the clause will depend on whether one wishes to go totally down the road pointed out by the Protocol. Appendix C is another sample clause to cover the keeping and submission of records. Finally, Appendix D graphically represents various situations involving concurrency, float and critical delays of various kinds.

2.7.7 Conclusion

The Protocol sets out ways of dealing with delays and disruption. Most of it is in line with what is generally understood to be the law on these matters. In some instances, the Protocol steps outside this boundary in order to suggest what it clearly considers to be a simpler or fairer way of dealing with the practicalities. All parties involved in construction contracts must be aware that the Protocol does not take precedence over the particular contract in use unless it is expressly so stated in the contract itself. Therefore, the recommendation should be viewed with caution. It will be of no avail for the architect, contract administrator or employer to argue that he has acted strictly in accordance with the Protocol if the contract prescribes action of a different sort. It is not yet known whether the incorporation of the Protocol in a standard form contract will have a good or adverse effect on tenders.

Chapter 3
Liquidated damages

3.1 The meaning and purpose of liquidated damages

Liquidated damages means a fixed and agreed sum as opposed to unliquid-ated damages which is a sum which is neither fixed nor agreed, but must be proved in court, arbitration or adjudication. A more comprehensive defin-ition of liquidated damages is given below. The addition of the words 'and ascertained' to 'liquidated damages' is not thought to be significant.

Litigation is generally recognised as being expensive and lengthy. In order to recover damages in matters involving breaches of contract it is necessary to prove that the defendant had a contractual obligation to the claimant, that there was a failure to fulfil the obligation wholly or partly and that the claimant suffered loss or damage thereby. Very often it is clear that there is damage, but it is difficult and expensive to prove it[90]. To avoid that situation, the parties may decide when they enter into a contract, that in the event of a breach of a particular kind the party in default will pay a stipulated sum to the other. This sum is termed liquidated damages.

In the building industry and elsewhere the terms 'liquidated damages' and 'penalty' are commonly used as though they were interchangeable. In fact, they are totally different in concept. Whereas liquidated damages are compensatory in nature and should be a genuine attempt to predict the damages likely to flow as a result of a particular breach, a penalty is a sum which is not related to probable damages, but rather stipulated *in terrorem*[91] – in other words, as a threat or even, in some instances, intended as a punishment. The courts will enforce the former but not the latter, though the parties may be no less agreed upon the matter in the first instance as in the second[92]. It is, therefore, of prime importance to establish into which category a particular sum will fall.

Building contracts usually include a date on which the contractor may take possession of the site and a further date by which he must have completed the building[93]. Alternatively, the contract may provide for a contract period which is triggered by a notice to commence[94]; or in some other way the building contract will provide a means of fixing the date on which building operations must be finished. It is established that the em-

---

[90] *Clydebank Engineering Co v. Don Jose Yzquierdo y Castenada* [1905] AC 6.
[91] *Cellulose Acetate Silk Co Ltd v. Widnes Foundry* [1933] AC 20.
[92] *Watts, Watts & Co Ltd v. Mitsui & Co Ltd* [1917] All ER 501.
[93] For example, JCT 98 Appendix entries 23.1.1 and 1.3.
[94] For example, GC/Works/1 (1998) clause 34.

ployer must give the contractor possession of the site on the due date and if
he is in breach of that obligation, he is liable in damages[95]. Provided the
contractor is able to enter upon the site on the date stipulated for possession
and thus to commence building work, he must finish by the completion
date. If he fails to complete, the employer may recover such damages under
the principles set out in *Hadley* v. *Baxendale*[96] as he can prove were a direct
result of the breach.

In practice, it may be difficult to allocate damages; which damages directly and naturally flow from the breach and which damages do not so flow
but depend upon special knowledge which the contractor had at the time
the contract was made? The amount of the damage is seldom easy to
ascertain and prove.

For more than 100 years it has been the practice in the building industry to
include a provision for liquidated damages in building contracts to avoid
these difficulties. The way the provision is generally expressed is that the
contractor must pay a certain sum to the employer for every week by which
the original completion date is delayed. That sum must represent a genuine
pre-estimate of the loss which the employer is likely to suffer.

3.2 Liquidated damages or penalty

The rules for deciding whether a sum is to be considered liquidated damages or a penalty were formulated by Lord Dunedin in *Dunlop Pneumatic
Tyre Co Ltd* v. *New Garage & Motor Co Ltd*[97]. These are set out below with
comment.

> '(i) Though the parties to a contract who use the words penalty or liquidated
> damages may prima facie be supposed to mean what they say, yet the expression
> used is not conclusive. The court must find out whether the payment stipulated is
> in truth a penalty or liquidated damages.'

Since *Kemble* v. *Farren*[98], the courts have paid little attention to the terminology adopted by the parties. In that case, not only was the sum expressed
by the parties as liquidated damages, it was clearly stated that it was 'not a
penalty or penal sum'. Notwithstanding the clear words, the court had little
hesitation in finding that the sum was a penalty. In other cases, the courts
have held that sums stated as penalties are in fact liquidated damages:

> 'All the circumstances which have been relied on in the different reported
> cases, as distinguishing liquidated damages from penalty, are to be found
> here. The injury to be guarded against was one incapable of exact
> calculation. The sum to be paid is not the same for every default, for
> that which should occasion small as for that which should occasion great

---

[95] *Rapid Building Group* v. *Ealing Family Housing Association Ltd* (1984) 1 Con LR 1.
[96] (1854) 9 Ex 341.
[97] [1915] All ER 739.
[98] [1829] All ER 641.

inconvenience, but one increasing as the inconvenience would become more and more pressing, and finally, the payments are themselves secured by the penalty of a bond.'[99]

Most modern forms of contract eschew the use of 'penalty' in favour of 'liquidated damages', but the term is often to be found in correspondence, site minutes and occasionally in forms of contract drafted by construction professionals. The term 'delay damages' which for no obvious good reason has been adopted by the NEC contract, seems to be equivalent to liquidated damages.

> '(ii) The essence of a penalty is a payment of money stipulated as in terrorem of the offending party; the essence of liquidated damages is a genuine covenanted pre-estimate of damage'[100]

A sum may be liquidated damages although it is not a genuine pre-estimate; for example, if the sum is agreed at a lower figure. Some examples will be mentioned later.

> '(iii) The question whether a sum stipulated is a penalty or liquidated damages is a question of construction to be decided upon the terms and inherent circumstances of each particular contract, judged of as at the time of making the contract, not as at the time of the breach.'[101]

This rule is in two parts. First that the decision whether a sum is liquidated damages or penalty will hinge not only on the terms of a particular contract, but also on the inherent circumstances of that contract. The second part of the rule is that the terms and inherent circumstances to be considered are those existing at the time the contract was made, not when the term was breached. This is of importance when considering whether a sum is a genuine pre-estimate of loss, particularly when the likely damages were difficult or impossible to forecast at that time, but perfectly clear later. In looking at a sum, it should be considered in the worst possible light just as, if there are several possible breaches, 'the strength of the claim must be taken at its weakest link'[102]. Therefore, if a sum would not normally be considered to be a penalty, but under certain circumstances it would be penal, then it is to be treated as penal in its entirety and the court will not sever any part.

Stanor Electric Ltd v. *R. Mansell Ltd*[103] provides a good example; liquidated damages expressed as a single sum for failure to complete two houses normally would present no problem. It was only the fact that one house was completed and taken into possession before the other that made the sum penal since it was not capable of division. The employer unsuccessfully attempted to deduct half the sum to represent one of the two houses. Had

---

99 *Ranger* v. *Great Western Rail Co* [1854] All ER 321 at 332 per Lord Cranworth LC.
100 *Clydebank* v. *Castenada* [1905] AC 6 was cited as authority for this proposition.
101 *Public Works Commissioner* v. *Hills* [1906] AC 368, [1906] All ER 919, and *Webster* v. *Bosanquet* [1912] AC 394 were cited as authorities for this proposition.
102 *Dunlop Pneumatic Tyre Co Ltd* v. *New Garage & Motor Co Ltd* [1915] All ER 739 at 743.
103 (1988) CILL 399.

both dwellings been delayed by an equal amount, the sum would not have been penal.

The principle is noticeable in the court's approach to hire purchase agreements. Very often, the sum to be paid on breach of the agreement by the hirer is not penal unless the breach occurs near the beginning of the hire period. Unless the sum is a genuine pre-estimate under all circumstances, it will not be upheld[104]. Two Hong Kong cases are instructive, because, although they indicate the same principle, they were overturned on appeal[105]. In each case, the liquidated damages provision in the contract was expressed in a complex form. At first instance, they were held to be void for uncertainty, because it was not easy to calculate the sum to be deducted at any particular stage, and the calculation could result in the sum being penal. The lesson appears to be that where complexities may arise, they should be severed from the primary liquidated damages provision.

In *Dunlop*, Lord Dunedin proceeded to set out tests which could prove helpful or even conclusive:

'(a) It will be held to be a penalty if the sum stipulated for is extravagant and unconscionable in amount in comparison with the greatest loss which could conceivably be proved to have followed from the breach.'

This is probably the most important of the tests. It has been explained thus:

'I do not think the word "unconscionable" there has any reference to the fact that the parties were on an unequal footing. It does not bring in at all the idea of an unconscionable bargain. It is merely a synonym for something which is extravagant and exorbitant.'[106]

The fact that the sum stipulated as liquidated damages bears no relation to the contract sum is not relevant.[107]

The correct burden of proof has been stated like this:

'The onus of showing such a stipulation is a "penalty clause" lies upon the party who is sued upon it. The terms of the clause may themselves be sufficient to give rise to the inference that it is not a genuine estimate of damage likely to be suffered but is a penalty. Terms which give rise to such an inference are discussed in Lord Dunedin's speech in *Dunlop Pneumatic Tyre Co* v. *New Garage & Motor Co* [1915] AC 79 at 87. But it is an inference only and may be rebutted. Thus it may seem at first sight that the stipulated sum is extravagantly greater than any loss which is liable to result from the breach in the ordinary course of things, i.e. the so-called "first rule" in *Hadley* v. *Baxendale* (1854) 9 Exch 341. This would give rise to the prima facie inference that the stipulated sum was a penalty. But the plaintiff may be able to show that owing to special

---

[104] *Landom Trust Ltd* v. *Hurrell* [1955] 1 All ER 839.
[105] *Arnold & Co Ltd* v. *Attorney General of Hong Kong* (1989) 5 Const LJ 263, (1990) 47 BLR 129 CA (Hong Kong) and *Philips Hong Kong Ltd* v. *Attorney General of Hong Kong* (1990) 50 BLR 122, (1991) 7 Const LJ 340 CA (Hong Kong).
[106] *Imperial Tobacco Co* v. *Parsley* [1936] 2 All ER 515 at 521 per Lord Wright MR.
[107] *Imperial Tobacco Co* v. *Parsley* [1936] 2 All ER 515 at 524 per Lord Slesser LJ.

circumstances outside "the ordinary course of things" a breach in those special circumstances would be liable to cause him a greater loss of which the stipulated sum does represent a genuine estimate.'[108]

There appears to be no case where a sum stipulated as liquidated damages in respect of breach of a single obligation has been held to be a penalty on these grounds. The importance of this test lies in its application to multiple breaches or to breaches of multiple obligations.

'(b) It will be held to be a penalty if the breach consists only in not paying a sum of money, and the sum stipulated is a sum greater than the sum which ought to have been paid.'

Lord Dunedin referred to this test as 'one of the most ancient instances'. An example of the operation of this principle is to be found in *Kemble* v. *Farren*[109], where a comedian was engaged to appear on certain nights for £3 6s 8d per night. The contract provided that if either party failed to fulfil the agreement or any part, the party in default would pay the other a sum of £1000. Tindall CJ said:

'But that a very large sum should become immediately payable in consequence of the non-payment of a very small sum, and that the former should not be considered as a penalty, appears to be a contradiction in terms; the case being precisely that in which the courts of equity have always relieved . . . '[110]

Although of little application to building contracts, it probably relies on the fact that where the breach lies in failure to pay a known sum of money, the likely damages are capable of precise calculation, being the sum itself together with, in certain circumstances, interest. Therefore, any greater sum must be a penalty.

'(c) There is a presumption (but no more) that it is a penalty when "a single lump sum is made payable by way of compensation, on the occurrence of one or more or all of several events, some of which may occasion serious and others but trifling damages".[111]'

The application of this principle is clearly to be seen in *Ariston SRL* v. *Charley Records Ltd*[112], where a sum of money claimable if certain manufacturing parts were not returned within 10 working days was held to be a penalty, because the same sum was payable whether the whole or just a few of the parts were late and in the latter case, the sum would be extravagant in relation to the greatest likely loss. It is suggested that the principle is the key to some other decisions in relation to building contracts where there have been no proper provisions for dividing a single sum, expressed as liquid-

---

[108] *Robophone Facilities Ltd* v. *Blank* [1966] 3 All ER 128 at 142 per Diplock LJ.
[109] [1829] All ER 641.
[110] [1829] All ER 641 at 642.
[111] *Lord Elphinstone* v. *Monkland Iron & Coal Co* (1886) 11 App Cas 332 at 342 per Lord Watson.
[112] (1990) The Independent 13 April 1990.

ated damages, to allow for the completion and taking into possession of part of a building[113].

> '(d) It is no obstacle to the sum stipulated being a genuine pre-estimate of damage that the consequences of the breach are such as to make precise pre-estimation almost an impossibility. On the contrary, that is just the situation when it is probable that pre-estimated damage was the true bargain between the parties.'[114]

This principle is of overriding importance in situations where the other tests have produced an inconclusive result.

There is a strong inference that a sum is liquidated damages where the parties have agreed a sum or sums as liquidated damages and the sum claimed is not excessive in relation to the actual loss suffered. It has been neatly summed up:

> 'The fact that the issue has to be determined objectively, judged at the date the contract was made, does not mean what actually happens subsequently is irrelevant. On the contrary it can provide valuable evidence as to what could reasonably be expected to be the loss at the time the contract was made.'[115]

More recently, the Court of Appeal set out four principles to differentiate liquidated damages from a penalty:[116]

(1) The parties' intentions must be identified by examining the substance rather than the form of words used.
(2) A sum would not be a penalty where a genuine pre-estimate of loss had been carried out.
(3) The contract should be construed at the time the contract was made, not at the time of the breach.
(4) It would be a penalty if the amount was extravagant or unconscionable compared to the greatest foreseeable loss.

It may be useful at this stage to summarise the effect of the rules and tests in the light of other judicial decisions:

(1) Where there is a single event and the pre-estimate of likely loss is relatively easy, a sum will be a penalty if it is greater than such loss, but otherwise liquidated damages[117].
(2) Where there is a single event and the pre-estimate of likely loss is difficult, the sum is more likely to be liquidated damages as the difficulty of pre-estimation increases[118].

---

[113] See, for example, *Stanor Electric* v. *R. Mansell* (1988) CILL 399 and *M. J. Gleeson* v. *Hillingdon Borough Council* (1970) 215 EG 165.
[114] *Clydebank* v. *Castenada* [1905] All ER 251 and *Webster* v. *Bosanquet* [1912] AC 394 were cited as authority for this proposition.
[115] *Philips Hong Kong Ltd* v. *The Attorney General of Hong Kong* (1993) 61 BLR 41 at 59 per Lord Woolf repeated in *Ballast Wiltshire PLC* v. *Thomas Barnes & Sons Ltd*, 29 July 1998, unreported at paragraph 50 per Judge Bowsher.
[116] *Jeancharm Ltd* v. *Barnet Football Club Ltd* (2003) CILL 1987 at 1989 per Gibson LJ.
[117] *Kemble* v. *Farren* [1829] All ER 641.
[118] *Dunlop Pneumatic Tyre Co Ltd* v. *New Garage & Motor Co Ltd* [1915] All ER 739.

(3) Where there are several events and the pre-estimate of likely loss in respect of any one of them is relatively easy, a sum will be a penalty if it is greater than such loss, but liquidated damages otherwise.
(4) Where there are several events and the pre-estimate of likely loss is difficult, the sum is likely to be liquidated damages, but other factors must be taken into account:

  (a) If one sum is payable in respect of several events which result in different kinds and amounts of loss, it is likely to be a penalty[119].
  (b) If one sum is payable in respect of several events and the damage is the same in kind, but giving rise to differing amounts of loss, it may be liquidated damages[120].
  (c) If one sum is expressly stated to be an average of the pre-estimated loss resulting from each of several events, it is likely to be liquidated damages[121].
  (d) Where different sums are payable in respect of different events, they are likely to be liquidated damages[122].

The two important considerations are the extent to which an accurate pre-estimate of loss can be carried out and the existence of different events, each of which are said to give rise to liquidated damages. But the decision of the Privy Council of the House of Lords in *Philips Hong Kong Ltd* v. *Attorney General of Hong Kong* is significant. The Law Lords held that hypothetical situations cannot be used to defeat a liquidated damages clause. The court will take a pragmatic approach:

> 'Whatever the degree of care exercised by the draftsman it will still be almost inevitable that an ingenious argument can be developed for saying that in a particular hypothetical situation a substantially higher sum will be recovered than would be recoverable if the plaintiff was required to prove his actual loss in that situation. Such a result would undermine the whole purpose of the parties to a contract being able to agree beforehand what damages are to be recoverable in the event of a breach of contract. This would not be in the interest of either of the parties to the contract since it is to their advantage that they should be able to know with a reasonable degree of certainty the extent of their liability and the risks which they run as a result of entering into the contract.'[123]

3.3 Liquidated damages as limitation of liability

It appears that a sum will be classed as liquidated damages if it can be said of it that it is a genuine pre-estimate of the loss or damage which would probably arise as a result of the particular breach[124]. The figure inserted in

---

[119] *Ford Motor Co* v. *Armstrong* (1915) 31 TLR 267.
[120] *Dunlop Pneumatic Tyre Co Ltd* v. *New Garage & Motor Co Ltd* [1915] All ER 739.
[121] *English Hop Growers* v. *Dering* [1928] 2 KB 174.
[122] *Imperial Tobacco Co* v. *Parsley* [1936] 2 All ER 515.
[123] (1993) 61 BLR 41 at 54 per Lord Woolf.
[124] *Philips Hong Kong Ltd* v. *Attorney General of Hong Kong* (1993) 61 BLR 41PC.

the contract must be a careful and honest attempt to accurately calculate the loss or damage which will be suffered and it must be a pre-estimate in the sense that it must be an estimate at the time the contract is made, not at the time of the breach[125]. It appears that the courts have made exceptions to this rule: the sum may be less than that which would represent an accurate forecast of probable loss perhaps because a party wishes to limit his liability:

> 'I agree that it is not a pre-estimate of actual damage. I think it must have been obvious to both parties that the actual damage would be much more than £20 a week, but it was intended to go towards the damage, and it was all that the sellers were prepared to pay. I find it impossible to believe that the sellers, who were quoting for delivery at 9 months without any liability, undertook delivery at 18 weeks, and in so doing, when they engaged to pay £20 a week, in fact made themselves liable to pay full compensation for all loss.'[126]

Clauses inserted as limitations of liability must now be examined in the light of the Unfair Contract Terms Act 1977. Section 3, in part, states that:

> 'This section applies as between contracting parties where one of them deals as a consumer or on the other's written standard terms of business.... As against that party, the other cannot by reference to any contract term.... when himself in breach of contract, exclude or restrict any liability of his in respect of the breach.... except in so far as (in any of the cases mentioned above in this subsection) the contract term satisfies the requirement of reasonableness.'

Where it can be shown that the lower sum was inserted as a limitation of liability and where one party deals on the other's standard written terms of business, the term must satisfy the requirements of reasonableness set out in section 11 and Schedule 2. In most building industry cases, the limitation of liability should be easily attributable to the application of sound business principles. It should not be ruled out that, in some instances, the limitation could be shown to be unreasonable and, therefore, unenforceable. In most cases, the liquidated damages are inserted into the contract by the employer and they are thought of as being for his benefit. Therefore, if they are less than one might expect, the employer must have had his own good reasons for it.

3.4 Sums greater than a genuine pre-estimate

It also seems that the courts are willing to countenance sums which are greater than that which would constitute a genuine pre-estimate in certain limited circumstances. The point was considered in *The Angelic Star*[127]. The

---

[125] *Public Works Commissioners* v. *Hills* [1906] All ER 919, [1906] AC 368.
[126] *Cellulose Acetate Silk Co Ltd* v. *Widnes Foundry (1925) Ltd* [1933] AC 20 at 23 per Lord Atkin.
[127] [1988] 1 Lloyds Rep 122.

case arose in connection with repayment of a substantial loan, on which the borrowers defaulted. A term of the agreement required immediate repayment of the whole sum of the outstanding balance on default. The Court of Appeal held that the provision was not a penalty, per Gibson LJ:

> 'Parties to a contract are free expressly to stipulate not only the primary obligations and rights under the contract but also the secondary rights and obligations, i.e. those which arise upon non-performance of any primary obligation by one of the parties to the contract.'[128]

This was subject to the rule of public policy against enforcing sums which the court is not satisfied are genuine estimates of the loss likely to be sustained. The court appears to have looked upon the repayment provision as a form of liquidated damages. The principles of genuine pre-estimate apply where the damages are expressed other than as money; for example, if a breach of obligations is to give rise to a transfer of property as liquidated damages.

3.5 Liquidated damages as an exhaustive remedy

A question often arises whether a party to a contract containing a liquidated damages clause can sue for actual damages suffered or whether the party is restricted to the sum expressed as liquidated damages. In principle, where parties enter into a contract, it must be assumed that they know what they are doing and that the contract is an expression of their intentions[129]. It follows that if parties agree that in the event of a particular kind of breach liquidated damages are payable by the party in breach, that agreement will be upheld by the courts and they will be allowed no other or alternative damages but the damages liquidated in the contract.

The sum expressed as liquidated damages was held to be exhaustive of the remedies available to the claimant for late completion in *Temloc Ltd* v. *Errill Properties Ltd* where the amount of liquidated damages was stated to be '£nil'[130]. It was held that the parties had agreed that, in the event of late completion, no damages should be applied. Even if a rate had been stated, the court considered that the rate would have represented an exhaustive agreement as to damages which were or were not to be payable by the contractor in event of his failure to complete on time. That, of course, does not preclude the employer from recovering as unliquidated damages other losses not directly caused by the breach of obligation to complete, but which may be connected to such breach. In *M. J. Gleeson plc* v. *Taylor Woodrow Construction Ltd*[131] a similar conclusion was reached when the court refused to allow the set-off of sums for which a liquidated damages figure

---

[128] *The Angelic Star* [1988] 1 Lloyds Rep 122 at 127.
[129] *Liverpool City Council* v. *Irwin* [1976] 2 All ER 39.
[130] (1987) 39 BLR 30.
[131] (1989) 49 BLR 95.

had been inserted in the contract and already deducted. This was on the straightforward principle that damages cannot be recovered twice for the same breach of contract. This view is supported by earlier decisions[132].

This principle should be distinguished from the situation where the defendant is in breach of two or more obligations, for one of which the stipulated remedy is liquidated damages and for the other(s) the remedy is to sue for unliquidated damages. A related situation is where there is but one breach which gives rise to a loss which may be said to trigger a remedy in liquidated damages and a separate kind of loss for which other damages are appropriate. The former situation is illustrated in *E. Turner & Sons Ltd* v. *Mathind Ltd*[133], where a number of flats were to be completed in stages and there was a final completion date for the whole development. Liquidated damages were stipulated only for failure to meet the final completion date. Although expressed obiter, it was the court's view that the liquidated damages clause, standing alone, was not an effective exclusion of any right to damages for earlier breaches of obligation. There was every reason to suppose that the parties intended the staging provisions to be contractual, possibly leading to a higher contract price. Without a specific overriding provision, breach of such provisions results in damages. The decision was curious because the development was carried out on the basis of the Standard Form of Building Contract 1963 Edition (JCT 63). It had a provision in clause 12(1) which was similar to the current clause 2.2.1 of JCT 98 to the effect that nothing in the bills of quantities was to override, modify or affect in any way whatsoever the application or interpretation of the 'Conditions'. This provision, although contrary to the normal rule that 'type prevails over print' has been upheld by the courts[134]. The staging provision was to be found only in the bills of quantities.

The only other case in point had been *M. G. Gleeson (Contractors) Ltd* v. *Hillingdon Borough Council*[135], where the argument really seems to have been about whether the single sum of liquidated damages could be distributed over the stages noted in the bills of quantities. The court had held that, on the basis of clause 12(1), such an interpretation could not be upheld. It is interesting to speculate whether the claimant in that case would have met with more success had he argued on the basis that there were two distinct breaches: breach of the obligation to complete the whole development on a fixed date for which the remedy was a sum set as liquidated damages; and breach or breaches of the obligation to comply with a set of intermediate dates for which the remedy was unliquidated damages.

In this context, it is useful to look at *Ford Motor Company* v. *Armstrong*[136], where the claimants agreed with the defendant that he should sell their

---

[132] See *Diestal* v. *Stevenson* [1906] 2 KB 345 and *Talley* v. *Wolsey-Neech* (1978) 38 P & CR 45, where the courts prevented the claimants from recovering amounts greater than those stipulated by way of liquidated damages.

[133] (1986) 5 Const LJ 273, CA.

[134] See, for example, *English Industrial Estates Corporation* v. *George Wimpey & Co Ltd* [1973] 1 Lloyds Rep 51.

[135] (1970) 215 EG 165.

[136] (1915) 31 TLR 267.

motor cars. He was to pay the claimants the sum of £250 if he should breach the agreement in any of the three following ways: by selling cars or parts at below the list price; by selling cars to persons or firms engaged in the motor car industry; and by exhibiting cars at any exhibition without the claimants' written permission. A majority of the Court of Appeal held the provision to be a penalty, and therefore unenforceable, on the basis that the breaches were not of the same kind. The claimants had argued that the reasoning behind the provision was to guard against damage to their business by the wholesale undercutting of the list prices. To that extent, the argument was the same as was put forward in *Dunlop* v. *New Motor*. The argument failed, because the breaches in *Dunlop*, although different in degree, were of the same type and each of the breaches could clearly be seen to have the same ultimate effect. In *Ford*, the breaches were quite different, in degree, type and result. The deciding factor appears to have been the fact that the same figure of £250 could not be considered as a genuine pre-estimate in respect of each of the three sets of breaches. On the basis of the judgment, it appears that the claimants could have avoided trouble by fixing different sums in respect of the three types of breaches. Alternatively, they could have fixed a sum of liquidated damages for the breach of selling below list price and sued in respect of the other breaches to obtain whatever damages they could prove.

It is debatable whether there were two breaches or just one in the situation considered in *Aktieselskabet Reidar* v. *Arcos*[137]. This concerned delay in loading cargo for which demurrage was stipulated. The meaning of demurrage has been stated thus:

> 'The word "demurrage" no doubt properly signifies the agreed additional payment (generally per day) for an allowed detention beyond a period either specified in or to be collected from the instrument; but it has also a popular or more general meaning of compensation for undue detention . . .'[138]

This meaning is very close to liquidated damages, particularly as commonly encountered in a construction situation, i.e. for delay in completion. The court appears to have dealt with demurrage in exactly the same way as liquidated damages. The principles to be derived from this case are thought-provoking. In essence, the facts are simple. The defendants failed to load a cargo at the agreed rate and as a result the ship was detained beyond the time stipulated. The delay also meant that the ship was only allowed to carry a winter cargo instead of the heavier summer cargo and the claimants suffered loss of freight. The claimants brought the action to recover demurrage (liquidated damages) for the period the ship was detained in port beyond the lay (allowed) days together with, as damages, the difference between the amount of freight the claimants would have earned if the

---

[137] [1926] All ER 140.
[138] *Lockhart* v. *Falk* (1875) LR 10 Exch at 135 per Cleasby B.

defendants had loaded at the correct rate and the amount which they did earn.

The members of the Court of Appeal took different approaches to the problem. Banks LJ thought that the obligation to load a full and complete cargo and the obligation to load the cargo at a stipulated rate were separate obligations. He went on to say:

> 'At one time I was inclined to think that, where parties had agreed a demurrage rate, the contract should be construed as one fixing the rate of damages for any breach of the obligation to load or discharge in a given time. On further consideration I do not think that such a view is sound. I can find no authority on the point, and it is noticeable that in the *Saxon Steamship Case*[139] it was not suggested that the claim for demurrage excluded the additional claim for special damage arising from the detention of this vessel.'[140]

Atkin LJ held that the 'provisions as to demurrage quantify the damages not for the complete breach, but only such damages as arise from the detention of the vessel'[141]. Sargant LJ was of the opinion that 'the same delay in loading, which might give rise to a claim for detention, also resulted in a breach of the obligation to load a full cargo,'[142] but there was a definite separate loss.

There is no doubt that the defendants were in breach of their obligation to load at a specified rate. The breach caused them to overrun their allotted time. The result of that was an obligation to compensate the claimants by liquidated damages. If, by overrunning the time period, the defendants had been able to load the agreed amount of cargo, the claimant's only loss would have arisen from the overrun itself, and liquidated damages would have been adequate. That is clearly what the parties contemplated at the date of the charterparty. The overrun, however, resulted in the defendant's failure to load the specified cargo in order to comply with the regulations for winter cargo which only applied as a result of the overrun. So the claimants suffered the additional loss due to a smaller cargo. If the cargo had been loaded at a rather quicker rate, it might have been possible for the full cargo to have been loaded before the winter rate applied.

The rate of loading is significant in that when it fell below a particular figure, it triggered the second breach. *Aktieselskabet Reidar* was considered in the House of Lords where it was thought that, 'There was a breach separate from although arising from the same circumstances as the delay . . . '[143] and that 'there were in that case breaches of two quite independent obligations; one was demurrage for detention . . . , the other was a failure to load a full and complete cargo . . . '[144]

---

[139] *Saxon Steamship Co Ltd* v. *Union Steamship Co Ltd* (1900) 69 LJQB 907.
[140] *Aktieselskabet Reidar* v. *Arcos Ltd* [1926] All ER 140 at 145.
[141] *Aktieselskabet Reidar* v. *Arcos Ltd* [1926] All ER 140 at 145.
[142] [1926] All ER 140 at 147.
[143] *Suisse Atlantique, etc.* v. *N V Rotterdamsche Kolen Centrale* [1966] 2 All ER 61 at 77 per Lord Hodson.
[144] [1966] 2 All ER 61 at 83 per Lord Upjohn.

In *Total Transport Corporation* v. *Amoco Trading Co (The 'Altus')*, the judge considered *Aktieselskabet Reidar* v. *Arcos* and concluded:

> 'I must treat the ratio decidendi of the case as being that where a charterer commits any breach, even if it is only one breach, of his obligation either to provide the minimum contractual load or to detain the vessel for no longer than the stipulated period, the owner is entitled not only to the liquidated damages directly recoverable for the breach of the obligation to load (dead freight) or for the breach of the obligation with regard to detention (demurrage), but also for, in the first case, to the damages flowing indirectly or consequentially from any failure to load a complete cargo if there is such a failure.'[145]

He proceeded to hold that where the charterer was in breach of his obligation to provide the minimum load, the owner was entitled to the damages flowing directly or indirectly from the failure in addition to liquidated damages for the breach of such obligation. On the facts of the case he was probably right. The probable consequences of the breach were known to both parties. It is difficult to accept his analysis of *Aktieselskabet*. The correct analysis must surely be that one default gave rise to two breaches because of the particular circumstances. Liquidated damages are certainly due as a result of the first breach and further damages may be due as a result of the second breach depending upon the facts[146]. The essential principle to be extracted from these cases is that although a breach for which liquidated damages are specified will give rise to such damages, they may not be the limit of a party's entitlement to damage resulting from the breach. This view appears to be supported by the following:

> 'The damages payable in respect of late completion of the works are one head of the general damages which may be recoverable by an employer for the contractor's breach of a building contract.'[147]

It is similar to the situation which sometimes occurs in building when a development must be completed by a particular date or it is ineligible for a grant. There, the failure to complete by the completion date would attract liquidated damages. Failure to complete by the further cut-off date for grant purposes would deprive the employer of the grant, but whether he could recover that amount from the builder would depend on whether it could be brought within the principles of special damages[148], i.e. whether the parties could reasonably be supposed to have contemplated such a result from the breach at the time they made the contract.

3.6 Injunction

Although a party cannot opt for unliquidated damages if liquidated damages have been set out in the contract, it seems that, if appropriate, a party

---

[145] [1985] 1 Lloyds Rep 423 at 435 per Webster J.
[146] *Koufos* v. *Czarnikow Ltd (The Heron II)* [1969] 1 AC 350 HL.
[147] *Temloc Ltd* v. *Errill Properties Ltd* (1987) 39 BLR 30 at 30 per Nourse LJ.
[148] See the full discussion of special damages in Chapter 5, section 5.2.

may opt for an injunction instead. In *General Accident Assurance Corporation* v. *Noel*[149], it was held that where a party was in breach of a covenant in restraint of trade, the injured party could not have both an injunction to restrain further breaches and liquidated damages in respect of the breaches already committed. The court concluded that the claimants had an option to elect between, but could not have both remedies. It is suggested that this is the correct answer to the problem posed when a party commits this kind of breach. If it is assumed that the breach must cause the innocent party undoubted but not readily quantifiable harm, liquidated damages appears ideally suited to the situation. But if the award of damages, as in this case, is expressed as a single sum, it may be argued that if the damages are paid, the party in effect has a licence to carry on committing the breach, because the injured party can recover no more. The answer to that argument seems to be that a party had the opportunity to make an appropriate bargain. An appropriate bargain in this case might well have been to have stipulated not a single sum as liquidated damages, but a sum for every week that the breach continued or, as in *Dunlop Pneumatic Tyre Co Ltd* v. *New Garage & Motor Co Ltd*, for each separate breach. If a single lump sum is stipulated, it must be assumed that it is calculated on the basis that any breach, whether brief or protracted, would have the same overall effect on the claimant's trade. Although that may be questioned in theory, in practical terms there is much to commend that approach[150]. It has been said, although probably *obiter*:

> 'Where there are different breaches and the agreement provides for a particular sum of liquidated damages to be payable for each and every breach, there is no bar to awarding the liquidated damages amount for each breach which has occurred to date of trial and also awarding an injunction to restrain future breaches.'[151]

The judge went on to say that there was no double recovery, because the two remedies were referable to different breaches. This contention seems to ignore two principles. The first is that where liquidated damages are stipulated they are exhaustive of the remedies available for a breach, and the second is that an injunction is normally refused if damages would be an adequate remedy. By agreeing a figure, be it a single sum or a sum for each

---

[149] [1902] 1 KB 377: a case dealing with a covenant in restraint of trade.

[150] *English Hop Growers* v. *Dering* [1928] 2 KB 174. The decision in *General Accident* v. *Noel* should be compared to *The Imperial Tobacco Co Ltd* v. *Parsley*. This case has been quoted as authority for the proposition that a party can recover liquidated damages and obtain an injunction where the sum stipulated as liquidated damages is graded according to the extent of the breach. The case concerned a price maintenance agreement by which the defendant undertook to pay the claimants £15 for every sale in breach of the agreement. The trial judge granted an injunction to restrain further breaches, but he refused to enforce the series of £15 payments on the basis that they were really penalties. As there was no appeal on the injunction, but only on the question whether the payments were penalties or liquidated damages, this case appears shaky ground on which to found any contention that both liquidated damages and an injunction are available remedies in certain instances.

[151] *Lorna P. Elsley, Executrix of the Estate of Donald Champion Elsley* v. *J. G. Collins Insurance Agencies Ltd* (1978) 4 Const LJ 318 at 320 per Dickson J. This Canadian case concerned a covenant in restraint of trade on breach of which the defendant was to pay liquidated damages.

breach, the parties are accepting that it is a complete remedy. This is the very foundation of the principle of liquidated damages. Under normal circumstances, the point has little application to a building situation. Liquidated damages are normally expressed as being payable for failure to complete by the contract completion date. Such a breach is not susceptible to easy remedy by injunction. Special constructional works, however, may require particular provisions to which these principles may apply.

3.7 *Liquidated damages in relation to loss*

The next question is whether a claimant is entitled to recover the amount specified as liquidated damages if the damage actually suffered is less than the amount or nothing at all. Indeed, is he able to recover liquidated damages though it can be demonstrated that he has actually gained from the breach? It is settled that a party can recover liquidated damages without being put to proof of actual loss[152]. If that is correct, it seems obvious that in some instances the actual loss will be greater and sometimes less than the sum in the contract. Indeed, it follows that in certain instances there will be no loss whatever.

The principle was applied in *BFI Group of Companies Ltd* v. *DCB Integration Systems Ltd*[153]. There, on an appeal from the award of an arbitrator, it was found that, although the claimants had suffered no actual loss as a result of being unable to use two vehicle bays, because they had, in any event, to execute fit out works after possession before being able to attract revenue, they were entitled to liquidated damages. The form of contract was on MW 80 terms and provided for the payment of liquidated damages if completion was delayed beyond the completion date. The claimants were given possession by the contractor on the extended date for completion although the arbitrator found that practical completion had not taken place. Had they not been given possession, they would have been obliged to wait until practical completion was certified before being able to execute the fit-out works. Unlike other forms of contract, such as JCT 98 or IFC 98, in MW 98 there is no provision for possession of part of the works before practical completion and the possession granted to the claimants in this case was a concession. Therefore, the claimants were able to carry out work during the period within which they were receiving liquidated damages. This represented a considerable advantage to the claimant.

A similar point arose in *Golden Bay Realty Pte Ltd* v. *Orchard Twelve Investments Pte Ltd*[154], an appeal to the Privy Council concerning liquidated damages in an agreement for the sale and purchase of commercial property. Lord Oliver, speaking of the calculation of the damages, expressed the view of the Privy Council that it was

---

[152] *Clydebank* v. *Castenada* [1905] AC 6.
[153] (1987) CILL 348.
[154] [1991] 1 WLR 981.

'difficult to support as a genuine pre-estimate of the damage likely to be suffered from delay in completion in any case. Particularly this would be so in a case in which the building is complete at the date of the contract and the purchaser is let into possession under the terms of the contract.'[155]

The purpose intended in this instance was to enable the claimant, among other things, to commence the fit-out works. It is difficult to fault the conclusion in *BFI* v. *DCB* on the facts as found by the arbitrator. The clear words of most contracts allow liquidated damages for the period between the date when the works should have reached completion until the date of practical completion. Unless possession is taken by the employer strictly in accordance with the contract terms, unlawful possession by the employer is not a trigger for the end of liquidated damages, despite what many adjudicators appear to think[156].

3.8   *Where there is no breach of contract*

It seems that the question whether a sum stipulated for payment on the happening of a particular event is a penalty or liquidated damages will be irrelevant if the event does not constitute a breach of obligation on the part of one of the parties. The situation has frequently arisen in connection with, but it is not confined to, hire purchase agreements. In *Associated Distributors Ltd* v. *Hall and Hall*[157] the agreement provided that if the hirer wished, he could determine the agreement. The owner was also entitled to determine if the hirer was in default with payments. On determination for any reason, the hirer must pay certain sums of money to the owners. Slesser LJ summarised the position in this way:

'This is a case where the hirer has elected to terminate the hiring. He has exercised an option, and the terms on which he may exercise the option are those set out in clause 7. The question, therefore, whether these payments constitute liquidated damages or penalty does not arise in the present case for determination.'[158]

This approach appeals as a very straightforward solution to the problem. *Lombard North Central PLC* v. *Butterworth*[159] was a case dealing with hire of computer equipment. The parties had stipulated that certain terms were to be treated as conditions. One of these terms was that on the hirer's failure to pay any single instalment, the owner was entitled to recover the goods

---

[155] [1991] 1 WLR 981 at 986.
[156] In *Impresa Castelli SpA* v. *Cola Holdings Ltd* (2002) CILL 1904, the employer had occupied part of the works, but not under the partial possession clause of WCD 98. It was held that such occupation did not amount to partial possession, there was no mechanism to reduce liquidated damages and, therefore, the full amount of liquidated damages could be recovered.
[157] [1938] 1 All ER 511, CA.
[158] [1938] 1 All ER 511, CA at 513.
[159] [1987] QB 527.

together with arrears of rentals, all further rentals which would have fallen due and damages for breach of the agreement. It was held not to be a penalty.

The right of parties to make their own bargain within specified limits has long been sacred and that seems to be the key to unravelling these decisions. It has been said:

> '... one purpose, perhaps the main purpose, of the law relating to penalty clauses is to prevent a plaintiff recovering a sum of money in respect of a breach of contract committed by a defendant which bears little or no relationship to the loss actually suffered by the plaintiff as a result of the breach by the defendant. But it is not and never has been for the courts to relieve a party from the consequences of what may in the event prove to be an onerous or possibly even a commercially imprudent bargain.'[160]

The case concerned a number of interlocking contracts of great complexity. In essence, the matter for consideration was whether breach of an obligation by one party to another could give rise to payment by a third party or was such a payment to be considered a penalty and unenforceable. A significant statement was made in the course of the appeal:

> '...the mere fact that a person contracts to pay another person, on a specified contingency, a sum of money which far exceeds the damage likely to be suffered by the recipient as a result of that contingency does not by itself render the provision void as a penalty.'[161]

The House of Lords concurred with this view. These and other similar cases give food for thought in the context of building industry contracts[162]. In each of the cases, the sum of money is payable on the occurrence of an event. This event is the termination. The difficulties seem to have arisen due to the specified grounds for termination. In each case, termination may take place at the instance of either party and some of the grounds for termination by the owner are breaches by the hirer. In one of the cases, even trivial breaches were made conditions so as to enable the owner to terminate.

Lord Denning drew attention to what he termed the 'absurd paradox' that if a hirer under a hire purchase agreement lawfully terminated the agreement, he would not be able to say the sum then payable by him according to the terms of the agreement was a penalty, but he would be able to do so about the same term if the agreement was terminated as a result of his breach of contract[163].

The courts seem to be agreed that no question of liquidated damages or penalties arises unless a breach of contract is involved. A Hong Kong case

---

[160] *Export Credits Guarantee Department* v. *Universal Oil* [1983] 2 All ER 205 at 222 per Lord Roskill.
[161] [1983] 2 All ER 205 at 215 per Slade LJ.
[162] See also *Cooden Engineering Co Ltd* v. *Stanford* [1952] 2 All ER 915 CA; *Campbell Discount Co Ltd* v. *Bridge* [1962] AC 600; *Re Apex Supply Co Ltd* [1941] 3 All ER 473; *Alder* v. *Moore* (1961) 2 QB 57, CA.
[163] *Campbell Discount Co Ltd* v. *Bridge* [1962] AC 600 at 629.

put this line of reasoning into effect in a construction contract[164]. But where a sum is payable on one of several events, some being breaches and some simply options, the courts have been less sure. In some instances they have avoided the issue by concentrating on the precise matter before them and ignoring the wider connotations; such a case was *Associated Distributers* where only the hirer's option to terminate was considered.

This point gives rise to an interesting speculation with regard to the provision for liquidated damages in building contracts. It is common practice that a contractor, in pricing his tender, will take account of the stipulated amount of liquidated damages. If he considers that the period stated for completing the work is insufficient, he may decide upon the period he requires and calculate the difference in liquidated damages, adding the amount to his tender figure although almost certainly disguised. It could be said that such a contractor who completes the work in, say, 10 months instead of 9 months is exercising an option. In building contracts, time is not of the essence, because there is provision for extending time in certain specified instances. However, if the contractor fails to complete by the contract completion date, he is in breach of contract and there is now the authority of the Court of Appeal that liquidated damages are not an agreed price to permit the contractor to continue his breach of contract[165].

If an otherwise penal sum were to be inserted, the employer may be able to argue that the sum payable was not a penalty following a breach, but simply the figure agreed by the parties as payable on the contractor opting to complete later than the contract completion date. In the hiring cases, there is an express clause which permits either party to terminate. There is no express clause in building contracts to enable the contractor to exceed the stipulated contract period. Although the contractor cannot claim that liquidated damages is an agreed price enabling him to continue his breach of contract, there appears to remain the possibility that the employer could give the contractor that option. It is interesting to speculate that, on that argument, a single sum would not be struck out on the basis that it was 'extravagant and unconscionable'[166].

3.9   Calculation of liquidated damages

Pre-estimation of loss is seldom easy. The employer may have little idea how much loss he may suffer if the building is not completed by the due date, particularly if the contract period is to be counted in years rather than months. Although it has been held that liquidated damages are especially suited to situations where precise estimation is almost impossible[167], the employer should do his best to calculate as accurate a figure as possible. The

---

[164] *Icos Vibro Ltd* v. *SFK Construction Management Ltd* (1992) APCLR 305.
[165] *Bath & North East Somerset District Council* v. *Mowlem* (2004) CILL 2081.
[166] *Dunlop Pneumatic Tyre Co Ltd* v. *New Garage & Motor Co Ltd* [1915] All ER 739 at 742 per Lord Dunedin.
[167] *Dunlop Pneumatic Tyre Co Ltd* v. *New Garage & Motor Co Ltd* [1915] All ER 739.

employer should include every item of additional cost which he predicts will flow directly from the contractor's failure to complete on the due date; that is, the damages recoverable under the first limb of the rule in *Hadley* v. *Baxendale*. It seems that the sum can be increased to include amounts which would normally only be recoverable under the second limb if the employer can show that special circumstances were involved[168]. It remains unclear whether, in the case of liquidated damages, the special circumstances must be known to the contractor when the contract is made. It seems appropriate to reveal such circumstances at tender stage although it could be argued that the higher figure for liquidated damages is itself a sufficient prior notification.

From a purely practical point of view, an employer will very often reduce such a figure in order to make the proposed damages more palatable to prospective tenderers. The Association of Consultant Architects Form of Building Agreement (ACA 3) is alone among standard forms of main contract in providing for unliquidated damages as an alternative. Some local authorities and other public bodies make use of a formula calculation which basically depends upon a percentage of the capital sum. Whether that would constitute liquidated damages will depend on the precise circumstances and particularly the difficulty with which a precise calculation could be made. Use of a formula is a perfectly sensible approach where it is obvious that substantial loss will be suffered in the event of a delay, but it is virtually impossible to calculate precisely in advance what that loss would be[169].

In *Multiplex Constructions Pty Ltd* v. *Abgarus Pty Ltd and Another*[170] a specially drafted liquidated damages clause was held to be entirely valid and enforceable despite the absence of any specified sum. The damages were expressed as two parts. The first part was to be the interest calculated with reference to Trading Banks on daily balances of the total of items listed in the clause. The items included: 'Payments made by the Proprietor under any contract relating to the execution of the Works' and 'Reasonable costs and expenses incurred by the Proprietor in enforcing or attempting to enforce any contract relating to the execution of the Works'. The other items were equally imprecise. The second part was rates, statutory charges 'and other reasonable outgoings...'.

Although referred to in the contract and by the court as 'liquidated damages', it is difficult to see how such a clause can justify that description. An important aspect of liquidated damages is that it is a known amount at the time the parties enter into the contract. Although that does not preclude the damages being expressed as a method of calculation, such a method should be known to have a certain result in any given set of circumstances. In *Multiplex* the individual items could not always be ascertained. Works

---

[168] *Philips Hong Kong Ltd* v. *Attorney General of Hong Kong* (1993) 61 BLR 41. A format/checklist for calculating liquidated damages is to be found in Chappell, Marshall, Powell-Smith and Cavenders' *Building Contract Dictionary* 3rd edition, 2001, Blackwell Publishing.
[169] *Philips Hong Kong Ltd* v. *The Attorney General of Hong Kong* (1993) 61 BLR 41 PC.
[170] [1992] APCLR 252.

such as 'charges assessed', 'reasonable costs', 'reasonably necessary' and 'reasonable outgoings' introduce elements of judgment which have no place in the calculation of liquidated damages after the event. Employers introducing clauses of that kind are simply courting disputes.

3.10 Maximum recovery if a sum is a penalty

A practical problem concerns the employer's position if liquidated damages are held to be a penalty. Is he restricted to recovery of such amount as he can prove up to, but not greater than, the amount of the sum held to be penal? One eminent commentator has come to the conclusion that the amount stipulated as a penalty is not a ceiling on the amount of damages recoverable[171], while another thinks the question is still open, at least in so far as building contracts are concerned[172]. In an early judgment in the Court of Appeal, Kay LJ traced the effect of courts of equity on sums stipulated as penalties and noted that if the actual damages could easily be estimated, 'the penalty would be cut down and the actual damage suffered would be assessed'[173]. No qualification is placed upon the statement and, at face value, it could be taken as authority for the assessment of damage of any amount, even greater than the penalty sum itself. It would probably be going too far to construe the remarks in that way, since removing a penalty in favour of actual damages is hardly likely to have been equitable if it resulted in the sum payable being thereby increased[174].

A strong argument against the penal sum being a ceiling on possible damages is to be found in the following extract:

'Now where a contract contains a clause which is in form indisputably a penalty clause the position of the parties was thus described by Lord Mansfield in *Lowe* v. *Peers*[175]: "There is this difference between covenants in general, and covenants secured by a penalty or forfeiture. In the latter case the obligee has his election. He may either bring an action of debt for the penalty, and recover the penalty; (after which recovery of the penalty he cannot resort to the covenant, because the penalty is to be in satisfaction for the whole;) or, if he does not choose to go for the penalty, he may proceed upon the covenant, and recover more or less than the penalty toties quoties".'[176]

---

[171] Harvey McGregor, *McGregor on Damages*, 15th edition, 1988, Sweet & Maxwell at 283.
[172] Stephen Furst and Vivian Ramsey, *Keating on Building Contracts*, 7th edition, 2001, Sweet & Maxwell at 288.
[173] *Law* v. *Redditch Local Board* [1892] All ER 839 at 895.
[174] *Diestal* v. *Stevenson* [1906] 2 KB 345 at first sight appears to be authority that the penal sum is not a ceiling on what is recoverable, but in that case the judge used the words 'it is agreed' – a clear indication that he was not deciding the matter. In the event, the sum was held to be liquidated damages, despite being referred to as a penalty, and the judge had no further need to refer to the point.
[175] (1768) 4 Burr 2225 at 2228.
[176] *Wall* v. *Rederiaktiebolaget Luggude* [1915] 3 KB 66 at 72 per Bailhache J, a judgment which was affirmed in glowing terms by the House of Lords in *Watts, Watts & Co Ltd* v. *Mitsui & Co Ltd* [1917] All ER 501.

The effect appears to be that, where a sum is held to be a penalty, a party may take action on the penalty and obtain judgment, but the court will only allow execution of the judgment up to the penal sum. However, the party may opt to disregard the penalty, in which case he may sue for and recover the full amount of damages suffered even if they exceed the penalty figure. Because one definition of a penalty is that it is 'extravagant and unconscionable in comparison with the greatest loss which could conceivably be proved to have followed from the breach', it will be rare that actual damages exceed the penalty figure[177].

However, there will be some situations where a sum is held to be a penalty because it consists of one sum payable on the happening of a number of different breaches, some resulting in substantial and others in only trifling amounts of damage. The result of some of these breaches is that the actual damage will exceed the penalty. It is unlikely that these cases establish the power of a party to opt for or against the penalty and the possibility of no limit on the amount of damages recoverable, for two reasons. First, the judgment of Bailhache J in *Wall* (much relied upon in *Watts, Watts & Co Ltd* v. *Mitsui & Co Ltd*[178]) relies upon a very old case modified by the application of a now defunct Act[179]. Not only is the ratio in *Watts, Watts* easily distinguishable, it is open to question whether it now has any application at all. Second, both *Wall* and *Watts, Watts* were concerned with charterparties and with a very common type of penalty clause in contracts of that kind. Indeed, the main thrust of argument was whether a slight amendment which had been made to the clause was sufficient to change it into an enforceable provision for liquidated damages. There is no such tradition regarding a penalty clause in common form in the building industry. More recently, it has been said:

> 'Where the court refuses to enforce a "penalty clause" of this nature, the injured party is relegated to his right to claim that *lesser measure of damages* to which he would have been entitled at common law for the breach actually committed if there had been no penalty clause in the contract.'[180] (emphasis added)

and later, in the same case:

> 'Again, it is by no means clear that "penalty clauses" are simply void, like covenants in unreasonable restraints of trade. There are dicta either way, and in *Cellulose Acetate Silk Co Ltd* v. *Widnes Foundry (1925) Ltd*[181] Lord Atkin expressly left open the question whether a penalty clause in a contract, which fixed a single sum as payable on breach of a number of different terms of the contract, some of which breaches may occasion only trifling damage but others damage greater than the stipulated sum,

---

[177] *Dunlop Pneumatic Tyre Co Ltd* v. *New Garage & Motor Co Ltd* [1915] All ER 742 per Lord Dunedin.
[178] [1917] All ER 501.
[179] Statute of William III 1697, 8 & 9 Will 3 c 11, repealed by the Statute Law Revision Act 1948.
[180] *Robophone Facilities* v. *Blank* [1966] 3 All ER 128 at 142 per Diplock LJ considering whether a sum was penal.
[181] [1932] All ER 567 at 570.

would be treated as imposing a limit on the damages recoverable in an action for a breach in respect of which it operated to reduce the damages which would otherwise be recoverable at common law.'[182]

What Lord Atkin actually said was:

'I desire to leave open the question whether, where a penalty is plainly less in amount than the prospective damages, there is any legal objection to suing on it, or in a suitable case ignoring it and suing for damages.'[183]

Lord Atkin in refusing to pass an opinion on the principle held in *Wall* and affirmed in *Watts, Watts*, appeared to be indicating that the House of Lords was disengaging itself from its earlier decision by refusing to apply it in general terms to all penal sums.

3.11   *Maximum recovery if liquidated damages do not apply*

Where the amount inserted in the contract is held to be a penalty, invariably such a holding will be against the wishes of the employer who has inserted the sum in the hope and perhaps expectation of getting it. However, where liquidated damages are held not to apply, that is usually because the employer has taken, or omitted to take, some action which destroys the right to such damages; possibly in an effort to recover greater damages.

In *The Rapid Building Group Ltd* v. *Ealing Family Housing Association Ltd*[184], the court affirmed the judge's holding that it was not open to the defendants to counterclaim the amount of liquidated damages. This was because they had been partly responsible for part of the delay in achieving the completion date. Since there was not adequate provision to allow the defendant to extend time for that particular reason, the liquidated damages clause did not apply[185]. The court accepted that the defendant could pursue a claim for unliquidated damages, but it refused to be drawn on the proposition that the claim would have a ceiling equal to the amount of liquidated damages. However, the Supreme Court of Canada has said:

'If the actual loss turns out to exceed the penalty, the normal rules of enforcement of contract should apply to allow recovery of only the agreed sum. The party imposing the penalty should not be able to obtain the benefit of whatever intimidating force the penalty clause may have in inducing performance, and then ignore the clause when it turns out to be to his advantage to do so. A penalty clause should function as a limitation on the damages recoverable, while still be ineffective to increase damages

---

[182]   *Robophone Facilities* v. *Blank* [1966] 3 All ER 128 at 142.
[183]   *Cellulose Acetate Silk Co Ltd* v. *Widnes Foundry* (1925) Ltd [1933] AC 20 at 570.
[184]   (1984) 1 Con LR 1.
[185]   *Peak Construction (Liverpool) Ltd* v. *McKinney Foundations Ltd* (1970) 1 BLR 114.

above the actual loss sustained when such loss is less than the stipulated amount.'[186]

This statement probably represents the modern approach to this problem where the stipulated sum is held to be a penalty rather than liquidated damages, but it is not clear whether it necessarily represents the position following the failure of the liquidated damages clause for every reason. Where parties have agreed a figure to represent estimated damages and the mechanism for putting their wishes into effect has been contractually disabled, can it be said that recovery of whatever damages can be proven should be allowed, even if they exceed the liquidated damages figure? A penalty is always a sum which is extravagant in relation to the damages likely to be incurred, but liquidated damages can operate as a limitation on damages[187].

In considering the question, it must be remembered that in the case of liquidated damages in a building contract, no default on the part of the contractor can prevent the application of the clause. The clause can only fail as a result of a default on the part of an employer. A contractor who enters into a contract with an employer, which includes a relatively small sum for liquidated damages, will have a valuable advantage. The employer will be equally and oppositely disadvantaged, but both parties will have agreed on the arrangement as part of the distribution of risk inherent in that particular contract.

Part of the employer's implied obligations will be that he will not prevent the contractor from due performance[188]. Among his express obligations will be, personally or through his agent the architect, to grant proper extensions of time at the right time. If the employer is so minded, it is possible for him to disable the liquidated damages clause by causing the architect to fail to grant an extension of time in appropriate circumstances and then he would be entitled to claim whatever amount of unliquidated damages he could prove[189]. If the sum stipulated in the contract is not a ceiling on what can be claimed in those circumstances, it would be open to the employer to effectively alter the distribution of risk and, as a result of his own default, be entitled to a greater sum in damages than if he had properly performed his part of the bargain. Purely on the principle that a party cannot profit by his own contractual breach to the detriment of the other party, there is a strong argument that the liquidated damages sum must be a ceiling on recovery[190]. On this analysis, the argument for a ceiling on recoverable damages is probably stronger where the liquidated damages are irrecoverable due to the employer's default than because they are held to be a penalty.

---

[186] *Lorna P. Elsley* v. *J. G. Collins Insurance Agencies Ltd* (1978) 4 Const LJ 318 at 320 per Dickson J.
[187] *Cellulose Acetate Silk Co Ltd* v. *Widnes Foundry* (1925) Ltd [1932] All ER 567.
[188] *London Borough of Merton* v. *Stanley Hugh Leach Ltd* (1985) 32 BLR 51.
[189] *Peak Construction (Liverpool) Ltd* v. *McKinney Foundations Ltd* (1970) 1 BLR 114.
[190] *Alghussein Establishment* v. *Eton College* [1988] 1 WLR 587.

3.12 *Bonus clauses*

Few contracts make provision for bonus clauses as a standard option[191].
Bonus clauses are usually written into contracts if the employer wishes to
provide an incentive for the contractor to finish early. They provide for the
payment of a sum of money for every day or week difference between
the date the contractor achieves practical completion, or the equivalent,
and the contract completion date. It is the reverse of a liquidated damages
provision. Bonus clauses need have no relation to liquidated damages. They
may be greater or less than the liquidated damages sum or there may be no
bonus clause at all. It is *not* true that where there is a liquidated damages
clause there must also be a bonus clause for the same amount. Exactly how a
bonus clause is structured depends on the requirements of the employer
and the ingenuity of the draftsman. Commonly, such a clause may provide
for a relatively modest payment if the contract completion date is beaten by
a few days stepping up to significantly larger sums as the contractor succeeds in achieving earlier completion dates.

A disagreeable feature of bonus clauses is that lost opportunity to achieve
a bonus will feature in many claims relating to contracts where a bonus is on
offer. It is worthwhile considering the effect of a bonus clause on such
clauses as 5.4 of JCT 98. The architect will usually comply with clause 5.4
if, put broadly, he provides information to the contractor in accordance with
the information release schedule or otherwise in such time that the contractor is able to complete the works by the contract date for completion.
Where a bonus clause is inserted, the contractor will doubtless call for
information much earlier than usual on the basis that he needs it earlier if
he is to earn the bonus. It will be difficult to resist this argument and clauses
5.4 of JCT 98 and 1.7 of IFC 98 would require redrafting accordingly. Indeed,
it is difficult to see how the information release schedule can sit happily
beside a bonus clause.

What amounts to a bonus clause may perhaps arise without the parties
being entirely aware of it. In *John Barker Construction Ltd* v. *London Portman
Hotels Ltd*, the parties entered into an acceleration agreement. Among other
things, the agreement stipulated that the contractor would be paid additional sums of £50,000 to be included in the valuation on 20 July 1994,
£20,000 to be included in the valuation on 3 August 1994 and £20,000 on
completion on 26 August 1994. Considering the final payment of £20,000
on 26 August 1994, Mr Recorder Toulson said:

> 'I conclude that the £20,000 was agreed to be a performance related
> payment if the plaintiffs completed by the 26 August 1994...
>
> It was also an express term of the acceleration agreement that the
> defendants would supply the plaintiffs with all outstanding information
> by the end of 12 July 1994, and I accept that there were implied

---

[191] A notable exception is the Engineering and Construction Contract (NEC), see Chapter 16.

non-hindrance terms as pleaded in paragraph 7 of the re-amended statement of claim.

By reason of the numerous changes made after the acceleration agreement the defendants were in breach of those implied terms, if not also of
the express term. The latter point turns on whether the express duty was
conditional upon receipt of a specific request for information from the
plaintiffs. I doubt that it was, but the point is academic.

It is impossible to tell whether, as a matter of probability, the plaintiffs
would or would not have finished by 26 August 1994, but for those
changes. They would have had a reasonable opportunity of doing so,
but they could easily have failed for all manner of reasons. In those
circumstances I would hold that the plaintiffs are entitled to damages
for loss of that chance equal to 50 per cent of the agreed performance
bonus, or £10,000.'[192]

Although based on a specially worded acceleration agreement, nevertheless, this part of the judgment gives useful guidance on the way in which the
courts may decide whether and to what extent a contractor has been
deprived of the opportunity to earn a bonus by the actions or defaults of
the employer and architect. In this instance the judge took a robust, if
somewhat rough and ready, approach.

---

[192] (1996) 83 BLR 31 at 64 per Mr Recorder Toulson.

Chapter 4
Basis for common law claims

4.1 General

Some of the most frequent common law claims are considered below. No standard form contract is a self-sufficient document; it must be read against the background of the general law, and common law claims can arise under any of the standard form contracts in current use.

The claims discussed in other chapters are those that arise under specific clauses in the contract and, in general, this specific right to claim, for example for direct loss and/or expense under JCT 98, clause 26, is in addition to any right to damages that the contractor possesses at common law for breach of contract or, for example, under the law of tort. Indeed, JCT 98, clause 26 makes this clear:

> The provisions of Clause 26 are without prejudice to any other rights or remedies which the Contractor may possess.

(The last sentence of clause 4.11 of IFC 1998 is to the same effect.)

The practical effect of this is that JCT 98 and IFC 98 (and, indeed, probably other current standard forms) allow additional or alternative claims for damages based on the same facts as those presented as part of a contractual claim, so that the contractor may pursue his remedy under the express contract term without prejudicing his right to claim at common law. The contractor is not obliged to make a claim under such a clause in respect of those grounds specified in the clause which are also breaches of contract, such as the architect's failure to provide information in due time. He may prefer to wait until completion and join the claim for damages with other claims for damages for breaches of other obligations under the contract. Common law claims may be based on implied terms in the contract. Of course, the contractor can only recover his loss once but a common law claim may avoid any restrictions imposed upon the contractual claim by the terms of the contract itself[193]; for example, the requirement for the contractor's application to be submitted within a reasonable time of the loss and/or expense becoming apparent. The contractor has the right to pursue both claims in tandem to the extent that he can make good such claims[194]. Moreover it should be noted that settlement of a claim made under contractual terms will not preclude the contractor from pursuing a claim for

---

[193] *London Borough of Merton* v. *Stanley Hugh Leach Ltd* (1985) 32 BLR 51
[194] *Fairclough Building Ltd* v. *Vale of Belvoir Superstore Ltd* (1992) 9-CLD-03-30.

damage arising from the same facts provided only that *additional* damages are claimed and that there is no element of double recovery[195].

There is an important distinction between contractual and common law claims. Where the contractor makes a claim under a specific contract provision, such as clause 26 of JCT 98, he is asking that the machinery provided for by the contract be operated in his favour. Accordingly, he must observe the procedural provisions in the contract relating to the giving of notices and so on. The wording of the particular provision relied upon may also circumscribe his remedy – and it usually does by limiting the amount of his claim by excluding what is usually called 'consequential loss'[196]. It is important to remember that not all the matters which under a standard form contract will trigger a claim for direct loss and/or expense are breaches of contract. For example, architect's instructions may entitle a contractor to loss and/or expense, but the giving of an instruction cannot be a breach, because it is expressly empowered by the contract. Therefore, a failure on the part of the contractor to operate the provisions of a 'claims clause' will sometimes leave him without any remedy at all.

Sometimes a contractor will submit a claim which relies partly on the contractual machinery and partly on damages for breach of contract. If the contractor has properly separated the contractual and common law parts of the claim, the architect will be able to deal with the one and decline the other. However, very often the two elements will be mixed together. In that case, the architect must request the contractor to disentangle his claim and submit only the contractual part.

Under the JCT forms it is only claims that arise under specific contract clauses that can be ascertained and certified by the architect and paid for under the normal contract provisions for payment. The architect has no power – express or implied – to certify payments in respect of common law claims. Such claims are governed by the general law. Once the adjudication or the arbitration procedure are invoked, however, the adjudicator or the arbitrator has power to deal with common law claims, but such claims must be expressly pleaded in the points of claim submitted by the claimant. Under JCT terms of contract, it is plain that the arbitrator may decide a claim based on tort alone (as well as breach of contract) and the limitation is that the tort must arise out of the transaction which is the subject matter of the contract[197].

It has been held that, on the true interpretation of clause 35 of JCT 63, the arbitrator has power to award damages for misrepresentation and mistake as well as power to order rectification of the contract.

'I have no doubt that a dispute between the parties based upon alleged mistake at the time this contract was entered into, and upon an alleged misrepresentation or negligent misstatement are ones "arising

---

[195] *Whittal Builders Ltd* v. *Chester-Le-Street District Council* (1996) 12 Const LJ 356 (reporting the 1985 *Whittal* case).
[196] *Saintline Ltd* v. *Richardson, Westgarth & Co Ltd* [1940] 2 KB 99.
[197] *Re Polemis and Furness, Withy & Co* [1921] 2 KB 560; The Arbitration Act 1996.

in connection" with that contract and thus within the scope of the arbitration clause ... there is ... nothing in that clause which would preclude an arbitrator from granting rectification or awarding damages if he finds such allegations made out.'[198]

JCT arbitration agreements now make express provision for the arbitrator to order rectification. Under the ACA terms (see Chapter 15) it is clear that the architect has power to assess common law claims as well, albeit only those claims against the employer and not claims against the contractor: clause 7.1 of the ACA contract refers specifically to claims resulting from 'any act, omission, default or negligence of the Employer or of the Architect'; and the contractor is 'entitled to recover the same in accordance with the provisions of this clause'. The provisions of clauses 7.2, 7.3 and 7.4 must be observed if the architect is to deal with such claims. Clause 7.5 makes it clear that failure to comply with those procedural provisions is not fatal to the contractor's claim; it merely delays the time of settlement. Consequently, it seems that the contractor's common law rights are unaffected by the express provisions of the clause.

The introduction of clause 26.2.11 into JCT 98 (see Chapter 12, end of section 12.2.5) and similarly worded clauses into the loss and/or expense provisions of other JCT contracts (but not MW 98) have effectively made it possible for the architect to deal, as loss and/or expense, with many if not all situations of employer default which previously would have had to have been dealt with as common law claims.

Finally, unless expressly precluded by the terms of the contract (such as the conclusiveness of the final certificate under JCT 98 and IFC 98), common law claims will only be subject to the Limitation Act 1980; and the contractor may pursue such common law claims at any time within the period of limitation, i.e. 6 years for actions based on simple contract or tort or 12 years if the contract is a deed. The limitation period is usually taken to run from practical completion so far as the contractor is concerned[199].

4.2   *Implied terms in building contracts*

The question of implied terms is fully dealt with in all the standard texts on general contract law. Implied terms usually fall into one of the following categories:

- By statute, for example under the Housing Grants, Construction and Regeneration Act 1996 and the Supply of Goods and Services Act 1982
- By local custom
- Particular trade usage[200]

---

[198] *Ashville Investments Ltd* v. *Elmer Contractors Ltd* (1987) 37 BLR 55 at 66 per May LJ.
[199] *Tameside Metropolitan Borough Council* v. *Barlows Securities Group Services Ltd* [2001] BLR 113 CA. This part of the first instance judgment was not overturned on appeal.
[200] *Symonds* v. *Lloyd* (1859) 141 ER 622.

- To give business efficacy
- If a term is the presumed intention of the parties which 'goes without saying'
- If there is a 'course of dealing' between the parties, similar terms will be implied into a new contract made on the same basis[201]
- At common law.

In practice, in the context of building contract claims, the concern is with those terms which will be implied into the contract by the courts, in order to make the contract commercially effective[202].

Where the express terms of the contract are clear and unambiguous, the courts will not imply a term simply to extricate a party from difficulties. The position has been put thus:

> 'An unexpressed term can be implied if and only if the court finds that the parties must have intended that term to form part of their contract: it is not enough for the court to find that such a term would have been adopted by the parties as reasonable men if it had been suggested to them: it must have been a term that went without saying, a term *necessary* to give business efficacy to the contract, a term which, although tacit, formed part of the contract which the parties made for themselves'[203].

In *London Borough of Merton* v. *Stanley Hugh Leach Ltd*[204] one of the many points at issue concerned the implication of certain implied terms. Vinelott J considered the general approach to be adopted as to the implication of terms in a contract in JCT 63 form, referring to the cases in which the courts have considered whether a term should be implied into a contract and which cover a wide variety of situations.

Many common law claims arising under building contracts are based on implied terms relating to the employer failing to co-operate with the contractor or interfering with the contractor's progress. In *Merton* v. *Leach* the contractor sought that a number of terms be implied into a JCT contract. The first two of these were:

(1) The employer would not hinder or prevent the contractor from carrying out its obligations in accordance with the terms of the contract and from executing the works in a regular and orderly manner.
(2) The employer would take any steps reasonably necessary to enable the contractor so to discharge its obligations and to execute the works in a regular and orderly manner.

Vinelott J unhesitatingly held that the first term ought to be implied:

> 'The implied undertaking not to do anything to hinder the other party from performing his part of the contract may, of course, be qualified by a term

---

[201] *McCutcheon* v. *David McBrayne Ltd* [1964] 1 WLR 125.
[202] *The Moorcock* (1889) 14 PD 64.
[203] *Trollope & Colls Ltd* v. *North West Metropolitan Regional Hospital Board* (1973) 9 BLR 60 at 70 per Lord Pearson.
[204] (1985) 32 BLR 51.

express or to be implied from the contract and the surrounding circumstances. But the general duty remains so far as qualified ... It is difficult to
conceive of a case in which this duty could be wholly excluded.'[205]

As regards the second term – the employer's duty to do all that is reasonably
necessary to bring the contract to completion – the judge was equally
emphatic. He accepted that it is well settled law that the courts will imply
a duty to do whatever is necessary to enable the other party to perform his
obligations under the contract[206]. He noted that there are limitations on the
principle, and referred with approval to the views expressed in *Mona Oil
Equipment Co* v. *Rhodesia Railway Co*:

> 'I can think of no term that can properly be implied other than one based
> on the necessity for co-operation. It is, no doubt, true that every business
> contract depends for its smooth working on co-operation, but in the
> ordinary business contract, and apart, of course, from express terms,
> the law can enforce co-operation only in a limited degree – to the extent
> that it is necessary to make the contract workable. For any higher degree
> of co-operation the parties must rely on the desire that both of them
> usually have that the business should be done'.[207]

Any building contract requires close co-operation between the contractor
and the architect. The JCT contracts contain many examples of the co-
operation which the architect is required to give to the contractor; for
example, by issuing instructions during the progress of the works and by
providing further drawings to explain or amplify the contract drawings.
Without that co-operation the contract cannot be completed expeditiously
and efficiently. However, the exact parameters of the duty of co-operation
are unclear, but this second implied term

> 'extends to those things which the architect must do to enable the con-
> tractor to carry out the work and that the building owner is liable for any
> breach of this duty on the part of the architect.'[208]

Another case in which a similar implied term was conceded is *Holland
Hannen & Cubitts* v. *WHTSO*[209] where the contract was also in JCT 63
form. The term there was to the effect that 'the employer would do all things
necessary to enable the contractor to carry out and complete the works
expeditiously, economically and in accordance with the contract'. The
important point is that the employer will only be liable for breach in respect
of those functions performed by the architect acting as the employer's agent
and not when he is acting as certifier because in the latter case the employer
is not vicariously liable. The position under a JCT contract has been
summarised in this way:

---

[205] (1985) 32 BLR 51 at 80 per Vinelott J.
[206] *Mackay* v. *Dick* (1881) 6 App Cas 251.
[207] [1949] 2 All ER 1014 at 1018 per Devlin J.
[208] *London Borough of Merton* v. *Stanley Hugh Leach Ltd* (1985) 32 BLR 51 at 81 per Vinelott J.
[209] (1981) 18 BLR 80.

'Under the standard conditions, the architect acts as the servant or agent of the building owner in supplying the contractor with the necessary drawings, instructions, levels and the like and in supervising the progress of the work and ensuring that it is properly carried out ... To the extent that the architect performs these duties the building owner contracts with the contractor that the architect will perform them with reasonable diligence and with reasonable skill and care. The contract also confers on the architect discretionary powers which he must exercise with due regard to the interests of the contractor and the building owner. The building owner does not undertake that the architect will exercise his discretionary powers reasonably; he undertakes that although the architect may be engaged or employed by him, he will leave him free to exercise his discretion fairly and without proper interference by him ... It is now clear that insofar as the architect exercises discretionary powers and the exercise of his discretion can be reviewed by the arbitrator, the arbitrator stands in the shoes of the architect and does not exercise a purely arbitral role ... But to the extent that the architect acts as agent or servant of the building owner by the contract his acts are not subject to review by the arbitrator – though they may found a claim for damages for breach of contract the extent of which will fall to be determined by the arbitrator'[210].

More recently the following useful analysis has been given:

'[The contract administrator], although employed by [the employer], was given authority by the parties to the contract to form and express the opinions and issue the certificates as and when required by its terms. He was not the agent for [the employer] in so acting so that [the employer] was liable as principal to [the contractor] for what he did or did not do in his capacity as certifier. On the other hand [the employer] was the party who could control him if he failed to do what the contract required. Since the contract is not workable unless the certifier does what is required of him, [the employer] as part of the ordinary implied obligation of co-operation, was under a duty to call [the contract administrator] to book ... if it knew that he was not acting in accordance with the contract ... the duty does not arise until the employer is aware of the need to remind the certifier of his obligations ... A mere failure by the certifier to act in accordance with the contractual timetable is not a failure on the part of the employer to discharge an implied obligation positively to co-operate and cannot be a breach of contract by the party whose employee is the certifier. On the facts set out in the award [the employer] could not therefore have been in breach of contract. In arriving at this conclusion I bear in mind the argument that the existence of an arbitration clause which confers on the arbitrator wide powers to open up etc. means that a failure to issue a final certificate can be put right ...'[211]

---

[210] *London Borough of Merton* v. *Stanley Hugh Leach Ltd* (1985) 32 BLR 51 at 78 per Vinelott J.
[211] *Penwith District Council* v. *V. P. Developments Ltd*, 21 May 1999, unreported at paragraph 36 per Judge Lloyd.

Apart from the two foregoing implied terms, it is an implied term of every building contract that the employer will give possession of the site to the contractor in sufficient time to enable him to complete the works by the due date[212]. All the standard form building contracts in common use contain express terms providing for the giving of possession of the site. In that case, of course, the employer's failure to give possession as provided in the contract is a breach for which the contractor will be entitled to damages in respect of any resultant loss. None of the standard form contracts in current use displaces the first two implied terms mentioned, and thus the employer is liable at common law for breach of them. This can have important practical implications, and consideration must be given to one of the most common grounds of common law claim arising under a building contract.

Other terms are commonly implied into building contracts that the contractor will carry out his work in a good and workmanlike manner and he will provide good and proper materials[213]. If there is no other designer, the work will be fit for the purpose so far as that has been made known to the contractor[214].

Bacal Construction (Midlands) Ltd v. *Northampton Development Corporation*[215] shows the possibility of the contractor bringing an action at common law for breach of an implied term or warranty that the ground conditions are as described in the tender documents. The tender documents in that case for the design and construction of some 518 dwellings and ancillary buildings included a statement that the site was a mixture of Northamptonshire sand and upperlias clay, but subsequently tufa was discovered. The employers had told the contractors in writing at the time of the tender that their design was to assume the soil conditions disclosed at the boreholes, which gave no warning of the presence of tufa. The contractors recovered damages for breach of an implied term or warranty that the ground conditions would accord with the hypotheses on which they were instructed to design. The contract was an amended form of JCT 63, and among other things the Court of Appeal held that the necessary redesigning of the foundations and the additional work occasioned by the discovery of the tufa did not rank as variations for the purposes of the contract. It was said:

> 'Bacal have submitted that there are strong commercial reasons for implying such a term or warranty in the contract as they have suggested. First, before designing the foundations of any building, it is essential to know the nature of the site conditions. Secondly, where the contract is for a comprehensive development of the kind here in question, the contractor must know the soil conditions at the site of each projected block in order to be able to plan his timetable and estimate his requirements for materials. These are matters which relate directly to the contract price.

---

[212] *Freeman & Son* v. *Hensler* (1900) 64 JP 260. Possession is dealt with in more detail in section 4.6 of this chapter.
[213] *Test Valley Borough Council* v. *Greater London Council* (1979) 13 BLR 63.
[214] *Viking Grain Storage* v. *T. H. White Installations Ltd* (1985) 3 Con LR 52.
[215] (1976) 8 BLR 88.

Thirdly, if the work is interrupted or delayed by unforeseen complications, the contractor is unlikely to be able to complete his contract in time.'[216]

The Court of Appeal, affirming the court's decision at first instance, accepted this analysis.

4.3 Variation of contract

It is important to distinguish between a variation expressly authorised under the terms of the contract (e.g. variation of the 'Works' under JCT contracts) and 'variation of contract'. The latter phrase is used to mean a change in the terms of the contract itself.

It is, of course, open to the parties to a contract to agree any changes to the terms of that contract which they please, provided that such changes are not illegal or do not render the contract void in some way. The important point is that the parties must both agree to the change, and the change then becomes, in effect, another contract between them. It follows that, unless the change is made by deed, which is always advisable in such circumstances, there must be consideration for the change – that is, there must be a return to one party for any benefit conferred upon the other party by the change.

An attempted variation of contract is illustrated in the case of *D & C Builders Ltd* v. *Rees*[217]. There the claimants had carried out work to the value of £482, but the defendant failed to pay the account. The claimants pressed for payment for several months. Finally, the defendant's wife, knowing that the claimants were in financial difficulties, offered £300 in settlement, saying that if the offer were not accepted, nothing would be paid. This was a clear attempt to vary the terms of the contract which specified the contract sum. The claimants accepted a cheque for £300 and gave a receipt 'in completion of the account'. Later, they sued for the balance. The Court of Appeal held that the claimants were not barred from recovering the balance. Quite apart from the fact that the claimant's consent had been obtained under pressure, there was lack of consideration, there was no accord and satisfaction and there was no equitable ground disentitling the claimants to recover.

Unless variations are specifically authorised by the terms of the contract, any change in the work will become a variation of contract. Also, if the power to order variations is in any way limited, any changes beyond the limits specified or implied will also be variations of contract. For instance, insofar as the power of the architect to order variations under the JCT forms is limited, and there are changes that need to be made to the works beyond those powers, they must then become the subject of a separate agreement between the employer and contractor. The architect will have no authority

---

[216] (1975) 8 BLR 88 at 100 per Buckley LJ.
[217] [1965] 3 All ER 837.

to deal with such variations unless both parties agree to give him that authority. The architect has no implied authority to vary the contract on behalf of the employer nor to vary the works without an express provision in the contract[218]. The terms upon which the contractor may agree to carry out such changes will also become a matter of negotiation directly between the contractor and the employer unless both agree to authorise the architect or the quantity surveyor to negotiate for the employer.

4.4   Omission of work to give it to others

It has always been a recognised principle of English contract law that (in the absence of express conditions to the contrary in the contract), once a man has contracted to do a certain quantity of work, he has the right to do it if it is to be done at all; if the contract so provides, the work may be omitted, but only if it is not to be done at all, not in order to give it to someone else. The case most generally quoted as authority for this proposition is Australian[219]. There, the contract provided that the employer would supply structural steel that would be fabricated and erected by the contractor. After a substantial delay in starting the work, due to the employer's failure to give possession of the site, the contractor was told that the employer had made other arrangements with another firm for the fabrication of the steel, which was to be omitted from the contract. This, taken together with the employer's failure to give possession of the site, was considered by the Australian court to be sufficient grounds for the contractor to treat the contract as having been repudiated, as they showed that the employer had no serious intention of considering himself bound by its terms.

There is an American case which deals with a contract which compares more closely to JCT contracts[220]. The contract, in a similar way to JCT forms, provided for the omission of work without vitiating the contract and provided that such omissions should be valued and deducted from the contract sum. Notably, the American appeal court held that the word 'omission' meant only work not to be done at all, not work to be taken from the contractor under the contract and given to another to do. There are now two English cases to the same effect[221].

Exactly the same principle applies with regard to the division of work within the contract. The original contract works are often divided up between work to be done and materials to be supplied by the main contractor and work to be done by nominated sub-contractors and materials to be supplied by nominated suppliers. That division, once fixed by the contract, cannot be changed unilaterally by the employer acting through the

---

[218] *Sharpe* v. *San Paulo Railway* (1873) 8 Ch App 597.
[219] *Carr* v. *J. A. Berriman Pty Ltd* (1953) 27 ALJR 273 and also *Commissioner for Main Roads* v. *Reed & Stuart Pty Ltd and Another* (1974) 12 BLR 55.
[220] *Gallagher* v. *Hirsch* (1899) NY 45.
[221] *Vonlynn Holdings Ltd* v. *Patrick Flaherty Contracts Ltd*, 26 January 1988, unreported; *AMEC Building Contracts Ltd* v. *Cadmus Investments Co Ltd* (1997) 13 Const LJ 50.

architect. The main contractor has a right to do that work which is set out in the contract documents for him to do, and it cannot be taken away from him in order that it be done by a nominated sub-contractor. Conversely, if work is set out to be done by a nominated sub-contractor, the main contractor cannot be forced to do it instead; nor, for that matter, can he insist upon doing it himself[222].

Where a main contractor brings others on to the site to supplement the labour of a sub-contractor, already lawfully on site, without that sub-contractor's consent, the main contractor is in breach of contract which may be repudiatory in nature so as to entitle the sub-contractor to leave site[223].

4.5 Extra work

The contractor is not entitled to payment simply because he carries out work which is additional to that which he originally contracted to execute. This is widely misunderstood. Moreover, a contractor will often claim extra payment on the grounds that he has provided better quality than the architect specified. He is not entitled to payment and indeed, in both instances, he is in breach of contract[224]. Although the law will not allow an employer without payment to gain a benefit from work which he has instructed to be carried out, the contractor is not entitled to payment for work which has not been instructed.

Occasionally there may be a dispute, because the architect insists that certain work is included in the contract, but the contractor refuses to accept it and demands that it be treated as an extra. Groundworks sometimes fall into this category. If the architect refuses to issue an instruction requiring a variation and the work is substantial, the contractor's remedy may be to refuse performance unless an instruction is issued and, in the absence of such instruction, to treat the contract as repudiated and sue for damages[225]. This could be a dangerous course of action with the risk of huge losses if the contractor is wrong. If the contractor proceeds with the work and attempts to make a claim at a later date, he may find that a court or arbitrator will find that he has acted in accordance with the architect's view of the contract and that he is not entitled to extra payment. Sometimes, it may be held that the employer has implicitly promised to pay if the work is, as a matter of law, additional to the contract[226].

On the other hand, if the contractor alleges that certain work is extra to the contract and refuses to carry it out unless the employer agrees to pay for it, and the employer does so agree, but it is subsequently found that the work

---

[222] This was an important part of the *ratio decidendi* in *T. A. Bickerton & Son Ltd* v. *North West Metropolitan Regional Hospital Board* [1970] 1 All ER 1039, concerning the architect's obligation to nominate a new sub-contractor under the JCT Form, 1963 edition, where the original nominee could not complete.

[223] *Sweatfield Ltd* v. *Hathaway Roofing Ltd* (1997) CILL 1235.

[224] *Holland Hannen & Cubitts* v. *Welsh Health Technical Services Organisation* (1981) 18 BLR 80.

[225] *Peter Kiewit Sons' Company of Canada Ltd* v. *Eakins Construction Ltd* (1960) 22 DLR (2d) 465.

[226] *Molloy* v. *Liebe* (1910) 102 LT 616.

was included in the original contract, the employer will not be bound to pay the extra price, as he will have received no value in return for the promise to pay. This is illustrated by the case of *Sharpe* v. *San Paulo Railway*[227]. There, the contractor submitted a lump-sum price for the construction of a railway, based upon a quantified specification produced by the engineer, but there was no indication in the contract that the quantities were guaranteed to be correct. It turned out that the engineer had made an error of 2 million cubic yards in his calculation of earthworks. The engineer promised to make other changes in the works which would reduce the cost of the work as a whole so as to compensate the contractor for the effect of the error. However, he did not make the changes promised and, at the end of the contract, certified for payment the original contract sum with no allowance for the extra earthworks. The Court of Appeal held that, since the contractors had undertaken simply to build the railway from terminus to terminus for the lump sum stated, and the engineer's quantities did not form part of the contract, the contractors must be held to have contracted to carry out whatever work was necessary for the construction in return for the lump sum and could not recover any extra payment for the additional earthworks; nor was the engineer's undertaking to make compensating savings enforceable.

An architect may only instruct such variations as the contract expressly provides. He has no automatic right to order variations[228]. Although he acts as agent for the employer, he has limited authority. So far as the contractor is concerned, the authority is limited to what is stated in the contract. Where the power to instruct is not expressed in precise terms, for example in MW 98, it will be implied that the architect can only issue instructions which are within the scope of the contract[229]. If an architect issues instructions which are not empowered by the contract, the contractor should not comply. If he does comply with unauthorised instructions, he is in breach of contract. Moreover, if the contractor does comply, it is conceivable that the architect may become personally liable to the contractor for the price. How the contractor would make such a claim against the architect, however, is unclear, because, to the contractor's clear knowledge, the instruction would concern works which are the property of the employer. Where the architect appears to have the employer's authority to order a variation, even if unauthorised by the employer, the employer will normally be liable to the contractor for the price and the employer in turn will look to the architect for reimbursement.

If the employer gives a direct instruction to the contractor, it would not be authorised under most standard forms, which reserve the power to issue instructions to the architect or the contract administrator (of whatever discipline). It is often suggested that the giving and receiving of such an instruction would create a separate little contract by which the contractor

---

[227] (1873) 8 Ch App 597.
[228] *Cooper* v. *Langdon* (1841) 9 M & W 60.
[229] *Sir Lindsay Parkinson & Co Ltd* v. *Commissioners of Works* [1950] 1 All ER 208.

would be entitled to payment on a *quantum meruit* basis. That, indeed, may be one analysis of the position. Perhaps the better view is that the employer and the contractor have agreed to vary the original contract and the contractor would be entitled to payment in accordance with the existing contract terms.

4.6 Possession of site

In every building contract it is an implied term that the employer will give possession of the site to the contractor within a reasonable time, i.e. in sufficient time to enable the contractor to complete the works by the contractual date. Under JCT terms, there is specific provision for the contractor to be given possession on the date specified in the Appendix.

JCT 98, clause 23.1.1 provides:

> 'On the Date of Possession, possession of the site *shall* be given to the Contractor who shall thereupon begin the Works...'. (emphasis added)

Any failure by the employer to give possession on the date named is a breach of contract and, indeed, if extended in time, it is a breach of a major term ('condition') of the contract, because without possession it is clear that the contractor cannot execute the works. It is a breach not only of the express terms of the JCT contracts but also a breach of the term that would be implied at common law in the absence of an express term. Since default in giving possession is a breach of a major term of the contract, protracted failure to give possession, and acceptance by the contractor of the employer's breach, entitle the contractor to accept the repudiation and to commence an action for damages, which would include the loss of the profit that he would otherwise have earned[230]. Few contractors would wish to take such a drastic course, and accordingly may elect to treat the breach as a breach of warranty only and to claim damages at common law for any loss actually incurred. But, at the very least, it is clear the contractor is entitled to damages for breach[231]. In practice, such damages may be slight if the failure lasts no more than a few days.

The right to possession of the site on the date given in the Appendix is an absolute one. In *London Borough of Hounslow* v. *Twickenham Garden Developments Ltd*, it was said:

> 'The contract necessarily requires the building owner to give the contractor such possession, occupation and use as is necessary to enable him to perform the contract, but whether in a given case the contractor in law has possession must, I think, depend at least as much on what is done as on what the contract provides ...'[232]

---

[230] *Wraight Ltd* v. *P. H. & T. Holdings Ltd* (1968) 13 BLR 27.
[231] *London Borough of Hounslow* v. *Twickenham Garden Developments Ltd* (1970) 7 BLR 81.
[232] (1970) 7 BLR 81 at 107 per Megarry J.

It is sometimes argued that this is authority for what is sometimes referred to as 'sufficient possession' and that, therefore, the employer need give only that degree of possession which is necessary to enable the contractor to carry out work. However, the statement is clearly *obiter*, and must be treated with caution because Megarry J had already said at the beginning of the previous paragraph that 'I do not think that I have to decide these or a number of other matters relating to possession'. A phrase from *Keating* is also often called in aid in this connection:

> 'Provided that the Contractor has sufficient possession, in all the circumstances, to enable him to perform, the Employer will not be in breach of contract.'[233]

Keating's only authority appears to be a Canadian case in the footnote[234]. It does not appear to support the statement. The contract referred to did not contain a clear possession clause:

> '...s.52 merely stipulates that the site of the work is to be provided by the appellant; it does not provide for the degree of possession of the site that was to be afforded to the respondent. It is obvious that in order to be able to perform his obligations under a construction contract, the contractor must have access to the site of the work and must also have, at least to a certain extent, possession of that site.'[235]

Again:

> '...the appellant failed to observe an implied term that the respondent would have a sufficient degree of uninterrupted and exclusive possession of the site to permit it to carry out its work unimpeded and in the manner of its choice.'[236]

In the context of a contract which does not contain a possession clause, that may be a correct statement of the law. However, Canadian decisions are not binding in England and they may not even be persuasive, particularly when there is authority within the English jurisdiction.

In *Freeman & Son* v. *Hensler* it was stated:

> 'I think there was an implied condition on the part of the defendant that he would hand over the land to the plaintiffs to enable them to carry out what they had contracted to do, and that it applied to the whole area.'[237]

This concerned a contract in which nothing was said about possession. The court considered the matter so important that they were prepared to imply a term that possession of the whole site must be given.

---

[233] Stephen Furst and Vivian Ramsey, *Keating on Building Contracts*, 7th edition, 2001, Sweet & Maxwell at 711.
[234] *The Queen* v. *Walter Cabot Construction* (1975) 21 BLR 42.
[235] (1975) 21 BLR 42 at 50 per Pratte J.
[236] (1975) 21 BLR 42 at 50 per Urie J.
[237] (1900) 64 JP 260 at 261 per Collins LJ.

Although it appears that the case itself is not in point, the commentary to *Walter Cabot* contains the following helpful observation:

> 'English standard forms of contract, such as the JCT Form, proceed apparently on the basis that the obligation to give possession of the site is fundamental in the sense that the contractor is to have exclusive possession of the site. It appears that this is the reason why specific provision is made in the JCT Form for the employer to be entitled to bring others on the site to work concurrently with the contractor for otherwise to do so would be a breach of the contract. No such right could be implied, at least on the wording of the standard form. This right is circumscribed since if completion of the works is delayed by the activities of those engaged by the employer or if the progress of the work is materially affected then the contractor may be entitled to an extension of time or compensation or both, as the case may be...'[238]

It is thought that this is a correct view.

In any event, whether or not the contractor has been given sufficient possession is a matter of fact. In *The Rapid Building Group Ltd* v. *Ealing Family Housing Association Ltd*[239], which arose under a contract in JCT 63 form, at the time when, by clause 21, the defendants were bound to give the claimants possession of the site, they were unable to do so because its north-east corner (an area of some size) was occupied by squatters. The defendants took eviction proceedings, but it was at least 19 days before the site was cleared of squatters so as to enable the contractors to take possession of the whole of the site. The Court of Appeal (affirming the decision of Judge John Newey QC, a very experienced Official Referee) held that the defendants were in clear breach of clause 21 because of their failure, for whatever reason, to remove the squatters until an appreciable time after they were bound to give the claimants possession. Although the contractors entered on the site, the trial judge found that they were unable to clear it and so the breach caused appreciable delay and disruption, which entitled the contractors to damages[240].

The phrase 'possession of the site' was considered in *Whittal Builders* v. *Chester-Le-Street District Council*[241]. It was held that the phrase meant possession of the whole site and that, in giving piecemeal possession, the employer was in breach of contract so as to entitle the contractor to dam-

---

[238] (1975) 21 BLR 42 at 44.

[239] (1984) 1 Con LR 1.

[240] This case should be contrasted with *Porter* v. *Tottenham Urban District Council* [1915] 1 KB 776, another decision of the Court of Appeal, where the contractor was wrongfully excluded from the site by a third party for whom the contractor was not responsible in law and over whom he had no control. There was no clause 21, and the court held that there was no implied warranty by the council against wrongful interference by a third party – an adjoining owner – with the only access to the site. The *Rapid Building* case is distinguishable from *LRE Engineering Services Ltd* v. *Otto Simon Carves Ltd* (1981) 24 BLR 127, where the point at issue was whether main contractors were in breach of a sub-contract term requiring that 'access ... shall be afforded', and this was denied to the sub-contractors because of unlawful picketing during a steel strike.

[241] (1987) 40 BLR 82 (the second case).

ages. The words of Mr Recorder Percival in the first *Whittal* case are also in
point:

> 'Taken literally the provisions as to the giving of possession must I think
> mean that unless it is qualified by some other words the obligation of the
> employer is to give possession of all the houses on 15 October 1973.
> Having regard to the nature of what was to be done that would not
> make very good sense, but if that is the plain meaning to be given to
> the words I must so construe them.'[242]

These words appear to be a very clear and precise statement of the law. He
proceeded to look at the contract and found that the Appendix had been
amended to refer to '...18 dwellings at any one time'. The Appendix is, of
course, part of the printed form which takes precedence.

Under JCT terms (both in 1963 and 1980 editions, before the 1987 amendment) there was no power in the architect to postpone the giving of possession of the site. This problem is less likely to arise under a JCT 98 contract
because clause 23 now has a sub-clause which provides:

> 23.1.2   Where clause 23.1.2 is stated in the Appendix to apply the Employer may
> defer the giving of possession for a period not exceeding six weeks or
> such lesser period stated in the Appendix calculated from the Date of
> Possession.

There is an appropriate Appendix entry and clauses 25 (extensions of time)
and 26 (loss and expense caused by matters affecting regular progress of the
works) contain an appropriate 'relevant event' and 'matter' respectively.

A similar provision appears in the JCT Intermediate Form, 1998 edition,
(clauses 2.2, 2.4.14, 4.11(a) and the Appendix). If it applies, it enables the
employer to defer giving the contractor possession of the site for a period
not exceeding 6 weeks or such shorter period as he has inserted in the
Appendix.

If the employer fails to give possession and the deferment provision is not
stated to apply or if the employer fails to operate the provision or if the
failure lasts longer than stipulated by the provision, the employer will be in
serious breach as if the deferment provision had not been included.

The position is straightforward under the terms of the ACA Contract, in
all its editions, because clause 11.1 provides:

> '... the Employer shall give to the Contractor possession of the site, or such part
> or parts of it as may be specified, on the date or dates stated in the Time Schedule
> ...'

Failure to give possession is, of course, a breach of contract on the
employer's part, but it seems that the architect's powers under that contract
are sufficiently wide to enable him to postpone the giving of possession.
Moreover, under ACA terms, the architect can give the contractor an
extension of time for late possession under clause 11.5 (in either of its
alternatives).

---

[242] (1985) 11 Con LR 40 at 51.

The wording is sufficiently wide to cover failure to give possession in accordance with clause 11.1, and disturbance to regular progress and loss and expense involved can be dealt with under clause 7.1, which again refers to 'any act' disrupting regular progress.

The position appears to be the same under GC/Works/1(1998): see clauses 34, 36(2)(b) and 46(1)(b) and the discussion of claims arising under GC/Works/1(1998) in Chapter 14.

4.7 Site conditions

Claims in respect of site conditions may arise in two principal ways: first, where the contractor is given misinformation by the employer about site conditions, and second in relation to the particular provisions of JCT 98, clause 2.2.2.1 of which requires that the bills have been prepared in accordance with the Standard Method of Measurement, 7th Edition. A statement, supposedly of fact, made by the employer or the architect or quantity surveyor on behalf of the employer and which it is intended that the contractor will act upon is a 'representation'. Such statements are commonly made in tender documents. If the statement is untrue, it is a 'misrepresentation'. A misrepresentation may be the basis of a claim if it is made part of the contract or if it was an inducement to the contractor to enter into the contract. Otherwise, it has no relevance. A misrepresentation can be fraudulent, negligent, innocent or under statute. The significance lies in the remedies available.

The contractor's claim for negligent misrepresentation and/or breach of warranty and/or under the Misrepresentation Act 1967, as amended, may arise for misrepresentations made by or on behalf of the employer. As a result of the Misrepresentation Act 1967, the remedies which were formerly restricted to cases of fraud or recklessness apply to all misrepresentations unless the party who made the representation can prove 'that he had reasonable ground to believe and did believe up to the time the contract was made that the facts represented were true'. It is thought that liability for misrepresentation is unaffected by the general common law rule that the employer does not warrant that the drawings, bills of quantities or specification are accurate or that the site is fit for the works or that the contractor will be able to construct on the site[243]. Most of the cases on which such common law principles are founded are based on nineteenth century cases which must be treated with caution in the light of the 1967 Act. Architects are personally liable at common law for any fraudulent or negligent misstatement or representation[244] and also under the 1967 Act and it is thought that the scope of such liability is increasing[245]. Section 3 of that Act restricts the employer's power to exclude liability for misrepresentation[246].

---

[243] *Appleby* v. *Myers* (1867) LR 2 CP 651; *Thorn* v. *London Corporation* (1876) 1 App Cases 120.
[244] *Hedley Byrne & Co* v. *Heller & Partners Ltd* [1964] AC 465.
[245] See the fuller discussion of this point in Chapter 1, section 1.4.
[246] Substituted by section 8(1) of the Unfair Contract Terms Act 1977.

There is no doubt that, in an appropriate case, the contractor may have a claim against the employer for misrepresentations about site and allied conditions made during pre-contractual negotiations. For example, in the Australian case of *Morrison-Knudsen International Co Inc* v. *Commonwealth of Australia*, the contractor claimed that basic information provided to him at pre-tender stage 'as to the soil and its contents at the site...was false, inaccurate and misleading...the clays at the site, contrary to the information, contained large quantities of cobbles'. On a preliminary issue (since the effect of the documents could not be finally determined until all the relevant facts were established), it was concluded:

> 'The basic information in the site document appears to have been the result of much technical effort on the part of a department of the defendant. It was information which the plaintiffs had neither the time nor the opportunity to obtain for themselves. It might even be doubted whether they could be expected to obtain it by their own efforts as a...tenderer. But it was indispensable information if a judgment were to be formed as to the extent of the work to be done...'[247]

In *Holland Hannen & Cubitts (Northern) Ltd* v. *Welsh Health Technical Organisation*[248], a case which ended in a settlement in favour of the contractors, one of the claims made by the contractors against the employers was for 'damages for negligent misrepresentations and/or breach of warranty and/ or pursuant to the Misrepresentation Act 1967 arising out of representations made or warranties given by or on behalf of' the employer. These related, among other things, to statements in the preliminaries section of the bills of quantities about the sequence of operations, letters from the architects, and statements made at pre-contractual meetings. In an appropriate case, therefore, an action would lie in respect of misleading statements about site conditions.

A factual misrepresentation made during pre-contractual negotiations by or on behalf of the employer and relied on by the contractor may give rise to an action under the 1967 Act. The possibilities are illustrated by the decision of the Court of Appeal in *Howard Marine & Dredging Co Ltd* v. *A. Ogden & Sons (Excavations) Ltd*[249] where barge owners were held liable to dredging contractors who had hired a barge for dredging works on the faith of a misrepresentation of the dead weight of the barge. The defendant's marine manager had stated that the barge's payload was 1600 tonnes, whereas in fact it was only 1055 tonnes. His misstatement was based on his recollection of an incorrect figure given in Lloyd's Register.

A representation followed by a warning that the information given may not be accurate will not usually be sufficient to protect the employer, because it is a clear intention to circumvent section 3 of the Act which provides:

---

[247] *Morrison-Knudson International Co Inc* v. *Commonwealth of Australia* (1972) 13 BLR 114 at 121 per Barwick CJ.
[248] (1981) 18 BLR 80.
[249] (1977) 9 BLR 34.

'3. If a contract contains a term which would exclude or restrict –
>   (a) any liability to which a party to a contract may be subject by reason of any misrepresentation made by him before the contract was made; or
>   (b) any remedy available to another party to the contract by reason of such a misrepresentation that term shall be of no effect except in so far as it satisfies the requirement of reasonableness as stated in section 11(1) of the Unfair Contract Terms Act 1977; and it is for those claiming that the term satisfies that requirement to show that it does.'

Indeed, such a statement may convert the representation into a misrepresentation[250]. In many instances the representation is made in the preliminaries section of bills of quantities or specification, as a statement about the subsoil, underground services or aspects of the site. When the contract is executed, such representations become terms of the contract and if they are incorrect, the contractor may have a claim for damages for breach of contract.

If the representation is fraudulent, i.e. 'knowingly, without belief in its truth, or recklessly, careless whether it is true or false'[251], the contractor may be able to recover more substantial damages under section 2(1) of the 1967 Act, which probably entitle him to recover all losses even those which are unforeseeable so long as they are not too remote. A very clear example of fraudulent misrepresentation is to be found in the old case of *Pearson* v. *Dublin Corporation*[252]. This concerned a complex project for the construction of an outfall sewer and associated works for a lump sum price. A key point was that an existing wall continued to a depth of 9 feet below a datum. This was shown on information provided by the employer's engineers. The true situation was that the wall scarcely reached 3 feet in depth. This caused the contractor considerably more expense than expected. It emerged that the engineers had carried out no proper survey and themselves doubted the accuracy of the information they provided. The court held that this amounted to fraud. The contract required the contractor to satisfy himself regarding the dimensions of the existing works and stated that the employer was not responsible for the accuracy of statements it had made about the existing structure. However, none of this was enough to provide a defence to the allegation of fraud. Had it not been fraud, it might have then succeeded. Whether such a defence would be enough now is doubtful.

The second possibility was originally suggested by Dr John Parris in relation to contracts in JCT 80 form where clause 2.2.2.1 provided that 'the Contract Bills ... are to have been prepared in accordance with SMM6'.

> 'This seems to require the employer to provide the contractor with information in his possession about potentially difficult site conditions. Other provisions require the employer to provide specific information. The contractor may have a claim against the employer, should site conditions not be as assumed ... SMM7 states that information regarding trial pits or bore holes is to be shown on location drawings under 'A. Preliminaries/

---

[250] *Cremdean Properties Ltd* v. *Nash* (1977) 244 EG 547.
[251] *Derry* v. *Peek* (1889) 14 App Cas 337.
[252] [1907] AC 351.

General Conditions' or on further drawings which accompany the bills of quantities or stated as assumed. Rock is classified separately.'[253]

This seems to obligate the employer to provide the contractor with information in his possession about potentially difficult site conditions. Other provisions require the employer to provide specific information.

In such circumstances, it seems that the contractor will have a claim against the employer, should site conditions *not* be as assumed. Support for this view is to be found in *C. Bryant & Son Ltd* v. *Birmingham Hospital Saturday Fund*[254], where Bryant contracted to erect a convalescent home in the RIBA form of contract, 1931 edition, but incorporating the 1909 edition as well, clause 11 of which provided:

> 'The quality and quantity of the work included in the contract sum shall be deemed to be that which is set out in the bills of quantities, which bills, unless otherwise stated, shall be deemed to have been prepared in accordance with the standard method of measurement of building works last before issued by the Chartered Surveyors' Institution. Any error in description or in quantity in, or omission of items from, the bills...shall not vitiate this contract, and shall be rectified and treated as an extra or omission [and valued accordingly].'

The relevant SMM provided that 'Where practicable the nature of the soil shall be described and attention shall be drawn to any existing trial holes. Excavation in rock shall be given separately'. Excavation in rock was not shown separately, although the contractor was required by the bills to satisfy himself as to site conditions and so on. On a case stated, Lewis J agreed with the arbitrator that Bryant was entitled to the extra cost of excavating the rock, the existence of which was known to the architect, although not shown in any of the contract documents. However, somewhat confusingly it has been held that a contractor is not entitled to rely on a ground investigation report if it was merely referred to on a drawing. The court held that it was not incorporated into the contract, but merely noted to identify a source of relevant information for the contractor. This was insufficient to override a clause in the contract which placed on the contractor the obligation to satisfy itself about the nature of the site and the sub-soil[255].

---

[253] John Parris, *The Standard Form of Building Contract: JCT 80*, 2nd edition, 1985, reproduced in the 3rd edition, p. 306 (2002) with the necessary changes, Blackwell Publishing.
[254] [1938] 1 All ER 503.
[255] *Co-operative Insurance Society* v. *Henry Boot Scotland Ltd* (2003) Const LJ 109.

Chapter 5
Direct loss and/or expense

5.1 Introduction

In most standard form contracts, the clauses which entitle the contractor to apply for additional money, as a result of disruption or prolongation, use the phrase 'direct loss and/or expense'. It is essential to understand the meaning of the phrase, otherwise neither the party claiming nor the party claimed against will be in a position to understand their rights. The term is used in clauses 26.1 and 34.3 of JCT 98; it is also used in clause 4.38.1 of the related JCT Nominated Sub-Contract Conditions (NSC/C), in clause 4.24 of the Domestic Sub-Contract Form (DSC), in clause 14.1 of IFC 98 Named Sub-Contract Conditions (NAM/SC) and in clause 14 of the IFC 98 Domestic Sub-Contract Conditions (IN/SC). The JCT Intermediate Form of Building Contract (IFC 98) uses the same expression in clause 4.11, as does the JCT Agreement for Minor Building Works (MW 98) in clause 3.6 and the JCT Works Contract Conditions (Works Contract/2) in clauses 4.45, 4.49 and 4.50.

Where other similar phrases such as 'direct loss and/or damage' or 'direct loss and expense' are found in other standard forms, it is suggested that their meaning is to all intents and purposes the same. Slightly different wording is used in Government Conditions (GC/Works/1 (1998), clause 46(1)) which speaks of the contractor 'properly and directly' incurring any expense which results in regular progress being materially disrupted or prolonged. The common feature of all such phrases is that a distinction is drawn between those losses and/or expenses which the law regards as 'direct' and those regarded as being indirect or too remote. This is a distinction of particular importance which will be considered later. The phrase 'direct loss and/or damage' has been judicially considered[256]:

> 'what is its usual, ordinary and proper meaning in the law: one has to ask whether any particular matter or items of loss or damage claimed have been caused by the particular matter.... if it has been caused by it, then one has to go on to see whether there has been some intervention or some other cause which prevents the loss or damage from being properly described as being the direct consequence of the [matter].'[257]

---

[256] In *Wraight Ltd* v. *P. H. & T. (Holdings) Ltd* (1968) 13 BLR 27.
[257] (1968) 13 BLR 27 at 33 per Megaw J.

It may, therefore, be said that references to 'direct loss and/or expense' or 'direct loss and/or damage' are the equivalent to references to damages at common law[258].

Direct loss and/or expense can usefully be considered separately as 'direct loss' and 'direct expense'. The word 'loss' is often used by the courts to include both loss of money which ought to have been received and expenditure of money which ought not to have been expended. The addition of the 'and/or' conjunction seems to have been made to remove any possible doubt regarding the scope of the contractor's entitlement. Thus, the contractor is entitled to claim his losses or his expenses or both together. It also suggests a distinction between the two expressions.

Effectively, the use of the dual phrase gives the contractor two separate avenues of claim:

- Actual losses incurred as a direct result of the circumstances giving rise to the entitlement.
- Actual expenditure occasioned as a direct result of the same circumstances.

This is important, because although 'loss' may be held to encompass both loss and expense, 'expense' is not wide enough to include loss[259].

5.2 Direct versus indirect

The law draws a distinction between direct and indirect (or consequential) loss or damage. A party in breach is not liable to pay all the damage suffered by the injured party. The courts limit the damages to what is reasonable in the circumstances by ruling out all damage said to be too remote. The rule has been stated as follows:

> 'Where two parties have made a contract which one of them has broken, the damages which the other party ought to receive in respect of such breach of contract should be such as may fairly and reasonably be considered either arising naturally, i.e. according to the usual course of things from such breach of contract itself, or such as may reasonably be supposed to have been in the contemplation of both parties at the time they made the contract, as the probable result of the breach of it.'[260]

The rule is said to have two 'limbs'. The first limb refers to damages 'arising naturally'. Such damages are often referred to as 'general damages'. They are the kind of damages that anyone would expect to be the result of the breach. The second limb refers to damages 'in the contemplation of both parties at the time they made the contract'. These kind of damages depend

---

[258] See *F. G. Minter Ltd* v. *Welsh Health Technical Services Organisation* (1980) 13 BLR 7.
[259] See Hodson LJ in *Chandris* v. *Union of India* [1956] 1 All ER 358 at 363: 'the primary meaning of the word "expense" is actual disbursement'.
[260] *Hadley* v. *Baxendale* (1854) 9 Ex 341 at 354 per Alderson B.

upon the knowledge of the parties of special circumstances and they are often referred to as 'special damages'. For example, suppose Ms A buys a car from Acme Used Cars and drives it away intending to use it immediately to drive to Southampton, there to start a pre-booked cruising holiday. Further suppose the car breaks down on the way to Southampton so that Ms A does not reach the ship in time before its departure. Ms A can certainly claim the cost of necessary repairs to the car, but she cannot claim the cost of the lost holiday, because Acme knew nothing of the projected holiday or the consequences of a mechanical breakdown. If all those facts had been made known to Acme before or at the time of the sale contract for the car, Ms A might have been able to claim the cost of the holiday also. In practice, it is unlikely that Acme would accept such a liability and there may well be a clause in its sale contract to deal with that eventuality, but the principle remains.

It is only the remoteness or the entitlement of a party to damages which is considered here; the amount of damages is a separate issue which will be considered later.

A useful explanation of this decision and of the authorities generally was given in *Victoria Laundry (Windsor)* v. *Newman Industries, Coulson & Co*[261]. It took the form of propositions which may be summarised as:

(1) The purpose of damages is to put the injured party in the same position, so far as money can, as if his rights had been observed, but to pursue that purpose would provide him with a complete indemnity and it is considered to be too harsh.

(2) The injured party may only recover loss reasonably foreseeable at the time of the contract.

(3) Foreseeability depends on the knowledge of the party committing the breach.

(4) Knowledge is of two kinds: (a) all reasonable people are assumed to know the kind of loss which is liable to result from a breach in the ordinary course of things; (b) actual knowledge of special circumstances which may cause greater loss.

(5) The contract breaker will be liable provided that, if he had to ask himself, he would have concluded, as a reasonable man, that the loss was liable to result from that breach.

(6) It is enough if the loss could be seen as likely to result.

These propositions were considered by the House of Lords in *The Heron II*[262]. Although they rejected proposition 6 as being too broad, they did not agree the test which should be substituted. Taking the opinions together, *it is enough if the loss was a serious possibility* appears to be a reasonable consensus. The crucial point which lies at the foundation of the rule in *Hadley* v. *Baxendale*, is the knowledge possessed by the contract breaker at the time the contract was entered into. The point is whether he had just

[261] [1949] 1 All ER 997 at 1002 per Asquith LJ.
[262] *Koufos* v. *Czarnikow Ltd (The Heron II)* [1969] 1 AC 350.

ordinary knowledge as an average person in that situation, or whether he had special knowledge which, had he considered the matter, would have led him to the conclusion that greater (or perhaps less) than the ordinary loss would result from the breach.

It seems that it is enough if the type of loss is within the reasonable contemplation of the parties even though the *extent* of such loss is far greater than they could have contemplated. In *H. Parsons (Livestock) Ltd* v. *Utley Ingham & Co Ltd*[263] a herd of pigs became ill with a serious disease as a result of a faulty pig-nut silo. In the ordinary course of events only a relatively minor illness might have been expected to result. The court held that it would not limit the defendants' liability because it was reasonable to suppose that the parties would have contemplated at the time of contract that, in the event of the silo being defective, there was a serious possibility of illness and even death among the pigs.

In practice, these concepts often lead to real difficulties of interpretation. In *Saintline Ltd* v. *Richardsons, Westgarth & Co Ltd*[264], Atkinson J discussed the distinction between direct and consequential loss or damage. He had there to consider the meaning of a contract clause providing that a manufacturer's liability should not 'extend to any indirect or consequential damages whatsoever'. The contract was for the supply of engines to a ship, and the manufacturers broke the contract. The buyers claimed damages for loss of profit for the time they were deprived of the use of the ship, together with the expenses of wages etc., and for the fees paid to experts for superintendence. The judge held that all these items were recoverable as being a direct and natural consequence of the breach of contract. In giving judgment he said:

> 'What does one mean by "direct damage"? Direct damage is that which flows naturally from the breach without other intervening cause and independently of special circumstances ... The words "indirect or consequential" do not exclude liability for damages which are the direct and natural result of the breaches complained of ... [The clause protects] the respondents from claims for special damages which would be recoverable only on proof of special circumstances and for damages contributed to by some supervening cause.'[265]

A construction industry case where the same distinction was drawn is *Croudace Construction Ltd* v. *Cawoods Concrete Products Ltd*[266]. The claimants were main contractors for the erection of a school and they contracted with the defendants for the supply and delivery of masonry blocks. It was a term of the supply contract that the defendants were 'not under any circumstances ... liable for any consequential loss or damage caused or arising by reason of late supply or any fault, failure or defect in any materials or goods supplied by us ... '.

---

[263] [1977] 2 Lloyds Rep 522.
[264] [1940] 2 KB 99.
[265] [1940] 2 KB 99 at 103 per Atkinson J.
[266] (1978) 8 BLR 20.

The claimants claimed against the defendants in respect of alleged late delivery and defective materials. They sought to recover:

(1) Loss of productivity
(2) Inflation costs resulting from delay
(3) The amount of a sub-contractor's claim for delay to his work.

The actual question in the case was whether such damages were to be regarded as 'consequential' loss. The Court of Appeal held that the items claimed were not excluded as 'consequential'. They were heads of direct loss. Their lordships quoted with approval: 'on the question of damages, the word "consequential" has come to mean "not direct" ...'[267].

Croudace Construction Ltd was one of the cases considered by the Court of Appeal in *F. G. Minter Ltd* v. *Welsh Health Technical Services Organisation*[268], which affirms the view that, in the JCT contracts, the term 'direct loss and/or expense' (and equivalent phrases in other contracts) in effect means that what is recoverable is substantially the same as the damages recoverable at common law according to the ordinary principles of damages under the first limb of the rule in *Hadley* v. *Baxendale*[269].

These principles apply to claims for direct loss and/or expense arising under JCT contracts and under other contracts where similar phraseology is used. It should be noted, however, that because the recovery of direct loss and/or expense is a specific term of the contract, the recovery of certain heads of claim may be permitted which might not be recoverable at common law[270].

---

[267] *Millar's Machinery Co Ltd* v. *David Way & Son* (1934) 40 Com Cas 204 at 210 per Maugham LJ.
[268] (1980) 13 BLR 7.
[269] (1854) 9 Ex 341.
[270] Some of the grounds for such contractual claims are not breaches of contract as an examination of the 'matters' in JCT 98 clause 26.2 will confirm.

Chapter 6
Potential heads of claim

6.1 Basic principles

Although the legal principles involved in formulating and ascertaining claims for direct loss and/or expense are well settled, the application of those principles in practice is far from easy.

6.1.1 Measure of damages

The principal function of damages is to put the injured party into the same position, so far as money can, as if the contract had been performed without breach[271]. This is the basic principle – the right of the injured party to receive precisely what he paid for, but it is not applied rigorously, for example, if some other lesser remedy would suffice and if the cost of strict entitlement is out of proportion to the benefit to be gained thereby[272]. Reimbursement of loss and/or expense is simply a means of putting the contractor back in the position in which he would have been but for the delay or disruption. Therefore, the loss and/or expense is not some notional or estimated figure, but the actual amount lost or spent by the contractor.

6.1.2 Burden of proof

The burden of proof usually lies with the party making a claim. There are some instances where the facts supporting a claim appear so obvious that the burden shifts to the other party to, in effect, prove that the claim is valueless (for example see section 6.1.3 below). Such instances are rare in building contracts and the general rule is that the contractor must prove his claim. Many architects and even contractors will state that the contractor's task is to convince them that the claim is both viable and worth as much as the contractor states. There is an old adage that 'he who asserts must prove'[273]; however, the courts have clearly set down the standard of proof in different circumstances. There is some misconception surrounding the standard of proof which a contractor must bring to his claim. Crucially, in

---

[271] *Wertheim* v. *Chicoutimi Pulp Co* [1911] AC 301 PC.
[272] See *Forsyth* v. *Ruxley Electronics & Construction Ltd and Others* (1995) 73 BLR 1.
[273] *Joseph Constantine Steamship Line* v. *Imperial Smelting Corporation Ltd* [1942] AC 154 at 174 per Viscount Maugham.

criminal cases, the Crown must prove its case 'beyond a reasonable doubt'. In civil cases, however, the claimant must prove 'on the balance of probabilities' – a very much less onerous standard. That is the standard required of a contractor in regard to his claim. In simple terms, the architect must be satisfied that it is more likely than not that the contractor has suffered the loss for the reasons he states.

6.1.3 *Res ipsa loquitur*

Literally: 'the thing speaks for itself'. There are some situations where the facts so clearly point one way that the burden of proof is placed upon the defendant to show that he is not at fault[274]. Rarely, some elements of the contractor's claim may fall into this category. Such a case might be where scaffolding is erected next to a public highway and a passer-by is found on the floor with head injuries. A blood-stained brick of the type being stacked on the scaffolding is found beside the injured person. The facts clearly indicate that a brick has fallen from the scaffolding on to the head of the passer-by, causing the injuries. In any legal action by the injured person, the burden of proof would probably lie with the contractor to show that he was not responsible.

6.1.4 Mitigation of loss

There is much confusion about the principles of mitigation of loss. They are quite simply stated:

(1) A party cannot recover damages resulting from the other party's breach of contract if it would have been possible to avoid any damage by taking reasonable measures.
(2) A party cannot recover damages which he has avoided by taking measures, even if such measures were greater than what might be considered reasonable.
(3) A party can recover the cost of taking reasonable measures to avoid or mitigate (reduce) his potential damages.

This is said to give rise to a duty to mitigate[275] although a failure to mitigate will not give rise to a legal liability; it will simply reduce the damages recoverable to what they would have been had mitigating measures been taken.

That is not to say that the claimant must do everything possible. He need not do anything other than an ordinary prudent person in the course of his business would do[276].

---

[274] *Scott* v. *London & St Katherine's Docks Co* (1865) 3 H & C 596.
[275] *British Westinghouse* v. *Underground Railways Company* [1912] AC 673.
[276] *London & South England Building Society* v. *Stone* [1983] 1 WLR 1242.

The application of this principle is illustrated by the example of plant standing idle as a result of a variation order. The contractor would not be entitled simply to accept the situation, but would be obliged to make reasonable endeavours to use the plant productively elsewhere or to persuade the plant owner to accept an early return. In the first instance, the costs of, say, moving the plant to another site so that it might be so used would be recoverable as a part of a direct loss claim, provided of course that this sum did not exceed the costs which would have been otherwise incurred. However, although the injured party must only take reasonable measures and not unreasonable measures, the courts usually will not look too critically in hindsight at his actions in attempting to mitigate. The crucial question is whether, in attempting to mitigate, the injured party acted reasonably[277]. Even if the actions of the injured party resulted in an increase in loss, the cost will be recoverable if he acted reasonably[278].

The position is, in fact, spelled out specifically in the extension of time clauses of many standard form contracts (for example see JCT 98, clause 25.3.4) in relation to delay. The clause expressly requires the contractor to use his best endeavours to prevent delay occurring and to mitigate the effects of a delay once encountered. So far as the money claims provisions are concerned, the general law which imposes a duty to take all reasonable steps and prevent claims for damages which have resulted purely from a failure to take such steps applies. It is for the party receiving the claim to show that the claimant has failed to mitigate[279]. Therefore, when the contractor submits an application for direct loss and/or expense, it is for the architect to show that the contractor has not mitigated his losses. But although that burden falls on the architect, he is entitled to seek relevant information from the contractor so that he can form an opinion about the matter.

6.2   Two basic claims situations

Many potential heads of claim are common to both. The first situation is delay in completion of the contract works beyond the date when they otherwise would have been completed or the date the contractor has undertaken to complete them; this is often known as a *prolongation claim* and, sometimes – inaccurately – as a claim for extended preliminaries. The second is a claim which is not related to the date for completion of the contract works. This is commonly referred to as a *disruption claim*. A disruption claim does not depend upon prolongation of the contract period and it may arise where the works are completed within the contract period originally specified, where they are completed within a shorter period programmed by the contractor, or where they are completed after the end of

---

[277] *Banco de Portugal* v. *Waterlow & Sons Ltd* [1932] AC 452.
[278] *Melachrino* v. *Nicholl & Knight* [1920] 1 KB 693.
[279] *Garnac Grain Co Inc* v. *Faure & Fairclough* [1968] AC 1130.

the contract period. An example of disruption would be if delays are caused to non-critical activities which, therefore, do not affect the overall period of time required to carry out the works. Not every such example will allow a contractor to recover loss and expense of course. Disruption claims are notoriously difficult to prove and every case will depend on the surrounding circumstances.

It is often contended that there is no such thing as a claim for disruption as part of loss and expense. Rather, it is said, such a claim is properly the subject of valuation of variations which standard form contracts provide for by adjustment of rates and even the creation of new rates to deal with changes of circumstances. There is some merit in such arguments, but they do not successfully deal with all types of disruption and especially cannot cover disruption which is not the result of a variation[280].

6.2.1 Prolongation

A prolongation claim is probably the commonest claim. It is also the simplest and most straightforward to prepare and to understand. In essence it simply states that the employer or architect has acted in such a way, or failed so to act, that the contractor has been unable to complete the works by the contractual date for completion, but has been obliged to complete later, i.e. the contract period has been prolonged. Moreover, the actions or inactions fall under one of the matters in clause 26 (where JCT 98 is involved). The claim then proceeds to quantify the loss and/or expense involved. Despite, or perhaps because of, its simplicity, many contractors are slovenly in its preparation. A 'claim' for extension of time is the usual precursor to a prolongation claim, to the extent that many contractors and architects believe that unless the contractor is given an extension of time first, he is not eligible for a prolongation payment. Nothing could be further from the truth[281]. Notwithstanding that, it is often convenient for the contractor to get his extension of time first, because the evidence in support of entitlements for extension of time will often be the same as the evidence required to establish an entitlement to loss and/or expense, although it will not establish the quantum.

Once the architect has been satisfied about the duration of the prolongation period, the contractor has merely to establish the loss and expense suffered as a result. Because it is actual loss and actual expense which is to be ascertained, the practice of using the priced preliminaries in the bill of quantities to arrive at the figure is not acceptable. The actual cost to the contractor of being on site for the extra period of time must be established. Although the period of prolongation tends, for obvious reasons, to be measured as starting from the contractual completion date and extending on as appropriate, this is not necessarily the time frame within which the

---

[280] See also Chapter 13.
[281] *H. Fairweather & Co Ltd* v. *London Borough of Wandsworth* (1987) 39 BLR 106.

costs should be ascertained, because the true costs resulting from a delay will usually follow the delay itself. The few weeks at the beginning and end of a contract will be characterised by a build-up then a reduction respectively of site-related costs.

Therefore, if a contract is prolonged for a period of 6 weeks, which the contractor can establish is a direct result of some clause 26 matters, the 6 weeks may be made up of several delaying occurrences taking place at differing times during the contract period. The task of identifying each delay and its monetary consequences is not always easy, but it must at least be attempted. Many architects will agree to take a representative slice of the appropriate number of weeks prolongation from somewhere in the middle of the contract period. This seems, at first sight, to be a reasonably good empirical method of establishing appropriate costs, but it is not an ascertainment in the proper sense and it is difficult to substantiate if challenged. In practice, if both employer advised by the architect and quantity surveyor and the contractor find that or some other approximate system is acceptable, there is nothing more to be said.

6.2.2 Disruption

Disruption is usually claimed separately from prolongation. It may be present with or without prolongation. Disruption has always been very difficult to establish with any precision and even more difficult to ascertain in monetary terms. Traditionally, a contractor's claim for disruption has relied both for substantiation of the fact of disruption and the ascertainment of its costs on the comparison of anticipated against actual labour costs. This bald approach does not bear consideration and it has been roundly condemned[282]. There may be many reasons for the actual costs of labour being greater than the contractor anticipated other than reasons for which the employer or the architect are responsible.

Commonly, disruption amounts to delays in non-critical parts of a project, but not to the extent that those parts become critical in programming terms. For example, the task of fitting external balcony railings on the front of a hotel project may not be critical. The contractor may have anticipated and priced for it to take 3 weeks during a 5-week available time slot. Therefore, if the work is delayed by one of the clause 26 matters so that it takes 4 instead of 3 weeks, there will be no resultant prolongation of the contract period, but the contractor will no doubt incur additional costs. That is a relatively simple example and, provided the contractor has kept proper records, there should be little difficulty in identifying the costs involved. Other instances are more complex.

The classic method of evaluating disruption is to compare the value to the contractor of the work done per man during a period of no disruption with the value per man doing the disrupted period and then to apply the ratio to

---

[282] *London Borough of Merton* v. *Stanley Hugh Leach Ltd* (1985) 32 BLR 51.

the total cost of labour[283]. In order for the method to work, it must be possible to identify a period free from disruption and the compared outputs must relate to similar work.

6.3 Foreshortened programme

Clause 23.1.1 of JCT 98 places an obligation upon the contractor to complete the works 'on or before' the completion date, i.e. the date for completion stated in the Appendix or any later date fixed by an extension of time under clause 25. It is therefore clear that the contractor may complete the works before the date fixed under the contract, if he wishes and can reasonably do so, while complying with all the other terms of the contract. It follows that the only significance of the completion date is to fix the date from which the employer may be entitled to recover liquidated damages. The contractor is not obliged to remain on site until that date. Indeed, by clause 17.1 of JCT 98 he is entitled to a certificate of practical completion whenever he in fact has finished the works, whether it is before or after the contract date for completion.

Therefore, when the cost of remaining on site for a longer period than would otherwise have been necessary is a factor to be included in the ascertainment of direct loss and/or expense, it is sometimes argued that the period of time which will form the basis of the ascertainment will not necessarily be measurable from the contract completion date, but may be measurable from an earlier or later date, i.e. the date when the contractor would otherwise have been able to complete. This contention is sometimes rebutted on the basis that the contractor has undertaken to stay on site until the completion date stated in the contract. That rebuttal seems to be based on an irrelevance. The contract completion date is modified by clause 23.1.1, noted above, which allows the contractor to finish early. Therefore, the contractor does not undertake to stay on site until the contract completion date, but only until the works have reached practical completion, which may be earlier. Therefore, the question is simply whether the contractor suffers loss and/or expense as a direct result of being kept on site longer than he needed to be on site.

However, there is one event listed in clause 26.2 when the use of an earlier date than the contractual date for completion would not be appropriate, and that is clause 26.2.1, dealing with late instructions. While the employer or architect must not prevent the contractor from meeting an earlier date, which he would otherwise have been able to meet, they are not obliged positively to assist him to do so. The architect's obligation is either to provide necessary information, instructions, etc. in accordance with an information release schedule (very rarely used in practice) or in due time, that is at a time that will enable the contractor to meet his obligations as to completion. Accordingly, if the architect has provided necessary informa-

---

[283] *Whittal Builders Co Ltd* v. *Chester-Le-Street District Council* [1996] 12 Const LJ 356.

tion at dates that have permitted the contractor (without undue effort or additional cost) to meet the contract completion date, then the architect has by definition fulfilled his obligations and the contractor will have no claim. This position is unaffected by the architect 'accepting' or 'approving' a programme from the contractor showing an earlier date for completion than that set out in the Appendix, because the architect has no power to vary the terms of the contract[284]. It may be a different matter if the shortened programme is accepted by the employer.

Judge James Fox-Andrews, sitting as an Official Referee, in the case of *Glenlion Construction Ltd* v. *The Guiness Trust*[285] considered the similar provisions in JCT 63. One of the questions the judge had to consider was:

> 'whether there was an implied term of the contract . . . that, if and in so far as the programme showed a completion date before the date for completion [stated in the contract] the employer by himself, his servants or his agents should so perform the said agreement as to enable the contractor to carry out the works in accordance with the programme and to complete the works on the said completion date'

i.e. information etc. should have been supplied at times which would have enabled the contractor to meet the earlier completion date. The judge's answer to the question was 'no'. As he said,

> 'it is not suggested by Glenlion that they were both entitled *and* obliged to finish by the earlier completion date. If there is such an implied term it imposed an obligation on the Trust but none on Glenlion.'[286]

This was manifestly wrong, said the judge, and there was, therefore, no obligation on the Trust or its architect to provide information at any times earlier than those necessary to enable the contractor to complete by the contract date. This principle is now enshrined in JCT 98 clause 5.4.2 and IFC 98 clause 1.7.2.

It is, therefore, clear that if the contractor intended to finish earlier than the contract date for completion, he would not have a claim for loss and/or expense associated with being kept on site longer than anticipated if the reason he could not finish on the earlier date was simply that the architect had provided information at such time as to allow completion by the contractual completion date. However, it is sometimes contended that if the contractor, as a matter of fact, had received all the information to allow him to finish at his earlier intended date, he may have a valid claim if he is prevented from finishing on that earlier date by one of the 'matters' in clause 26, even though he still finished before or on the contract completion date.

The arguments in favour of that contention are that, if it was not for the 'matter', the contractor would have completed the works by his intended

---

[284] See the commentary to *Glenlion Construction Ltd* v. *The Guiness Trust* at (1987) 39 BLR 93.
[285] (1987) 39 BLR 89.
[286] (1987) 39 BLR 89 at 63 per Judge Fox-Andrews.

date and made a certain profit. He was denied that profit by a 'matter' which under the terms of clause 26 entitles the contractor to reimbursement of loss and/or expense. Moreover, the 'matters' in clause 26 are in no way tied to a prolongation of the contract period beyond the contract completion date. The only criteria are:

(1) Whether regular progress has been materially affected: there can be no doubt that it has been, because if it was not for the 'matter' the contractor would have finished by his earlier date
(2) Whether the contractor has suffered direct loss and/or expense: he must have suffered loss and/or expense by being kept on site longer than would otherwise have been the case
(3) Whether he has been reimbursed under some other provision of the contract: he has not been reimbursed under any other provision for the profit he would have made if he had finished at the earlier date.

The arguments against the contention are that, under the terms of the contract, the contractor is assumed to have allowed for being on site until the contract completion date. His earlier date is not a contractually agreed date, but simply an attempt by the contractor to try to make more profit. Therefore, he has not suffered any loss and/or expense by staying on site until the completion date. The contractor has already been reimbursed under the provisions of the contract which provide for payment to the contractor of the agreed contract sum.

   The situation may be further complicated if the contractor has proposed a foreshortened programme at the commencement of the project and the employer, rather than the architect, has accepted or agreed the programme. In such a case, it may be convincingly argued that the parties have effectively varied the contract by agreeing a new completion date to which all are bound to work instead of the former date.

6.4 The 'knock-on' effect

This is often referred to as a 'winter working' claim, but although the principle is probably most commonly encountered in connection with winter working, it could apply to any other situation where a delay arises which inevitably causes the works to be carried out in a situation which is less felicitous than originally envisaged. In the case of winter working, the problem for the contractor is that something which causes him delay and which entitles him to loss and/or expense directly resulting from the delay may also move the programme of work so that site operations needing good weather to execute may have to be carried out during a period of bad weather. At the extreme, activities programmed for the summer months have to be carried out in the winter. There will be occasions, of course, in which the contractor has had to allow in his tender for work to be carried out at a difficult time of year and a delay caused by the employer may actually improve the working conditions. The employer is not entitled to argue that the contractor should reimburse some money in consequence, but it may be appropriate for the architect to take it into account, depending on the facts, when ascertaining loss and/or expense for the delay.

Under JCT 98 and IFC 98 standard forms, a contractor is entitled to an extension of time only for *exceptionally* adverse weather conditions. The weather conditions, although not what was envisaged, may not be exceptionally adverse in normal circumstances. Even if they are sufficiently bad to warrant an extension of time, bad weather is not in itself grounds for a claim for loss and/or expense. Whether or not the contractor can found a claim for loss and/or expense in the circumstances will depend on whether causation can be established.

The case which is much relied upon in this situation is the Canadian case of *Ellis-Don* v. *The Parking Authority of Toronto*[287]. The basic facts of the case were simply that a contract period of 52 weeks was delayed by a further 32 weeks. Of this delay, 7 weeks were caused by the employer who failed to obtain the appropriate permit so that the project commenced 7 weeks late. This led to further delay in starting up and a total of 17.5 weeks was laid at the door of the employer. Part of the work had to be carried out in winter and the contractor was successful in claiming additional payment.

In essence, however, the principle is simply damages for breach of contract[288]. In another case, the judge postulated a knock-on scenario thus:

'Assume the following facts: A contract is entered into in this form of contract in May for 1 year for completion on 31 July of the next. The work is of a tunnelling nature. No tunnelling can be carried out from 1 November to 31 March for seasonal reasons but during that period the contractor will have expensive equipment lying idle. In early April when the works were on course for completion on 31 July the architect issues an instruction under 11(1) requiring a variation the execution of which will add 3 months to the contract period. At the same time on the contractor's application he grants an extension of time for completion to 31 October. A fortnight before 31 October when the works as varied are on course for completion in due time a strike occurs which continues until 31 March. The contractor recommences work on 1 April but because he had no opportunity to protect his machinery during the 6 months period it then takes the contractor 2 months not 2 weeks to complete. There has been no fault on either party.

If the architect grants an extension of time of eight months only under 23(d) I can see no reason why the contractor under the contract cannot still recover all his direct loss and expense under 11(6).'[289]

There are two things immediately to remark about this extract. The first is that the reference is to JCT 63. The principle, however, would be the same

---

[287] (1978) 28 BLR 98.

[288] See Chapter 5, section 5.2 and the view of the House of Lords in *Koufos* v. *Czarnikow Ltd* (*The Heron II*) [1969] 1 AC 350.

[289] *H. Fairweather & Co Ltd* v. *London Borough of Wandsworth* (1987) 39 BLR 106 at 118 per Judge Fox-Andrews.

under JCT 98. The second thing is that the mathematics in the extract is wrong in a number of places. For example, it is surprising that a 1-year contract entered into in May is to be completed on 31 July the following year and the '6 months period' is actually 5 months. The principle, however, remains unaffected.

In the commentary to this case, the editors of *Building Law Reports* note that it is a useful example to demonstrate why there is no necessary link between the grant or the refusal of an extension of time and the success of an application for loss and/or expense. They proceed, however, to disagree with the judge. In their view the contractor's costs flowed directly from the strike, not from the variation. They point out that a contractor takes the risk that a strike may occur not only during the original contract period, but also during any period of extension. They agree that the contractor would be entitled to an extension of the contract period for the strike. This example very clearly highlights the difficulties in considering knock-on claims. The question to be asked in each case is 'What is the direct cause of the winter working?'[290].

The contractor must, of course, be able to prove a direct link between the cause of the delay and the working in worse conditions, with no compensating saving[291]. This view is supported by *Bush* v. *Whitehaven Port & Town Trustees*[292].

In *Bush*, Bush contracted with the Trustees to lay a 15 inch water main from Ennerdale Water to Whitehaven for £1335. The contract was made in June, and the Trustees contracted to be ready at all times to give Bush sufficient possession of the site to enable him to proceed with the works. In fact, through the fault of the Trustees the whole of the site was not available until October. As a result of this delay, the contract was thrown into the winter months, and Bush was put to heavy extra expense for which he sued the defendants.

His claim was successful, despite the fact that there was an express term saying that 'non-delivery of the site ... shall not ... entitle the contractor to any increased allowance in respect of money ...'.

In one sense the decision is unsatisfactory, since the court put forward an alternative base for its decision, namely that a summer contract having, by implication, been in the contemplation of the parties when the contract was made, Bush was entitled to a *quantum meruit* (as much as he has earned) or damages in respect of the increased expenditure.

In *Sir Lindsay Parkinson Ltd* v. *Commissioners of Works*[293], the decision was explained as being based on an implied term about the circumstances in which the contract works were to be done. This certainly seems to have been the view of Lord Caleridge CJ in the *Bush* case itself. He dealt with the question whether a term could be implied for breach of which the plaintiff

---

[290] Causation is discussed in Chapter 7.
[291] The principles of causation are discussed in Chapter 7.
[292] (1888) 52 JP 392. This case was disapproved in part in *Davis Contractors* v. *Fareham Urban District Council* [1956] 2 All ER 145, but not on this point.
[293] [1950] 1 All ER 208.

could recover damages, and was inclined to answer it in the affirmative, and referred several times to the contract being made 'with reference to certain anticipated circumstances' and becoming inapplicable to the actual situation. What was to have been a summer contract had been turned into a winter contract:

> 'It was turned into a winter contract when wages were different...when days were short, instead of long; when weather was bad, instead of good; when rivers which had to be dealt with, and had to be crossed by the pipes, were full not empty; and when, in fact,...a great many most important circumstances under which the contract was to be executed, had...changed from those which...were in the contemplation of the parties when the contract was entered into. The contract, nevertheless, was carried on and completed...with the knowledge of the defendants...the contractor [completed the works] under the altered conditions...'[294]

This view has been echoed in Ellis-Don[295]. Essentially, a knock-on claim flows naturally from the breach, whatever it was, which caused the delay. It is by no means easy to identify the chain of causation correctly and contractors should not rely upon the kind of example put forward in the decision in *H. Fairweather & Co Ltd* v. *London Borough of Wandsworth*[296].

A contractor is obliged to take responsibility for those delays which he has caused, but he is not bound to take the unforeseeable into account. Prolongation of a contract which means working through an additional winter period almost inevitably results in 'direct loss and/or expense' to the contractor. There are occasions, of course, when a delay during the progress of the works may have the result of pushing work into a summer period to the contractor's advantage.

6.5 *The more common heads of loss*

The following are not intended to be exhaustive heads of loss, but simply those that most generally apply. The basic principle to be borne in mind is that, subject to the restrictions of directness and foreseeability, the contractor should be put into the financial position which he would have been in had the delay or disruption not occurred. If this general principle is borne in mind, there should be no difficulty in judging or putting forward other heads of loss where the particular circumstances permit.

Loss and expense is the equivalent of damages at common law. The measure of such damages can be quite complex, but the starting position is to put the injured party in the same position, so far as money can do it, as if the contract had been correctly performed[297]. In recovering such damages,

---

[294] (1888) 52 JP 392 at 393 per Coleridge CJ.
[295] *Ellis-Don* v. *Parking Authority of Toronto* (1978) 28 BLR 98.
[296] (1987) 39 BLR 106 at 118 per Judge Fox-Andrews. See the commentary on p. 110 of the judgment.
[297] *Robinson* v. *Harman* (1848) 1 Ex 850.

the law will allow only the recovery of losses actually suffered or expense actually incurred. The contractor should recover his actual costs if he has the records to substantiate them in preference to his tender costs, even though his tender costs are part of his tendered and accepted price.

6.5.1 On-site establishment costs

These are often called site overheads or commonly simply 'preliminaries' because the prices are normally found in the preliminaries section of the bills of quantities. The bills of quantities prices are not normally to be used to calculate the loss and/or expense. Actual costs should be used. On-site establishment costs are perhaps the easiest head of claim to establish because the data should be readily available once the period of delay has been settled. They will consist of the supervisory and administrative staff engaged upon the site of the particular contract, site accommodation, plant and tools, telephone and electricity charges, costs of welfare and sanitary facilities, light and heat where not covered by electrical charges, and the like. Not all of these items are time related. Some are clearly dependent on work or value and care must be taken that inappropriate items are excluded.

All these costs should be readily ascertainable from the contractor's cost records. A note of caution is necessary. In the normal course of events the contractor will be running down his site establishment as the contract work approaches its end. Simply to take these costs from the date when the work would have been completed to the date of actual completion may, therefore, be unfair to the contractor and it is the site establishment during the period of delay which is relevant. In most instances that will be at full strength.

On very large contracts it is desirable for the person actually responsible for the ascertainment to check that the staff on site are working exclusively in connection with the particular contract and are not carrying out any work in connection with other contracts. Efficient contractors will require their staff to complete time-sheets and the architect or quantity surveyor should call for, and examine, them. It is not unknown for a contractor to keep a supervisor on site longer than necessary simply because he has no other project on which to place him. Supervisors who are not required on one job are sometimes placed on another if it seems that it may be possible to include them in a claim for site establishment.

6.5.2 Head office overheads

The position is succinctly put in *Keating on Building Contracts*:

> 'A contractor's overheads are commonly taken to be recovered out of the income from his business as a whole and ordinarily where completion of one contract is delayed the contractor claims to have suffered a loss arising from the diminution of his income from the job and hence the

turnover of his business. But he continues to incur expenditure on overheads which he cannot materially reduce or, in respect of the site, can only reduce, if at all, to a limited extent. But for the delay, the workforce would have had the opportunity of being employed on another contract which would have had the effect of contributing to the overheads during the overrun period.'[298]

Where a contractor was kept on site longer than the contract period, it used to be taken for granted that he would be able to recover his overhead costs for the period of delay. That is no longer the presumption and recovery of head office overheads is becoming far more difficult.

It is usually argued that the contractor must be able to show that he had other work which he could have done during the delay period, otherwise, in the absence of any delay, there would have been no contribution during the period and, therefore, no loss[299]. An exception has been made where a contractor was able to show that he carried out one project at a time[300].

Essentially, it seems that on any contract, delay or disruption may lead to some increase in direct head office administrative costs, relating not only to any period of delay but also to the involvement of staff in dealing with the problems caused by disruption, e.g. contract managers spending more time in organising additional labour, recasting programmes, etc., buying staff in, ordering additional materials, arranging plant hire and the like. If, however, they would not have been gainfully employed, but for the delay, the contractor may face difficulties in recovering such costs which, it will be argued, he would have incurred as part of his head office expenditure in any event. It has been said:

> '... it is for [the contractor] to demonstrate that he has suffered the loss which he is seeking to recover...it is for [the contractor] to demonstrate, in respect of the individuals whose time is claimed, that they spent extra time allocated to a particular contract. This proof must include the keeping of some form of record that the time was excessive, and that their attention was diverted in such a way that loss was incurred. It is important, in my view, that [the contractor] places some evidence before the court that there was other work available which, but for the delay, he would have secured, but which, in fact, he did not secure because of the delay; thus he is able to demonstrate that he would have recouped his overheads from those other contracts and thus, is entitled to an extra payment in respect of any delay period awarded in the instant contract.'[301]

---

[298] Stephen Furst and Vivian Ramsey, *Keating on Building Contracts*, 7th edition, 2001, Sweet & Maxwell at 267.
[299] *Finnegan* v. *Sheffield City Council* (1988) 43 BLR 124.
[300] *Alfred McAlpine Homes North Ltd* v. *Property & Land Contractors Ltd* (1995) 76 BLR 65.
[301] *AMEC Building Ltd* v. *Cadmus Investments Co Ltd* (1997) 13 Const LJ 50 at 56 per Mr Recorder Kallipetis. See also *City Axis Ltd* v. *Daniel P. Jackson* (1998) CILL 1382, *Norwest Holst Construction Ltd* v. *Co-operative Wholesale Society Ltd*, 17 February 1998, unreported, and *Beechwood Development Company (Scotland) Ltd* v. *Stuart Mitchell (t/a Discovery Land Surveys)* (2001) CILL 1727 where the criteria for head office overheads are set out. The importance of the availability of other work is common to all.

Significantly, the problem, identified by the arbitrator and confirmed by the judge, was that the delay was not sufficient to deter a building contractor of the size and standing of the contractor in this case from tendering for other work. The recovery of head office overheads as part of prolongation costs is likely to be difficult in future where large contractors are concerned. Indeed, it is always difficult for a contractor to show that he has been prevented from using his workforce on another project, because the current project is delayed. In practice, much if not all of the workforce will be sub-contracted. In any event, the types of operatives engaged during a period of delay at the end of a contract are finishing trades and not the groundworkers and other early trades needed for a new project. Even the supervisors will often be finishing foremen.

Efficient contractors will require their staff to keep time records and where this is done the direct costs involved should be readily ascertainable.

Head office overheads include not only costs of staff engaged upon individual contracts but also such general items as rent, rates, light, heating, cleaning, etc. and also clerical staff, telephonists, etc. and general costs such as stationery, office equipment, etc. It is important to distinguish between these two elements of overhead costs however calculated. One set of overhead costs is costs which are expended in any event: rates, electricity and the like. The other is managerial time which is directly allocatable to the project and to no other.

The use of formulae for calculating head office overheads and profit was not approved by the High Court in *Tate & Lyle Food Distribution Co. Ltd* v. *Greater London Council*[302], especially if other more accurate systems are available but the contractor fails to take advantage of them. This case throws doubt on the legitimacy of charging a percentage to represent head office or any managerial time spent as a result of delay or disruption, at least in the absence of specific proof. Here, it was held that expenditure of managerial time in remedying an actionable wrong done to a trading company was claimable at common law as a head of 'special damage'. The claim failed because the company had kept no record of the amount of managerial time actually spent on remedying the wrong and, accordingly, there was no proof of the claim. The High Court refused to speculate on quantum by awarding a percentage of the total of the other items of the claim, although it was expressly invited to do so by counsel, relying on an old line of cases in Admiralty matters. This was not a case involving building contracts, but it is suggested that the principles enunciated are of general application. Although the case went to the House of Lords, the reasoning of the High Court was not at issue.

> 'I have no doubt that the expenditure of managerial time in remedying an actual wrong done to a trading concern can properly form the subject-matter of a head of claim. In a case such as this it would be wholly unrealistic to assume that no such additional managerial time was in

---

[302] [1982] 1 WLR 149.

fact expended. I would also accept that it must be extremely difficult to quantify. But modern office arrangements permit of the recording of the time spent by managerial staff on particular projects. I do not believe that it would have been impossible for the plaintiffs...to have kept some record to show the extent to which their trading routine was disturbed...In the absence of any evidence about the extent to which this has occurred the only suggestion...is that I should award a percentage on the total damages...While I am satisfied that this head of damage can properly be claimed I am not prepared to advance into an area of pure speculation when it comes to quantum. I feel bound to hold that the plaintiffs have failed to prove that any sum is due under this head.'[303]

There is some authority to the effect that, if all other methods of calculating loss fails, then provided that it is clear that some loss has been sustained, a court will accept an estimate which in some instances may be little more than speculation[304].

In order to make a claim involving either overhead levels or profit levels (or both), it appears that actual overheads and profits must be identified – not merely theoretical or assumed levels. If 'direct loss and/or expense' is the equivalent of what is claimable as damages for breach of contract at common law, the common law principles must apply. Force has been given to this argument by the courts:

'Managers are of course employed to sort out problems as they arise. If, however, the magnitude of the problem is such that an untoward degree of time is being spent on it then their costs are recoverable. Looking at the hours recorded, I am quite satisfied that is the position in this case. The costs of course go beyond those of managers and represent staff cost that would not have been incurred but for the defendant's breach. The plaintiffs might have provided an alternative quantification by reference to the additional costs to them of employing others but I do not consider that they are obliged to do so if they can satisfactorily demonstrate the cost to them of time unnecessarily spent and therefore lost. It is for the defendants to show that the losses *prima facie* incurred are not the correct measure of damage and this [the defendants] failed to do.'[305]

In a further case, the judge has made observations of general importance, albeit there were some special circumstances which the arbitrator, from whom the case was heard on appeal, had taken into account on the basis that the claimants only carried out one major project at a time and, therefore, all their overheads were referable to that project. The judge noted with approval some 'clear and sensible conclusions' of the arbitrator:

---

[303] *Tate & Lyle Food & Distribution Co Ltd* v. *Greater London Council* [1982] 1 WLR 149 at 152 per Forbes J.
[304] *Chaplin* v. *Hicks* [1911] 2 KB 786. See also the Canadian case of *Wood* v. *Grand Valley Railway Co* (1913) 16 DLR 361 and the more recent *Middlesborough Football & Athletic Company (1986) Ltd* v. *Liverpool Football & Athletic Grounds plc* (2003) CILL 1959.
[305] *Babcock Energy Ltd* v. *Lodge Sturtevant Ltd* (1994) 41 Con LR 45.

'Efficient contractors normally require their staff to keep accurate time records which allow actual costs related to projects to be ascertained. This is a duty commonly left to the respective quantity surveyors, although the use of a formula would perhaps be appropriate where no such records are available or where there is an agreement between the parties that the "broad brush" approach would be acceptable. Practitioners are generally sceptical about the application of such formulae on the grounds that it is the actual loss and expense which is admissible and that the contractor must specify precisely the expense which has been incurred. It is clear in my mind that this was the intention of the JCT standard form in respect of clause 26.'[306]

Later the judge added:

'The requirements that the loss or expense should be "direct", that it should not "be reimbursed by a payment under any other provision in [the] contract" and that the architect or quantity surveyor is to "ascertain the amount of such loss and/or expense", all suggest strongly that the amount of direct "loss and/or expense" will not exceed what might have been recoverable as damages. In particular the requirement that the amount should not be reimbursed under another provision of the contract is likely further to limit the occasions on which a formula might be appropriate, (although like the use of preliminaries to measure prolongation costs a formula is not infrequently agreed by the contracting parties to be a convenient short cut even though it would not otherwise have been legitimate). Furthermore "to ascertain" means "to find out for certain" and it does not therefore connote as much use of judgment or the formation of opinion had "assess" or "evaluate" been used. It thus appears to preclude making general assessments as have at times to be done in quantifying damages recoverable for breach of contract.'[307]

For all these reasons it appears that the use of formulae cannot generally be justified as a method of ascertaining the contractor's entitlement under the contract terms unless backed up by supporting evidence and subject to the limitations indicated. This reference to ascertainment was considered in a later case dealing with an appeal from an arbitrator's award where the duty to 'ascertain' was softened to allow some measure of judgment to be used:

'A judge or arbitrator who assesses damages for breach of contract will endeavour to calculate a figure as precisely as it is possible to do on the material before him or her. In some cases, the facts are clear, and there is only one possible answer. In others, the facts are less clear, and different tribunals would reach different conclusions. In such cases, there is more scope for the exercise of judgment. The result is always uncertain until the damages have been assessed. But once the damages have been

---

[306] *Alfred McAlpine Homes North Ltd v. Property & Land Contractors Ltd* (1995) 76 BLR 65 at 70 per Judge Lloyd.
[307] *Alfred McAlpine Homes North Ltd v. Property & Land Contractors Ltd* (1995) 76 BLR 65 at 88 per Judge Lloyd.

assessed, the figure becomes certain: it has been ascertained. In my view, precisely the same situation applies to an arbitrator who is engaged on the task of "ascertaining" loss or expense under one of the standard forms of building contract. Indeed, it would be strange if it were otherwise, since a number of the events which give rise to recover loss or expense under the contract would also entitle the claimant to be awarded damages for breach. I would hold, therefore, that, in ascertaining loss or expense, an arbitrator may, and indeed should, exercise judgment where the facts are not sufficiently clear, and that there is no warrant for saying that his approach should differ from that which may properly be followed when assessing damages for breach of contract.

Thus in cases such as the present, the arbitrator must decide *inter alia* whether the costs built into the tender rates were realistic on the footing that the contract proceeded without delay or disruption. That decision inevitably involves an element of judgment, just as the tendering process itself involves an element of judgment. There is no place for pure speculation in the ascertainment of loss or expense, any more than there is in the assessment of damages. Moreover, I think that an arbitrator should not readily use typical or hypothetical figures, but it would be wrong to say that they can never be used.[308]

However, in *Ellis-Don Ltd* v. *Parking Authority of Toronto*[309], the Supreme Court of Ontario applied the Hudson formula[310] somewhat uncritically in respect of a contractor's claim for damages for off-site overheads and profit. Importantly, however, it should be noted that O'Leary J expressly found that (1) the overheads and profit would have been capable of being earned elsewhere had it not been for the delay caused by the employer and (2) 'on this project without taking into account the results of this law suit Ellis-Don made 4 per cent of the contract price to be applied against overhead and as profit', the contractor having claimed that 3.87 per cent of the contract price had been included for these items[311].

It is essential to remember that formulae assume a healthy construction industry and that the contractor has finite resources so that, if delayed on a project, he will be unable to take on other work. In a period of recession, if workload for a particular contractor is not heavy, or if, as in the *AMEC* case noted above, the contractor is of significant size, he will have difficulty in showing that a delay caused him to lose the opportunity to carry out other work. Indeed, as noted earlier, there may be other reasons why being delayed on one project would not prevent the contractor undertaking another. When the construction industry is buoyant or booming at the material time, a formula approach may be acceptable[312].

---

[308] *How Engineering Services Ltd* v. *Lindner Ceilings Partitions plc* [1999] 2 All ER (Comm) 374 at 383 per Dyson J.
[309] (1978) 28 BLR 98.
[310] See below.
[311] See also *Shore & Horwitz Construction Co Ltd* v. *Franki of Canada Ltd* [1964] SCR 589.
[312] *St Modwen Developments Ltd* v. *Bowmer & Kirkland Ltd* (1996) 14 CLD-02-04.

Care must be taken to avoid double-recovery in respect of directly engaged administrative staff if some kind of formula is used to deal with the element of general overhead costs. If some or all of the prolongation period is caused by additional work, the contractor will have recovered an appropriate proportion of overheads. Even if he has not so recovered, the amount is 'reimbursable' under the valuation of variation clause and, therefore, cannot be recovered under the loss and/or expense clause which covers only loss and/or expense for which he 'would not be reimbursed by a payment under any other provision'[313]. Note the reference is not to reimbursement which has been made, but which could have been made.

Before a decision is made to use a formula, it is essential to ensure that it does not overstate the actual loss to the contractor, and the formula should be backed up by supporting evidence, e.g. the tender make-up, head office and project records and accounts, showing actual and anticipated overheads before, during and after the period of delay. Any formula should be used with caution.

The Hudson formula

The best-known formula is called the Hudson formula, set out on page 1076 of *Hudson's Building and Engineering Contracts*[314], which takes the allowance made by the contractor for head office overheads and profit in his original tender, divides it by the original contract period and multiplies the result by the period of contract overrun.

The formula is as follows:

$$\frac{HO/Profit\%}{100} \times \frac{Contract\ sum}{Contract\ period} \times Period\ of\ delay\ (in\ weeks)$$

Loss of profit is dealt with as a separate head of claim (see section 6.5.3). The formula links head office overheads and profit together on the reasonable assumption that contractors normally add a single percentage to their prices to cover both. However, the author says of this calculation that 'in the case of a *delayed* contract, where the concern is to ascertain the "profit" which the delayed contract organisation might have expected to earn *elsewhere in the market on other contracts*, it is this necessary combined operating margin of profit and fixed overhead which, in appropriate market conditions, the contractor's enterprise will have lost as a consequence of the period of owner caused delay...'. This is not necessarily correct. If a contract has overrun, the contractor has not actually lost overheads or, indeed, profits. What he has lost is the opportunity to earn these two elements on other work during the overrun period. The formula is related entirely to an

---

[313] Part of clause 26.1 of JCT 98. Other JCT contracts have wording to much the same effect (see Chapter 12).
[314] I. N. Duncan Wallace, *Hudson's Building and Civil Engineering Contracts*, 11th edition, 1995, London, Sweet & Maxwell, where the strengths and some weaknesses of this and the Eichleay formula are discussed.

overrun of contract time, and to apply it would mean that a contractor would receive additional sums under the contract in respect of head office overheads and profit. This appears to be wrong.

It is wrong because the percentages are based upon the contractor's annual accounts prior to and during the contract 'or other available information'[315] and may never have been achievable on a particular project. In addition, the formula as set out above contains a mathematical inaccuracy because if used in that form it allows a claim for overheads and profit on the amount of overheads and profit already included in the contract sum and if used at all the formula would be:

$$\frac{\text{HO/Profit\%}}{100} \times \frac{\text{Contract sum} - \text{o/heads profit}}{\text{Contract period}} \times \text{Period of delay}$$

Furthermore the formula itself cannot be used as it is printed unless adjustment is made to the various factors to allow for delays in completion for which recovery is not permitted (e.g. delays due to contractor's own inefficiency or to matters for which extensions of time may be granted on grounds for which there is no similar grounds entitling the contractor to make application for loss and/or expense) and also for the effect of variations and fluctuations upon the contract sum. The formula assumes that there has been no change in the contract price. Where the value of the work carried out by the contractor has increased, e.g. because of variation, the pricing of those variations would normally include an allowance for overheads and profit by the application of the rates in the contract bills or under the 'fair valuation' rule. Fluctuations being recoverable net (except, of course, where formula price adjustment is used), must have the effect of reducing the percentage of overhead recovery on actual cost. The formula also ignores the fact that the contractor should make realistic efforts to deploy his resources elsewhere during a period of delay[316].

The Emden formula

There is another formula published in a leading legal textbook, *Emden's Building Contracts and Practice*[317], which runs as follows:

$$\frac{h}{100} \times \frac{c}{cp} \times pd$$

where *h* equals the head office percentage arrived at by dividing the total overhead cost and profit of the contractor's organisation as a whole by the total turnover; *c* equals the contract sum in question; *cp* is the contract period and *pd* equals the period of delay, the last two being calculated in the same units, e.g. weeks.

---

[315] *Hudson*, para 8.179.
[316] *Peak Construction (Liverpool) Ltd* v. *McKinney Foundations Ltd* (1970) 1 BLR 114.
[317] *Emden's Building Contracts and Practice*, 8th edition, vol. 2, p.N/46.

This approach is open to some of the same criticisms as the Hudson formula. It can be useful as an approach where actual costs of head office staff directly engaged upon the individual contract are not obtainable. In that case, the proportion of the contractor's overall overhead costs that can be shown from his accounts to be spent upon staff directly engaged on contracts can be substituted for the element h to obtain a rough and ready approximation of the cost of staff engaged on the particular contract during the period of delay. However, clearly this approach does not make an allowance for the cost of greater staff involvement during the original contract period due to disruption.

This formula has been accepted in *Finnegan v. Sheffield City Council*[318] and *Beechwood Development Company (Scotland) Ltd v. Stuart Mitchell (t/a Discovery Land Surveys)*[319] if there is no practicable means of otherwise calculating the amount. In both instances, the court incorrectly referred to the formula as 'Hudson's'.

The Eichleay formula

This is an American version of the Hudson formula, and is a three-stage calculation which applies daily rates. It is usually expressed as follows:

$$(1) \quad \frac{\text{Contract billings}}{\substack{\text{Total contractor billings} \\ \text{for contract period}}} \times \substack{\text{Total HO overheads for} \\ \text{contract period}} = \substack{\text{Allocable} \\ \text{overhead}}$$

$$(2) \quad \frac{\text{Allocable overhead}}{\text{Days of performance}} = \text{Daily contract HO overhead}$$

$$(3) \quad \text{Daily contract HO overhead} \times \text{Days of compensable delay}$$
$$= \text{Amount of recovery}$$

Once again, the formula gives only a rough approximation; it does not require any proof from the contractor of his actual increased overhead costs from the delay and there is the possibility of double-recovery. In order to allow for this it is necessary at least to deduct any head office overhead recovery allowed under the normal valuation rules. Furthermore, although it is widely used in American Federal Government Contracts and has been adopted in some other cases, it has been subject to judicial criticism and it is not universally accepted[320]. A very sensible view is expressed by the editors of *Building Law Reports*:

> 'Criticism of the formula is not of course criticism of the proposition that in a period of reduced activity on site a contractor will incur off-site overheads for which payment is not being recovered from revenue gen-

---

[318] (1988) 43 BLR 124.
[319] (2001) CILL 1727.
[320] See the comments of the USA Court in *E. Berley Industries v. City of New York* (1978) 385 NE (2d) 281.

erated at site. However, unintelligent use of the formula will demonstrate its inherent weakness.'[321]

Later they add, in relation to the Eichleay formula, that 'it will, for example, be necessary for the contractor still to establish, as a matter of fact, that the services provided at head office were underfunded in the period of delay because they were not designated as general services'[322].

Application of formulae

There is a use for formulae in appropriate situations, usually as a last resort where it is clear there has been a loss, but where there is a complete lack of proper evidence. However, their uncritical use without regard to available facts and without supporting evidence is to be deplored. It appears that there are two distinct situations:

(1) A disruption situation, in which case the *Tate & Lyle* principle applies and the 'management time' and overhead element are recoverable on proof as indicated by Forbes J.
(2) A prolongation situation, i.e. where there is a delay, in which case a modified formula approach may sometimes be used. It has been, rightly, described as a claim for 'loss of business'[323].

In either situation, it is probable that no formula is suitable for general application. The best summary of the position is contained in Abrahamson's *Engineering Law*:

'It is claims for loss of overhead return and profit that are most likely to produce a dividend for the contractor beyond his actual losses. The theory is that the contractor's site and management resources are his revenue earning instrument, and that insofar as they are delayed on a contract by delay he will lose the earnings he would have made with them on some other contract out of which he would have paid his overheads and pocketed his profit. The reality is that in many cases, particularly where the delay affects a small part of a large contractor's total resources, the contractor's organisation has sufficient flexibility to cope with the extra time on site without sacrificing other contracts that may be available, so that the contractor's total overhead and profit is not in fact adversely affected on foot of the usual mathematical formula. On the other hand where a large part of his resources are tied down on a site because of a delay, the ultimate length of which is not known, the contractor genuinely may be inhibited from tendering for other work at competitive rates: the edge may be taken off his tendering in a way not susceptible of very clear proof. The difficulty is to establish the real facts

---

[321] 28 BLR at 103.
[322] 28 BLR at 105.
[323] *Hudson* at para 8.186.

and some indication of the proof necessary has been given [in *Peak* v. *McKinney Foundations*].[324]

This seems to be the true position.

6.5.3. Profit

Loss of profit, which the contractor would otherwise have earned but for the delay or disruption, is an allowable head of claim. It is recoverable under the first part of the rule in *Hadley* v. *Baxendale*[325]. It is only the profit normally to be expected which can be claimed and if, for example, the contractor was prevented from earning an exceptionally high profit on another contract, this special profit would not be recoverable unless, at the time the delayed or disrupted contract has been entered into, the employer was aware of the exceptional profit. This is the second part of the rule.

Victoria Laundry (Windsor) Ltd v. *Newman Industries Ltd*[326] is a case in point (although not in the construction industry) where the plaintiffs, who were launderers, contracted to buy a boiler from the defendants, who knew that the boiler was wanted for immediate use. Delivery of the boiler actually took place 5 months after the date specified in the contract. The Court of Appeal held that the plaintiffs were entitled to recover the profit which might reasonably have been expected to result from the normal use of the boiler during the 5 months in question, but that no account could be taken of the exceptionally profitable character of some of the contracts that they lost.

A profit percentage is not invariably recoverable as a head of claim arising from disruption and delay. The better view is that such a claim is allowable only where the contractor is able to demonstrate that he has been prevented from earning profit elsewhere in the normal course of his business as a direct result of the disruption or prolongation, e.g. he has been prevented from taking up other work available to him. The position is similar to that discussed above in section 6.5.2 regarding overheads. Indeed, for convenience, claims for loss of profit are often grouped together with loss of overheads. To some extent the success of such a claim may, therefore, depend on the economic climate at the time since it may be difficult for a contractor to show when there is a shortage of work that any actual loss of this kind has been suffered. Indeed, it may be that a contractor is operating at no profit or even a small loss. But it will also depend on the extent to which the prolongation is the result of additional work, the value of which contains an appropriate proportion of profit.

There is a possible argument to the effect that loss of the overhead and profit-earning capacity of additional resources devoted to a contract because of delay or disruption is to be assumed without necessity of proof. For the reasons stated earlier, this kind of argument is no longer acceptable.

---

[324] Max Abrahamson, *Engineering Law*, 1985, 4th edition, Elsevier Applied Science Publishers, p.369.
[325] (1854) 9 Ex 341.
[326] [1949] 1 All ER 997.

Indeed, the proposition that loss of profit-earning capacity must be proved is supported by *dicta* in *Peak Construction (Liverpool) Ltd* v. *McKinney Foundations Ltd*[327], which is also helpful in giving some indication of the sort of evidence needed to substantiate such a claim. In that case work was suspended on a contract for some 58 weeks, and although the case arose from a non-standard form of contract and involved some special facts, the principles are of general application:

> 'The way in which the claim for loss of profit was dealt with below has caused me some anxiety. The basis upon which the claim was put in the pleadings was that for 58 weeks no work had been done on this site. Accordingly, a large part of the plaintiffs' head office staff, and what was described as their site organisation, was either idle or employed on non-productive work during this period, and the plaintiffs accordingly suffered considerable loss of gross profit... When the matter came before the court below the matter was put rather differently. The case was put on the basis that in the time during 1966 and 1967 when they were engaged in completing the construction of the East Lancashire Road project they were unable to take on any other work, which they would have been free to do had the East Lancashire Road project been completed on time, and they lost the profit which they would have made on this other work. When the case was argued in this court it seemed to me that the plaintiffs were a little uncertain about which basis they were opting for.
>
> In the end however I think they came down in favour of the second basis: that is, the one that was argued before the court below...
>
> Possibly some evidence as to what the site organisation consisted of, what part of the head office staff is being referred to and what they were doing at the material times could be of help. Moreover, it is possible, I suppose, that a judge might think it useful to have an analysis of the yearly turnover from, say, 1962 right up to, say, 1969, so that if the case is put before him on the basis that work was lost during 1966 and 1967 by reason of the plaintiffs being engaged upon completing this block and, therefore, not being free to take on any other work, he would be helped in forming an assessment of any loss of profit sustained by the plaintiffs.'[328]

Later in the same case it was said:

> 'Under this head (i.e. loss of profit) the plaintiffs were awarded the sum of £11,619. The defendants submit that this sum should be wholly disallowed, no loss of profit having been established. This outright denial is, in my judgment, probably untenable, it being a seemingly inescapable conclusion from such facts as are challenged that the plaintiffs suffered some loss of profit. The sum awarded was arrived at on the basis of a gross profit calculated at nine per cent of the main contract figure of

---

[327] (1970) 1 BLR 114.
[328] (1970) 1 BLR 114 at 122 per Salmon LJ.

£232,000. Whether this was a satisfactory method of approach need not be decided now, though I have substantial doubts on the matter.'[329]

A case involving a successful claim for loss of profit on a somewhat different basis is *Wraight Ltd* v. *P. H. & T. (Holdings) Ltd*[330], where the contractors properly determined their own employment under clause 26 of JCT 63 and, therefore, correctly claimed as part of the direct loss and/or damage the profit they would have earned had they been able to complete the contract work. The judge had little hesitation in finding it to be a valid claim. He said:

> 'In my judgment, the position is this *prima facie*, the claimants are entitled to recover, as being direct loss and/or damage, those sums of money which they would have made if the contract had been performed, less the money which has been saved to them because of the disappearance of their contractual obligation.'[331]

In referring to 'those sums of money which they would have made' the judge went to the nub of the matter. What is recoverable is the actual profit on that contract. The profit which might usually be obtained in such circumstances is not relevant. This situation is distinct from that referred to above, where the contractor is prevented from obtaining an exceptional profit on another contract[332]. The difficulty here is in determining the level of profit the contractor would have made if he had been allowed to do the work. It is probably not enough for the contractor simply to demonstrate the profit he put into his tender, because such profit may not have been realisable. The *Wraight* case must also be distinguished from normal claims for direct loss and/or expense arising from delay or disruption, because the profit lost in this instance was that which would have been earned on work that the contractor was not permitted to carry out[333] rather than work which was delayed or carried out under different conditions than those originally anticipated.

6.5.4 Inefficient or increased use of labour and plant

Delay and disruption can lead to increased expenditure on labour and plant in two ways. It may be necessary to employ additional labour and plant or the existing labour and plant may stand idle or be underemployed. The latter is sometimes referred to as 'loss of productivity'. Although this is an allowable head of claim, it can be difficult if not impossible to establish the amount of the actual additional expenditure involved. Contractors commonly attempt to overcome this problem by presenting the claim on a total cost basis. In other words, they maintain an entitlement to be remunerated

---

[329] (1970) 1 BLR 114 at 126 per Edmund Davies LJ.
[330] (1968) 13 BLR 27.
[331] (1968) 13 BLR 27 at 36 per Megaw J.
[332] See *Parsons (Livestock) Ltd* v. *Utley Ingham & Co Ltd* [1978] 1 All ER 525.
[333] See Chapter 4, section 4.4, 'Omission of work to give it to others'.

for all the work they have done and for all the resources they have expended. This type of claim has been roundly condemned[334]. It could only be sustained if the contractor could show that he was absolutely blameless and that none of the resource time was occupied other than by carrying out the work. Loss of productivity is demonstrated by reference to original tender figures to establish the anticipated percentage productivity, then actual labour figures are used to show the fall in productivity. Usually a new percentage is calculated to form the basis of calculation of the claim. A contractor should be able to establish the actual costs incurred, but it will clearly be impossible to prove as a matter of fact what the costs would have been had the delay or disruption not occurred. The tender breakdown is irrelevant. The contractor's intended use of labour and plant by reference to the original programme of work is unlikely to be an accurate forecast.

The problem is that the intended use may be inadequate. Some of the additional labour and plant time may be the difference between the contractor's wrongly estimated proposed resources and what he would have had to use even if the contract had proceeded without delay or disruption. In appropriate cases it is possible to demonstrate the true loss by ignoring the tender breakdown showing intention and simply comparing a period of normal working with a period when disruption is present. Assuming that the building work is of a fairly repetitive nature, this method produces a fairly convincing ratio for application throughout the project[335].

A further difficulty is that of relating particular items of additional expenditure under these or indeed other heads to particular events. Contractors seldom keep cost records in such a detailed form as to enable this to be done, particularly where there may be several concurrent causes of delay and disruption, some of which may be claimable and others not[336]. All that can be said is that the architect or quantity surveyor must do his best to arrive at a reasonable conclusion from whatever evidence is available. It must be a reasonable assumption that some loss will have been suffered in these respects where delay or disruption has occurred and the architect or quantity surveyor cannot resist making some reasonable assessment simply on the grounds that the contractor cannot prove in every detail the loss he has suffered.

This is, in fact, one aspect of what is called 'the global approach', which is discussed in Chapter 8. In *London Borough of Merton* v. *Stanley Hugh Leach Ltd*[337] one of the issues before the court was whether the terms of a JCT 63 contract allowed the contractor to recover direct loss and/or expense 'when it is not possible for the contractor to state in respect of any such alleged event the amount of loss and/or expense attributable' to the specific event. Following *J. Crosby & Son Ltd v Portland UDC*[338], the question was answered

---

[334] *London Borough of Merton* v. *Stanley Hugh Leach Ltd* (1985) 32 BLR 51 at 112 per Vinelot J.

[335] The method was approved by Mr Recorder Percival in the 1985 *Whittal Builders* v. *Chester-Le-Street District Council* (1996) 12 Const LJ 356. There were two cases by this name (one in 1987).

[336] Concurrency and the inherent difficulties are discussed in Chapter 2, section 2.4.

[337] (1985) 32 BLR 51.

[338] (1967) 5 BLR 121.

affirmatively. Where there is more than one head of claim, then provided that the contractor has not unreasonably delayed in making the claim and so created the difficulty himself, the architect or quantity surveyor must ascertain the global loss which is directly attributable to the various causes. However he must disregard any loss or expense which would have been recoverable had the claim been made under one head in isolation and which would not have been recoverable under the other head, also considered in isolation. The learned judge, however, added an important reservation:

> 'It is implicit [in *Crosby*] that a rolled-up award can only be made in a case where the loss or expense attributable to each head of claim cannot in reality be separated and secondly that a rolled-up award can only be made where apart from that practical impossibility the conditions which have to be satisfied before an award can be made have been satisfied in relation to each head of claim'.[339]

6.5.5 Winter working

One other factor that can lead to a claim which is effectively one for loss of productivity is the carrying out of work in less favourable circumstances, e.g. excavation work carried out in winter rather than in summer. In such circumstances, there is potentially a claim in respect of the additional costs caused by working in winter when, but for the delay, the work would have been completed during the summer period. The principle behind this type of claim is discussed in section 6.4. Clearly there will be no, or at least little, chance of a claim on this basis where work scheduled to be carried out in winter is pushed into spring or summer.

6.5.6 Plant

Plant is actually part of site overheads, but there are certain considerations which must be taken into account when considering costs related to prolongation. It is important to identify plant which the contractor has hired and separate it in the reckoning from his own plant.

Plant hired

If the plant is hired from an external source there is no problem. The amount claimable is the loss which has actually been incurred: the sums payable to the plant owner under terms of the hire contract, subject to the normal principles regarding mitigation of loss.

If there is a minimum period of hire, that is the minimum period for which the contractor is entitled to claim. Obviously if it is clear that the

---

[339] (1985) 32 BLR 51 at 102 per Vinelott J.

disruption will be prolonged, the contractor must mitigate his loss by either trying to use the plant elsewhere or terminating the plant hire contract and returning the plant to its owner in accordance with the terms of the hire.

Questions often arise if the contractor is part of a group and one of the companies in the group hires out plant and equipment to the others. Once it is established that there is a claim in principle, the question to be answered refers to the amount actually lost or expended by the contractor in hiring the plant. Does the plant hire sister company hire out plant to the contractor at the same rates as it would apply to other contractors? It is likely that there would be a discount. It is for the contractor to prove that he actually has to pay the hire charge. If the contractor is a separate limited company, in other words a separate legal entity, whatever charge has been paid will be claimable unless there is some arrangement between the companies which allows the contractor to recover the outlay in another way.

Sometimes a contractor will argue that he is entitled to claim hire charges even though it is his own plant, because he operates a plant hire business, hiring out spare plant to other contractors. In such circumstances, the contractor would have to show that, if the particular plant was not being used by the contractor, he would have hired it out. He must also show that he had an opportunity to do so.

Contractor's own plant

Where the contractor is using his own plant, he is not entitled to claim any notional hiring charge and arriving at the true cost of his plant standing idle is more difficult. The case of *B. Sunley & Co Ltd* v. *Cunard White Star Ltd*[340] is illustrative of the principles involved. There, a machine had to be transported from Doncaster to Guernsey for use on a contract, but was delayed by 1 week in its arrival. While delayed in Doncaster it worked for 1 day and earned £16. The Court of Appeal held that in the absence of evidence as to actual loss of profit, the measure of damage was depreciation in the value of the machine during the period of delay, interest on the money invested in its purchase, costs of maintenance, and some wages thrown away. Assumptions were made as to the values involved and in the absence of firm evidence a figure of £30 was arrived at, less the £16, making a net award of £14. It is notable that, in arriving at the figure for depreciation, the Court of Appeal took into account that the machine would not depreciate as much while standing idle as it would when working, but the court's apparently harsh attitude on this point was possibly influenced by the fact that the plaintiff's original claim had been for £577 on grounds which the court considered wholly unsustainable, and by the plaintiff's inability to produce any real evidence. A certain testiness is discernible in the judgment. The ruling of the Court of Appeal is summarised in the head note:

---

[340] [1940] 2 All ER 97. See also the Canadian case *Shore & Horwitz Construction Co Ltd* v. *Franki of Canada Ltd* [1964] SCR 589 which followed the same principles.

'*Held:* The machine was a chattel of commercial value and there were four possible heads of damage: (i) Depreciation; (ii) interest on money invested; (iii) cost of maintenance; and (iv) expenditure of wages thrown away. The plaintiffs could not complain that they were refused relief under heads (ii), (iii) and (iv) for they had not given any evidence on these legitimate topics. £20 would be allowed for depreciation, £10 for interest, maintenance and wages – in all £30 – against which receipts for interest, maintenance and wages – in all £30 – against which receipts for £16 must be deducted leaving a net sum for damages of £14.'

At first instance[341] Hallet J held that the proper measure of damages was the amount the plaintiffs would have made by the use of the machine during the period it was idle, but the Court of Appeal took the view that, in the absence of proof of special damage, the plaintiffs could only recover nominal damages based largely upon a mathematical calculation of the rate of depreciation of the machine.

It appears that two consequences flow from this decision in the normal claims situation. *First,* if the contractor can prove his actual loss by detailed calculations based on actual cost records, he is entitled to recover that amount. Second, if the contractor cannot prove actual damage, he is entitled to recover only a nominal amount, which is normally limited to depreciation. There is 'no authority for the proposition that if the owner of a profit-earning chattel does not prove the loss he has sustained the judge may make a fortuitous guess and award him some arbitrary sum'[342]. The only real evidence in the *Sunley* case was that the capital cost of the machine had been £4500 and its life was 3 years. The Court of Appeal took the view that on a pure sinking-fund basis the depreciation for the claim period would be £29 for the week, but 'We cannot believe...that a machine exposed to such working strains in use that it will last only 3 years depreciates as much when idle as it does when working and we, therefore, think that £20 for the week is as much as ought to be allowed for depreciation'. Despite these strictures, it seems that in assessing the loss (in the absence of actual proof) the architect must work on the same assumptions as did the Court of Appeal and must make a reasonable assumption that some loss will have been suffered where delay and disruption have occurred. In a more recent case, the court has held that ascertainment should take into account the substantiated cost of capital and depreciation, but not the elements normally included in hire rates on the basis that plant will only be profitable for some of the time:

'... in ascertaining direct loss or expense under clause 26 of the JCT conditions in respect of plant owned by the contractor the actual loss or expense incurred by the contractor must be ascertained and not any hypothetical loss or expense that might have been incurred whether by way of assumed or typical hire charges or otherwise.'[343]

---

[341] [1939] 3 All ER 641.
[342] [1940] 2 All ER 97 at 101 per Clauson LJ.
[343] *Alfred McAlpine Homes North Ltd* v. *Property & Land Contractors Ltd* (1995) 76 BLR 65 at 93 per Judge Lloyd.

Hudson's Building and Engineering Contracts sets out the position succinctly as follows:

> 'Where plant is delayed in the sense of being delayed on site for too long because of owner breaches, claims may be validly advanced in a number of ways which will vary according to whether it is hired or owned by the contractor. Thus where there is a disturbance but no overall delay, there may be a loss of productivity claim based on standing or idle time and driver's wages thrown away if hired, or on an interest and depreciation basis if owned. If there is overall delay (or extension of the period during which particular plant is programmed to be on site) hire payments will be extended in the one case (as, for example, hired site accommodation) or interest or depreciation or maintenance if the plant is owned.'[344]

As noted earlier, there are intermediate positions between the two extremes of hired in and contractor's own plant. Each of these positions must be examined carefully, it being remembered that the key point is that a contractor can only recover, as direct loss and/or expense, what he has actually lost or spent.

6.5.7 Increased costs

Additional expenditure on labour, materials or plant due to increases in cost during a period of delay is an allowable head of claim. Claims may also be sustainable where disruption has resulted in labour-intensive work being delayed and carried out during a period after an increased wage award. It should be noted that, where a claim of this kind is being made in respect of a delay in completion, it is not only the period of delay that should be considered. The correct measure would be the difference between what the contractor would have spent on labour, materials and plant and what he has actually had to spend over the whole period of the work as a result of the delay and disruption concerned. In making this calculation, of course, due allowance must be made for any recovery of increased costs under the appropriate fluctuations clauses in the contract.

Contractors may often seek to make such calculations easier by using some kind of formula or notional percentage to produce a result. That approach is not acceptable. The contractor must show that the increases in costs have been the inevitable consequence of the cited occurrence. This can be quite complicated in the case of materials, because the contractor must demonstrate that he could not reasonably have placed his order earlier to avoid the increases. Above all, it must not be assumed that all work and all materials after the period of delay or during a prolongation period after the contract completion date, will automatically suffer a price increase.

---

[344] I. N. Duncan Wallace, *Hudsons Building and Engineering Contracts*, 11th edition, 1995, Sweet & Maxwell, para 8.195.

6.5.8 Financing charges and interest

Financing charges

Whatever the position may be at common law about interest on outstanding debts and claims, it is now settled law that under the 'direct loss and/or expense' provisions of the JCT Forms – and, it is submitted, under similarly worded provisions in other forms – finance charges by way of interest expended are allowable as a head of claim. Indeed, it is the loss of interest that might have been earned on the money diverted from investment, i.e. compensation for the loss of use of money. The contractor is to be compensated for the financial burden arising from the fact that primary loss or expense would have been incurred some time before ascertainment and certification. 'Direct loss and/or expense' covers the financial burden to the contractor of being stood out of his money; it is *not* interest on a debt but a constituent part of the loss and/or expense. This principle was first established by the decision of the Court of Appeal in *F.G. Minter Ltd* v. *Welsh Health Technical Services Organisation*[345], which recognised the realities of the financing situation in the construction industry and gave a sensible and practical interpretation to the claims provisions.

> '[In] the building and construction industry the "cash flow" is vital to the contractor and delay in paying him for the work he does naturally results in the ordinary course of things in his being short of working capital, having to borrow capital to pay wages and hire charges and locking up in plant, labour and materials capital which he would have invested elsewhere. The loss of the interest which he has to pay on the capital he is forced to borrow and on the capital which he is not free to invest would be recoverable for the employer's breach of contract within the first rule in *Hadley* v. *Baxendale*[346] without resorting to the second, and would accordingly be a direct loss, if an authorised variation of the works, or the regular progress of the works having been materially affected by an event specified . . . has involved the contractor in that loss.'[347]

In that case, the plaintiff was engaged to construct a hospital in Wales. The contract was in JCT 63 form. As the work progressed substantial numbers of variations were ordered, and these affected regular progress of the works as a whole and a nominated sub-contractor's work was materially affected by a lack of necessary instructions. Claims were made and paid under clauses 11(6) and 24(1)(a) (now JCT 98, clause 26). The plaintiff challenged the amounts paid because they had not been certified until long after the losses and expenses had been incurred. They claimed that finance charges which they had incurred as a result of being kept out of their money were 'direct

---

[345] (1980) 13 BLR 7.
[346] (1854) 9 Ex 341.
[347] (1980) 13 BLR 7 at 15 per Stephenson LJ.

loss and/or expense'. The Court of Appeal agreed, and rejected the employer's argument that they were not 'direct' but were 'a naked claim for interest'.

The decision of the Court of Appeal can be summarised in this way:

(1) The words 'direct loss and/or expense' must be interpreted as covering those losses and expenses which arose naturally and in the ordinary course of things, i.e. the same criteria applicable whenever common law damages are being considered. A particular item of loss or expense is regarded as 'direct' if it can fairly be said to fall within the first branch of the rule in *Hadley* v. *Baxendale*[348].

(2) In the construction industry, in the ordinary course of things, a contractor who is required to finance particular operations will be required to utilise capital. Either it must be borrowed (in which case he will have to pay borrowing charges) or else he must find it from his own resources. In that case, he will forgo the interest which might otherwise have been earned. Therefore, if a contractor has to carry out varied work or if regular progress has been materially affected by one of the specified events, he should expect to be fully reimbursed by interim payments for work properly executed as provided for by the contract. To the extent that he has been involved in laying out money which is not so reimbursed, the amount paid for or lost by obtaining the use of the necessary capital is a part of the loss or expense which arises in the ordinary course of things.

(3) In the context of JCT 63, the financing charges are implicitly part of the recoverable direct loss and/or expense or what was being claimed 'is not interest on a debt, but a debt which has as one of its constituent parts interest charges which have been incurred'[349].

(4) On the wording of the JCT 63 provisions, recovery was restricted to direct loss and/or expense incurred up to the date of the contractor's application, i.e. for the period between the loss and/or expense being incurred and making of the written application for reimbursement.

Consequently, under JCT 63, subsequently incurred charges were recoverable only under successive applications. This is because JCT 63 did not deal with *continuing or future losses*; it referred only to losses and expenses *which have already been incurred*. Under JCT 63 the cut-off point was the date of the contractor's notice.

There is no such limitation under JCT 98, clause 26 or IFC 98, clause 4.11, where only one application or notice from the contractor is required to cover loss and/or expense *'that he has incurred or is likely to incur'*. In fact, proper application of the JCT 98 and IFC 98 provisions *should* ensure that little in the way of finance charges is incurred by the employer. The matter has been taken a stage further by the decision of the Court of Appeal in *Rees & Kirby*

---

[348] (1854) 9 Ex 341.
[349] (1980) 13 BLR 7 at 23 per Ackner LJ.

Ltd v. *Swansea City Council*[350] where the court extended the *Minter* principles. Finance charges, should be calculated on a compound interest basis.

> '... [It] seems to me, we must adopt a realistic approach. We must bear in mind, moreover that what we are considering is a debt due under a contract... [It is] a claim in respect of loss or expense in which a contractor has been involved by reason of certain specified events. The [contractors]... operated over the relevant period on the basis of a substantial overdraft at their bank, and their claim in respect of financing charges consists of a claim in respect of interest paid by them to the bank on the relevant amount during that period. It is notorious that banks do themselves, when calculating interest on overdrafts, operate on the basis of periodic rests: on the basis of [the *Minter* principle], which we here have to apply, I for my part can see no reason why that fact should not be taken into account when calculating the [contractor's] claim for loss or expense ...'[351]

In *Rees & Kirby Ltd*, in 1972 the plaintiffs were engaged to construct a housing estate for the defendants, under a fixed-price contract in JCT 63 terms. The contract completion date was in July 1973. A large number of variations were ordered and there were delays in issuing instructions and information. The works were not in fact practically completed in July 1974. By a letter of 4 July 1973 the contractors notified the Corporation of the causes of delay, giving further additional details and notified a claim for reimbursement of loss and/or expense by a letter in general terms dated 18 December 1973.

During 1972 there was a sharp escalation in construction industry wage rates and it soon appeared that this would convert the contract into a substantial loss-maker. In October 1973, the Minister for Housing and Construction issued a statement that local authorities could in appropriate cases consider making *ex gratia* payments to contractors working on fixed-price contracts. In light of this, the plaintiffs' contractual claims were put aside while the parties negotiated for an *ex gratia* settlement or alternatively for the contract to be converted into a fluctuating price one. By the end of 1976 the plaintiffs concluded that no settlement was possible. On 24 December 1976 they wrote to the defendants indicating that their losses were being 'aggravated by interest charges on monies outstanding for our contractual claim'. On 11 February 1977 it became clear that negotiations had failed. The plaintiffs' detailed formal claim was finally submitted in June 1978, but despite chasing letters, the architect did not respond until February 1979. He then granted the full extension of time claimed (52 weeks), but said that the interest claim was not a matter for his decision.

Between February and August 1979 the architect issued various interim certificates in respect of the direct loss and/or expense claimed. None of the sums certified included any element of interest. The contractors signed the

---

[350] (1985) 5 Con LR 34.
[351] (1985) 5 Con LR 34 at 51 per Robert Goff LJ.

draft final account on 10 August 1979, but expressly reserved the claim for interest. The final certificate was issued without prejudice to the contractor's right to press that outstanding claim.

The ruling of the Court of Appeal was important and worth consideration in some detail:

(1) The contractor's application under clauses 11(6) and 24(1)(a) of JCT 63 need not be in any particular form, but must contain a reference sufficient to make it clear that it includes some loss or expense suffered because the contractor has been kept out of his money. Robert Goff LJ emphasised that, even under JCT 63, he did

> 'not consider that more than the most general reference is required, sufficient to give notice that the contractor's application does include loss or expense incurred by him by reason of his being out of pocket in respect of the relevant variation or delayed instruction, or whatever may be the relevant event giving rise to a claim under the clause.'[352]

The position is, it seems, different under both JCT 98 and IFC 98, because of the revised wording.

> '[The] clauses in question have since been revised to allow for applications in respect of loss or expense which the contractor has incurred or is *likely* to incur; and it may be (though I express no opinion upon it) that, under clauses so revised there need be no express reference to interest as financing charges.'[353]

To avoid uncertainty, the wise contractor's application will always include reference to finance charges.

(2) Normally, the contractor's notice must be given within a reasonable time of the loss or expense having been incurred as the clause makes plain. However, on the facts the Council could not rely on the delay between practical completion and the formal application of June 1978 since they were estopped from enforcing their strict legal rights under the principle of promissory estoppel first stated by Lord Cairns LC in *Hughes* v. *Metropolitan Railway Co*[354]. This is an important point, but if negotiations for a settlement are taking place, it is sensible practice for the contractor expressly to reserve his legal rights, both as to interest or otherwise: see later.

(3) There is no cut-off point at the date of practical completion as the Council had argued.

> 'I can see no reason why the financing charges should not continue to constitute direct loss or expense in which the contractor is involved by reason of, for example a variation, until the date of the last application made before the issue of the certificate issued in respect of the primary

---

[352] (1985) 5 Con LR 34 at 48 per Robert Goff LJ.
[353] (1985) 5 Con LR 34 at 48 per Robert Goff LJ.
[354] (1877) 2 App Cas 439.

loss or expense incurred by reason of the relevant variation. At the date of the issue of the certificate, the right to receive payment in respect of the primary loss or expense merges in the right to receive payment under the certificate within the time specified in the contract, so that from the date of the certificate, the contractor is out of his money by reason either (1) that the contract permits time to elapse between the issue of the certificate and its payment, or (2) that the certificate has not been honoured on the due date, but I can for my part see no good reason for holding that the contractor should cease to be involved in loss or expense in the form of financing charges simply because the date of practical completion has passed.'[355]

(4) The delay in payment which occurred because of negotiations for an *ex gratia* settlement, which was between the date of practical completion and 11 February 1977 when it became clear that the plaintiffs' claim must be advanced under the terms of the contract, was delay which was attributable to an independent cause, and not to the ordering of variations or the giving of late instructions. The financing charges incurred by the contractor during this period were not direct loss and/or expense for the purposes of the contract.

If only for this reason, it is advisable for the contractor to reserve his rights to finance charges if negotiations are taking place or, alternatively, to give notice of arbitration. This part of the Court of Appeal's reasoning involves difficult problems of causation and has caused some comment. It raises potential legal and practical difficulties[356]. The best that can be said is that the facts of *Rees & Kirby Ltd* were rather special.

(5) The contractors were entitled to recover finance charges for the period from 11 February 1977 (when the architect notified them that in effect the claim must be processed under the terms of the contract) until 10 August 1979 when they appended their signature to the draft final account. These finance charges were to be calculated on the basis of compound interest.

A connected, but rather different, question concerns the date at which finance charges start to run. Is it the date on which the contractor makes application and the architect has his first intimation that there is a matter to consider under the terms of the contract? Is it rather the date by which the architect has sufficient information to consider the point? Consider, for example, the situation where the contractor makes the briefest of applications and, despite a detailed and precise request from the architect, he is very slow in providing the information reasonably necessary to enable the architect to ascertain the amount of loss and/or expense. Who is to bear the financing charges for the intervening period? One view is that, whatever may be the reason for the architect's inability to ascertain, if the claim is

---

[355] (1985) 5 Con LR 34 at 49 per Robert Goff LJ.
[356] See the commentary on this case in *Building Law Reports*, vol 30, pp. 5-6.

found ultimately to be valid, the contractor is bearing the financing charges, while the money remains in the employer's pocket and, therefore, the contractor is entitled to reimbursement. Another approach is to say that the period commences when the architect has received all the information he requires; before that the contractor is the author of his own misfortune. On balance, it is thought that the latter is the better view and receives some support from *Rees & Kirby Ltd*. In any particular case it will be a matter of fact to be taken into account how much of the delay between making application and providing full information is the responsibility of the contractor.

It appears that, if the principles enunciated in *F. G. Minter Ltd* v. *Welsh Health Technical Services Organisation* and *Rees & Kirby Ltd* v. *Swansea Corporation* are to be followed, the resulting additional financing charges must be considered to result directly from the delay and disruption concerned, and consequently are recoverable. It should be said that this principle and, indeed, the *Minter* decision itself, at first sight appears to conflict with a decision of the House of Lords in a case known as *The Edison*[357]:

> 'the appellants' actual loss, in so far as it was due to their impecuniosity arose from that impecuniosity as a separate and concurrent cause, extraneous to and distinct in character from the tort; the impecuniosity was not traceable to the respondents' acts, and, in my opinion, was outside the legal purview of the consequences of these acts'.[358]

However, as Dr Parris has observed, 'this decision of the House of Lords has in fact long been ignored by the courts, if it has ever indeed been followed'[359], and cites in support *Dodd Properties (Kent) Ltd* v. *The City of Canterbury*[360]. It therefore appears that *The Edison* has no application to the type of situation envisaged above[361]. It has recently been held that interest is recoverable as damages since it was within the parties' contemplation at the time the contract was entered into that such charges might be incurred[362].

Rees & Kirby Ltd v. *Swansea City Council* has silenced the debate as to whether financing charges are a proper head of claim under contracts in JCT terms. The principles it establishes are applicable to other similarly worded forms of contract. Indeed, in 1984 the Property Services Agency formally accepted, in principle, that in appropriate circumstances financing charges were recoverable under Edition 2 of GC/Works/1. The principle was re-affirmed in the Scots Court of Session following legal argument which essentially proceeded from first principles[363].

---

[357] (1933) AC 449.
[358] (1933) AC 449 at 460 per Lord Wright.
[359] John Parris, *The Standard Form of Building Contract*, 2nd edition 1985, Blackwell Science, section 10.05, Interest as 'direct loss and expense'.
[360] (1979) 13 BLR 45.
[361] (1979) 13 BLR 45 at 54 per Megaw LJ and at 61 per Donaldson LJ.
[362] *AMEC Process & Energy Ltd* v. *Stork Engineers & Contractors BV* (2002) CILL 1883.
[363] *Ogilvie Builders Ltd* v. *The City of Glasgow District Council* (1994) 68 BLR 122.

Rate of interest

Some problems still remain, for example as regards deciding the rate of interest on money diverted from investment. Where a contractor is, in fact, operating on borrowed money the position is simple, since supporting evidence as to the additional financial charges incurred can readily be obtained from his bankers, subject to what is said below. However, where what the contractor is claiming is loss of investment potential, it may be necessary for him to show the manner in which he customarily invests his money and the interest that a contractor in his position could have expected to have earned[364].

Special difficulties will be encountered where a contractor in a small or medium way of business wavers from a credit to a debit situation and back again during the period of the contract. It may become very difficult to establish at any particular point whether what he is entitled to claim is interest on overdraft or loss of investment potential. In such cases it is suggested that some reasonable average be drawn, having regard to the realities of the investment market. Indeed, the actual percentages involved may differ very little.

With the abolition of minimum lending rate, overdraft rates tend to fluctuate at short intervals and will also vary as between contractors as, indeed, has always been the case. Where a wholly exceptional and possibly penal rate of interest is being charged to the particular contractor, his recovery would be limited to a 'normal' rate, because that must be presumed to be what was in the contemplation of the employer as a foreseeable consequence of the matter in regard to which the contractor is claiming.

There is authority for the view that any interest or finance charges allowed as part of direct loss and/or expense should be assessed at a rate equivalent to the cost of borrowing (or normal investment, as the case may be), disregarding any special position of the contractor. One looks

> '... at the cost to the plaintiff of being deprived of the money which he should have had. I feel satisfied that in commercial cases the interest is intended to reflect the rate at which the plaintiff would have had to borrow money to supply the place of that which was withheld. I am also satisfied that one should not look at any special position in which the plaintiff may have been; one should disregard, for example, the fact that a particular plaintiff could only borrow money at a very high rate or, on the other hand, was able to borrow money at specially favourable rates. The correct thing to do is to take the rate at which plaintiffs in general could borrow money. This does not, however,... mean that you exclude entirely all the attributes of the plaintiff other than that he is a plaintiff... [It] would always be right to look at the rate at which plaintiffs with the general attributes of the actual plaintiff in the case (though not, of course,

---

[364] *Tate & Lyle Food & Distribution Co Ltd* v. *Greater London Council* [1982] 1 WLR 149.

with any special or peculiar attribute) could borrow money as a guide to
the appropriate interest rate . . . '[365]

This seems to be a sensible approach under JCT terms. The alternative is to
take the rate at which the contractor is borrowing money, or the rate of
investment interest he is receiving if he has a deposit, and ask him to
produce an auditor's or banker's certificate in support. Many architects
and quantity surveyors prefer this approach because the calculation is
based on actual costs, for the contract refers to 'the amount of such loss
and/or expense which has been or is being incurred by the contractor'. But
if the particular contractor was paying for his overdraft at above the normal
rate, it is at least arguable that the excess should be disallowed.

Either method of arriving at the interest rate is reasonable in practice,
although if the claimant is paying a specially high rate of interest (or is
receiving above-average investment interest), the judicial approach is preferable. The rate at which an ordinary commercial borrower can borrow
money is probably the safest guide.

The financial pages of the quality daily newspapers will be a useful
source of reference and, in practice, it may well be acceptable to average-
out rates of interest over a period rather than to go through the tedious
mathematical exercise of detailed calculation on the basis of rates from day
to day.

The interest allowable on an overdraft is that actually payable by the
contractor (subject to what has been said above) and would therefore be
compounded at the normal intervals adopted by the contractor's bank, e.g.
quarterly or half-yearly.

Interest

The law relating to interest payments is complex and it is still developing. It
still retains some of the medieval abhorrence of usury which seems strange
in today's commercial environment. At common law it is well settled that
debts do not carry interest so that, for example, a person who pays late on an
invoice discharges his obligation by paying the sum certified[366]. The rule
has been affirmed by the House of Lords[367]. Their lordships thought that the
rule was unsatisfactory, and expressed the view that the position should be
altered by statute, as had been recommended by the Law Commission in its
report on interest in 1978[368].

The general position, when considering the position of a contractor who is
trying to recover interest, is that the recovery may fall into one or more of
five categories:

---

[365] *Tate & Lyle Food & Distribution Co Ltd* v. *Greater London Council* [1981] 3 All ER 716 at 722 per Forbes J.
[366] *London Chatham & Dover Railway Co* v. *South-Eastern Railway Co* [1893] AC 429.
[367] *President of India* v. *La Pintada Cia Navegacion SA* [1984] 2 All ER 773.
[368] Cmnd 7229. Eventually this has been put into effect, at least in part, by the Late Payment of
Commercial Debts (Interest) Act 1998, see (5) below.

(1) *If there is an express term of the contract providing for interest in specific circumstances*

Where applicable, this category is generally unmistakable. The JCT standard forms and the ICE Conditions of Contract do have such provision under which the employer must pay interest on overdue payments.

(2) *If the contract can be construed as giving a contractual right to interest payment*

Interest and financing charges as part of direct loss and/or expense fall into this category. They have been discussed already.

(3) *If interest is awarded by the court on judgment for a debt*

There has to be a debt – more than a mere assertion that money is due, before judgment will be given and interest will be awarded. The court is empowered to award interest provided the money was outstanding at the time proceedings were commenced although it may have been paid subsequently[369]. Arbitrators have a like power which provides:

> (1) The parties are free to agree on the powers of the tribunal as regards the award of interest.
> (2) Unless otherwise agreed by the parties the following provisions apply.
> (3) The tribunal may award simple or compound interest from such dates and with such rests as it considers meets the justice of the case –
>> (a) on the whole or part of any amount awarded by the tribunal, in respect of any period up to the date of the award;
>> (b) on the whole or part of any amount claimed in the arbitration and outstanding at the commencement of the arbitral proceedings but paid before the award was made, in respect of any period up to the date of payment.
> (4) The tribunal may award simple or compound interest from the date of the award (or any later date) until payment, at such rates and with such rests as it considers meets the justice of the case, on the outstanding amount of any award (including the award of interest under subsection (3) and any award as to costs).
> (5) References in this section to an amount awarded by the tribunal include an amount payable in consequence of a declaratory award by the tribunal.
> (6) The above provisions do not affect any other power of the tribunal to award interest.[370]

The normal practice is that interest will be awarded from the date on which payment should have been made, but there may be reasons why a lesser period will be awarded, for example unreasonable delay on the part of the claimant.

Both judgment debts and sums directed to be paid by an arbitrator's award carry interest at the prescribed statutory rate as from the date of judgment or the award.

---

[369] See section 15 of the Administration of Justice Act 1982 (amending section 35A of the Supreme Court Act 1981).
[370] Section 49 of the Arbitration Act 1996.

(4) *If it can be shown that there are special circumstances*
This is a promising category for interest seekers. Where a creditor is able to establish that he has suffered special damage, for example by himself having to pay interest on an overdraft, as a result of the debtor's late payment of a debt, the creditor is entitled to claim that special damage, provided that he can bring himself within the second part of the rule in *Hadley* v. *Baxendale*[371]. This follows from the decision of the Court of Appeal in *Wadsworth* v. *Lydall*[372], which was approved by the House of Lords in *La Pintada*[373].

In *Wadsworth* v. *Lydall*, the plaintiff contracted to sell a property to the defendant, and in anticipation of receiving the purchase price of £10,000 from his purchaser, contracted to buy another property. The defendant (the purchaser) defaulted, and in fact paid a lesser sum very late. As a result the plaintiff had to pay his vendor interest on the unpaid purchase price of the other property, and also incurred the cost of raising a mortgage to meet the balance. Subsequently he sued the defaulting purchaser, claiming, among other things, £355 in respect of interest which he had to pay his vendor for late completion and £16.20 for mortgage costs.

The Court of Appeal allowed these two items as special damage, i.e. as an application of the second limb of the rule in *Hadley* v. *Baxendale*.

'The defendant knew or ought to have known that if the £10,000 was not paid to him the plaintiff would need to borrow an equivalent amount or would have to pay interest to his vendor or would need to secure financial accommodation in some other way. The plaintiff's loss in my opinion is such that it may reasonably be supposed that it would have been in the contemplation of the parties as a serious possibility, had their intention been directed to the consequences of a breach of contract.'[374]

The Court of Appeal distinguished the *London Chatham and Dover Railway Co* case.

'In my view the court is not constrained (i.e. in relation to interest) by the decision of the House of Lords. In *London Chatham and Dover Railway Co* v. *South-Eastern Railway Co* the House of Lords was not concerned with a claim for special damages. The action was an action for an account. The House was concerned only with a claim for interest by way of general damages. If a plaintiff pleads and can prove that he has suffered special damage as a result of the defendant's failure to perform his obligation under a contract, and such damage is not too remote, on the principle of *Hadley* v. *Baxendale*, I can see no logical reason why such special damage should be irrecoverable merely because the obligation on which the defendant defaulted was an obligation to pay money and not some other type of obligation . . .'[375]

This case seems, therefore, to open up the way to such claims for interest as 'special damage' in arbitration proceedings arising under building contracts

---

[371] (1854) 9 Ex 341.
[372] [1981] 2 All ER 401.
[373] *President of India* v. *La Pintada Cia Navegacion SA* [1984] 2 All ER 773.
[374] *Wadsworth* v. *Lydall* [1981] 2 All ER 401 at 405 per Brightman LJ.
[375] *Wadsworth* v. *Lydall* [1981] 2 All ER 401 at 405 per Brightman LJ.

generally. It has been held that a contractor may be entitled to interest or financing charges if he pleads and proves them as special damage[376].

An interesting application of the principle of *Wadsworth* v. *Lydall is* to be found in the decision of the High Court of Justice in Northern Ireland in *Department of the Environment for Northern Ireland* v. *Farrans (Construction) Ltd*[377], which arose under the JCT Form (1963 edition) and adds to the debate in the construction industry about the recovery of interest or financing charges in building contract cases. Under clause 22, JCT 63 (clause 24.2.1, JCT 98), once the architect has issued his certificate of delay the employer is entitled to deduct liquidated and ascertained damages at the prescribed rate from moneys due or to become due to the contractor. If, subsequently, the architect grants an extension of time so that a refund is due to the contractor, is the contractor entitled in law to make any claim against the employer for interest upon, or financing charges which are attributable to, the moneys refunded?

This was the point at issue in the Ulster decision, but it should be noted that Murray J was not asked to decide the question whether, under JCT 63, the architect can issue more than one certificate under clause 22. For the purposes of the case, the parties agreed that he could.

Murray J decided, on the facts before him, that where several certificates were issued, with the result that sums previously deducted as liquidated damages fell to be repaid, the contractor was entitled to interest thereon. He held that clause 22 was to be construed as meaning that when the employer received the first or any subsequent certificate it was open to him to start deducting liquidated damages. The employer did this at his own risk in the sense that if a later certificate was issued, which had the effect of vitiating the earlier certificate, the employer was without protection against a claim for breach of contract in failing to pay on the due dates the amounts shown in the relevant interim certificates. In those circumstances the contractor was entitled to the remedy appropriate for a common law claim for breach of contract. His lordship applied *Wadsworth* v. *Lydall* and held that the arbitrator had power to award damages, which could include interest incurred or lost as a foreseeable consequence of the employer's breach of contract.

However, the wording of the clause makes it plain that the employer is not making good an earlier breach of contract. Clause 24.2.1 of JCT 98 confers on him the right to deduct liquidated damages once the architect has issued a certificate under clause 24.1 and the employer has given written notice of intended deduction to the contractor. Once these two conditions are satisfied the employer has a contractual right to deduct liquidated damages – and clause 24.2.2 deals with what is to happen if the completion date is later altered in the contractor's favour. The employer cannot be in breach of contract by doing that which the contract expressly empowers him to do.

---

[376] *Holbeach Plant Hire Ltd* v. *Anglian Water Authority* (1988) 14 Con LR 101. The court was applying the proposition laid down in *President of India* v. *Lips Maritime Corporation* [1987] 3 All ER 110 at 116 per Lord Brandon. The position under JCT forms is dealt with in section 6.5.8, 'Financing charges and interest'.
[377] (1982) 19 BLR 1.

(5) *If it is a commercial debt*

The Late Payment of Commercial Debts (Interest) Act 1998 has been fully in force from November 2002. As the name implies, it only deals with commercial debts; consumers are excluded. Broadly, if invoices are outstanding for longer than the prescribed period (usually about 30 days), the creditor is entitled to claim interest at 8% above the Bank of England Base rate current at the previous end of June or end of December as the case may be. In addition and depending upon the size of the debt, a modest lump sum is to be added to the amount of interest.

6.6 Costs of claim

It is generally accepted that the contractor is not entitled to reimbursement for any costs he has incurred in preparing the claim, since he is not required to prepare a claim as such, but merely to make a written application to the architect, backed up by supporting information, as explained in subsequent chapters. Most certainly, fees paid to claims specialists or to independent quantity surveyors or other professional advisers are not in principle allowable as a head of claim at law. Where a claim proceeds to arbitration or litigation, of course, the contractor is entitled to claim his costs and the arbitrator's award or judgment of the court can condemn the employer in costs. In *James Longley & Co Ltd* v. *South West Regional Health Authority*[378] on a summons to review taxation of costs of an arbitration which was settled during the hearing, the fees of a claims consultant in respect of work done in preparing the contractor's case for arbitration (the preparation of three schedules annexed to the Points of Claim) were allowed as those of a potential expert witness in the arbitration. The practice of the High Court is that 'costs follow the event', i.e. in the ordinary way, the successful party will receive his costs, and an arbitrator must follow the same principle[379].

However, it has been held that the expenditure of managerial time in remedying an actionable wrong done to a trading company can properly form the subject matter of a claim for 'special damage' in an action at common law[380]. It seems that in light of this decision there can in principle be a claim for the cost of managerial time spent on preparing a claim, assuming that this is not covered by any claim for head office overheads, in appropriate circumstances and subject to proof that the time had been spent in a manner in which it would not have been spent otherwise. Under the same principle, the cost of employing an outside expert might be recoverable as damages for breach of contract. Clearly, if the claim is prepared as part of arbitration or litigation, the costs can be recovered as part of the costs of the action.

---

[378] (1984) 25 BLR 56.
[379] Refer to the Arbitration Act 1996 which provides the arbitrator with important new powers in regard to costs.
[380] *Tate & Lyle Food & Distribution Co Ltd* v. *Greater London Council* [1982] 1 WLR 149.

Chapter 7
Causation

7.1 Theory

Causation is the relationship between cause and effect. It is an extremely important concept in the context of liability. A wrongful act may trigger a series of events which eventually results in damage being suffered. This is called the 'chain of causation'(see Figure 7.1).

The loss and/or expense must be direct in the sense of remoteness and also in the sense of the chain of causation, that is, the relationship between cause and effect. The matter on which the contractor seeks to rely must be linked, without interruption, to the loss suffered. Therefore if the cause is not the matter, but some intervening event[381], there will be no liability and no claim. To put the situation another way, the loss and/or expense must have been caused by the breach or act relied on and not merely be the occasion for it[382].

Two simple examples may be contrasted. In the first one, a variation is ordered which necessitates plant lying idle for some days. The plant is needed for the original work, but at a very late stage the work is varied and so the plant is not needed. Suppose the plant is hired in. The contractor's hire charges, subject to any reletting or the plant owner accepting an early return, would be a direct loss and, therefore, reimbursable. In the second example, a variation substitutes slates for roof tiles. After the contractor has ordered the new slates, problems are encountered at the slate quarry, which mean that the supply of slates is interrupted so that the supplier is in breach of the supply contract. The delay and disruption to the contract works consequent upon the interruption of supply are clearly a direct consequence of the supplier's breach of the supply contract and only an indirect consequence of the variation. It is in fact the direct consequence of an intervening event – the supplier's breach. In such a case, it is for the contractor to look, if at all, to the supplier for recompense. The principles of causation have been set out in classic statements:

> 'It seems to me that there is no abstract proposition, the application of which will provide the answer in every case, except this: one has to ask oneself what was the effective and predominant cause of the accident that happened, whatever the nature of that accident may be.'[383]

---

[381] The legal term used to be expressed as *novus actus interveniens* – a new act coming in between.
[382] *Weld-Blundell* v. *Stevens* [1920] AC 956.
[383] *Yorkshire Dale Steamship Co Ltd* v. *Minister of War Transport* [1942] AC 691 at 698 per Viscount Simon LC.

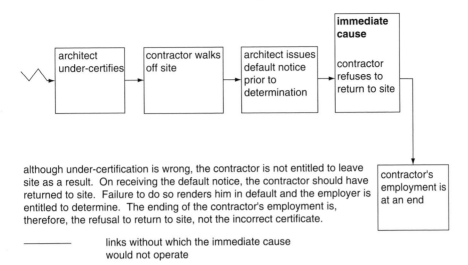

although under-certification is wrong, the contractor is not entitled to leave site as a result. On receiving the default notice, the contractor should have returned to site. Failure to do so renders him in default and the employer is entitled to determine. The ending of the contractor's employment is, therefore, the refusal to return to site, not the incorrect certificate.

——————   links without which the immediate cause
                    would not operate

Figure 7.1   The chain of causation.

'This choice of the real or efficient cause from out of the complex of facts must be made by applying commonsense standards. Causation is to be understood as the man in the street, and not as either the scientist or the metaphysician, would understand it.'[384]

In *P & O Developments Ltd* v. *Guy's & St Thomas' National Health Service Trust*, Judge Bowsher aptly summarised the position:

'The test is what an informed person in the building industry (not the man in the street) would take to be the cause without too microscopic analysis but on a broad view.'[385]

Everything depends on the facts and circumstances. Some situations are very complex and it will be important to identify the damage from which it was intended to protect a party[386]. A graphic example of the concept of causation is to be found in a case where negligent architects issued defective interim certificates and the contractors withdrew from site[387]. The contractors lost their claim against the negligent architects, because they broke the chain of causation by persisting in suspension of the works despite the service by the employer of a preliminary notice of default prior to determination. They alone, not the architects, were responsible for the termination of the contract. Although the architect's negligence was the source of the events, it was overtaken and overwhelmed by the contractor's serious breach of contract.

---

[384] *Yorkshire Dale Steamship Co Ltd* v. *Minister of War Transport* [1942] AC 691 at 706 per Lord Wright.
[385] [1999] BLR 3 at 9 per Judge Bowsher.
[386] *Skandia Property (UK) Ltd* v. *Thames Water* [1999] BLR 338.
[387] *Lubenham Fidelities & Investment Co* v. *South Pembrokeshire District Council and Wigley Fox Partnership* (1986) 6 Con LR 85.

Where there appear to be concurrent causes, one being the responsibility of one party and the other being the responsibility of the other party, the correct test to apply is not whether one of the causes is the sole cause or the dominant cause. The correct test has been held to be whether a cause is an effective cause[388]. Where there are several possible causes, the burden of proof on the contractor is to show that one cause is more likely than the others[389].

7.2 *Use of networks*

Computers are very commonly used to generate graphical information to assist in presenting a claim. Such graphics cannot usually be said to 'support' the claim in the same way that hard evidence, such as correspondence and site minutes, will support it, but if the graphics are used sensibly, they can pictorially represent what the documentary evidence proves happened on site so as to make it easier to understand. Architects, trained to visualise, usually find that such things as histograms, graphs and pie charts explain what the contractor is trying to say better than thousands of words. A particularly useful tool is the computer planning programme. There are many versions on the market. Both architects and contractors will find them helpful in preparing programmes for construction works and in analysing the programmes and the effects of delays. They have a part to play in extensions of time and the analysis of loss and expense, whether prolongation or disruption.

The courts have shown themselves ready to accept such analysis if properly carried out[390]. Of course, there is nothing magical about computers. They simply do at great speed what would take the ordinary mortal a considerable time to achieve. The particular tool used to programme and analyse is the network or the precedence diagram also called the PERT (Performance Evaluation and Review Technique) chart. All these charts provide a way of connecting together the operations on site in a series of logic links (e.g. pouring concrete cannot commence until trenches are dug, etc.). They also provide the means of delaying some activities and bringing forward others. For example, pouring concrete can start before trench digging is entirely completed. They enable the critical path or paths to be identified and delays to be introduced. Not least, resources can be added. This is not the place to venture into even a brief description of the preparation of a network and there are many excellent books on the topic. Particularly to be commended are those books published to assist in understanding unfathomable official software manuals. Most of them include excellent explanations of the theory behind programming.

---

[388] *Loftus-Brigham* v. *Ealing LBC* (2004) 20 Const LJ 82.
[389] *Plater* v. *Sonatrach* (2004) CILL 2073.
[390] *John Barker Construction Ltd* v. *London Portman Hotels Ltd* (1996) 12 Const LJ 277.

All architects and project managers should use computerised programmes to monitor progress and assist in analysing claims. Contractors
should routinely submit detailed programmes on disk as well as in hard
copy. If this was done, all parties would be assisted in making prompt
claims and speedy responses, claim making and understanding would be
eased and disputes avoided or at least made less frequent. Programmes can
be prepared to show as-built compared to intended progress, and known
employer-generated delays can be taken out to examine the likely situation
had those delays not occurred. The reverse operation can be tried. These
techniques are sometimes known as the 'subtractive' or 'additive' methods.
Provided accurate records are available, the only limit to possible methods
of analysis are the limits to the architect's or the contractor's ingenuity.
Because the logic links will determine effect, the most important part is to
ensure they are properly represented.

On the other hand, it should be borne in mind that computer programmes
are not the solution to all the ills which afflict contractor's claims and it is
easy to be seduced by the slick visuals in a typical software package into
thinking otherwise.

Balfour Beatty Construction Ltd v. *The Mayor and Burgesses of the London
Borough of Lambeth*[391] concerned an application for summary judgment
following an adjudication decision. Its interest lies in the references to the
use of programmes for estimating extensions of time. As part of its submission to the adjudicator, Balfour Beatty referred to the 'most widely recognised and used' delay analysis methods:

> '(I) Time Impact Analysis (or "time slice" of "snapshot" analysis). This
> method is used to map out the impacts of particular delays at the point in
> time at which they occur permitting the discrete effects of individual
> events to be determined.
> (II) Window analysis. For this method the programme is divided into
> consecutive time "windows" where the delay occurring in each window
> is analysed and attributed to the events occurring in that window.
> (III) Collapsed as-built. This method is used so as to permit the effect of
> events to be "subtracted" from the as-built programme to determine
> what would have occurred but for those events.
> (IV) Impacted plan where the original programme is taken as the basis of
> the delay calculation, and delay faults are added into the programme to
> determine when the work should have finished as a result of those
> delays.
> (V) Global assessment. This is not a proper or acceptable method to
> analyse delay.'[392]

Later the judge said:

> 'By now one would have thought that it was well understood that, on a
> contract of this kind, in order to attack, on the facts, a clause 24 certificate

---

[391] [2002] BLR 288.
[392] [2002] BLR 288 at 292 per Judge Lloyd.

for non-completion (or an extension of time determined under clause 25), the foundation must be the original programme (if capable of justification and substantiation to show its validity and reliability as a contractual starting point) and its success will similarly depend on the soundness of its revisions on the occurrence of every event, so as to be able to provide a satisfactory and convincing demonstration of cause and effect. A valid critical path (or paths) has to be established both initially and at every later material point since it (or they) will almost certainly change. Some means has also to be established for demonstrating the effect of concurrent or parallel delays or other matters for which the employer will not be responsible under the contract.'[393]

Although it is possible to agree in principle with this statement, it does not mean that a programme, adjusted as indicated, must be used in just that way on every occasion. It is perfectly possible to determine a critical path in words and to deduce the effect of delays by applying reason rather than computer technology. However, that may only be practical in relatively simple delay situations.

It is very easy for a contractor to make a mistake in calculating elements of his claim, whether it be a matter of extra time or money or a combination of both. In *McAlpine Humberoak Ltd* v. *McDermott International Inc (No 1)*, the Court of Appeal identified flaws in the methodology adopted by the very experienced civil engineer engaged by the contractor to prepare its claim. The engineer's approach assumed that if one man was working for one day on a particular variation order, the whole contract was delayed for that day. Thus in one instance an inspection took no more than an hour and £39 was claimed, but the engineer allowed a day's delay to the whole of the work. A more serious defect was that the claim assumed that the whole of the workforce planned for a particular activity was engaged continuously on that activity from start to finish although that situation was hardly likely[394].

7.3 Float

'Float' is a term often used in connection with programming, especially with network analysis. Essentially, it is the time difference, if any, between the time required to perform a task and the time available in which to do it. If an activity has 5 days of float, it means that the activity could be extended by up to 5 days without affecting the completion date of the project. Alternatively, the activity could start up to 5 days late without any overall delaying effect. One of the definitions of a 'critical activity' is that it has no float. In other words, there is no scope for any delay at all before the completion date of the project is affected.

---

[393] [2002] BLR 288 at 302 per Judge Lloyd.
[394] (1992) 58 BLR 1 at 25 per Lloyd LJ.

Much debate rages about the 'ownership' of float in a programme. Contractors will usually claim it for themselves, sometimes to the disadvantage
of sub-contractors[395]. A contractor may argue that an extension of time is
due even if a non-critical activity is delayed. No one owns float. The
argument is sometimes extended to the effect that if a contractor programmes to complete a 10-week contract in 9 weeks, the extra week is the
contractor's float and if the project is delayed by a few days, an extension of
time will be due, even if the contractor finishes before the completion date.
That is manifestly wrong. The better view is that no one owns the float. If an
activity has a float of 3 days and this float is used, because the architect is
late in providing information, the contractor has no entitlement to an extension of time. That is not to say, of course, that the contractor has no claim to
loss and/or expense due to disruption, but that is a different matter.

In *Ascon Constructing Ltd* v. *Alfred McAlpine Construction Isle of Man Ltd*,
the judge made a very useful analysis of the concept of float:

'Before addressing those factual issues I must deal with the point made
by McAlpine as to the effect of its main contract "float", which would in
whole or in part pre-empt them. It does not seem to be in dispute that
McAlpine's programme contained a "float" of 5 weeks in the sense, as I
understand it, that had work started on time and had all sub-
programmes for sub-contract works and for elements to be carried out
by McAlpine's own labour been fulfilled without slippage the main
contract would have been completed 5 weeks early. McAlpine's argument seems to be that it is entitled to the "benefit" or "value" of this float
and can therefore use it at its option to "cancel" or reduce delays for
which it or other sub-contractors would be responsible in preference to
those chargeable to Ascon.

In my judgment that argument is misconceived. The float is certainly of
value to the main contractor in the sense that delays of up to that total
amount, however caused, can be accommodated without involving him
in liability for liquidated damages to the employer or, if he calculates his
own prolongation costs from the contractual completion date (as
McAlpine has here) rather than from the earlier date which might have
been achieved, in any such costs. He cannot, however, while accepting
that benefit as against the employer, claim against the sub-contractor as if
it did not exist. That is self-evident if total delays as against sub-
programmes do not exceed the float. The main contractor, not having
suffered any loss of the above kinds, cannot recover from sub-contractors
the hypothetical loss he would have suffered had the float not existed,
and that will be so whether the delay is wholly the fault of one sub-
contractor, or wholly that of the main contractor himself, or spread in
varying degrees between several sub-contractors and the main contractor. No doubt those different situations can be described, in a sense,
as ones in which the "benefit" of the float has accrued to the defaulting

[395] *Ascon Contracting Ltd* v. *Alfred McAlpine Construction Isle of Man Ltd* (2000) 16 Const LJ 316.

party or parties, but no-one could suppose that the main contractor has, or should have, any power to alter the result so as to shift that "benefit". The issues in any claim against a sub-contractor remain simply breach, loss and causation.

I do not see why that analysis should not still hold good if the constituent delays more than use up the float, so that completion is late. Six sub-contractors, each responsible for a week's delay, will have caused no loss if there is a 6 weeks' float. They are equally at fault, and equally share in the "benefit". If the float is only 5 weeks, so that completion is a week late, the same principle should operate; they are equally at fault, should equally share in the reduced "benefit" and therefore equally in responsibility for the one week's loss. The allocation should not be in the gift of the main contractor.

I therefore reject McAlpine's "float" argument. I make it clear that I do so on the basis that it did not raise questions of concurrent liability or contribution; the contention was explicitly that the "benefit", and therefore the residual liability, fell to be allocated among the parties responsible for delay and that the allocation was entirely in the main contractor's gift as among sub-contractors, or as between them and the main contractor where the latter's own delay was in question.'[396]

This supports the view that float is owned by no one. A useful and neat summary of the position has been set out by Nicholas Carnell:

'In fact consideration of the role of float from first principles shows that the debate is less complex than might be supposed.
(1) In the majority of standard form contracts, the programme is not a contract document. The contractor's obligation is to carry out and complete the works by the completion date, rather than by any specific activity date.
(2) Accordingly, unless the effect of delaying a particular activity is to cause delay to the completion date of the works, the programme is to be regarded as a planning tool and no more.
(3) Within the constraints of the need to complete the works by the date for completion, the contractor can programme the works as he wishes.
(4) Similarly, if the employer's conduct causes the contractor to use up some or all of the float without causing delay to the works, the consequences may be disruption if the contractor can identify the need to deploy additional resource, but it will not entitle him to any extension of time.'[397]

The judgment in *How Engineering Services Ltd* v. *Lindner Ceilings Partitions plc* gives support to this view[398].

A more recent case appears to have thrown some doubt on that interpretation:

---

[396] (2000) 16 Const LJ 316 at 338 per Judge Hicks.
[397] Nicholas J Carnell, *Causation and Delay in Construction Disputes* (2000) Blackwell Science, p.185.
[398] 17 May 1995, Unreported.

'Under the JCT conditions, as used here, there can be no doubt that if an architect is required to form an opinion then, if there is then unused float for the benefit of the contractor (and not for any other reason such as to deal with pc or provisional sums or items), then the architect is bound to take it into account since an extension is only to be granted if completion would otherwise be delayed beyond the then current completion date. This may seem hard to a contractor but the objects of an extension of time clause are to avoid the contractor being liable for liquidated damages where there has been delay for which it is not responsible, and still to establish a new completion date to which the contractor should work so that both the employer and the contractor know where they stand. The architect should in such circumstances inform the contractor that, if thereafter events occur for which an extension of time cannot be granted, and if, as a result, the contractor would be liable for liquidated damages then an appropriate extension, not exceeding the float, would be given. In that way the purposes of the clause can be met: the date for completion is always known; the position on liquidated damages is clear; yet the contractor is not deprived permanently of "its" float.'[399]

The rationale behind this statement is not immediately apparent. It is certainly *obiter*, because the judge said later that it was not certain that there was any float in the programme under consideration. It seems that the judge was referring to the kind of float which a contractor may put in his programme at the end of all activities, to give himself a cushion if it takes him rather longer than expected to complete the works. Where it is clear that the contractor has placed that kind of float in his programme, it is difficult to discern the difference in law from the situation where the contractor simply attempts to finish early. That situation has already been considered in Chapter 6, section 6.3.

Essentially float is simply the space before or after individual activities, when a group of activities is put together in the form of a programme. Whether it actually exists at all depends on the extent to which the programme mirrors reality.

---

[399] *Royal Brompton Hospital NHS Trust* v. *Hammond and Others (No 8)* (2002) 88 Con LR 1 at 187 per Judge Lloyd.

Chapter 8
The global approach

In general, it is necessary for the contractor to establish each and every head of claim, by means of supporting documentation and other evidence. *London Underground Ltd* v. *Kenchington Ford plc and Others*[400] concerned, among other things, a claim for delay due to the alleged excessive number of requests for information (RFIs) which had to be made. The judge neatly summarised the position:

> 'In the manner of pointing a blunderbuss at a target it is maintained that there were many RFIs, and there was considerable delay. The delay in part can be explained by other causes but a balance is left which must be caused by the volume of RFIs. And by reason of the volume of them negligence must be concluded. It is termed a global claim. It can properly be described as a global claim in the sense that it is the antithesis of a claim where the causal nexus between the alleged wrongful act or omission of the defendant and the loss of the plaintiff has been clearly spelt out and pleaded.'[401]

In another case, not concerned with construction, such claims were termed 'total-total', where the totality of the losses is attributable to the totality of misrepresentations, breaches of contract and acts of negligence. This is compared to a 'cumulative-total' case where the cumulative effect of all the breaches led to the totality of the losses claimed[402]. This concept was explained later in an extract from the particulars of claim:

> ' "Accordingly, it is not primarily the plaintiff's case that particular aspects of the losses are individually attributable to any particular misrepresentation, breach of contract or act of negligence. The plaintiff's case is that in respect of each of the allegations of misrepresentation, breach of contract and negligence the cumulative effect thereof led to the totality of the losses claimed in the action. Further, as set out in more detail hereunder, some of the individual acts of misrepresentation, breach of contract and negligence were sufficient to cause the entirety of the plaintiff's losses.
>
> Generally, however, it is not possible directly to attach individual losses to individual allegations of misrepresentation, breach of contract or negligence in the manner implied by the form of request. The plaintiff's case is that the totality of the losses claimed arise from the totality of

---

[400] (1999) CILL 1452.
[401] (1999) CILL 1452 at 1453 per Judge Wilcox.
[402] *GAB Robins Holdings Ltd* v. *Specialist Computer Centres Ltd* (1999) 15 Const LJ 43.

the wrongful acts established to have been committed by the defendant. This is not to be taken as advancing a case that no loss is shown unless all the pleaded wrongful acts are proved but as advancing a case that all of the claimed losses resulted from the deficiencies shown to have existed in the defendant's conduct and the resultant software ..."

The last two paragraphs in those particulars seek to establish a comprehensive claim on the part of the plaintiff which seeks to cover every eventuality and ensure out of an abundance of caution that no aspect of the claim is lost in support of the overall claim for damages of £2.8 million. One can understand how the defendant is somewhat alarmed by this approach. However, bearing in mind that this is a general statement of the plaintiff's claim, I do not consider that a case has been made out which would warrant the striking out of either paragraph 31 in its original form, or any part of the particulars under headings (A), (B) and (C).'[403]

In some circumstances a 'global' apportionment of the claim may be admissible and this applies to claims for extensions of time as well as claims for money. This global approach was recognised in *J. Crosby & Sons Ltd* v. *Portland UDC*[404] which was decided under the ICE Conditions of Contract (4th Edition) and which established the acceptable criteria for a global claim.

One of the points at issue was that the contractors had made a general claim for delay and disorganisation. Completion had been delayed by 46 weeks by a combination of matters, some of which entitled the contractors to additional time and/or money and some of which did not. The court upheld the arbitrator's view that the contractor was entitled to compensation on a global basis in respect of 31 weeks of the overall delay and rejected the employer's argument that the arbitrator must necessarily build up the sum by finding amounts due under each of the individual heads of claim upon which the contractor relied in support of his overall claim for delay and disruption.

The parameters of this approach are closely confined. The global approach is only justified where a claim depends 'on an extremely complex interaction in the consequences of various denials, suspensions and variations' and where 'it may well be difficult or even impossible to make an accurate apportionment of the total extra cost between the several causative events'. In those limited circumstances it seems that there is no reason why an architect, engineer or arbitrator 'should not recognise the realities of the situation and make individual awards in respect of those parts of individual items of the claim which can be dealt with in isolation and a supplementary award in respect of the remainder of those claims as a composite whole'[405].

This does not, of course, relieve the contractor of producing substantiating evidence and proving each head of claim. What it does is to enable the architect or quantity surveyor to adopt a common-sense method of

---

[403] *GAB Robins Holdings Ltd* v. *Specialist Computer Centres Ltd* (1999) 15 Const LJ 43 at 47 per Otton LJ.
[404] (1967) 5 BLR 121.
[405] (1967) 5 BLR 121 at 136 per Donaldson J.

ascertaining certain complex claims where it is impossible or totally impracticable to prove the cost resulting from each individual item. The general
and limiting qualification under JCT forms is well expressed by the
following:

> 'The events which are the subject of the claim must be complex and
> interact so that it is difficult if not impossible to make an accurate
> apportionment. It is very tempting to take the easy course and to lump
> all the delaying events together in order to justify the total overrun or
> total financial shortfall. That argument is justifiable only if the alternative
> course is shown to be impracticable.'[406]

In other words, the *Crosby* approach is very much a long-stop and cannot be
relied upon in the vast majority of cases. It cannot be invoked where in
reality a contractor cannot establish a valid claim but can merely show a
financial shortfall, and the dangers of this approach are illustrated by the
American case of *Bruno Law* v. *US*[407], where a contractor claimed over $US
1 million extra on a contract, apparently on the theory that it was sufficient
to take the overrun between the original and actual completion dates, point
to a number of individual delays for which the employer was allegedly
responsible and which contributed to the overall delay, and then arrive at
the conclusion that the entire overrun time was attributable to the employer.
The trial commissioner pointed out that, upon the evidence:

> 'Many of the incidents relied on by plaintiff were isolated and non-
> sequential and therefore could not possibly have caused any significant
> delay in the overall progress of the contract. Furthermore, with respect
> to the great bulk of such incidents, plaintiff has failed to prove, or
> indeed even to attempt to prove, the crucial factors of the specific
> extent of the alleged wrongful delay to the project operations caused
> thereby'.[408]

The court upheld the trial commissioner's award of approximately one-fifth
of the amount claimed by the contractor on the basis of approximately one-
tenth of the alleged specific items of delay.

The point was made again in the case, already much quoted, of *London
Borough of Merton* v. *Stanley Hugh Leach Ltd*[409]. The type of calculation put
forward by Leach as a claim under the contract was well described by the
arbitrator in a passage from his interim award as follows:

> 'The calculation commences with the "direct site costs", which I can only
> interpret as being the total expenditure incurred by [Leach] on all labour,
> plant and materials involved in the construction works. From the very
> limited information available to me I can interpret the word "direct" as
> indicating that the costs relate to [Leach's] own expenditure and that of

---

[406] (1967) 5 BLR 121 see the commentary at 123.
[407] (1971) 195 Ct Cl 370.
[408] (1971) 195 Ct Cl 370.
[409] (1985) 32 BLR 51.

his direct sub-contractors to the exclusion of expenditure through nominated sub-contractors and suppliers.

From this total site cost, [Leach] deduct the assessment for fluctuations which under clause 31A of the conditions are to be adjusted on a net basis. A percentage for profit and overheads is then added to the total site costs excluding fluctuations and finally the net fluctuations are added back to arrive at the alleged remunerable total cost to the contractor of £3,721,970.

If one could imagine a building contract which proceeded to completion without any hitch, delay or variation whatsoever this calculation would provide [Leach] at line (5) with a direct comparison with his tender figure. However as that ideal situation is rarely, if ever, met and certainly was not met in the instant contract, the figures in line (1) (and so those in lines (4) and (7)) must include the costs to [Leach] of all the 'hitches' of whatever nature that occurred on the site.'

The judge commented on that in this way:

'I find it impossible to see how [this calculation] can be treated as even an approximation for a claim, whether or not rolled up (as in *Crosby*), under clause 11(6) or 24(1) [of JCT 63]. As the arbitrator points out in the passage I have cited, the calculation in effect relieves Leach from the burden of additional costs resulting from delays in respect of which Leach is not entitled to any extension of the completion date.'[410]

He might have added: and to any recovery of 'direct loss and/or expense'.

Claims such as that submitted by Leach (a claim in that case for a sum well into seven figures and contained on one side of a sheet of A4 paper albeit backed up with a vast mass of other material) are not uncommon.

Contractors are faced with difficult problems where the facts are truly interconnected in such a complex way that to unravel them into the classic approach of particularised cause and effect is impossible. On the other hand, it is clearly inequitable if an employer responsible for just one occurrence giving rise to a delay, and for which a clear causal nexus can be demonstrated, is more likely to be made to suffer the consequences than an employer guilty of a large number of interconnected occurrences.

Despite the fact that the Judicial Committee of the Privy Council stated quite categorically that the judgment involved no question of general importance, the Privy Council decision in *Wharf Properties Ltd* v. *Eric Cumine Associates*[411] was seized upon by some commentators as ringing the death knell on global claims. The case was brought against a firm of architects and the question was whether the pleadings as presented by the plaintiffs established an essential link between the breaches and the damages claimed. It was asserted by the defendants that the pleadings disclosed no reasonable cause of action or that they were so embarrassing as to warrant them being struck out as an abuse of the court process. The committee held

---

[410] (1985) 32 BLR 51 at 112 per Vinelott J.
[411] (1991) 52 BLR 1.

that, although the plaintiffs would face 'extraordinary evidential difficulties', there were no grounds for saying that the pleadings disclosed no reasonable cause of action, but they were an abuse of the process. The most useful part of the decision is as follows:

> '... the pleading is hopelessly embarrassing as it stands and their Lordships are wholly unpersuaded by Counsel for Wharf's submission that the two cases of *J. Crosby and Sons* v. *Portland Urban District Council* and *Merton* v. *Leach* provide any basis for saying that an unparticularised pleading in this form ought to be permitted to stand. Those cases establish no more than this, that in cases where the full extent of extra costs incurred through delay depend upon a complex interaction between the consequences of various events, so that it may be difficult to make an accurate apportionment of the total extra costs, it may be proper for an arbitrator to make individual financial awards in respect of claims which can conveniently be dealt with in isolation and a supplementary award in respect of the financial consequences of the remainder as a composite whole. This has, however, no bearing upon the obligation of a plaintiff to plead his case with such particularity as is sufficient to alert the opposite party to the case which is going to be made against him at the trial. ECA [Eric Cumine Associates] are concerned at this stage not so much with quantification of the financial consequences – the point with which the two cases referred to were concerned – but with the specification of the factual consequences of the breaches pleaded in terms of periods of delay. The failure even to attempt to specify any discernible nexus between the wrong alleged, and the consequent delay, provides, to use the phrase of Counsel for ECA, "no agenda" for the trial.'[412]

This decision does not overturn *Crosby* and *Merton*; quite the reverse, it upholds them. Essentially, it was the link between cause and effect – the liability – with which the Privy Council was concerned. The decision was followed in *Mid Glamorgan County Council* v. *J. Devonald Williams & Partner*[413] although in that case the pleadings were not struck out on the facts. There, the position was analysed as follows:

> '56.1   A proper cause of action has to be pleaded.
> 56.2   Where specific events are relied upon as giving rise to a claim for moneys under the contract then any preconditions which are made applicable to such claims by the terms of the relevant contract will have to be satisfied, and satisfied in respect of each of the causes of events relied upon.
> 56.3   When it comes to quantum, whether time based or not, and whether claimed under the contract or by way of damages, that proper nexus should be pleaded which relates each event relied upon to the money claimed.

---

[412] (1991) 52 BLR 1 at 20 per Lord Oliver.
[413] (1993) 8 Const LJ 61. See also *Imperial Chemical Industries PLC* v. *Bovis Construction Ltd, GMW Partnership and Oscar Faber Consulting Engineers* (1992) 8 Const LJ 293.

56.4   Where, however, a claim is made for extra costs incurred through delays as a result of various events whose consequences have a complex interaction that renders specific relation between event and time/money consequence impossible and impracticable, it is permissible to maintain a composite claim.'[414]

The logic of a 'total cost' claim has been explained like this:

'(a) the contractor might reasonably have expected to perform the work for a particular sum, usually the contract price;
(b) the proprietor committed breaches of contract;
(c) the actual reasonable cost of the work was a sum greater than the expected cost.

The logical consequence implicit in this is that the proprietor's breaches caused that extra cost or cost overrun. This implication is valid only so long as, and to the extent that, the three propositions are proved and a further unstated one is accepted: the proprietor's breaches represent the only causally significant factor responsible for the difference between the expected cost and the actual cost... The unstated assumption underlying the inference may be further analysed. What is involved here is two things: first, the breaches of contract caused some extra cost; secondly, the contractor's cost overrun is this extra cost... It is the second aspect of the unstated assumption... which is likely to cause the more obvious problem because it involves an allegation that the breaches of contract were the material cause of all of the contractor's cost overrun. This involves an assertion that, given that the breaches of contract caused some extra cost, they must have caused the whole of the extra cost because no other relevant cause was responsible for any part of it.'[415]

The problems inherent in this kind of analysis are obvious. Nevertheless, the courts have continued to allow claims to be made on a global basis[416]. Duncan Wallace provides a useful, if slightly provocative, summary of the global claim position[417], but his conclusion that global claims are always embarrassing has been questioned in some recent cases[418]. It seems that it may be enough if the contractor sets out the claim in sufficient detail that the employer knows what is being claimed and, in some instances, it may be that the employer is in a perfectly good position to calculate the amount the contractor should be paid without the contractor being obliged to separate

---

[414] *Mid Glamorgan v. Devonald Williams* (1993) 8 Const LJ 61 at 69 per Mr Recorder Tackaberry.
[415] *John Holland Construction & Engineering Pty Ltd v. Kvaerner R. J. Brown Pty Ltd* (1996) 82 BLR 83 at 85 per Byrne J.
[416] See for example *Nauru Phosphate Royalties Trust v. Matthew Hall Mechanical & Electrical Engineers Pty Ltd and Another* (1992) 10 BCL 178; *British Airways Pension Trustees Ltd v. Sir Robert McAlpine & Sons Ltd and Others* (1994) 72 BLR 102; *Bernard's Rugby Landscapes Ltd v. Stockley Park Consortium Ltd* (1997) 82 BLR 39; *John Doyle Construction Ltd v. Laing Management (Scotland) Ltd* [2002] BLR 393.
[417] I. N. Duncan Wallace, *Hudson's Building and Engineering Contracts*, 11th edition, 1995, Sweet & Maxwell, vol. 1, pp.1086–9.
[418] See *British Airways Pension Trustees Ltd v. Sir Robert McAlpine & Sons Ltd and Others* (1994) 72 BLR 102 and *Inserco Ltd v. Honeywell Control Systems Ltd*, 19 April 1996, unreported.

the claim into its various parts for the purpose of allocating value[419]. The position seems to be that the contractor is entitled to put his claim in any rational way albeit that putting forward a claim on a global basis may present particular evidential difficulties[420]. A particular difficulty is the establishment of a definable connection between the alleged wrong and the consequent delay and damage[421]. The position was re-emphasised in an Australian case. It held that where the connection between cause and loss is not apparent, each aspect of the connection must be set out unless the probable existence of the connection can be demonstrated by evidence or argument, or unless it can be shown that it is impossible or impracticable for it to be itemised further[422].

The topic of global claims was examined again from first principles by the Scottish courts in *John Doyle Construction Ltd* v. *Laing Management (Scotland) Ltd* and it is instructive to read what was said:

'The logic of a global claim demands, however, that all the events which contribute to causing the global loss be events for which the defender is liable. If the causal events include events for which the defender bears no liability, the effect of upholding the global claim is to impose on the defender a liability which, in part, is not legally his. That is unjustified. A global claim, as such, must therefore fail if any material contribution to the causation of the global loss is made by a factor or factors for which the defender bears no legal liability...

...The point has on occasions been expressed in terms of a requirement that the pursuer should not himself have been responsible for any factor contributing materially to the global loss, but it is in my view clearly more accurate to say that there must be no material causative factor for which the defender is not liable...

...Failure to prove that a particular event for which the defender was liable played a part in causing the global loss will not have any adverse effect on the claim, provided the remaining events for which the defender was liable are proved to have caused the global loss. On the other hand, proof that an event played a material part in causing the global loss, combined with a failure to prove that the event was one for which the defender was responsible, will undermine the logic of the global claim. Moreover, the defender may set out to prove that, in addition to the factors for which he is liable founded on by the pursuer, a material contribution to the causation of the global loss has been made by another factor or other factors for which he has no liability. If he succeeds in proving that, again the global claim will be undermined.'[423]

---

[419] *British Airways Pension Trustees Ltd* v. *Sir Robert McAlpine & Sons Ltd and Others* (1994) 72 BLR 102.

[420] *GMTC Tools & Equipment Ltd* v. *Yuasa Warwick Machinery Ltd* (1994) 73 BLR 102.

[421] *Ralph M. Lee Pty Ltd* v. *Gardner & Naylor Industries Pty Ltd* (1997) 12 Const LJ 125.

[422] *John Holland Construction & Engineering Pty Ltd* v. *Kvaerner R. J. Brown Pty Ltd* (1996) 82 BLR 83. This judgment was considered with apparent approval in *Bernard's Rugby Landscapes Ltd* v. *Stockley Park Consortium Ltd* (1997) 82 BLR 39.

[423] [2002] BLR 393 at 407 per Lord McFadyen.

This thoughtful judgment emphasises the necessity for the employer to be responsible for all the major causative factors before a global claim can succeed, a point that is often overlooked when such claims are made. This change in emphasis starkly highlights the difficulties involved in successfully making such claims.

An earlier excellent judgment usefully set out in some detail what is required of the contractor when making a claim[424]. A summary of the court's findings can be expressed as follows:

- The claimant must set out an intelligible claim which must identify the loss, why it has occurred and why the other party has an enforceable obligation recognised at law to compensate for the loss.
- The claim should tie the breaches relied on to the terms of the contract and identify the relevant contract terms.
- Explanatory cause and effect should be linked.
- There is no requirement that the total amount of loss must be broken down so that the sum claimed for each specific breach can be identified. But an 'all or nothing' claim will fail in its entirety if a few causative events are not established.
- Therefore, a global claim must identify two matters:
  - The means by which the loss is to be calculated if some of the causative events alleged have been eliminated. In other words, what formula or device is put forward to enable an appropriate scaling down of the claim to be made?
  - The means of scaling down the claim to take account of other irrevocable factors such as defects, inefficiencies or events at the contractor's risk.

This still seems to be an exceptionally clear exposition of the current position.

---

[424] *How Engineering Services Ltd v. Lindner Ceilings Partitions PLC* 17 May 1995, unreported.

Chapter 9
Substantiation of claims

9.1 Principles

Many contractors seem to approach contract claims in the same way as the claimants in *Bruno Law* v. *US*[425] and in *London Borough of Merton* v. *Stanley Hugh Leach*[426], and many a potentially valid claim founders because of lack of proper substantiating evidence and the contractor's inability to (a) establish the direct link between cause and effect with regard to delays, and (b) provide sufficient particulars of the resulting loss and/or expense. It has been rightly said:

> 'General words intended to arouse sympathy or other emotion favourable to the contractor's interest should be used sparingly lest they excite a suspicion that the contractor really thinks his claim is founded in mercy and not in law. Further, it is rarely useful to say that the contractor has suffered loss due to causes outside his control or which he did not reasonably anticipate. It is elementary law that such factors are not grounds in law for extra payment unless, which is rare, they are expressly so made. If this is put forward as the sole ground of a claim the reader may well assume that there is no other ground.'[427]

In fact, emotion of this kind is out of place in a claims situation, and so is emotive language. What the claimant must do is to state the facts giving rise to his alleged entitlement, point to the contract term under which the claim originates, establish that he has satisfied any conditions precedent to his claim, and set out the financial basis of his claim. In relation to this last requirement, clause 26.1.3 of JCT 98 requires the contractor to submit, upon request, such details of the loss and/or expense incurred as are reasonably necessary to enable the architect or quantity surveyor to carry out the ascertainment. Although not strictly required, it is probably in the contractor's best interests that this should be a comprehensive document setting out fully what the contractor considers to be his entitlement, supported by appropriate evidence. The contractor would be well advised to take this approach even under clause 34.3 of JCT 98, although there is no specific requirement of the contractor to provide even the details called for in clause 26.1.3.

---

[425] (1971) 195 Ct C1 370.
[426] (1985) 32 BLR 51.
[427] Stephen Furst and Vivian Ramsey, *Keating on Building Contracts*, 7th edition, 2001, Sweet & Maxwell, p. 552.

If the contractor is to provide substantiating evidence he must have adequate records. His obligation is to provide the architect and/or quantity surveyor with all the information necessary for them to ascertain the direct loss and/or expense. This information will vary according to the project and the claim, but might well include cost records, charts showing actual progress measured against programme, original tendering calculations showing level of anticipated profit, possibly even profit and loss accounts and balance sheets if head office overheads are being claimed.

It is clear that a heavy burden is placed upon the contractor, but unless detail of this kind is provided no proper ascertainment can be made. That is not just a question of the contract provision, but one of practicality. In *Peak Construction (Liverpool) Ltd* v. *McKinney Foundations Ltd*[428], when dealing with the sort of evidence required to substantiate a head office overheads and loss of profit claim, it was said that not only was the evidence of the contractor's auditor as to profitability relevant, but also:

> 'some evidence as to what the site organisation consisted of, what part of the head office staff is being referred to, and what they were doing at the material times...'[429]

It was also suggested that it would be useful to have an analysis of the contractor's yearly turnover for a period of some 7 years in order to establish the level of profitability and the effect upon it of the particular contract disruption and overrun.

It is very much in the interest of the contractor to keep detailed records of all cost factors related to individual contracts and be prepared to abstract them for the purpose of substantiating any contract claim.

The foregoing is best illustrated by an example. Where there is a claim for direct loss and/or expense arising under JCT 98, clause 26.2.7, in respect of a variation, the contractor's claim must establish the following:

(1) That an instruction amounting to a variation has been properly issued and carried out by the contractor
(2) The date of issue and receipt of the variation instruction
(3) The content of the variation
(4) The time at which it was *necessary* to carry out the work contained in the instruction
(5) That the carrying out of the variation instruction unavoidably affected regular progress of the works
(6) The extent to which the work was affected as a direct result of the carrying out of the variation which must be more than trivial
(7) Appropriate evidence of the direct loss and/or expense suffered or incurred directly resulting from the variation
(8) That the contractor made a proper and timely written application to the architect in respect of direct loss and/or expense.

---

[428] (1970) 1 BLR 114.
[429] (1970) 1 BLR 114 at 122 per Salmon LJ.

Item (1) may seem straightforward, but the difficulty which even this can involve for the contractor is illustrated by *M. Harrison & Co (Leeds) Ltd* v. *Leeds City Council*[430], where an architect issued a document headed 'Variation Order No. 1', which read as follows:

'You are hereby authorised to execute the following work involving a variation on your Contract, in accordance with the terms thereof, namely...

Omit Item A page 159 The PC Sum for Structural Steel work £77,000.

Add Place an order with [XYZ Ltd] for the supply and erection of structural steelwork at the above in accordance with the fixed price quotation enclosed ... in the sum of £98,084 ls 6d.'

Despite this apparently clear wording, it was held that this instruction did not amount to a variation within the meaning of clause 11 of JCT 63 (now clause 13, JCT 98) for, as Megaw LJ remarked in his judgment, 'the definition in that clause is narrow'.

Item (2) is important because the period of notice of the variation will have a bearing upon the contractor's ability to accommodate it by reasonable adjustment of his programme.

Item (3) is self-explanatory: the variation should be described.

The emphasis under item (4) must be on the word *necessary* since the contractor must demonstrate that he executed the variation at a time that reasonably mitigated the amount of any direct loss and/or expense incurred. If the timing of the execution of the work was forced upon the contractor, say by the timing of the issue of the instruction itself, then the contractor should give details.

So far as item (5) is concerned, the contractor must establish that the effect upon regular progress was a direct result of the instruction and that it was unavoidable by any reasonable means.

It is obviously essential to show how and to what extent regular progress was affected by the variation (item (6)). The JCT 98 contract provides in clause 26.1 that the regular progress must be 'materially' affected. In other words, substantially and not trivially.

Under JCT 98, clause 26.1.3, the contractor is not obliged to supply particulars of his loss and/or expense (item (7)) until asked for them, but it is clearly in his interests to provide them, whether requested or not and, as previously indicated, a fully calculated statement (backed by supporting evidence) is desirable.

Item (8) is of great importance, because for sound practical as well as legal reasons noted elsewhere, the contractor's entitlement to reimbursement under the contract provisions depends upon his having satisfied the administrative requirements.

It is not necessary to prepare the claim in any particular form, but it is prudent to prepare every claim with the possibility of eventual adjudication, arbitration or litigation in mind. It is therefore advisable that it should be in

---

[430] (1980) 14 BLR 118.

such a form that it can be readily used for the purpose of formal proceedings. Not only will that save time if the matter does eventually come before a tribunal, but the discipline involved in preparing formal claim documents for legal proceedings (formerly referred to as 'pleadings') is aimed at clarifying the basis and logic of the claim together with its supporting evidence.

Just because the contractor is quite unable to provide substantiation, he will not necessarily be deprived of the right to some reimbursement for loss and expense, and some estimate must be made[431]. In practice, however, without proper evidence he is likely to receive substantially less than he might think is his due.

9.2 Nature of supporting evidence

9.2.1 Introduction

Claims submitted by contractors are often lacking supporting evidence. Some consist merely of broad and sweeping assertions of incompetence on the part of the design team or generalised allegations of disruption and loss without any hard foundation. Others, while marginally more detailed and substantiated, on examination fail to meet the criteria set out here because, although superficially they link particular disruption and cost to particular causes of delay, they totally fail to establish that the extent of the disruption and cost inevitably and directly arose from the cause alleged and not from some other cause.

Contractors sometimes use averages or generalisations or even figures plucked out of the air (typically percentages for lack of productivity). It is the *actual* loss and/or *actual* expense which is required and not figures taken from the bills of quantities or from government or other indices.

The nature of the back-up evidence will obviously depend on the type of claim, but in almost every case detailed cost records and comparative programme/progress schedules will be necessary, together with references to correspondence, records of site meetings, site diaries and the like. The phrase 'contemporary records' was considered in a Falkland Islands case[432] which was concerned with the FIDIC Conditions of Contract 4th Edition which contained a clause (53) requiring contemporary records to be produced. The court concluded that it meant original or primary documents or copies of them. Importantly, they should have been prepared at or about the time the claim arose. The court considered that it would be perverse to allow a contractor who failed to comply with the terms of the contract to introduce non-contemporary records including witness statements to support a claim. However, a witness statement may be used to clarify contemporary records. The court's analysis is generally applicable to all contractual claim situations.

---

[431] *Chaplin* v. *Hicks* [1911] 2 KB 786.
[432] *Attorney General for the Falkland Islands* v. *Gordon Forbes (Falklands) Construction Ltd* (2003) 19 Const LJ T1 49.

In putting together a claim, the contractor or his advisers often paraphrase parts of such correspondence, site meetings, etc, the better to make their point. Naturally, the best possible gloss is put on such evidence. However, it is essential that such paraphrasing is done with complete accuracy. The recipient of the claim will certainly refer to the original documents and, if it is found that the paraphrasing represents what the contractor wishes was in the documents rather than what is actually there, the whole basis of the claim is compromised. Unfortunately, this scenario is fairly common to a greater or lesser degree in many claims. Substantial inaccuracies came to light in a recent adjudication, in which the contractor was claiming nearly £1,000,000, with disastrous results for the contractor, who recovered nothing.

9.2.2 Cost records

Ideally, cost records should be capable of pin-pointing the precise cost effects of particular events, but this is a counsel of perfection, which is rarely fulfilled. None the less this is the ideal at which to aim and the increasing use of computers, adequately programmed, should make this task very much easier. Despite this development, it will still remain the fact that the adequacy and accuracy of cost records will always ultimately depend upon the keeping of detailed time-sheets by workpeople and particularly by site supervisory staff. Currently, time-sheets for manual workers are of a general nature, and care should be taken to ensure that they are in such a form that relevant information to support a particular claim can be readily abstracted. In the case of a claim arising from a variation, for example, supervisory staff should take care to direct operatives actually carrying out the work involved to record separately the time actually spent on the variation work. This applies similarly to those executing work not itself varied, but nevertheless affected, by the introduction of the variation.

Similar considerations apply to records of plant time, while records of materials should present no particular problem. The essential point is that the cost records relied upon should be clearly referable to the particular disruption or other event.

9.2.3 Programme/progress schedules

Clause 5.3.1.2 of JCT 98 requires the contractor 'within 14 days of any decision by the Architect' with regard to the completion date under the contract to provide the architect with two copies of any amendments and revisions to the master programme to take account of that decision. While useful for general monitoring purposes, neither the original nor amended master programme is likely to be sufficient to substantiate a monetary claim arising from any particular cause of delay or disruption. It is certainly evidence, however, and should be referred to.

What is needed to substantiate a money claim is a chart showing the progress that could have been achieved had the particular cause of delay or disruption not arisen, compared with actual achievement. This is best achieved by the use of a computer-generated network[433]. The architect can make life very much simpler for all concerned if the master programme to be provided under clause 5.3.1.2 is required by the preliminaries to be in network form, fully resourced and indicating the logic links. As much annotation as possible should be provided to demonstrate precisely the manner in which progress was affected. The architect should make full use of every modern aid in dealing with claims and an architect has been criticised for failing in this respect:

> '[The architect] did not carry out a logical analysis in a methodical way of the impact which the relevant matters had or were likely to have on the plaintiffs' planned programme.'[434]

9.2.4 Correspondence

Copies of any letters, memoranda, etc. relevant to the claim should be annexed, including of course, copies of the relevant applications, etc. required by the terms of the contract. Particular paragraphs in the correspondence can be referenced by numbers and referred to in the claim document and cross-referenced to cost records, progress schedules and so on. However, it should be noted that the inclusion of large amounts of copy correspondence is no substitute for careful linking of cause and effect. Letters must be read with care and always in the context of other letters and documents. They may have been written especially to support a future claim. They may not mean what they appear to mean at first sight.

9.2.5 Records of site meetings

Records of site meetings can provide useful supporting evidence. Usually, such records are prepared by one party without reference to the other. On receipt of a copy, the other party should check it carefully and challenge any alleged inaccuracies, ambiguities or misrepresentations at that time and ensure that the relevant corrections are recorded both in correspondence and in the record of the next meeting. If such inaccuracies are not challenged contemporaneously, with the passage of time it will become increasingly difficult (though not impossible) to establish that they are not an accurate record of what actually happened. Such records form the mainstay of many claims, because matters are recorded there which are not referred to elsewhere. Moreover, they usually contain very useful progress reports. A contractor who is recorded as regularly reporting a project as being on

---

[433] *John Barker Construction Ltd* v. *London Portman Hotels Ltd* (1996) 12 Const LJ 277.
[434] (1996) 12 Const LJ 277 at 286 per Mr Recorder Toulson.

time may find it difficult to argue later that he was in delay, when seeking an extension of time. Minutes that are not recorded as agreed in later minutes have a reduced value, somewhat similar to personal notes.

9.2.6 Site diaries

Site diaries are useful, but they sometimes contain entries that may be embarrassing to the person putting them forward. Diaries kept by each side, e.g. by the contractor and the clerk of works, may well differ in their recording of the same events. The court's attitude in such cases is shown by the approach of Judge Sir William Stabb in *Oldschool* v. *Gleeson (Construction) Ltd*, where there was a conflict between the diaries kept by a consulting engineer and the site agent. The judge said:

> 'I found [the agent's] diary entries unsatisfactory, in the sense that they do not record warnings and complaints when, as I believe, these were given. I cannot believe that the district surveyor could have written in these terms on 21 March if [his assistant] had not in fact given the same warning, and yet not a hint of it appears in [the agent's] diary; which, together with other instances, leads me to think that he was not anxious to record criticisms or complaints when they were made, and where such appear in [the engineer's] diary and not in the diary of [the agent], I am bound to say that I have no hesitation in accepting [the engineer's] contemporaneous record as being the accurate one.'[435]

Site diaries are, however, often useful as substantiating evidence as to progress. The evidence of the diary may be quite crucial if there is no other evidence.

9.2.7 The narrative

At one time every claim used to start with a 'narrative' telling the story of the claim – a piece of writing of dubious value, because it tended to be loosely assembled and worded and poorly substantiated. Such text cannot stand up to rigorous examination. One such example stated:

> 'The work was becoming increasingly complex.'

That kind of vague statement invites the architect to ask questions which may be impossible to answer: how complex was the work originally? How do you measure complexity? At what rate did the complexity increase? How complex was it when it had stopped increasing? Regrettably, narratives are still employed in order to achieve, by emotive means, what really needs detailed evidential substantiation. If a narrative is used at all, it should be concise and merely for the purpose of setting the scene.

---

[435] (1976) 4 BLR 103 at 117 per Sir William Stabb J.

Reference Number	Nature of Variation	Claimant's		Respondent's		Amount in dispute	Arbitrator's comments
		Details	Amount claimed	Comments	Amount agreed		

Figure 9.1   A typical Scott Schedule.

The putting together of a claim requires all the discipline of legal argument or a piece of scholarly research, analytical in nature, leaving nothing out, taking nothing for granted and citing appropriate authority or evidence for every statement.

9.3 'Scott Schedule'

A 'Scott' or Official Referee's Schedule is a formal document used in litigation, arbitration and adjudication, setting out the issues in dispute in tabular form, with space for the contentions of the opposing parties to be set out against each other for easy reference. While it is inappropriate to submit a claim in such a form initially, the claim should be prepared so that the factual information in it can, if necessary, easily be transferred to such a schedule.

Particularly where a claim is based on a large number of variations or even the final account, a simplified version of the schedule can be useful to give an overview and keep control over which parts of the claim are agreed and which disputed.

There is no standard form for a Scott Schedule and every case will generate its own schedule. Figure 9.1 is an example of a typical schedule relating to a contractor's claim resulting from variations arising under clause 13 of JCT 80.

PART II

Chapter 10
Extension of time under JCT standard form contracts

10.1 Extension of time position under JCT 98

10.1.1 Clause 25

The full text of clause 25 is as follows:

25   **Extension of time**

25.1   In clause 25 any reference to delay, notice or extension of time includes further delay, further notice or further extension of time.

   .1   .1   If and whenever it becomes reasonably apparent that the progress of the Works is being or is likely to be delayed the Contractor shall forthwith give written notice to the Architect of the material circumstances including the cause or causes of the delay and identify in such notice any event which in his opinion is a Relevant Event.

   .1   .2   Where the material circumstances of which written notice has been given under clause 25.2.1.1 include reference to a Nominated Sub-Contractor, the contractor shall forthwith send a copy of such written notice to the Nominated Sub-Contractor concerned.

25.2   .2   In respect of each and every Relevant Event identified in the notice given in accordance with clause 25.2.1.1 the Contractor shall, if practicable in such notice, or otherwise in writing as soon as possible after such notice:

   .2   .1   give particulars of the expected effects thereof; and

   .2   .2   estimate the extent, if any, of the expected delay in the completion of the Works beyond and Completion Date resulting therefrom whether or not concurrently with delay resulting from any other Relevant Event

and shall give such particulars and estimate to any Nominated Sub-Contractor to whom a copy of any written notice has been given under clause 25.2.1.2.

25.2   .3   The Contractor shall give such further written notices to the Architect, and send a copy to any Nominated Sub-Contractor to whom a copy of any written notice has been given under clause 25.2.1.2, as may be reasonably necessary or as the Architect may reasonably require for keeping up-to-date the particulars and estimate referred to in clauses 25.2.2.1 and 25.2.2.2 including any material change in such particulars or estimate.

25.3 .1 If, in the opinion of the Architect, upon receipt of any notice, particulars and estimate under clauses 25.2.1.1, 25.2.2 and 25.2.3,

    .1 .1 any of the events which are stated by the Contractor to be the cause of the delay is a Relevant Event and

    .1 .2 the completion of the Works is likely to be delayed thereby beyond the Completion Date

the Architect shall in writing to the Contractor give an extension of time by fixing such later date as the Completion Date as he then estimates to be fair and reasonable. The Architect shall, in fixing such new Completion Date, state:

    .1 .3 which of the Relevant Events he has taken into account and

    .1 .4 the extent, if any, to which he has had regard to any instructions issued under clause 13.2 which require as a Variation the omission of any work or obligation and/or under clause 13.3 in regard to the expenditure of a provisional sum for defined work or for Performance Specified Work which results in the omission of any such work,

and shall, if reasonably practicable having regard to the sufficiency of the aforesaid notice, particulars and estimate, fix such new Completion Date not later than 12 weeks from receipt of the notice and of reasonably sufficient particulars and estimate, or, where the period between receipt thereof and the Completion Date is less than 12 weeks, not later than the Completion Date.

If, in the opinion of the Architect, upon receipt of any such notice, particulars and estimate it is not fair and reasonable to fix a later date as a new Completion Date, the Architect shall if reasonably practicable having regard to the sufficiency of the aforesaid notice, particulars and estimate so notify the Contractor in writing not later than 12 weeks from receipt of the notice, particulars and estimate, or, where the period between receipt thereof and the Completion Date is less than 12 weeks, not later than the Completion Date.

25.3 .2 After the first exercise by the Architect of his duty under clause 25.3.1 or after revision to the Completion Date stated by the Architect in a confirmed acceptance of a 13A Quotation in respect of a Variation the Architect may in writing fix a Completion Date earlier than that previously fixed under clause 25 or than that stated by the Architect in a confirmed acceptance of a 13A Quotation if in his opinion the fixing of such earlier Completion Date is fair and reasonable having regard to any instructions issued after the last occasion on which the Architect fixed a new Completion Date

    — under clause 13.2 which require or sanction as a Variation the omission of any work or obligation; and/or

    — under clause 13.3 in regard to the expenditure of a provisional sum for defined work or for Performance Specified Work which result in the omission of any such work.

Provided that no decision under clause 25.3.2 shall alter the length of any adjustment to the time required by the Contractor for the completion of the Works in respect of a Variation for which a 13A Quotation has been given and which has been stated in a confirmed acceptance of a 13A Quotation or in respect of a Variation or work for which an adjustment to the time for completion of the Works has been accepted pursuant to clause 13.4.1.2 paragraph A7.

25.3   .3   After the Completion Date, if this occurs before the date of Practical Completion, the Architect may, and not later than the expiry of 12 weeks after the date of Practical Completion shall, in writing to the Contractor either

    .3   .1   fix a Completion Date later than that previously fixed if in his opinion the fixing of such later Completion Date is fair and reasonable having regard to any of the Relevant Events, whether upon reviewing a previous decision or otherwise and whether or not the Relevant Event has been specifically notified by the Contractor under clause 25.2.1.1; or

    .3   .2   fix a Completion Date earlier than that previously fixed under clause 25 or stated in a confirmed acceptance of a 13A Quotation if in his opinion the fixing of such earlier Completion Date is fair and reasonable having regard to any instructions issued after the last occasion on which the Architect fixed a new Completion Date

        — under clause 13.2 which require or sanction as a Variation the omission of any work or obligation; and/or

        — under clause 13.3 in regard to the expenditure of a provisional sum for defined work or for Performance Specified Work which result in the omission of any such work; or

    .3   .3   confirm to the Contractor the Completion Date previously fixed or stated in a confirmed acceptance of a 13A Quotation.

25.3   .4   Provided always that:

25.3   .4   .1   the Contractor shall use constantly his best endeavours to prevent delay in the progress of the Works, howsoever caused, and to prevent the completion of the Works being delayed or further delayed beyond the Completion Date;

    .4   .2   the Contractor shall do all that may reasonably be required to the satisfaction of the Architect to proceed with the Works.

25.3   .5   The Architect shall notify in writing to every Nominated Sub-Contractor each decision of the Architect under clause 25.3 fixing a Completion Date and each revised Completion Date stated in the confirmed acceptance of a 13A Quotation together with, where relevant, any revised period or periods for the completion of the work of each Nominated Sub-Contractor stated in such confirmed acceptance.

25.3   .6   No decision of the Architect under clause 25.3.2 or clause 25.3.3.2 shall fix a Completion Date earlier than the Date for Completion stated in the Appendix.

25.4   The following are the Relevant Events referred to in clause 25:

25.4   .1   force majeure;

25.4   .2   exceptionally adverse weather conditions;

25.4   .3   loss or damage occasioned by any one or more of the specified perils;

25.4   .4   civil commotion, local combination of workmen, strike or lock-out affecting any of the trades employed upon the Works or any of the trades engaged in the preparation, manufacture or transportation of any of the goods or materials required for the Works;

25.4   .5   compliance with the Architect's instructions

    .5   .1   under clauses 2.3, 2.4.1, 13.2 (except for a confirmed acceptance of a 13A Quotation), 13.3 (except compliance with an Architect's instruction for the expenditure of a provisional sum for defined work or of a provisional sum for Performance Specified Work), 13A.4.1, 23.2, 34, 35 or 36; or

    .5   .2   in regard to the opening up for inspection of any work covered up or the testing of any of the work, materials or goods in accordance with clause 8.3 (including making good in consequence of such opening up or testing) unless the inspection or test showed that the work, materials or goods were not in accordance with this Contract;

25.4   .6   .1   where an Information Release Schedule has been provided, failure of the Architect to comply with clause 5.4.1;

    .6   .2   failure of the Architect to comply with clause 5.4.2;

25.4   .7   delay on the part of Nominated Sub-Contractors or Nominated Suppliers which the Contractor has taken all practicable steps to avoid or reduce;

25.4   .8   .1   the execution of work not forming part of this Contract by the Employer himself or by persons employed or otherwise engaged by the Employer as referred to in clause 29 or the failure to execute such work;

    .8   .2   the supply by the Employer of materials and goods which the Employer has agreed to provide for the Works or the failure so to supply;

25.4   .9   the exercise after the Base Date by the United Kingdom Government of any statutory power which directly affects the execution of the Works by restricting the availability or use of labour which is essential to the proper carrying out of the Works or preventing the Contractor from, or delaying the Contractor in, securing such goods or materials or such fuel or energy as are essential to the proper carrying out of the Works;

25.4   .10 .1   the Contractor's inability for reasons beyond his control and which he could not reasonably have foreseen at the Base Date to secure such labour as is essential to the proper carrying out of the Works; or

    .10 .2   the Contractor's inability for reasons beyond his control and which he could not reasonably have foreseen at the Base Date to secure such goods or materials as are essential to the proper carrying out of the Works;

25.4 .11   the carrying out by a local authority or statutory undertaker of work in pursuance of its statutory obligations in relation to the Works, or the failure to carry out such work;

25.4 .12   failure of the Employer to give in due time ingress to or egress from the site of the Works or any part thereof through or over any land, buildings, way or passage adjoining or connected with the site and in the possession and control of the Employer, in accordance with the Contract Bills and/or the Contract Drawings, after receipt by the Architect of such notice, if any, as the Contractor is required to give, or failure of the Employer to give such ingress or egress as otherwise agreed between the Architect and the Contractor;

25.4 .13   where clause 23.1.2 is stated in the Appendix to apply, the deferment by the Employer of giving possession of the site under clause 23.1.2;

25.4 .14   by reason of the execution of work for which an Approximate Quantity is included in the Contract Bills which is not a reasonably accurate forecast of the quantity of work required;

25.4 .15   delay which the Contractor has taken all practicable steps to avoid or reduce consequent upon a change in the Statutory Requirements after the Base Date which necessitates some alteration or modification to any Performance Specified Work;

25.4 .16   the use or threat of terrorism and/or the activity of the relevant authorities in dealing with such use or threat;

25.4 .17   compliance or non-compliance by the Employer with clause 6A.1;

25.4 .18   delay arising from a suspension by the Contractor of the performance of his obligations under the Contract to the Employer pursuant to clause 30.1.4;

25.4 .19   save as provided for in clauses 25.4.1 to 25.4.18 any impediment, prevention or default, whether by act or omission, by the Employer or any person for whom the Employer is responsible except to the extent that it was caused or contributed to by any default, whether by act or omission, of the Contractor or his servants, agents or sub-contractors.

10.1.2 Commentary

'If and whenever it becomes reasonably apparent' – This is a key phrase, and it should be noted that the contractor is to give notice not only when the progress of the works is being delayed, but also when it becomes reasonably apparent that it is likely to be delayed in the future. It has to be reasonably apparent that the progress of the works is being or is likely to be delayed. It is the actual progress and not the contractor's planned-for or hoped-for progress which is relevant, as the wording makes clear. Apparent means 'manifest' and once it becomes reasonably apparent that the

progress is actually being delayed or is likely to be delayed, the contractor must notify the architect in writing.

'*The Contractor shall forthwith give written notice*' – As soon as it is reasonably apparent to the contractor that the progress of the works is or is likely to be delayed the contractor must give written notice to the architect. The notice must specify the cause of delay. Although often overlooked, it is important for the contractor to identify the precise activity (or activities) which is (are) delayed together with its (their) relation to the project's critical path. This notice was held not to be a condition precedent to giving an extension of time by the architect under JCT 63[436] and the decision probably applies to JCT 98 also. In any event, under JCT 98 revised wording, the architect has power to give an extension in the absence of such written notice; failure by the contractor to give written notice merely means that the architect does not need to make a decision on extensions until a later date, i.e. on his review of the completion date not later than the expiry of 12 weeks from the date of practical completion 'whether or not the relevant event has been specifically notified by the Contractor': clause 25.3.3. It is less clear whether he is entitled to give an extension of time before practical completion in the absence of written notice. On balance, that he cannot do so appears to be the better view. There seems to be no obstacle to making such a notice a condition precedent if that is what the parties agree in clear words. In *City Inn Ltd* v. *Shepherd Construction Ltd*[437], the court considered JCT 80 with amendments. One such amendment stated:

> '13.8.5   If the Contractor fails to comply with one or more of the provisions of Clause 13.8.1, where the Architect has not dispensed with such compliance under Clause 13.8.4, the Contractor shall not be entitled to any extension of time under Clause 25.3.'

Clause 13.8.1 required the contractor to submit within a specified timescale written details after receipt of an architect's instruction. It was argued by the defendants that clause 13.8.5 constituted a penalty. This was rejected by the court:

> 'It was perfectly legitimate for the employer to require and the contractor to accept that, in relation to architect's instructions, the employer should be forewarned of anticipated consequential delay, and for it to be agreed that, in the event of the contractor failing to provide such forewarning in accordance with clause 13.8.1, the risk of loss through delay should shift from the shoulders of the employer to those of the contractor. Such provision did not constitute a penalty.'[438]

A factor taken into account by the court was that the amount specified as liquidated damages was genuinely pre-estimated and could not be considered 'extravagant, penal or oppressive'.

---

[436] *London Borough of Merton* v. *Stanley Hugh Leach Ltd* (1985) 32 BLR 51.
[437] 17 July 2001, unreported; the decision upheld on appeal: [2003] BLR 468.
[438] 17 July 2001, unreported, at paragraph 8 per Lord MacFadyen.

'... *of the material circumstances including the cause or causes of the delay'* −
The contractor's notice is to state not just the 'cause or causes' of the delay; it
must also state 'the material circumstances'. The notice should go into some
detail as to why and how the delay is occurring or is likely to occur. The
'material circumstances' will include, for example, the stage the contract has
reached, the proposed order of works, and so on. The fact that the delay is
likely to make a summer contract into a winter one, with consequent likely
further disruption, would also be a 'material circumstance'. The duty is not
limited to notifying the causes of delay listed as relevant events; it is a duty
to give notice of delay, howsoever caused. This is so even if it is uncertain
whether the current completion date will be affected. The phrase 'cause of
the delay' is probably best interpreted in practice as meaning 'circumstances
giving rise to the delay'. Anecdotal evidence suggests that it is rare, virtually
unknown, for a contractor to give notice of delays unless he believes that a
relevant event giving rise to an extension of time is involved. The purpose of
the notice is simply to warn the architect of the situation, and it is then up to
him to monitor it[439]. He may, if necessary, take remedial action, possibly
eradicating the delay completely, and forewarn the employer.

In other words, the contractor must give prior notice of delays which it is
reasonable for him to anticipate. This gives the architect the opportunity,
once he has been notified of any impending delay, to take steps to rectify the
situation and bring the contract back on schedule. One of the things the
architect can do so as to reduce the time needed to complete the project is to
omit work under clause 13.2 if it is practicable to do so and if, of course, the
employer agrees. Further, if the contractor fails to give notice of a delay
which he clearly should have been able to anticipate, the architect can in fact
say that the contractor has not used his best endeavours to prevent delay in
progress, which he is bound to do by the terms of clause 25.3.4.1. If the
contractor fails to give notice (even of his own delays), he is in breach of
contract and the architect is entitled to take such breach into account when
giving a future extension of time.

A simple example will make clear how the principle works[440]. Take the
case of an architect who has been given a specific written request for information by the contractor at the right time, neither too early nor too late. The
architect has overlooked the request. A week before the information is
required, it is reasonably apparent to the contractor that it is not likely to
arrive, but he does nothing. On the day it is required, it has still not arrived,
but it is not until a week later and the job is seriously delayed that the
contractor sends the architect a notice of delay. By the time the architect has
diverted staff onto the task and the information is produced, a further
2 weeks have passed making a total of 3 weeks delay. The contractor is
clearly in breach, because he failed to notify the architect that the work was
likely to be delayed a week before the delay occurred. In fixing a new
completion date, the architect is entitled to take the contractor's breach

---

[439] *London Borough of Merton* v. *Stanley Hugh Leach Ltd* (1985) 32 BLR 51.
[440] *London Borough of Merton* v. *Stanley Hugh Leach Ltd* (1985) 32 BLR 51.

into account by asking himself what would have been the length of delay had he received notification at the correct time. It would still have taken 2 weeks to produce the information, but the preparation would have been able to start a week before it was required. The contractor would have been delayed one week and that is all the architect need consider in fixing a new completion date. Events are rarely quite as straightforward as that and, in practice, there may be numerous other factors to be considered.

'... *identify in such notice any event which in his opinion is a Relevant Event'* – This phrase acknowledges that the notice of delay may contain causes of delay which are not relevant events. The notice must state those listed causes of delay (or cause of delay) which, in the contractor's opinion, entitle him to an extension of time. The 'Relevant Event' so identified must be one (or more) of those listed in clause 25.4 and it is for the contractor to specify those causes of delay.

'... *notice to the nominated sub-contractor concerned'* – Clause 25.2.1.2 introduces a further and important requirement, and that is that a copy of the contractor's original notice must be sent to any nominated sub-contractor to whom reference is made in it. One of the 'Relevant Events' listed is 'delay on the part of nominated sub-contractors or nominated suppliers which the contractor has taken all practicable steps to avoid or reduce': see clause 25.4.7. 'Delay on the part of' a nominated sub-contractor has a very limited meaning as will be seen later. Delay by the employer in making a timely nomination of a new sub-contractor to replace a failed nominated sub-contractor is within clause 25.4.5 and the main contractor is entitled to an extension of time to cover such delay[441]. The purpose of giving a copy of the notice to affected nominated sub-contractors is to forewarn the nominated sub-contractor so that he may in turn, if necessary, make application for extension of time to the main contractor under clause 2.2 of the Nominated Sub-Contract Form NSC/C.

'*The Contractor shall ... give particulars of the expected effects'* – Clause 25.2.2.1 imposes an additional obligation on the contractor. Either in his original notice or, where that is not practicable, as soon as possible after the notice, the contractor must state in writing to the architect *particulars of the expected effects* of each and every relevant event identified in the notice, i.e. particulars of the expected effects on progress, and each and every relevant event must, for this purpose, be considered in isolation. No doubt, the contractor must go into sufficient detail to enable the architect to make a judgment.

'... *estimate the extent'* – This is a provision of vital significance. The contractor must give his own estimate of the expected delay in completion of the works beyond the completion date 'whether or not concurrently with delay resulting from any other Relevant Event'. This is a particularly onerous task. The contractor must address each delay separately and its effects even if two or more delays are acting together. Contractors will commonly, indeed invariably, promote their views of the delay, but rarely split so as to

---

[441] *Percy Bilton Ltd* v. *Greater London Council* (1982) 20 BLR 1.

deal with each delay as a separate item. The particulars and estimate must be 'reasonably sufficient' to enable the architect to form a judgment: see clause 25.3.1.

'*... give such particulars and estimate to any Nominated Sub-Contractor'* – A copy of the contractor's particulars and estimate must be given to any nominated sub-contractor to whom a copy of the original notice was given under clause 25.2.1.2.

'*... shall give such further written notices'* – The contractor must, by the terms of clause 25.2.3, keep each notice of delay under review and revise his statement of particulars and estimate and/or give whatever further notices 'may be reasonably necessary or as the architect may reasonably require'. Notice that the duty extends to 'any material change' in the particulars, etc. The contractor must keep the architect up to date with developments as they occur and the contractor's duty is not dependent upon the architect's request. The architect's time limit for dealing with applications for delay only starts to run when he has received 'reasonably sufficient particulars and estimates' from the contractor: see clause 25.3.1. The clear intention of these provisions is to provide the architect with sufficient information reasonably to form his own judgment on the matter. He may not have been on the site at the time of the delay, though he must use whatever records he has as well[442].

'*... copy to any Nominated Sub-Contractor'* – This is self-explanatory: affected nominated sub-contractors must also be kept informed.

'*... If, in the opinion of the Architect'* – It is entirely for the architect to decide whether, in his opinion, a delay in the contract completion date is likely to occur or has occurred and also whether the cause of delay is one of those listed in clause 25.4 and therefore one for which he should grant an extension.

Once the architect is notified by the contractor of delay, it is for him to monitor the position. The position has been summarised neatly:

'Clause 23 imposes on the architect the duty of considering whether completion of the works is likely to be or has been delayed beyond the date for completion by way of the causes there set out and if it has whether any and if so what extension should be granted. That duty is owed both to the contractor and the building owner. The architect is entitled to rely on the contractor to play his part by giving notice when it has become apparent to him that the progress of the works is delayed. If the contractor fails to give notice forthwith upon it becoming so apparent he is in breach of contract and that breach can be taken into account by the architect in deciding whether he should be given an extension of time. But the architect is not relieved of his duty by the failure of the contractor to give notice or to give notice promptly. He must consider independently in the light of his knowledge of the contractor's programme and the progress of the works and of his knowledge of other matters affecting or

---

[442] *London Borough of Merton v. Stanley Hugh Leach Ltd* (1985) 32 BLR 51.

likely to affect the progress of the works ... whether completion is likely to be delayed by any of the stated causes. If necessary he must make his own inquiries, whether from the contractor or others.'[443]

The extension of time clause under the form of contract there being considered was clause 23 of JCT 63 (clause 25 under JCT 98). If the contractor feels that the architect has been unreasonable in reaching his opinion, his recourse is to adjudication or arbitration. On receipt of the contractor's written notice the architect must decide if the cause of delay is covered by clause 25. If in his view it is not then, subject to the contractor's right to challenge that opinion by adjudication or arbitration, that is the end of the matter.

Of course, the architect must not arrive at his decision on a whim. He should carefully analyse the position and consider the effect of individual delays[444]. However, having reached his decision, that decision has a considerable status under the contract as indicated in the judgment in *Balfour Beatty* v. *London Borough of Lambeth*:

> 'Lambeth was in my view entitled to criticise BB's case without putting forward an alternative. Since BB had not justified its case Lambeth was not obliged to justify the architect's extensions of time or certificates of non-completion. It was entitled to rely on them as they were apparently valid decisions by the architect and the parties by adopting the JCT conditions have agreed to be bound by them (subject to review by an Adjudicator or arbitrator). BB had to persuade the Adjudicator that the architect's decisions were wrong. Lambeth were not obliged to prove that they were right (although it is often prudent to do so).'[445]

'upon receipt of any notice, particulars and estimate' – It is when, and only when he has received the notice, particulars and estimate from the contractor that the architect must consider them before practical completion. The initiative passes to the architect who must then decide: (1) Whether any of the causes of delay specified by the contractor in the notice is in fact a relevant event, i.e. he may disagree that the particular cause specified by the contractor is a relevant event, in which case the architect need not consider the next point. The contractor may, for example, have specified 'exceptionally adverse weather conditions' and gone on to give the necessary detail, and yet the architect may say that the details show 'adverse weather conditions', not *'exceptionally* adverse weather conditions'. (2) Whether completion of the works is in fact likely to be delayed *thereby* (i.e. by the specified relevant event) beyond the completion date. Then, and only then, does his duty to give an extension of time arise.

   '... *the completion of the Works is likely to be delayed* – The architect must then decide whether or not the delay is going to mean a likely failure to complete by the date for completion. In making up his mind on this point

---

[443] *London Borough of Merton v. Stanley Hugh Leach Ltd* (1985) 32 BLR 51 at 93 per Vinelott J.
[444] *John Barker Construction Ltd* v. *London Portman Hotels Ltd* (1996) 12 Const LJ 277.
[445] [2002] BLR 288 at 303 per Judge Lloyd.

the architect is entitled to consider the proviso to the clause, which begins 'provided always that the contractor shall constantly use his best endeavours to prevent delay'. The contractor's duty is to prevent delay, so far as he can reasonably do so, e.g. a delay in progress of the works at an early stage may be reduced or even eliminated by the contractor using his best endeavours.

A common situation is where the contractor is in advance of this programme, and a delaying event occurs.

The correct analysis appears to be as follows:

> 'Provided the contractor has given written notice of the cause of delay, the obligation to make an extension appears to rest on the architect without the necessity of any formal request for it by the contractor. Yet he is required to do this only if the completion of the works "is likely to be or has been delayed beyond the Date for Completion", or any extended time for completion previously fixed. If a contractor is well ahead with his works and is then delayed by a strike, the architect may nevertheless reach the conclusion that completion of the works is not likely to be delayed beyond the date of completion. Under condition 21(1), the contractor is under a double obligation: on being given possession of the site, he must "thereupon begin the works and regularly and diligently proceed with the same", and he must also complete the works "on or before the Date for Completion", subject to any extension of time. If a strike occurs when two-thirds of the work has been completed in half the contract time, I do not think that on resuming work a few weeks later the contractor is then entitled to slow down the work so as to last out the time until the date for completion (or beyond, if an extension of time is granted) if thereby he is failing to proceed with the work "regularly and diligently".'[446]

These observations were purely *obiter* but the analysis appears to be impeccable, and the consequence is that the architect is entitled to take into account the fact that the contractor is in advance of programme when considering what extension to grant; and he may also make use of the contractor's 'float' element in the contract programme[447]. Where there are overlapping causes of delay, the architect must consider each cause separately, so that if there is one ground justifying an extension and another not, the architect cannot deprive the contractor of any reasonable extension for the relevant event merely because there is an overlapping cause; but the cumulative effect on progress must be taken into account: it is delay to progress which is the important factor[448].

It is difficult to overemphasise the importance of the completion date. A delay which does not affect the completion date does not entitle the contractor to an extension of time. Delay to an activity which may or may

---

[446] *London Borough of Hounslow* v. *Twickenham Garden Developments Ltd* (1970) 7 BLR 81 at 113 per Megarry J.
[447] *How Engineering Services Ltd* v. *Lindner Ceilings Partitions PLC* 17 May 1995 unreported; *Ascon Contracting Ltd* v. *Alfred McAlpine Construction of the Isle of Man Ltd* (2000) 16 Const LJ 316 at 338.
[448] See Chapter 2, section 2.4 'Concurrency' for a detailed consideration of this point.

not have an effect on the completion date must be differentiated from delay to the completion date:

> '... had [the architect] directed its mind, when considering the question whether to grant a second extension of time on the Hydrotite Ground, to the issue whether the progress of the Works, as opposed to the activity "Wall and Floor Finishes", had been further delayed since the grant of the first extension of time on the Hydrotite Ground, it could only have concluded that it had not. It is thus, in my judgment, clear that in relation to the second grant of an extension of time on the Hydrotite Ground [the architect] negligently failed to direct its mind to the correct issue, and, if it had directed its mind to the correct issue, it could only have concluded that no further extension of time was appropriate[449].'

'*... shall in writing ... give an extension of time*' – The architect must give an extension of time to the contractor if his conclusions are positive, i.e. if in his opinion there is a relevant event and that relevant event is likely to delay the completion of the works.

Having received from the contractor all the requisite information, the architect's duty is to consider the information provided and to make his own assessment of the situation as to whether or not an extension should be granted at that time and, if so, what extension should be granted.

The problem has been put squarely:

> 'Perhaps the greatest difficulty which may be encountered will be in deciding whether or not the notice, particulars and estimates, which the contractor is required to provide, are sufficient for the architect to make his decision on extending the completion date. This requires close co-operation between contractor and architect and architects ought to be decisive in the matter and not use alleged insufficiency of particulars and estimates as an excuse for delaying the issue of extensions of time'[450].

In considering this question, the architect will have to take into account any overlapping of delays resulting from different relevant events and:

> 'where optional clause 5.3.1.2 is not deleted, the architect will be assisted by the contractual obligation on the contractor to provide and keep up to date a master programme for the execution of the works'[451].

In fact, there is no contractual obligation to provide a master programme. It is common and sensible practice for the architect to include such a requirement, and specify the type of programme required, in the bills or specification, in which case the requirement will become a contractual

---

[449] *The Royal Brompton Hospital National Health Service Trust* v. *Frederick Alexander Hammond and Others (No 7)* (2001) 76 Con LR 148 at 214 per Judge Seymour.
[450] The Aqua Group, *Contract Administration for the Building Team*, 8th edition, 1996, Blackwell Science, p.114.
[451] The Aqua Group, *Contract Administration for the Building Team*, 8th edition, 1996, Blackwell Science, p.114.

provision[452]. It would seem reasonable that, as a minimum, the architect should require a programme in bar chart *and* in network form with key dates and resources clearly shown.

Under clause 25 the architect is required to grant the extension of time by fixing as a new completion date for the works *'such later date ... as he then estimates to be fair and reasonable'*. It should be noted that the architect is only expected to estimate the length of extension and not to ascertain. Ascertainment would be impossible.

'The Architect shall, in fixing such new Completion Date, state' – The architect must inform the contractor in writing of the new completion date, and he must state two things, first, which of the relevant events he has taken into account, and second, the extent, if any, to which he has had regard to any omission instruction issued. If the architect has not issued any omission instruction, the only information he must give to the contractor is the new completion date and the relevant events taken into account. Many architects actually state the period of time attributable to each relevant event, but there is no obligation to do so and it is generally not in the employer's interests. The contractor will, of course, demand these details, because without them it is very difficult to challenge the architect's decision unless it is grossly wrong. The contractor will often mistakenly believe that an extension of time is a prerequisite to claiming loss and/or expense. The architect would be obliged to reveal his calculations during an adjudication or arbitration if he wanted to defend his decision, but by that time the contractor must have already made his decision to challenge, based on his own opinion and perhaps that of his expert. The relevant events which the architect has taken into account should be, of course, all the relevant events notified by the contractor even if the architect has discounted some of them. A clause which appears to give some credence to the argument that the period of time should be stated is clause 26.3. However, that has been judicially questioned[453].

It appears to be clear that the architect may take account of omissions when he decides on the first extension application that he grants because the definition of completion date in clause 1.3 is 'the Date for Completion as fixed and stated in the Appendix or any later date fixed' under the relevant provisions, and date for completion includes the original date. It follows that there is always a 'previous completion date' for the purposes of clause 25.3 and, indeed, it would be absurd if any other interpretation were to be placed upon this provision. The only proviso is that the architect cannot thereby fix a completion date earlier than the date for completion stated originally in the Appendix – something which is forbidden by clause 25.3.6. If he takes account of omissions, the architect must inform the contractor in writing when fixing the new completion date.

' ... shall, if reasonably practicable ... not later than 12 weeks' – The architect is given a time limit of 12 weeks from receipt of the contractor's notice of

---

[452] *Glenlion Construction Ltd* v. *The Guinness Trust* (1987) 39 BLR 89.
[453] *Methodist Homes Housing Association Ltd* v. *Messrs Scott & McIntosh* 2 May 1997, unreported; considered in Chapter 12, section 12.2.6.

delay and of 'reasonably sufficient particulars and estimate' from the contractor in which to reach a decision and, if he considers it appropriate, to give an extension of time. This is qualified by 'reasonably practicable'. The correct operation of these provisions really depends upon both architect and contractor being of one mind as to whether the information supplied by the contractor is 'reasonably sufficient' to enable the architect to form an opinion. From the employer's point of view it is important that the architect should decide in due time because of the fluctuations provisions: see clauses 38.4.8.2; 39.5.8.2; and 40.7.2.2, the effect of which is that, unless the architect carries out his duties timeously, the right of the employer to freeze the contractor's fluctuations on the due date for completion is lost.

However, if there are fewer than 12 weeks left between receipt of the contractor's notice, particulars and estimate and the currently fixed completion date, the architect must reach his decision and grant any extension no later than that date. The intention clearly is that the contractor should always have before him a date towards which he can work. If there is a very short period left before the completion date, it may not be 'reasonably practicable' for the architect to come to a decision in time. It appears that the architect has no power under clause 25.3.1 to make the decision after the completion date and the decision will have to be made as part of the review under clause 25.3.3. The contractor may be disappointed, but he is the author of his own misfortune.

If it seems that the architect would be able to make a decision if a further week or so was available, he may decide to make the best decision practicable (which may well be conservative) before the completion date and then use the additional period thus created to come to a more considered decision. Thus, an architect faced with making a decision just one week before completion date may decide he is able to give 2 weeks extension of time and the extra 2 weeks may enable him, on mature reflection, to give a further 1 week. However, the architect is not obliged to act in this way.

Some architects have adopted the practice of amending clause 25 so as to do away with the time limits. This is an unwise practice, if only because of the fluctuations provisions. Fluctuations are only to be frozen at completion date 'if the printed text of clause 25 is unamended and forms part of the Conditions': see clause 38.4.7; 39.5.7; and 40.7.1.1. In the absence of a fixed period, an adjudicator or arbitrator, if called upon, may decide that the architect ought to come to a decision in less than 12 weeks.

'. . . *not fair and reasonable to fix a later date as a new Completion Date*' – The final paragraph of clause 25.3.1 is important and its effect is that in respect of each application for an extension of time under clause 25.2 after the provision of any further particulars and estimates required, the architect must notify the contractor in writing where his decision is not to fix a later completion date as a new completion date. It is important, because the architect's decisions are required before the provisions restricting the level of fluctuations or formula adjustment (clauses 38.4.7, 39.5.7 and 40.7) can be operated if the contractor is in default over completion, a point which should not escape those using JCT 98. The architect is allotted the same time period in which to make his decision whether it is positive or negative. If it is not reasonably practicable for him to make a decision in the time available, even if that decision is ultimately going to be to decline the extension of time, it appears that such decision will be postponed until clause 25.3.3 applies.

'After the first exercise' – Clause 25.3.2 is much misunderstood. After the first extension of time that he gives or after a revision to the completion date stated in a confirmed acceptance of a clause 13A quotation or the contract-or's price statement under clause 13.4[454], the architect can use his powers under the clause. He cannot, in any case, fix any earlier date than the original completion date: clause 25.3.6. But if he has issued instructions which result in the omission of work or obligations under clause 13.2 or under clause 13.3 in regard to provisional sums for defined work or for performance specified work, he is entitled to take this into account and fix a completion date earlier than that previously fixed under clause 25 if in his opinion the fixing of such earlier completion date is fair and reasonable having regard to those instructions. Some commentators take the view that this is in conflict with clause 25.3.1.4. already discussed. The architect can, under clause 25.3.1.4, reduce extensions previously granted and even extinguish them completely so as to return to the original date for completion, but he can fix no earlier date than that, no matter how much work or obligations he omits. Under clause 25.3.2 he can only reduce extensions of time on account of omissions of work etc. instructed since he last granted an extension. Each extension is deemed to take into account all omissions instructed up to the date of the extension.

Clause 25.3.2 used to restrict the architect's right to fix an earlier completion date to those situations where he had issued variation instructions omitting work, so that variation instructions requiring the omission of obligations or restrictions (see clause 13.1.2) and provisional sum instructions under clause 13.3 requiring omissions of work, obligations or restrictions gave no grounds for the operation of clause 25.3.2. This lacuna seems to have been unintentional and the original wording has been replaced by 'having regard to any instructions'.

If architects wish to take advantage of this power to reduce extensions previously granted on account of omissions of work or obligations, it is suggested that they should take the decision and notify the contractor at the earliest possible moment – preferably when issuing the instruction – and not leave it until they next give an extension of time. The reason for this is that experience suggests that architects have tended to be conservative in giving extensions of time, knowing that they can, at the end of the day, grant a little more time: see below. To err too much in the direction of parsimony – and to be unrealistic in considering the effect of omissions – is not good contract practice. It should be noted that the architect's powers under clause 25.3.2 are not dependent upon the contractor's notice of delay.

---

[454] See Chapter 13, section 13.5.2.

'... *not later than then expiry of 12 weeks after the date of Practical Completion'*
– Clause 25.3.3 sets out the extension of time regime after the completion
date which is quite separate from what has gone before. It gives the architect
the opportunity to make a final decision on extensions of time. Some
commentators believe that in *Temloc Ltd* v. *Errill Properties Ltd* the Court of
Appeal held that the requirement to do so within 12 weeks is not mandatory. This is a wrong view of the judgment[455].

Clause 25.3.3 requires the architect to review the completion date in any
event; and he must do this in light of any relevant events whether or not
specifically notified to him by the contractor. The opening sentence makes it
clear that the architect must take account of any relevant events which have
occurred since the commencement of the contract. It is intended to be the
architect's final opportunity to consider extensions of time. It is at least
arguable, on a strict reading of clause 25.3.3, that the architect can exercise
this power only once. Therefore, if he chooses to do so after the completion
date but before practical completion, it may be that he cannot do it again
afterwards. In practice, an architect will usually wait until after practical
completion to act under this clause.

'... *shall in writing to the Contractor'* – The wording indicates that the
architect has no discretion; he must write to the contractor and do one of
three things:

(1) *Fix a completion date later than that previously fixed* – He must do this if in
    his opinion to do so is 'fair and reasonable having regard to any of the
    Relevant Events', i.e. those listed in clause 25.4, '*whether upon reviewing a
    previous decision or otherwise'* and 'whether or not the Relevant Event has
    been specifically notified by the Contractor'.
(2) *Fix a completion date earlier than that previously fixed* – He must do this if
    'in his opinion ... [it] is fair and reasonable having regard to' any
    omission instructions issued since he last granted an extension of time.
(3) *Confirm to the contractor the completion date previously fixed.*

In practice, the architect should write to the contractor soon after practical
completion, reminding him of the 12 weeks period and that the architect has
no power to make any extension of time after the expiry of the period. The
contractor should be given a date by which any final submissions should be
made; this is not the time for the submission of large numbers of weighty
lever arch files.

'... *the contractor shall use constantly his best endeavours to prevent delay* –
This seems to be no more than an express restatement of the contractor's
common law obligation and reinforcement of the express provisions of the
contract relating to completion and the contractor's obligation to proceed
regularly and diligently. This is a matter which the architect must take into
account when deciding upon extension of time under clause 25. The contractor *shall do all that may reasonably be required to the satisfaction of the
architect to proceed with the works.* This is the contractor's obligation in any

---

[455] See the discussion in Chapter 2, section 2.2 of this Chapter.

case, but the architect has no power to order that acceleration measures be taken (either under this provision or any other provision in the contract). If 'best endeavours' obliged the contractor to accelerate, there would be little need for an extension of time clause. Indeed, it is doubtful whether a contractor has any obligation at all to increase resources on a project over and above the level necessary to complete the work for which the contractor originally tendered. There would be no necessity for a relevant event dealing with architect's instructions requiring additional work if the contractor was obliged to increase his labour to carry out the additional work. The best legal view of the effect of the proviso is the following:

> 'This proviso is an important qualification to the right to an extension of time. Thus, for example, in some cases it might be the contractor's duty to re-programme the works either to prevent or reduce delay. How far the contractor must take other steps depends upon the circumstances of each case, but it is thought that the proviso does not contemplate the expenditure of substantial sums of money'[456].

However, so far as the first part of the proviso is concerned – that 'the contractor shall use constantly his best endeavours to prevent delay' – it has been said that when a contractor undertakes to use his best endeavours, he undertakes to do anything within his power to prevent delay to the progress of the works irrespective of extra cost.

There are supporters of this point of view, pointing to the use of the phrase obligating the contractor to 'do all that may reasonably be required' in the second part of the proviso, which is in contrast with the 'best endeavours' obligation[457]. There appears to be no relevant construction industry case, but in other contexts using best endeavours has been held to mean doing everything prudent and reasonable to achieve an objective[458]. The Court of Appeal held, in connection with the obtaining of planning permission, that 'best endeavours' obliged a person to take 'all those reasonable steps which a prudent and determined man, acting in his own best interests and desiring to achieve that result would take[459]. Clearly, it is a lesser obligation than to 'ensure' or to 'secure', which impart an absolute liability to perform the duty set out[460].

The addition of the word 'constantly' clearly increases the contractor's obligation to the extent that he must never cease to use his best endeavours. It may be said with some force that the obligations in clause 25.3.4 amount to a condition precedent and that the contractor's failure disqualifies him from any extension of time for the particular relevant event and in the circumstances concerned.

---

[456] Stephen Furst and Vivian Ramsey, *Keating on Building Contracts*, 7th edition, 2001, Sweet & Maxwell, p. 730.
[457] *Transfield Pty* v. *Arlo International* (1980) 30 ALR 201.
[458] *Victor Stanley Hawkins* v. *Pender Bros Pty Ltd* (1994) 10 BCL 111.
[459] *IBM (UK) Ltd* v. *Rockware Glass Ltd* [1980] FSR 335
[460] *John Mowlem & Co* v. *Eagle Star Insurance Co Ltd* (1995) 62 BLR 126.

'... *shall notify in writing every Nominated Sub-Contractor'* – Every nominated sub-contractor must be notified by the architect of each decision of his fixing a completion date under clause 25.3 and also after the acceptance of a clause 13A quotation which includes a revised date. It is not merely those nominated sub-contractors referred to in the contractor's notice of delay. Each time he fixes a new completion date under the main contract the architect must notify all nominated sub-contractors of the new date so fixed. This is obviously highly relevant information for sub-contractors as it will affect their own obligations under their nominated sub-contracts, or could potentially do so. This is so even if a nominated sub-contractor has finished his work and left site – even after he may have been paid for his work in full.

'*No decision of the architect'* – Clause 25.3.6 makes plain that no decision of the architect under clause 25.3 can fix a completion date earlier than that stated in the Appendix. That date is sacrosanct, no matter how much work is omitted.

10.1.3 Grounds for extension of time

JCT 98, clause 25.4 lists the grounds (relevant events) on which the architect is entitled to revise the date of completion by the contractor. The corresponding provision in IFC 98 is clause 2.4, which uses the term 'event' to describe such a ground. Similar provisions are to be found in WCD 98 (clause 25.4) and PCC 98 (clause 2.6). They divide into two groups:

(1) *Those which are the fault of neither party:* JCT 2003 clauses 25.4.1, 25.4.2, 25.4.3, 25.4.4, 25.4.9, 25.4.10, 25.4.11, 25.4.15, 25.4.16.
(2) *Those which are the responsibility of the employer or his architect or those for whom the employer is responsible in law:* JCT 2003, clauses 25.4.5, 25.4.6, 25.4.7, 25.4.8, 25.4.12, 25.4.13, 25.4.14, 25.4.17, 25.4.18, 25.4.19.

The grounds on which an extension of time may be given are as follows.

Force majeure: clause 25.4.1

Force majeure is a French law term which is wider in its meaning than the common law term 'Act of God', which is an overwhelming superhuman event[461]. In fact, under JCT contracts the term *force majeure* has a restricted meaning because many matters such as war, strikes, fire and exceptional weather are expressly dealt with later in the contract. There appear to be no reported cases dealing with the matter in the context of JCT contracts, and the English authority usually quoted is *Lebeaupin* v. *Crispin*, where McCardie J accepted that:

> 'This term is used with reference to all circumstances independent of the will of man, and which it is not in his power to control ... Thus war,

---

[461] *Oakley* v. *Portsmouth & Ryde Steam Packet Co* (1856) T1 Exchequer Reports 6 1F.

inundations and epidemics are cases of *force majeure;* it has even been decided that a strike of workmen constitutes a case of *force majeure* ... [But] a *force majeure* clause should be construed in each case with a close attention to the words which precede or follow it and with due regard to the nature and general terms of the contract. The effect of the clause may vary with each instrument.'[462]

In fact, it seems that the term *force majeure* as used in JCT contracts is of limited effect, and decisions on the meaning of the term when used in other forms of contract are of little assistance. The dislocation of business caused by the general coal strike of 1912 has been held to be covered by the term and also covered the breakdown of machinery, but not delay caused by bad weather, football matches or a funeral.

'These are the usual incidents interrupting work and the defendants, in making their contract, no doubt took them into account.'[463]

The best that can be said is that the event relied upon as *force majeure* must make the performance of the contract wholly impossible and, in this sense, the term is somewhat akin to the English law doctrine of frustration of contract.

Exceptionally adverse weather conditions: clause 25.4.2

This phrase is a frequent source of argument between architects and contract-ors and the change in wording in the 1980 form from 'inclement' to 'adverse' was intended to make it clear that the ground was intended to cover any kind of adverse conditions including unusual heat or drought. The emphasis is on the word *exceptionally* and the meaning of the phrase is to be found by considering two factors: first, the kind of weather that may be expected at the site at the particular time when the delay occurs; second, the stage which the works have reached. In regard to the first factor, reference to local weather records may be helpful in showing that the adverse weather was 'excep-tional' for that area, i.e. exceeding what may on the evidence of past years be reasonably expected. The dictionary meaning of adverse is 'contrary' or 'hostile', while 'exceptionally' means 'unusual'. In regard to the second factor, even if the weather is exceptionally adverse for the time of year it must be such that it interferes with the works at the particular stage they have reached when the exceptionally adverse weather occurs. It matters not that the works have been affected only because they have been delayed through the contractor's own fault[464]. If despite the weather, work could continue then the works have not been delayed by the exceptionally adverse weather. The contractor is expected to programme the works making due allowance for normal adverse weather, i.e. the sort of weather which is to be expected in

---

[462] [1920] 2 KB 714.
[463] *Matsoukis* v. *Priestman & Co Ltd* [1915] 1 KB 681.
[464] *Walter Lawrence* v. *Commercial Union Properties* (1984) 4 Con LR 37.

the area and at the time of year during the course of the works. His programme for those parts of the work which may be affected by rain, wind or frost should acknowledge the fact that interruption is likely to occur, and should allow for them. This view is supported by the following example:

> 'If it were known at the time that a contract was let that the work was to be carried out during the winter months, and if that work was delayed by a fortnight of snow and frost during January, such a delay could not be regarded as due to exceptionally inclement weather. If, however, such work was held up by a continuous period of snow or frost, from early January until the end of March, in this case an extension of time would probably be justified ... '465

On a strict reading of the relevant event, it is only the 'exceptional' aspect of the adverse weather which will attract an extension of time. Thus, if 10 days of snow in January is just on the borderline between usual and exceptional and a project suffers 15 days adverse weather, the contractor is not entitled to an extension of time for the consequences on the completion date of the whole 15 days, but only for the extra 5 days. On any view, the common practice whereby clerks of works keep records of 'wet time' so that every couple of months the architect can give an extension of time covering the total period of wet time, is insupportable.

Loss or damage occasioned by one or more of the specified perils: clause 25.4.3

This simply gives the contractor the necessary time to fulfil his obligations to repair damage caused by fire, lightning, explosion, storm, tempest, flood, bursting or overflowing of water tanks, apparatus or pipes, earthquake, aircraft or other aerial devices, or articles dropped therefrom, riot and civil commotion, but excluding what are called the 'Excepted Risks': These are defined in clause 1.3.

The only important practical question arising under this heading is whether or not the contractor is entitled to an extension of time where the events are caused by the default or negligence of the contractor's own employees. On the plain reading of the wording it would appear that the contractor is still entitled to an extension.

Strikes and similar events: clause 25.4.4

The full list of events is given in the clause. 'Civil commotion' means, for insurance purposes, 'a stage between a riot and a civil war'466. So far as strikes are concerned, it should be noted that extension may be given, not only for circumstances affecting the contractor himself and his work on site, but also those engaged in preparing or transporting any goods and mater-

---

465 The Aqua Group, *Contract Administration for the Building Team*, 8th edition, 1996, p. 109.
466 *Levy* v. *Assicurazioni Generali* [1940] 3 All ER 427 at 431 per Luxmore LJ, approving an extract from Welford and Otterbarry's *Fire Insurance*, 3rd edition, at p. 64.

ials required for the works. The wording covers both official and unofficial strikes, but it does not cover 'working to rule' or other obstructive practices which fall short of a strike. It has been held that a strike by workers employed by statutory undertakers directly engaged by the employer to execute work not forming part of the works was not covered by the fore-runner of this clause in JCT 63[467]. A strike or other event referred to in the sub-clause must be one in which the trades mentioned in it are directly involved.

'Local combination of workmen' is an antiquated phrase beloved of the draftsmen of insurance policies. Possibly today it might be held to cover obstructive activities falling short of a strike provided they were confined to one area or site, but the true position is far from clear.

It is thought that a situation where deliveries to site are delayed, not due to a strike but due to some form of secondary picketing, does not fall under this relevant event.

Architect's instructions: clause 25.4.5.1

The instructions referred to are:

(1) Clause 2.3 – Discrepancies in contract bills, etc.
(2) Clause 2.4.1 – Discrepancies between performance specified work statement and architect's instructions
(3) Clause 13.2 – Variations
(4) Clause 13.3 – Expenditure of provisional sums in bills and sub-contracts
(5) Clause 13A.4.1 – Valuation of variations
(6) Clause 23.2 – Postponement of any work to be executed under the contract
(7) Clause 34 – Antiquities
(8) Clause 35 – Nominated sub-contractors
(9) Clause 36 – Nominated suppliers.

Compliance with an architect's instruction for the expenditure of a provisional sum for defined work is expressly excluded[468]. That is because in both cases, the contractor has been given sufficient information to enable him to make appropriate allowance in planning his work at tender stage.

If there is a failure of a nominated sub-contractor, and there is unreasonable delay by the architect in making a renomination, the contractor may be entitled to an extension of time under clause 25.4.5.1, with its specific reference to clause 35; he is certainly so entitled under clause 25.4.6 which empowers the architect to extend time in such circumstances[469].

---

[467] *Boskalis Westminster Construction Ltd* v. *Liverpool City Council* (1983) 24 BLR 83.
[468] See Chapter 13, section 13.5.2 under the sub-heading 'Approximate quantities, defined and undefined provisional sums'.
[469] *Percy Bilton Ltd* v. *Greater London Council* (1982) 20 BLR 1. See also *Fairclough Building Ltd* v. *Rhuddlan Borough Council* (1985) 30 BLR 26.

Opening up for inspection and testing: clause 25.4.5.2

The contractor may be entitled to an extension of time if delay has been caused by work being opened up for inspection or materials and goods being tested in accordance with architect's instructions (clause 8.3) if, and only if, the work inspected or the materials or goods tested prove to be in accordance with the contract. Clause 8.4.4 has elaborate provisions for repeated testing in the event of a defect being shown in repetitive work. Here again, an extension of time may only be granted if the tests of other items of similar work show that it is in accordance with the contract. The contractor also would be entitled to an extension of time to the extent that instructions under clause 8.4.4 to open up or test are not reasonable in all the circumstances.

Late instructions and drawings: clause 25.4.6

The relevant event is in two parts. If an information release schedule is used, clause 25.4.6.1 refers to the failure of the architect to comply with clause 5.4.1. This means that if the architect does not provide the information as set out in the schedule, the contractor has a ground for extension of time provided other criteria are met. That is very straightforward, and easy to understand and to operate.

The second part of the relevant event is contained in clause 25.4.6.2. It refers to the failure of the architect to comply with clause 5.4.2. Clause 5.4.2 deals with the situation if an information release schedule has not been provided or if there is information required which is not listed on the schedule. Assessing delays under this ground is less easy than under clause 25.4.6.1, because there are no dates on which information should have been provided to act as benchmarks. It is rather a matter of judgment by the architect who must decide when he should have provided the information under clause 5.4.2 and whether or not he did so. To make matters more difficult, the architect's obligation is to provide the information to the contractor to enable him to carry out and complete the works by the completion date, but there are two qualifications.

If the contractor's rate of progress is such that he will not finish by the due date, the architect may have 'regard' to this fact. It appears that this means that the architect is entitled, if he so wishes, to slow down the rate of provision of information to match the contractor's progress. That is not a practice to be advocated, not least because months later it may be unclear whether the contractor's slow progress provoked the slower delivery of information or if it was a result of it. Moreover, if the contractor looks likely to finish before the completion date, the architect is not obliged to furnish information to allow this to happen[470].

---

[470] This echoes the judgment in *Glenlion Construction Ltd* v. *The Guinness Trust* (1987) 39 BLR 89.

The question of terms to be applied as to the time within which further drawings, details or instructions are to be given have been considered by the courts although not in relation to a JCT contract. It has been stated that such information must be given within a reasonable time, but it has been made clear that this is a limited duty. Although the case was concerned with the duty of an engineer, it is thought that the principle applies equally to architects:

'What is a reasonable time does not depend solely upon the convenience and financial interests of the [contractors]. No doubt it is in their interest to have every detail cut and dried on the day the contract is signed, but the contract does not contemplate that. It contemplates further drawings and details being provided, and the engineer is to have a time to provide them which is reasonable having regard to the point of view of him and his staff and the point of view of the employer as well as the point of view of the contractor.'[471]

This is a common-sense business approach. Under JCT contracts the architect does not control the order of the works and, therefore, in interpreting the phrase 'to enable the Contractor to carry out and complete the Works in accordance with the Conditions' the contractor's viewpoint is paramount. The architect must take into account the time necessary for the contractor to organise adequate supplies of labour, materials and plant and to execute or have executed any prefabrication or prepare materials in such time as to ensure that these things are available on site having regard to his obligation to complete the works in accordance with the contract.

The current date for completion must always be borne in mind when analysing the particular circumstances. Vinelott J had this to say of the provision of information clause in JCT 63 with its rather more substantial clause calling for a 'specific written application' to be made:

'What the parties contemplated by these provisions was first that the architect was not to be required to furnish instructions, drawings, etc., unreasonably far in advance from the date when the contractor would require them in order to carry out the work efficiently nor to be asked for them at a time which did not give him a reasonable opportunity to meet the request. It is true that the words "on a date" grammatically govern the date on which the application is made. But they are … capable of being read as referring to the date on which the application is to be met. That construction seems to me to give effect to the purpose of the provision – merely to ensure that the architect is not troubled with applications too far in advance of the time when they will be actually needed by the contractor … and to ensure that he is not left with insufficient time to prepare them. If that is right then there seems … to be no reason why an application should not be made at the commencement of the work for all the instructions etc. which the contractor can

---

[471] *Neodox v. Borough of Swinton & Pendlebury* (1958) 5 BLR 34.

foresee will be required in the course of the works provided the date specified for delivery of each set of instructions meets these two requirements. Of course if he does so and the works do not progress strictly in accordance with this plan some modification may be required to the prescribed timetable and the subsequent furnishing of instructions and the like ... It does not follow that the programme was a sufficiently specified application made at an appropriate time in relation to every item of information required, more particularly in light of the delays and the rearrangement of the programme for the work.'[472]

Clause 5.4.2 requires the architect to provide 'such further drawings or details which are reasonably necessary to explain and amplify the Contract Drawings and ... to enable the Contractor to carry out and complete the Works in accordance with the Conditions', i.e. by the completion date. It is, of course, clear that apart from the express terms of the contract there would be an implied term that the contractor will receive decisions, drawings, details, etc. from the architect in time to avoid disruption of the contract.

However, the last part of the clause stipulates that if the contractor is aware and has reasonable grounds for believing that the architect is not aware when the contractor should receive information, he must inform the architect, giving him sufficient time to prepare the information. Although the clause is governed by the phrase 'if and to the extent that it is reasonably practicable to do so', it is difficult to envisage many circumstances when it would not be practicable. To that extent, the fulfilling of this requirement by the contractor may prove a hurdle to some claims for delays under this head. However, a contractor who deliberately refrains from making specific requests for drawings, on the basis that time will become 'at large' because the architect cannot give an extension of time, is likely to be considered to have a wholly unmeritorious claim.

Delay on the part of nominated sub-contractors or nominated suppliers: clause 25.4.7

A limited meaning is to be given to this provision; 'delay on the part of Nominated Sub-Contractors' only means delay by the nominated sub-contractor during the execution of the sub-contract works, and if a nominated sub-contractor (or, for that matter, a nominated supplier) ostensibly completes his sub-contract work or his supply contract but later is found to be in breach, e.g. because defects appear in the work, and has to return to remedy the breach, that is not a 'delay' within the meaning of this sub-clause[473].

Where a nominated sub-contractor withdraws from site, the duty of the employer is limited to giving instruction for nomination of a replacement within a reasonable time after receiving a specific application in writing

---

[472] *London Borough of Merton* v. *Stanley Hugh Leach Ltd* (1985) 32 BLR 51.
[473] *Westminster Corporation* v. *J. Jarvis & Sons Ltd* (1970) 7 BLR 64, as explained in *Percy Bilton Ltd* v. *Greater London Council* (1982) 20 BLR 1.

from the main contractor[474]. It follows that, if a nominated sub-contractor drops out, the result is not that time becomes at large but that the date for completion remains unaffected by the delay.

The delay referred to by the sub-clause must, it is noted, be delay which the contractor has taken all practicable steps to avoid or reduce, and perhaps the observations of Viscount Dilhorne in *Bickerton* v. *North West Metropolitan Regional Hospital Board* as to the general legal situation will serve as a warning to main contractors. He said:

> 'I cannot myself see that the extent of the contractor's obligation … is in any respect limited or affected by the right of the architect to nominate the sub-contractors. He has accepted responsibility for the carrying out and completion of all the contract works including those to be carried out by the nominated sub-contractor. Once the sub-contractor has been nominated and entered into the sub-contract, the contractor is as responsible for his work as he is for the works of other sub-contractors employed by him with the leave of the architect.'[475]

However, it is clear that 'delay by the employer in making the timeous nomination of a new sub-contractor is within the express terms of clause 23(f)' of JCT 63 and, of course, within JCT 98, clause 25.4.6[476]. It is for the contractor to make application to the architect for a renomination in respect of failure by a nominated sub-contractor or nominated supplier.

Work not forming part of the contract: clause 25.4.8

There are two separate situations. The first situation is covered by clause 25.4.8.1, where the employer engages others to carry out part of the works under the provisions of clause 29. It should be noted in particular that there is no reference to 'delay on the part of'. The circumstances in which an extension of time can be given seem to be much broader than under the previous relevant event. The meaning of 'work not forming part of the contract' has been defined:

> 'For some purposes the work does form a part, literally, of the contract; but for other purposes it does not. It is not work which the employer can require the contractor to do. All that he can require is that the contractor affords attendance etc. on those who do the work … [and that] I take the pragmatic view that the relevant work is work not forming part of the contract.'[477]

The second situation is a curious ground for extension of time in the context of this contract and it is to be found in clause 25.4.8.2. There are three points worth noting. First, in both situations it is not merely the employer's failure,

---

[474] *Westminster Corporation* v. *J. Jarvis & Sons Ltd* (1970) 7 BLR 64, as explained in *Percy Bilton Ltd* v. *Greater London Council* (1982) 20 BLR 1.
[475] [1970] 1 All ER 1039 at 1054 per Viscount Dilhorne.
[476] *Percy Bilton Ltd* v. *Greater London Council* (1982) 20 BLR 1.
[477] *Henry Boot Construction Ltd* v. *Central Lancashire New Town Development Corporation* (1980) 15 BLR 8 at 19 per Judge Fay.

but also the employer's success in correctly performing his obligations which are grounds for extending time. Second, unlike the execution of work by others, there is no contractual provision which entitles the employer to provide materials or goods. The words 'has agreed' are inappropriate and 'has elected with the agreement of the contractor' would better indicate that invariably the supply will be initiated by the employer, possibly because he believes that he has located a source of supply that will result in a financial saving, but probably because he believes that, by reserving the supply to himself, he will be certain to get what he wants.

Third, an interesting scenario would be created if the materials subsequently were found to be defective. It can reasonably be argued that the supply of materials and goods can only refer to materials and goods which are in accordance with the contract. Therefore, the supply by the employer of materials and goods which are not in accordance with the contract amounts to a failure to supply, because they should not be used by the contractor, who should reject them. However, if the employer has elected to supply, the election will almost certainly have been taken prior to tendering and certainly prior to executing the contract. Other than stating that the employer will supply stated materials and goods, it is unlikely that they will be specified in detail. In the absence of a specification, and if they were to be supplied by the contractor, there are certain terms which would be implied, such as that such materials and goods will be appropriate for their purpose. Where the employer is responsible, through the architect, for specification, and if no specification is given, it is considered that the contractor would be not be liable if an inferior product was supplied, because the employer would be relying on the architect, not the contractor, in such matters[478]. Clearly, if the materials or goods supplied were so inferior that they would seriously jeopardise the project or even become a danger, the contractor would have a duty to warn the employer.

But a difficult problem would arise if, either the materials and goods appeared to be in accordance with the contract, or if there was no specification, they appeared to be satisfactory, and were built into the construction and subsequently they were found to be defective and required replacement causing delay and additional expense. The employer could not require the contractor to supply replacement materials or goods without first issuing an instruction through the architect which would have the effect of adding these items to the contract at a very late stage. The cost of the materials or goods could be dealt with by the variation clause (13) and all the other costs would amount to loss and/or expense under clause 26. An appropriate extension of time would be indicated under the category of architect's instructions requiring a variation or possibly late instructions. If there was a specification and the employer's materials or goods were shown not to comply, an extension under this ground would be appropriate. That would certainly be the situation if the employer did not require the architect to issue an instruction, but simply himself supplied replacement materials or goods. Alternatively, the

---

[478] *Rotherham Metropolitan Borough Council v. Frank Haslam Milan & Co Ltd* (1996) 59 Con LR 33.

contractor could bring a common law claim against the employer for breach of contract and simply recover all his loss and/or expense as damages.

Government action: clause 25.4.9

The action must be taken by the government after the base date. The 'Base Date' is that date written into the Appendix. In the case of JCT 80 before its amendment of 11 July 1987, the reference was to the 'Date of Tender', meaning '10 days before the date fixed for receipt of tenders by the employer' (clauses 38.6.1 and 39.7 in their original form), which did not always in practice provide a firm date if the date for receipt of tenders was amended.

This provision might, for example, be relied upon wherever the British Government exercises any statutory power in the sense described in the clause, e.g. under the Defence of the Realm Regulations. The real significance of this is that it is no longer to be covered by *force majeure*, and this relevant event prevents the contract being brought to an end by frustration. It is simply dealt with by an extension of time. A prolonged stoppage of work for this reason would not, therefore, be grounds for the contractor to determine his own employment under clause 28.

Inability to obtain labour and goods: clause 25.4.10

Under JCT 98 the clause is mandatory and not merely an optional ground for extension. Moreover, the date at which any shortage is to be unforeseeable is the 'Base Date'. There are two sub-clauses. One deals with labour, the other with materials. In order to qualify as a relevant event, not only must the shortage be unforeseeable, the inability to obtain labour or materials must be for reasons which are beyond the contractor's control. Although there will be certain fairly rare instances when the contractor will not be able to obtain certain materials no matter what measures he takes or what price he is prepared to pay, he will always be able to obtain labour. Sometimes he will have to pay a grossly inflated price or he may be obliged to bus them in to site from some distance away, but he will always be able to secure labour.

On that basis, this clause would never bite and it would be redundant. However, a contract must be construed so as not to defeat the parties' intentions[479]. The parties clearly intended the inability to obtain labour to be grounds for an extension of time. In order to make sense of this particular event it is necessary to make some implication regarding the availability of labour or materials at prices which could reasonably be assumed by the parties at the base date. This is a peculiarly difficult event to consider in practice.

Local authority or statutory undertaker's work: clause 25.4.11

These provisions cover delay caused by the carrying out by a local authority or statutory undertaker of work 'in pursuance of its statutory obligations' in relation to the works, or its failure to do so.

---

[479] *Hydrocarbons Great Britain* v. *Cammel Laird Shipbuilders* (1991) 53 BLR 84.

In *Henry Boot Construction Ltd* v. *Central Lancashire Development Corporation*, which arose from an award made in the form of a special case by an arbitrator, Judge Edgar Fay QC was concerned with the problem of whether 'statutory undertakers' were 'artists, tradesmen or others engaged by the Employer' for the purpose of JCT 63, clauses 23(h) and 24(1)(d)[480]. He was bound by the arbitrator's finding of fact that statutory undertakers carrying out particular work under particular circumstances were doing it by virtue of a contract with the employer. They were not doing the work because statute obliged them to do it but because they had contracted with the employer to do it and so, he found, were 'engaged by the employer in carrying out work not forming part of this contract'. It followed from this that extensions of time in respect of delays on their part should be granted under clause 23(h) rather than 23(l), and that the contractor could, therefore, claim direct loss and/or expense from the employer under clause 24(l)(d).

Despite some assertions to the contrary, this decision has made no difference whatever to the perfectly clear meaning to be attached to these words, which the arbitrator clearly understood very well. Extensions of time can only be granted under this head if the statutory undertakers are carrying out work that is a statutory obligation.

Statutory undertakers frequently do work that is not done under statutory obligation, even though only they can do it. In such a case, if they have been engaged by the employer, any extension of time would be made under clause 25.4.8.1, with a possible claim for direct loss and/or expense under clause 26.2.4.1. If, however, they were engaged by the contractor, it seems that no extension of time would be due.

Statutory undertakers may, of course, affect the work in other ways. A water supplier might, for example, be laying water mains in the public road which provides access to the site, not for the purposes of the particular contract works but perhaps for another site nearby. In that case, even though they might be under an obligation to lay the mains (and be carrying out the work as a matter of statutory obligation) they would not be doing so in relation to the works. In such a case, there could be no extension of time under clause 25.4.11 because the statutory undertaker would not be carrying out 'work ... in relation to the Works' and, indeed, it seems that there is no provision in JCT 98 under which an extension of time could be given. The suggestion that such activities amount to *force majeure* and, therefore, could be grounds for an extension of time on that basis does not bear scrutiny.

In practical terms, whether or under which sub-clause an extension of time should be given for delays caused by local authorities and statutory undertakers depends entirely upon the sort of work they are doing and the circumstances under which they are doing it.

---

[480] (1980) 15 BLR 8.

Failure to give ingress or egress: clause 25.4.12

The exact wording of this provision should be noted, because it is not as extensive as some architects and contractors appear to think. An extension of time can only be granted under this sub-clause where there is failure by the employer to provide access to or exit from the site of the works across any *adjoining or connected* 'land, buildings, way or passage' which is in his own *possession and control*. It does not, therefore, cover, for instance, failure to obtain a wayleave across an adjoining owner's property, or where, for example, access to the highway is obstructed. It would equally not extend to the situation where protestors impeded access to a site where contractors were carrying out work[481].

It should be made clear that there is no right in the contract to receive (or power in the architect to grant) an extension of time under this sub-clause for the employer's failure to give possession of the site itself on the date for possession. This is of particular importance where the employer has not taken power to defer the giving of possession of the site to the contractor or where the deferment exceeds the period allowed in the contract.

There is also a further limitation in the sub-clause itself. The wording refers to access, etc. 'in accordance with Contract Bills and/or the Contract Drawings', which suggests that the undertaking to provide such access must be stated in the bills or drawings, and in that case any extension of time will be dependent upon the contractor giving whatever notice may be required by the provision in the bill before access is to be granted. The clause is unusual, because it extends the architect's powers as agent to act for the employer in agreeing access. It refers to the failure of the employer to give such ingress or egress as otherwise agreed between the architect and the contractor. This would seem to give the architect the authority to reach such an agreement as agent on the employer's behalf so that the employer in effect becomes responsible for failing to honour such an agreement even though it may have been reached without his being consulted. This, if correct, seems an extraordinary and unwarranted extension of the architect's powers. The architect's authority as the employer's agent is a limited one in law and the current *Standard Form of Agreement for the Appointment of An Architect 1999 (SFA/99)* makes clear how limited his authority is.

Deferment of possession of site: clause 25.4.13

This ground was added to JCT 80 in July 1987 in response to widespread criticism, because in practice, failure by the employer to give possession of the site is quite common. Clause 23.1 was accordingly amended so as to enable the employer to defer giving the contractor possession of the site for a period of up to 6 weeks unless he has inserted a shorter period in the Appendix. He only has this power if it is expressly stated in the Appendix,

---

[481] *LRE Engineering Services Ltd* v. *Otto Simon Carves Ltd* (1981) 24 BLR 127.

the relevant entry being 'Clause 23.1.2 applies/does not apply' and one of the alternatives is to be deleted. In view of the often unpredictable nature of demolition, site clearance works and tenants, to say nothing of unlawful squatters, it seems risky to reduce this period. Indeed, in some instances, employers would be wise to increase the 6 weeks and make the relevant amendment to references to 6 weeks. Where the employer does defer the giving of possession, there will be entitlement to extension of time. It is considered that deferment is a positive activity which the employer should signal by giving written notice, although clause 23.1.2 does not expressly so state. It should be noted that, on a strict reading of clauses 23.1 and 25.4.13, the extension can only be given where the employer has actually exercised his right to defer. This relevant event probably does not apply if the employer, without deferring possession, has simply failed to give possession on the due date.

Approximate quantity not a reasonably accurate forecast: clause 25.4.14

This ground was added by Amendment 7 (issued July 1988) as part of the incorporation of reference to the 7th edition of *Standard Method of Measurement* (SMM7). It proceeds on the perfectly reasonable basis that a contractor will plan his work using, among other things, the quantities in the bills of quantities. Where such quantities are described as 'approximate', it is presumably because the architect and/or the quantity surveyor either does not know, or has not quite decided upon, the amount required. All the contractor can do is to use the approximate quantities as if they gave a reasonably accurate forecast of the quantities required. If they give a lower forecast, he will presumably need additional time to carry out the work.

Change in statutory requirements necessitating alteration or modification of performance specified work: clause 24.4.15

Provision for performance specified work and hence this relevant event was added to the 1980 version of the contract by Amendment 12 (issued July 1993). This relevant event is similar to clause 25.4.13 of what is now WCD 98. It applies to the situation when, after the base date stated in the Appendix, there is a change in statutory requirements. Such a change might be an amendment to the Building Regulations which obliges the contractor to revise his proposals. It seems that there would be two possible bases for extension: if the redesign involves delay or if the changed or additional work takes longer to execute. The contractor must have taken all practicable steps to avoid or reduce the delay and, of course, there are no grounds for extension of time if the change occurred before the base date and, therefore, could have been taken into account.

Use or threat of terrorism: clause 25.4.16

It is thought that the threat would have to be more substantial than just the fact that other terrorist incidents have occurred in the area. A specific terrorist threat directed at the project or a threat to an area which, if it was to be carried out, would affect the project would qualify. The activities of the relevant authorities which would qualify under this ground would include such measures as evacuation of premises and the restriction of access. This ground is not restricted to the site of the works and, therefore, it is likely that any such threat or action which affected the execution of the works in any way (such as the forced evacuation or destruction of the contractor's offices) would give entitlement to extension of time.

Compliance or non-compliance with clause 6A.1: clause 25.4.17

Clause 6A.1 refers to the employer's obligation to ensure that the planning supervisor carries out his duties under the *CDM Regulations 1994* and, if the contractor unusually is not the principal contractor under the regulations, to ensure that he carries out his duties also. The employer's obligation to 'ensure' is onerous. It should be noted that the ground encompasses 'compliance or non-compliance' so that the proper carrying out of duties may also attract an extension of time. The problem for the employer is that the planning supervisor has duties under the regulations which he may have to carry out after the issue of any architect's instruction. Therefore, each instruction may attract an extension of time under this ground even if it does not qualify under clause 25.4.5.

Suspension by the contractor of performance of his obligations: clause 25.4.18

This ground is included to comply with the last part of section 112 of the Housing Grants, Construction and Regeneration Act 1996 which entitles a contractor to suspend performance of his obligations on 7 days written notice if the employer does not pay a sum due in full by the final date for payment. The suspension part of section 112 is dealt with by clause 30.1.4. This relevant event covers section 112(4) which states:

'(4) Any period during which performance is suspended in pursuance of the right conferred by this section shall be disregarded in computing for the purposes of any contractual time limit the time taken, by the party exercising the right or by a third party, to complete any work directly or indirectly affected by the exercise of the right.'

The relevant event is more generous than the Act which apparently provides that if a party suspends performance for 6 days, the effective extension to the period for completing the work is to be 6 days. This ignores any time the contractor may need to get ready to recommence. The wording of clause 25.4.18, by referring to 'delay arising from a suspension', clearly requires the architect to consider all the delay including remobilisation and not just the actual period of suspension.

Impediment, prevention or default by the employer: clause 25.4.19

This was added to JCT 98 by Amendment 4 in January 2002. It is obviously intended as a catch-all clause to avoid any possibility of time becoming at large due to an act of prevention or the like on the part of the employer. A saving provision makes clear that this clause does not cover the content of any of the other relevant events. Therefore, although the employer's failure to give access is undoubtedly an act of prevention, the contractor cannot be given an extension of time for that reason under this clause. It also excludes such part of any act or omission of the employer as was caused or contributed to by the default of the contractor, his servants, agents or sub-contractors. The key words in this relevant event seem strange bedfellows. To impede is to retard or hinder. To prevent is to hinder or stop. A default has been variously defined. It certainly covers a breach of contract, but it may go further than that in some circumstances[482].

10.2 Extension of time position under IFC 98

10.2.1 Clauses 2.3–2.5

The provisions of IFC 98, clauses 2.3, 2.4 and 2.5, concerning extensions of time are as follows:

Extension of time

2.3 Upon it becoming reasonably apparent that the progress of the Works is being or is likely to be delayed, the Contractor shall forthwith give written notice of the cause of the delay to the Architect, and if in the opinion of the Architect the completion of the Works is likely to be or has been delayed beyond the Date for Completion stated in the Appendix or beyond any extended time previously fixed under this clause, by any of the events in clause 2.4, then the Architect shall so soon as he is able to estimate the length of delay beyond that date or time make in writing a fair and reasonable extension of time for completion of the Works.
If an event referred to in clause 2.4.5, 2.4.6, 2.4.7, 2.4.8, 2.4.9, 2.4.12, 2.4.15, 2.4.17 or 2.4.18 occurs after the Date for Completion (or after the expiry of any extended time previously fixed under this clause) but before Practical Completion is achieved the Architect shall so soon as he is able to estimate the length of the delay, if any, to the Works resulting from that event make in writing a fair and reasonable extension of the time for completion of the Works.

---

[482] See *Building Contract Dictionary*, 3rd edition, 2001, Blackwell Publishing, p.119 for a fuller consideration of the term.

At any time up to 12 weeks after the date of Practical Completion, the Architect may make an extension of time in accordance with the provisions of this clause 2.3, whether upon reviewing a previous decision or otherwise and whether or not the Contractor has given notice as referred to in the first paragraph hereof. Such an extension of time shall not reduce any previously made.

Provided always that the Contractor shall use constantly his best endeavours to prevent delay and shall do all that may be reasonably required to the satisfaction of the Architect to proceed with the Works.

The Contractor shall provide such information required by the Architect as is reasonably necessary for the purposes of clause 2.3.

Events referred to in clause 2.3

2.4   The following are the events referred to in clause 2.3:

2.4.1   force majeure;

2.4.2   exceptionally adverse weather conditions;

2.4.3   loss or damage caused by any one or more of the Specified Perils;

2.4.4   civil commotion, local combination of workmen, strike or lock-out affecting any of the trades employed upon the Works or any trade engaged in the preparation, manufacture or transportation of any of the goods or materials required for the Works;

2.4.5   compliance with the Architect's instructions under clauses

      1.4   *(inconsistencies)*, or
      3.6   *(Variations)*, or
      3.8   *(Provisional sums)*
          except, where the Contract Documents include bills of quantities, for the expenditure of a provisional sum for defined work included in such bills, or
      3.15   *(Postponement)*,
         or, to the extent provided therein, under clause
      3.3   *(Named sub-contractors)*;

2.4.6   compliance with the Architect's instructions requiring the opening up or the testing of any of the work, materials or goods in accordance with clauses 3.12 or 3.13.1 (including making good in consequence of such opening up or testing), unless the inspection or test showed that the work, materials or goods were not in accordance with this Contract;

2.4.7   .1   where an Information Release Schedule has been provided, failure of the Architect to comply with clause 1.7.1;

      .2   failure of the Architect to comply with clause 1.7.2;

2.4.8   the execution of work not forming part of this Contract by the Employer himself or by persons employed or otherwise engaged by the Employer as referred to in clause 3.11 or the failure to execute such work;

2.4.9   the supply by the Employer of materials and goods which the Employer has agreed to supply for the Works or the failure so to supply;

2.4.10 where this clause is stated in the Appendix to apply, the Contractor's inability for reasons beyond his control and which he could not reasonably have foreseen at the Base Date to secure such labour as is essential to the proper carrying out of the Works;

2.4.11 where this clause is stated in the Appendix to apply, the Contractor's inability for reasons beyond his control and which he could not reasonably have foreseen at the Base Date to secure such goods or materials as are essential to the proper carrying out of the Works;

2.4.12 failure of the Employer to give in due time ingress to or egress from the site of the Works or any part thereof through or over any land, buildings, way or passage adjoining or connected with the site and in the possession and control of the Employer, in accordance with the Contract Documents, after receipt by the Architect of such notice, if any, as the Contractor is required to give, or failure of the Employer to give such ingress or egress as otherwise agreed between the Architect, the Contract Administrator and the Contractor;

2.4.13 the carrying out by a local authority or statutory undertaker of work in pursuance of its statutory obligations in relation to the Works, or the failure to carry out such work;

2.4.14 where clause 2.2 is stated in the Appendix to apply, the deferment of the employer giving possession of the site under that clause;

2.4.15 by reason of the execution of work for which an Approximate Quantity is included in the Contract Documents which is not a reasonably accurate forecast of the quantity of work required;

2.4.16 the use or threat of terrorism and/or the activity of the relevant authorities in dealing with such use or threat;

2.4.17 compliance or non-compliance by the Employer with clause 5.7.1;

2.4.18 delay arising from a suspension by the Contractor of the performance of his obligations under this Contract to the Employer pursuant to clause 4.4A;

2.4.19 save as provided for in clauses 2.4.1 to 2.4.18 any impediment, prevention or default, whether by act or omission, by the Employer or any person for whom the Employer is responsible except to the extent that it was caused or contributed to by any default whether by act or omission, of the Contractor or his servants, agents or sub-contractors.

Further delay or extension of time

2.5   In clauses 2.3, 2.6 and 2.8 any references to delay, notice, extension of time or certificate include further delay, further notice, further extension of time, or further certificate as appropriate.

10.2.2 Significant differences

It will be seen that these provisions are essentially a cut down version of clause 25 in JCT 98, but there are some significant differences which should be highlighted.

(1) The mandatory requirement for the contractor to provide particulars of the expected effects of delays and his own estimate of the resulting delay in completion has been omitted. Instead there is an obligation to provide 'such information required by the Architect as is reasonably necessary'. There is still, therefore, an obligation upon the contractor to provide the architect with the information he will need in order to make a proper extension of time, but there is no obligation on the contractor to provide his own estimate of the extension he will need. The word 'required' does not appear to place an obligation on the architect to ask for the information or to specify what he needs in this instance, although 'required' can be used in that sense; if that had been intended the word 'requested' would presumably have been used.

(2) The time limit for the architect to deal with extensions of time is omitted and IFC 98 reverts to the old form of words: 'so soon as he is able to estimate the length of the delay'. Delays may be said to fall into two classes: those that are a single definable cause of delay with a finite result, which is immediately apparent, and those that are a continuing cause of delay extending over a considerable period or even over the whole currency of the contract. An example of the first case may be a single major variation, which must be carried out before further work can commence; the architect must, in such a case and provided that the contractor has given a written notice of delay, issue his extension of time, if not immediately, then within a reasonable time, and failure on his part to do so may lead to the contract being considered 'at large' as to time with the consequent forfeiture of the employer's right to deduct liquidated damages. An example of the second case might be a continuing inability to obtain sufficient labour, which may extend up to, or nearly up to, completion of the works. In that case the architect would not be unreasonable in saying that he could not estimate the extent of the delay until the whole of the relevant work was completed. The true position is, probably, as follows:

'I think it must be implicit in the normal extension clause that the contractor is to be informed of his new completion date as soon as is reasonably practicable. If the sole cause is the ordering of extra work, then in the normal course extensions should be given at the time of ordering, so that the contractor has a target for which to aim. Where the cause of delay lies beyond the employer, and particularly where its duration is uncertain, then the extension order may be delayed, although even then it would be a reasonable inference to draw from the ordinary extension clause that the extension should be given a reasonable time after the factors which will govern the exercise of the [architect's] discretion have been established. Where there are multiple causes of delay, there may be no alternative but to leave the final decision until just before the issue of the final certificate.'[483]

---

[483] *Fernbrook Trading Co Ltd* v. *Taggart* [1979] 1 NZLR 556 at 568 per Roper J. See also *Amalgamated Building Contractors Ltd* v. *Waltham Holy Cross UDC* [1952] 2 All ER 452.

These principles are sound common sense and good contract practice.

When the architect is considering the grant of an extension of time, the effect of the cause of delay is to be assessed at the time when the works are actually carried out and not when they were programmed to be carried out. This appears to be so even if the contractor is in culpable delay during the original or extended contract period.[484]

(3) The architect is not required to 'fix a completion date' but simply to 'make' in writing an extension of time, i.e. to specify a period of extension and not a date for completion. It is not thought that this is of significance on the current wording of the clause.

(4) The requirement for the architect to notify the contractor if he decides not to grant an extension of time is not repeated here, but obviously it is good practice.

(5) The power of the architect to reduce extensions previously granted if he has omitted work in the interim does not appear. It seems this would not affect his right to take such omissions into account when next granting an extension of time, but he will not be able actually to withdraw or reduce an extension already made.

(6) The architect is expressly given power to extend time if any delays which are the responsibility of the employer or of the architect occur after completion date, but before practical completion. It seems likely that he has that power in any event, but this provision puts the matter beyond doubt so far as this contract is concerned[485]. It is thought that the inclusion of such an express term precludes the architect from extending time if a 'neutral' relevant event occurs during this period. This approach appears to have received judicial approval[486]. Older authority from the Court of Appeal, however, suggests the contrary[487].

(7) The provision for review of extensions following practical completion of the works has been made discretionary and not mandatory by the use of the word 'may' instead of 'shall' in the paragraph equivalent to JCT 98 clause 25.3.3. It will usually be in the employer's interest for the architect to carry out such a review.

(8) While these provisions are in general a simplification of the JCT 98 provisions, it seems regrettable that IFC 98 has returned in some respects to the old unsatisfactory wording of JCT 63, particularly the use of the words 'so soon as he is able to estimate the length of the delay' in relation to the architect's time for granting extensions. This might, wrongly, be used by architects as an excuse for leaving matters until the end of the contract, and it would have been preferable to see some firmer wording used. However, if the contractor fulfils his obligation to provide the architect with 'such information … as is reasonably neces-

---

[484] *Walter Lawrence & Son Ltd* v. *Commercial Union Properties (UK) Ltd* (1984) 4 Con LR 37.
[485] *Balfour Beatty* v. *Chestermount Properties Ltd* (1993) 62 BLR 1, in which an amended JCT 80 was under consideration.
[486] *Balfour Beatty* v. *Chestermount Properties Ltd* (1993) 62 BLR 1, in which an amended JCT 80 was under consideration.
[487] *Amalgamated Building Contractors Ltd* v. *Waltham Holy Cross UDC* [1952] 2 All ER 452.

sary' in order for the architect to make a decision there will be no reason why the architect cannot estimate the length of the delay in completion and make his decision quickly.

(9) Under IFC 98 there is no provision for the nomination of sub-contractors. Instead sub-contractors may be 'named', either in the contract documents or by instructions for the expenditure of provisional sums. Except that the sub-contract must be on a prescribed standard form entered into after specified procedures, such sub-contractors become virtually domestic sub-contractors. There are no provisions for the certification of payments by the architect or for direct payment by the employer if the contractor defaults and there is no provision entitling the contractor to an extension of time for delay on their part. However, if the sub-contractor defaults in the performance of his work to the extent that his employment is determined, the contractor is to notify the architect who must then issue instructions either, (a) naming a replacement sub-contractor, or (b) instructing the contractor to make his own arrangements for the completion of the work, or (c) omitting the remaining work, in which event the employer may make his own arrangements for completion. Whichever instruction the architect issues, the contractor will be entitled to an extension of time for the delaying effect of the instruction. If the sub-contractor determines his own employment because of the contractor's default, the architect is still to issue an instruction, but such an instruction will not entitle the contractor to an extension of time.

Similarly, if the contractor finds that he cannot enter into a sub-contract with a sub-contractor named in the contract documents because of some problem over the particulars of the sub-contract as set out in those documents, the architect is to issue instructions either changing the particulars so as to remove the problem, or omitting the work or by substituting a provisional sum. Any sub-contractor subsequently named would be regarded as having been named in an instruction for the expenditure of that sum.

Where a sub-contractor is named in an instruction for the expenditure of a provisional sum, such an instruction will have been issued under clause 3.8 of the contract. If the instruction causes delay to the completion date for any reason, therefore, the contractor will be entitled to an extension of time.

10.3 Extension of time position under MW 98

10.3.1 Clause 2.2

The full text of clause 2.2 is as follows:

Extension of contract period

2.2 If it becomes apparent that the Works will not be completed by the date for completion inserted in clause 2.1 hereof (or any later date fixed in accordance with the provisions of this clause 2.2) for reasons beyond the control of the Contractor, including compliance with any instruction of the Architect under this Agreement whose issue is not due to a default of the Contractor, then the Contractor shall thereupon in writing so notify the Architect who shall make, in writing, such extension of time for completion as may be reasonable. Reasons within the control of the Contractor include any default of the Contractor or of others employed or engaged by or under him for or in connection with the Works and of any supplier of goods or materials for the Works.

10.3.2 Significant differences

The most striking thing about the extension of time clause under this form is that it is very brief. The main differences are as follows:

(1) The contractor is not obliged to give notice of every delay. He need only give notice of such delays as will prevent completion of the Works by the current completion date and resulting from reasons beyond the control of the contractor.

(2) The contractor must give notice in writing, but there is no provision for him to provide supporting information. Common sense dictates that he must give sufficient information to allow the architect to understand what the delay entails, but the architect cannot, as a right, demand any particular further information. In practice, an architect will ask for any information he requires and if the contractor refuses to provide it, it seems that the architect must do the best he can in the circumstances. The contractor's refusal would effectively preclude him from making any substantial criticism of the extension of time thereafter.

(3) No time limit is set on the architect's exercise of his duty (note 'the architect *shall* make ... ') to make an extension of time. It is reasonable to suppose that the duty must be performed as quickly as practicable, bearing in mind that the nature of this contract suggests that projects executed under it will be of short duration, and a term would be implied to that effect to give business efficacy to the contract.

(4) The architect's extension must be 'reasonable'. Other forms refer to 'fair and reasonable', but it is not thought that anything significant turns on the distinction.

(5) There is no list of delaying events. There is merely reference to reasons beyond the control of the contractor including compliance with any architect's instruction provided it is not issued as a result of the contractor's default. It is fundamental that the architect only has power to make extensions of time for the reasons set out in the contract and that those grounds will be interpreted very strictly particularly in regard to delays which are due to the employer or architect[488]. The phrase ' ... reasons beyond the control of the contractor ... ' has been held, under another earlier form of contract, not to be specific enough to include

---

[488] *Peak Construction (Liverpool) Ltd* v. *McKinney Foundations Ltd* (1970) 1 BLR 114.

employer delays[489]. That raises the possibility of a successful challenge by a contractor that the architect's inability to extend time for employer delays other than architect's instructions renders time at large whenever such a delay becomes apparent. The fact that no such challenge has appeared in the law reports probably says more for the low value of work intended to be carried out under MW 98 than for the drafting of the clause. It is another reason why MW 98 should never be used for projects whose parameters are broader than set out in the guidance note.

(6) Although it is a basic principle of law that a party who is permitted to sub-contract retains full responsibility for the performance of such sub-contractors, the point has not been without doubt[490]. The last sentence of this clause was inserted to clarify the position and the position has since been upheld[491].

(7) There is no provision for the architect to carry out any review of extensions of time after the date of practical completion and it appears that, unless the delay is ongoing almost to practical completion, he has no general power to do so.

10.4 Extension of time position under WCD 98

10.4.1 Clause 25

The full text of clause 25 is as follows:

25 **Extension of time**

25.1 In clause 25 any reference to delay, notice or extension of time includes further delay, further notice or further extension of time.

.1 If and whenever it becomes reasonably apparent that the progress of the Works is being or is likely to be delayed the Contractor shall forthwith give written notice to the Employer of the material circumstances including the cause or causes of the delay and identify in such notice any event which in his opinion is a Relevant Event.

.2 In respect of each and every Relevant Event identified in the notice given in accordance with clause 25.2.1 the Contractor shall, if practicable in such notice, or otherwise in writing as soon as possible after such notice:

.2 .1 give particulars of the expected effects thereof; and

.2 .2 estimate the extent, if any, of the expected delay in the completion of the Works beyond the Completion Date resulting therefrom whether or not concurrently with delay resulting from any other Relevant Event.

25.2 .3 The Contractor shall give such further written notices to the Employer as may be reasonably necessary for keeping up-to-date the particulars

[489] *Wells* v. *Army & Navy Co-operative Society* (1902) 86 LT 764.
[490] *Scott Lithgow Ltd* v. *Secretary of State for Defence* (1989) 45 BLR 1.
[491] *John Mowlem & Co* v. *Eagle Star Insurance Co Ltd* (1995) 62 BLR 126.

and estimate referred to in clauses 25.2.2.1 and 25.2.2.2, including any material change in such particulars or estimate.

25.3 .1 If

.1 .1 any of the events which are stated by the Contractor to be the cause of the delay is a Relevant Event and

.1 .2 the completion of the Works is likely to be delayed thereby beyond the Completion Date,

the Employer upon receipt of any notice, particulars and estimate under clauses 25.2.1, 25.2.2 and 25.2.3 shall make in writing to the Contractor such extension of time, if any, for completion of the Works beyond the Completion Date as is then fair and reasonable, by fixing a later date as the Completion Date. The Employer shall, in making such extension of time, state:

.1 .3 which of the Relevant Events he has taken into account and

.1 .4 the extent, if any, to which he has had regard to any instruction under clause 12.2 requiring the omission of any work issued since the fixing of the previous Completion Date

and shall, if reasonably practicable having regard to the sufficiency of the aforesaid notice particulars and estimate, fix such new Completion Date not later than 12 weeks from receipt of the notice and of reasonably sufficient particulars and estimate, or, where the period between receipt thereof and the Completion Date is less than 12 weeks, not later than the Completion Date.

If upon receipt of any such notice, particulars and estimate it is not fair and reasonable for the Employer to fix a later date as a new Completion Date, the Employer shall, if reasonably practicable having regard to the sufficiency of the aforesaid notice, particulars and estimate, so notify the Contractor in writing not later than 12 weeks from receipt of the notice, particulars and estimate. or, where period between receipt thereof and the Completion Date is less than 12 weeks, not later than the Completion Date.

25.3 .2 After first making an extension of time under clause 25.3.1, the Employer may fix Completion Date earlier than that previously fixed under clause 25 if the fixing of such earlier Completion Date is fair and reasonable having regard to the omission of any work or obligation instructed by the Employer under clause 12.2 after the last occasion on which the Employer made an extension of time.

Provided that no decision under clause 25.3.2 shall alter the length of any adjustment to the time required by the Contractor for the completion of the Works in respect of a Change or work for which an adjustment to the time for completion of the Works has been accepted pursuant to clause 12.4.2 paragraph A7.

25.3 .3 After the Completion Date, if this occurs before the date of Practical Completion, the Employer may, and not later than the expiry of 12 weeks after the date of Practical Completion, shall in writing to the Contractor either

.3 .1 fix a Completion Date later than that previously fixed if the fixing of such later Completion Date is fair and reasonable having regard to any of the Relevant Events, whether upon reviewing a previous decision or otherwise and whether or not the Relevant Event has been specifically notified by the Contractor under clause 25.2.1, or

.3 .2 fix a Completion Date earlier than that previously fixed under clause 25 if the fixing of such earlier Completion Date is fair and reasonable having regard to any instructions of the Employer effecting a Change requiring the omission of any work issued under clause 12.2 where such issue is after the last occasion on which the Employer made the extension of time, or

3 .3 confirm to the Contractor the Completion Date previously fixed.

25.3 .4 Provided always that:

.4 .1 the Contractor shall use constantly his best endeavours to prevent delay in the progress of the Works, howsoever caused, and to prevent the completion of the Works being delayed or further delayed beyond the Completion Date; and

.4 .2 the Contractor shall do all that may reasonably be required to the satisfaction of the Employer to proceed with the Works.

25.3 .5 No decision of the Employer under clause 25.3 shall fix a Completion Date earlier than the Date for Completion stated in Appendix 1.

25.4 The following are the Relevant Events referred to in clause 25:

25.4 .1 force majeure;

25.4 .2 exceptionally adverse weather conditions;

25.4 .3 loss or damage occasioned by any one or more of the Specified Perils;

25.4 .4 civil commotion, local combination of workmen, strike or lock-out affecting any of the trades employed upon the Works or any of the trades engaged in the preparation, manufacture or transportation of any of the goods or materials required for the Works, or any persons engaged in the preparation of the design of the Works;

25.4 .5 compliance with the Employer's instructions;

.5 .1 under clauses 2.3.1, 12.2, 12.3, 23.2, or 34, or

.5 .2 in regard to the opening up for inspection of any work covered up or the testing of any of the work, materials or goods in accordance with clause 8.3 (including making good in consequence of such opening up or testing) unless the inspection or test showed that the work, materials or goods were not in accordance with this Contract;

25.4 .6 the Contractor not having received in due time necessary instructions, decisions, information or consents from the Employer which the Employer is obliged to provide or give under the Conditions including a decision under clause 2.4.2 for which he specifically applied in writing provided that such application was made on a date which having regard to the Completion Date was neither unreasonably distant from nor unreasonably close to the date on which it was necessary for him to receive the same;

25.4 .7 delay in receipt of any necessary permission or approval of any statutory body which the Contractor has taken all practicable steps to avoid or reduce;

25.4 .8 .1 the execution of work not forming part of this Contract by the Employer himself or by persons employed or otherwise engaged by the Employer as referred to in clause 29 or the failure to execute such work;

25.4 .8 .2 the supply by the Employer of materials and goods which the employer has agreed to provide for the Works or the failure so to supply;

25.4 .9 the exercise after the Base Date by the United Kingdom Government of any statutory power which directly affects the execution of the Works by restricting the availability or use of labour which is essential to the proper carrying out of the Works, or preventing the Contractor from, or delaying the Contractor in, securing such goods or materials or such fuel or energy as are essential to the proper carrying out of the Works;

25.4 .10 .1 the Contractor's inability for reasons beyond his control and which he could not reasonably have foreseen at the Base Date to secure such labour as is essential to the proper carrying out of the Works; or

25.4 .10 .2 the Contractor's inability for reasons beyond his control and which he could not reasonably have foreseen at the Base Date to secure such goods or materials as are essential to the proper carrying out of the Works;

25.4 .11 the carrying out by a local authority or statutory undertaker of work in pursuance of its statutory obligations in relation to the Works, or the failure to carry out such work;

25.4 .12 failure of the Employer to give in due time ingress to or egress from the site of the Works or any part thereof through or over any land, buildings, way or passage adjoining or connected with the site and in the possession and control of the Employer, in accordance with the Employer's Requirements after receipt by the Employer of such notice, if any, as the Contractor is required to give, or failure of the Employer to give such ingress or egress as otherwise agreed between the Employer and the Contractor;

25.4 .13 delay which the Contractor has taken all practicable steps to avoid or reduce consequent upon a change in the Statutory Requirements after the Base Date which affect the Works as referred to in clause 6.3.1, or an amendment to the Contractor's Proposals to which clause 6.3.2 applies;

25.4   .14   where clause 23.1.2 is stated in Appendix 1 to apply, the deferment by the Employer of giving possession of the site under clause 23.1.1;

25.4   .15   the use or threat of terrorism and/or the activity of the relevant authorities in dealing with such use or threat;

25.4   .16   compliance or non-compliance by the Employer with clause 6A.1;

25.4   .17   delay arising from a suspension by the Contractor of the performance of his obligations under the Contract to the Employer pursuant to clause 30.3.8;

25.4   .18   save as provided for in clauses 25.4.1 to 25.4.17 any impediment, prevention or default, whether by act or omission, by the Employer or any person for whom the Employer is responsible except to the extent that it was caused or contributed to by any default, whether by actor omission, of the Contractor or his servants, agents or sub-contractors.

10.4.2 Significant differences

Clause 25 closely follows the extension of time provisions in JCT 98. The principal differences are as follows:

(1) Reference is to the employer fixing a new completion date not to the architect, because there is no architect named as such under WCD 98. An architect can, of course, act as employer's agent and this arrangement is quite common, but an architect acting as agent is not acting in an independent capacity[492]. Under article 3 the employer's agent acts for the employer except to the extent that the employer specifically notifies the contractor in writing. It will be rare for the employer to act personally under this clause, particularly if the employer's agent is an architect or other professional with experience of dealing with extensions of time.

(2) JCT 98 refers to *giving* an extension, WCD 98 (and MW 98) to *making* an extension. It is not clear why JCT have chosen different wording unless in an attempt to emphasise that an extension of time is a right and not a benefaction. In practice, an extension of time is often referred to as being *awarded* or *granted*, both of which misleadingly suggest something not entirely deserved.

(3) There are no references to clause 13A quotations, because that is a provision of JCT 98. WCD 98 has a very similar provision (supplementary provision S6), but it is not referred to in clause 25 and, read strictly, references in clauses 25.3.2 and 25.3.3.2 to completion dates previously fixed under clause 25 exclude completion dates fixed under S6. It may

---

[492] *J. F. Finnegan Ltd* v. *Ford Seller Morris Developments Ltd* (1991) 53 BLR 38.

be that clause S6.1 which states, among other things, that clause 25 shall have effect as modified by S6, will fill the gap, but it is by no means certain because S6 makes no specific reference to clauses 25.3.2 and 25.3.3.2. The result may be that the employer has no power to fix an earlier date under these clauses if S6 has been operated to fix a revised completion date[493].

(4) There are no references to defined work or approximate quantities, because except in the unlikely event that S5 has been operated, there will be no bills of quantities and, therefore, no SMM7. If the employer has prepared bills of quantities to which supplementary provision S5 applies, clause S5.1 requires the employer to state the applicable method of measurement and clause S5.2 makes clear that errors in description or quantity in the bills must be corrected by the employer and the correction is to be treated as a change in the Employer's Requirements. Such a change would be ground for an extension of time if the completion date was delayed thereby. It is possible that the Employer's Requirements may stipulate that the contract sum analysis must be in the form of bills of quantities and it is possible that SMM7 has been used, but errors in the contractor's own documents cannot form grounds for an extension of time.

(5) '... or any persons engaged in the preparation of the design of the Works': clause 25.4.4 This is an extension of the equivalent JCT 98 clause to provide for an extension of time if the design element in this contract is affected by civil commotion and other occurrences in this relevant event, although it would be unusual for strikes to affect an independent consultant engaged by the contractor to design the works.

(6) 'compliance with Employer's instructions ...': clause 25.4.5 Among other things, this relevant event covers the correction of divergencies between Employer's Requirements and the definition of the site boundary under clause 7, changes, provisional sums and the postponement of design or construction.

(7) 'delay in receipt of any necessary permission ... of any statutory body: clause 25.4.7 The requirement that the contractor must have taken all practicable steps to reduce the delay must be taken seriously. Realistically, this will probably amount to little more than that the contractor must have made any necessary applications in good time, replied promptly to queries and used his best endeavours to obtain the permissions or approvals. An architect in the position of making applications to statutory bodies cannot guarantee the result and neither can the contractor. The contractor will be entitled to an extension of time under this relevant event if he can show that the delay was not due to his fault. This relevant event refers to any kind of statutory permission or approval. Virtually all buildings require planning permission and they must satisfy the Building Regulations. There are, however,

---

[493] The effect of S6 is considered in detail in Chapter 13, section 13.8.4.

many other possible controls over such things as fire, water and enter-
tainment.

(8) '...*change in the Statutory Requirements after the Base Date*...': *clause
25.4.13*. The contractor must have taken all practicable steps to reduce
the delay. To qualify, the Contractor's Proposals must need amending
either because the change has affected the works under clause 6.3.1 or to
conform with the terms of any permission or approval for development
control requirement under clause 6.3.2. Essentially this relevant event
quite reasonably entitles the contractor to an extension of time if he has
to make some alteration to his proposals due to events outside either
party's control.

(9) '*compliance or non-compliance by the Employer with clause 6A.1*': *clause
25.4.16*. This relevant event will only take effect if the contractor does
not take on the role of planning supervisor and principal contractor
under the CDM Regulations. It will be seldom that he does not assume
the latter role, but the employer may often require the employer's agent
to act as planning supervisor. In any event, the contractor may find it
difficult to carry out this role, because his involvement from a very early
date would be required. Although the duties of the planning supervisor
may inevitably delay progress, the contractor is not entitled to an extension of time if he is carrying out those duties himself.

10.5 Extension of time position under PCC 98

10.5.1 Clauses 2.5–2.7

The full text of clauses 2.5–2.7 is as follows:

Extension of time (2.5.1 to 2.5.12)

Delay to progress – Contractor's obligations

2.5 .1 If and whenever (including the circumstances referred to in clause 2.5.8)
it becomes reasonably apparent that the progress of the Works is being or
is likely to be delayed the Contractor shall forthwith give written notice to
the Architect of the material circumstances including the cause or causes
of the delay and identify in such notice any event which in his opinion is
a Relevant Event.

.2 In respect of each and every Relevant Event identified in the notice
given in accordance with clause 2.5.1 the Contractor shall, if practicable
in such notice, or otherwise in writing as soon as possible after such
notice:

give particulars of the expected effects thereof; and
estimate the extent if any of the expected delay in the completion
of the Works beyond the Completion Date resulting there

from whether or not concurrently with delay resulting from any other Relevant Event.

.3   The Contractor shall give such further written notices to the Architect as may be reasonably necessary or as the Architect may reasonably require for keeping up-to-date the particulars and estimate referred to in clause 2.5.2 including any material change in such particulars or estimate.

.4   The Contractor shall use constantly his best endeavours to prevent delay in the progress of the Works howsoever caused and shall do all that may be reasonably required to the satisfaction of the Architect to proceed with the Works.

Review of progress

2.5   .5   Whenever the Architect considers it reasonably necessary, whether or not after a notice by the Contractor under clause 2.5.1 has been given, the Contractor shall review the progress of the Works with the Architect. Such review may include consideration of the extent to which, in the opinion of the Architect, any additional resources are required to maintain progress for which the cost would be included in the Prime Cost.

Fixing Completion Date

2.5   .6   If, in the opinion of the Architect, upon receipt of any notice, particulars and estimate under clauses 2.5.1 and 2.5.2 and, where applicable, clause 2.5.3

any of the events which are stated by the Contractor to be the cause of the delay is a Relevant Event and

the completion of the Works is likely to be delayed thereby beyond the Completion Date

the Architect shall in writing to the Contractor give an extension of time by fixing such later date as the Completion Date as he then estimates to be fair and reasonable. The Architect shall, in fixing such new Completion Date, state:

which of the Relevant Events he has taken into account and
the extent, if any, to which he has had regard to any instruction under clause 3.3.2 which requires the omission of any work or obligation shown or described in the Contract Documents or previously instructed under clause 3.3.2 and which has been issued since the fixing of the previous Completion Date,

and shall, if reasonably practicable having regard to the sufficiency of the aforesaid notice, particulars and estimate, fix such new Completion Date not later than 12 weeks from receipt of the notice and of reasonably sufficient particulars and estimate, or, where the period between receipt thereof and the Completion Date is less than 12 weeks, not later than the Completion Date.

If, in the opinion of the Architect, upon receipt of any such notice, particulars and estimate, it is not fair and reasonable to fix a later date as a new Completion Date, the Architect shall if reasonably practicable having regard to the sufficiency of the aforesaid notice, particulars and estimate so notify the Contractor in writing not later than 12 weeks from receipt of the notice, particulars and estimate, or, where the period between receipt thereof and the Completion Date is less than 12 weeks, not later than the Completion Date.

.7 After the first exercise by the Architect of his duty under clause 2.5.6 the Architect may in writing fix a Completion Date earlier than that previously fixed under clause 2.5 if in his opinion the fixing of such earlier Completion Date is fair and reasonable having regard to any instruction under clause 3.3.2 which requires the omission of any work or obligation shown or described in the Contract Documents or previously instructed under clause 3.3.2 and which has been issued by the Architect since the fixing of the previous completion Date.

.8 If a Relevant Event referred to in clauses 2.6.5.1, 2.6.5.2, 2.6.6, 2.6.10, 2.6.11 or 2.6.14 occurs after the Completion Date but before Practical Completion is achieved the Architect, so soon as he is able to estimate the length of the delay, if any, to the Works resulting from that Relevant Event and after receipt of such particulars and estimate of the kind referred to in clause 2.5.2 as the Architect may reasonably require shall in writing to the Contractor

either give an extension of time by fixing such later date as the Completion Date as he then estimates to be fair and reasonable

or state that in his opinion it is not fair or reasonable to fix a later date as the Completion Date.

.9 Not later than the expiry of 12 weeks from the date of Practical Completion the Architect shall in writing to the Contractor

either fix a Completion Date later than that previously fixed if in his opinion the fixing of such later Completion Date is fair and reasonable having regard to the Relevant Events whether upon reviewing a previous decision or otherwise and whether or not the Relevant Event has been specifically notified by the Contractor under clause 2.5.1;

or fix a Completion Date earlier than that previously fixed if in his opinion the fixing of such earlier Completion Date is fair and reasonable having regard to any instruction under clause 3.3.2 which requires the omission of any work or obligation shown or described in the Contract Documents or previously instructed under clause 3.3.2 and which has been issued by the Architect since the fixing of the previous Completion Date;

or confirm to the Contractor the Completion Date previously fixed.

.10 No decision of the Architect shall fix a Completion Date earlier than the Date for Completion stated in the Appendix unless the Employer and the Contractor otherwise agree in writing.

Nominated Sub-Contractors

.11 Where the material circumstances of which written notice has been given under clause 2.5.1 include reference to a Nominated Sub-Contractor the Contractor shall send to the Nominated Sub-Contractor concerned a copy

> of such notice; and
>
> of the particulars and estimate given under clause 2.5.2; and
>
> of any further written notices given under clause 2.5.3.

.12 The Architect shall notify in writing to every Nominated Sub-Contractor each decision of the Architect under clause 2.5 fixing a Completion Date.

Relevant Events

2.6   The following are the Relevant Events referred to in clause 2.5:

.1   force majeure;

.2   exceptionally adverse weather conditions;

.3   loss or damage occasioned by any one or more of the Specified Perils;

.4   civil commotion, local combination of workmen, strike or lock-out affecting any of the trades employed upon the Works or any of the trades engaged in the preparation, manufacture or transportation of any of the goods or materials required for the Works;

.5   compliance with the Architect's instructions

.1   .1   under clauses 1.10, 3.3.2 (except to the extent that the instructions are to carry out the items of work as described in the Specification and/or as shown upon the Contract Drawings, if any), 3.3.4, 3.16.2, 8A or 8B.

.1   .2   in regard to the opening up for inspection of any work covered up or the testing of any of the work, materials or goods in accordance with clause 3.11 (including making good in consequence of such opening up or testing) unless the inspection or test showed that the work, materials or goods were not in accordance with this Contract;

.6   the Contractor not having received in due time necessary instructions, drawings, details or levels from the Architect for which he specifically applied in writing provided that such application was made on a date which having regard to the Completion Date was neither unreasonably distant from nor unreasonably close to the date on which it was necessary for him to receive the same;

.7   delay on the part of Nominated Sub-contractors or Nominated Suppliers which the Contractor has taken all practicable steps to avoid or reduce;

.8   where this clause 2.6.8 is stated in the Appendix to apply, the Contractor's inability for reasons beyond his control and which he could not reasonably have foreseen at the Base Date to secure such labour as is essential to the proper carrying out of the Works;

.9   the Contractor's inability for reasons beyond his control and which he could not have foreseen at the Base Date to secure such goods or materials as are essential to the proper carrying out of the Works;

.10   the execution of work not forming part of this Contract by the Employer himself or by persons employed or otherwise engaged by the Employer as referred to in clause 3.13 or 3.14 or the failure to execute such work;

.11   the supply by the Employer of materials and goods which the Employer has agreed to supply for the Works or the failure so to supply;

.12   the exercise after the Base Date by the United Kingdom Government of any statutory power which directly affects the execution of the Works by restricting the availability or use of labour which is essential to the proper carrying out of the Works or preventing the Contractor from, or delaying the Contractor in, securing such goods or materials or such fuel or energy as are essential to the proper carrying out of the Works;

.13   the carrying out by a local authority or statutory undertaker of work in pursuance of its statutory obligations in relation to the Works, or the failure to carry out such work;

.14   failure of the Employer to give in due time ingress to or egress from the Site or any part thereof through or over any land, buildings, way or passage adjoining or connected with the Site and in the possession and control of the Employer in accordance with the Contract Documents, after receipt by the Architect of such notice, if any, as the Contractor is required to give, or failure of the Employer to give such ingress or egress as otherwise agreed between the Architect and the Contractor;

.15   where clause 2.1.2 is stated in the Appendix to apply, the deferment by the Employer of giving possession of the Site under clause 2.1.2;

.16   the use or threat of terrorism and/or the activity of the relevant authorities in dealing with such use or threat;

.17   compliance or non-compliance by the Employer with clause 5.19;

.18   delay arising from a suspension by the Contractor of the performance of his obligations under this Contract to the Employer pursuant to clause 4.3.4;

.19   save as provided for in clauses 2.6.1 to 2.6.18 any impediment, prevention or default, whether by act or omission, by the Employer or any person for whom the Employer is responsible except to the extent that it was caused or contributed to by any default, whether by act or omission, of the Contractor or his servants, agents or sub-contractors.

Further delay or extension of time etc.

2.7   In clause 2.5 any references to delay, notice, extension of time, or certificate include further delay, further notice, further extension of time, or further certificate as appropriate.

10.5.2 Significant differences

Although in structure and wording this extension of time provision is clearly based on clause 25 of JCT 98, there are some important differences as follows:

(1) The contractor must review progress whenever the architect considers it to be reasonably necessary under clause 2.5.5. It is not quite clear what is intended and the clause goes into no detail. The architect's right to have a review carried out does not depend on the contractor's notice of delay under clause 2.5.1. The clause stipulates that the review must be carried out with the architect. Therefore, there is no question of the contractor simply submitting a progress report. They must sit down together. During the review the architect may come to a conclusion about the amount of additional resources necessary to maintain progress. This clause does not actually give the architect power to accelerate the works. Indeed, it simply states that the cost of such additional resources *would* (presumably if used) be included in the prime cost. The key to this strangely worded provision lies in the philosophy of this particular contract. It is based on a rough estimate of cost for known work and contractors tender on the basis of recovery of the whole of the prime cost of the work together with a sum to represent overheads and profit. The architect must issue instructions for the carrying out of all work including work in the original specification and/or drawings. It seems that, under clause 3.3.2, he can instruct the contractor to employ additional resources on the job. Clause 2.5.5 makes clear that, in such an instance, the employer pays in the usual way. This clause must be read in conjunction with clause 1.5 which requires the contractor to carry out the work as economically as possible in all the circumstances, taking care not to engage more personnel than reasonably required. Seen in context, clause 2.5.5 provides a useful tool to enable the architect to review and improve the progress of the work if the contractor's original allowance is thought to be too low.
(2) Express provision is made for the architect to extend time for employer delay relevant events which occur after the completion date, but before practical completion. This particular provision is absent from JCT 98, but present in somewhat abbreviated form in IFC 98.
(3) Although compliance with architect's instructions is a relevant event, instructions given to carry out work described in the specification or shown on the contract drawings are excluded to allow for the fact that the architect must instruct all work (see (1) above).

10.6 Extension of time position under MC 98

10.6.1 Clauses 2.12–2.14

The full text of clauses 2.12–2.14 is as follows:

Extension of time (2.12 to 2.14)

2.12 .1 If and whenever it becomes reasonably apparent that the Completion Date is not likely to be or has not been achieved, the Management Contractor shall forthwith advise the Architect of the cause of the delay and if in the opinion of the Architect the completion of the Project is likely to be or has been delayed beyond the Completion Date by any of the Project Extension Items in clause 2.13 then the Architect shall as soon as he is able to assess the length of the delay beyond the Completion Date give in writing an extension of time by fixing such later date as the Completion Date which he considers to be fair and reasonable provided that no extension shall be made in the case of delay which the Management Contractor has not used his best endeavours to avoid or reduce. If in the opinion of the Architect upon receipt of such advice from the Management Contractor it is not fair and reasonable to fix a later date as a new Completion Date he shall so notify the Management Contractor.

.2 After the first occasion on which the Architect fixed a new Completion Date the Architect may in writing fix a Completion Date earlier than that previously fixed under clause 2.12.1 if in his opinion the fixing of such earlier Completion Date is fair and reasonable having regard to the omission of any work or obligations instructed under clause 3.4 after the last occasion on which the Architect fixed a new Completion Date.

2.13 The Project Extension Items referred to in clause 2.12.1 are:

.1 any cause which impedes the proper discharge by the Management Contractor of his obligations under this Contract including

any impediment prevention or default whether by act or omission of the Employer or any persons for whom the Employer is responsible in regard to the Project or

the Management Contractor not having received in due time necessary specifications or bills of quantities for Works Contracts or Instructions, drawings, details or levels from the Professional Team for which he specifically applied in writing provided that such application was made on a date which having regard to the Completion Date was neither unreasonably distant from nor unreasonably close to the date on which it was necessary for him to receive the same;

where clause 2.3.2 is stated in the Appendix to apply the deferment of the Employer giving the possession of the site under clause 2.3.1;

.2 any Relevant Event except the Relevant Event referred to in clause 2.10.7.1 of the Works Contract Conditions which entitles any Works Contractor to an extension of time under clause 2.3 and/or clause 2.7 of the Works Contract Conditions for completion of his Works.

Provided that no Project Extension item shall be considered to the extent that it was caused or contributed to by any default, whether by act or omission, of the Management Contractor his servants or agents or of any Works Contractor his servants or agents or sub-contractors.

2.14 The Management Contractor shall in accordance with clause 2.3 of the Works Contract Conditions notify the Architect of any proposed decision on extensions of the period or periods for completion of a Works Contract in sufficient time so that the Architect can express in writing to the Management Contractor any dissent from the proposed decision before the Management Contractor is required to notify the Works Contractor of his decision in accordance with the provisions of clauses 2.3 and 2.4 of the Works Contract Conditions. If the Architect wishes to dissent from the proposed decision of the Management Contractor he shall so notify the Management Contractor in writing before the Management Contractor is required under the aforesaid clauses of the Works Contract Conditions to notify the Works Contractor of his decision.

10.6.2 Significant differences

The wording of this clause reflects the structure of the management contract. This clause refers to the project as a whole, but demonstrates the relationship with the works contracts which together form the project. The procedures for notifying delays and giving extensions of time are generally similar in effect to JCT 98 although simpler in expression. The principal differences are as follows:

(1) The management contractor's obligation to notify the architect comes into effect if and whenever it becomes reasonably apparent that the completion date is not likely to be or has not been achieved, not when progress is likely to be delayed although it probably comes to much the same thing. The management contractor is to 'advise' the architect and it is not specified to be in writing although common sense suggests that it is essential to put every notice in writing.

(2) The management contractor is not required to make any estimation of the effect of the delay on the completion date.

(3) The architect must give the extension of time 'so soon as he is able'. That does not mean he can take as long as he wishes and the comments in section 10.2.2 (2) are relevant.

(4) There is no provision for the management contractor to give, nor the architect to request, further information. Under this procurement system, however, it is generally expected that the management contractor will work closely with the architect and the provision of information should not be a problem in practice.

(5) The requirement for 'best endeavours' is much more emphatic under this contract and there is little doubt that it is a condition precedent. Clause 2.12.1 states quite unequivocally that no extension must be made for any delay which the management contractor has not used his best endeavours to avoid or reduce. Note that, on a strict reading of the clause, failure by the management contractor does not simply reduce the amount of extension for any particular delay, it totally negates it and the architect has no power to use discretion in the matter. It remains to

be seen whether such a potentially draconian remedy will be held to be a penalty in appropriate cases.

(6) There is no express power to review extensions of time after the date of practical completion of the project.

(7) The grounds for delay are termed 'Project Extension Items'. There are only two grounds, but the first one is so broad that it could be argued that the second is superfluous:

(a) The first ground is *'any cause which impedes the proper discharge by the Management Contractor of his obligations under this Contract ... '*. This is stated to include any impediment, prevention or default, whether by act or omission of the employer or persons for whom he is responsible, late information applied for in writing in due time and the deferment of possession (if applicable). This ground could hardly be wider and it certainly includes all those employer-generated occurrences which could result in time becoming at large if the architect had no power to deal with them[494].

(b) The second ground is any relevant event under the works contract conditions which entitles any works contractor to an extension of time. The exception is the relevant event in clause 2.10.7.1 which entitles a works contractor to an extension of time due to delay by other works contractors. This is clearly something which the contract looks to the management contractor to sort out between works contractors.

There is an interesting proviso that no project extension item must be considered *to the extent* that it is caused or contributed to by any default of the contractor, his servants, agents or sub-contractors. It is rather difficult to unravel the precise nuances of this clause, particularly in the light of judicial opinion that an architect should not take into account the contractor's own delays[495]. It appears, however, that the architect is not to entirely discount a ground just because the management contractor has contributed to it by his default; rather the architect must still consider so much of the ground as is unaffected by the default. This promises to be a skilful balancing task.

(8) Clause 2.14 is a very curious clause indeed. It obliges the management contractor to notify the architect of any decision which the management contractor proposes to make in regard to an extension of time for a works contractor. The notice must give the architect sufficient time to dissent in writing before the management contractor has to notify the works contractor. What then? Other than that the management contractor must pass on the architect's dissent to the works contractor under the terms of the works contract, there is nothing to prevent the management contractor proceeding to give the extension of time as originally proposed. It is difficult to see what purpose is served by this clause. The commentary in Practice Note MC/2 merely implies that if

---

[494] *Peak Construction (Liverpool) Ltd* v. *McKinney Foundations Ltd* (1970) 1 BLR 114.
[495] *John Barker Construction Ltd* v. *London Portman Hotels Ltd* (1996) 12 Const LJ 277.

the management contractor disregards the architect's dissent, he may find that his own application for extension, based on the same facts, would be refused. It does not invariably follow, of course, that an extension of time under the works contract confers a right to an extension under MC 98. However, clause 2.13.2 states that such a relevant event becomes a project extension item and, therefore, the management contractor has a basic case for an extension of time for the project whether the architect dissents or not. As with every delay, the management contractor has to decide whether he is convinced that he is correct. If so, he must give the extension of time to the works contract and notify the architect, if appropriate under clause 2.12.1. It is then for the architect properly to carry out his duty.

10.7 Extension of time position under TC/C 02

10.7.1 Clauses 2.2–2.5

The full text of clauses 2.2–2.5 is as follows:

Delay – extension of the Works Contract time

Written notice by the Trade Contractor

2.2 .1 If and whenever it becomes reasonably apparent that the commencement, progress or completion of the Works or any part thereof is being or is likely to be delayed, the Trade Contractor shall forthwith give written notice to the Construction Manager of the material circumstances including, insofar as the Trade Contractor is able, the cause or causes of the delay and identify in such notice any event which in his opinion is a Relevant Event as described in clause 2.5.

Particulars, estimates and further written notices

2.2 .2 In respect of each and every Relevant Event identified in the notice given in accordance with clause 2.2.1, the Trade Contractor shall, if practicable in such notice, or otherwise in writing as soon as possible after such notice:

  1 give particulars of the expected effects thereof; and

  .2 estimate the extent, if any, in the expected delay in the completion of the Works beyond the Completion Period resulting therefrom whether or not concurrently with delay resulting from any other Relevant Event; and

  .3 give such further written notices to the Construction Manager as may be reasonably necessary or as the Construction Manager may reasonably require for keeping up to date the particulars and estimate referred to in clause 2.2.2.1 and 2.2.2.2 including any material change in such particulars or estimate.

Revising Completion Period

2.3 .1 If in the opinion of the Construction Manager, upon receipt of any notice, particulars and estimate under clause 2.2.2:

   .1 any of the events which are stated by the Trade Contractor to be the cause of the delay is a Relevant Event and

   .2 the completion of the Works is likely to be delayed thereby beyond the Completion Period,

the Construction Manager shall in writing to the Trade Contractor give an extension of time by fixing such revised or further revised Completion Period as he then estimates to be fair and reasonable. The Construction Manager shall, in fixing such revised Completion Period, state:

   .3 which of the Relevant Events he has taken into account and

   .4 the extent, if any, to which he has had regard to any instruction requiring as a Variation the omission of any work issued since the last occasion on which the Completion Period was revised

and shall, if reasonably practicable having regard to the sufficiency of the aforesaid notice, particulars and estimates, fix such revised Completion Period not later than 12 weeks from receipt by the Construction Manager of the notice and of reasonably sufficient particulars and estimates or where the time between receipt thereof and the expiry of the Completion Period is less than 12 weeks, not later than the expiry of the Completion Period.

If, in the opinion of the Construction Manager, upon receipt of any such notice, particulars and estimate it is not fair and reasonable to revise the Completion Period, the Construction Manager shall, if reasonably practicable having regard to the sufficiency of the aforesaid notice, particulars and estimate, so notify the Trade Contractor in writing not later than 12 weeks from receipt of the notice, particulars and estimate or where the time between receipt thereof and the expiry of the Completion Period is less than 12 weeks, not later than the expiry of the Completion Period.

.2 The Construction Manager may, in writing after the first exercise of his duty under clause 2.3.1 or after any revision of the Completion Period in a confirmed acceptance by him of a 3A Quotation in respect of a Variation, fix a Completion Period earlier than that previously fixed under clause 2.3.1 or stated in a confirmed acceptance of a 3A Quotation if in his opinion the fixing of such earlier Completion Period is fair and reasonable having regard to omission of any work or obligation instructed or sanctioned by him after the last occasion on which he fixed a revised Completion Period. Provided that no decision under clause 2.3.2 shall alter the length of any adjustment to the time required by the Trade Contractor for the completion of the Works in respect of a variation for which a 3A Quotation has been given and which has been stated in a confirmed acceptance of a 3A Quotation.

.3 On the expiration of the Completion Period, if this occurs before the date of practical completion of the Works, the Construction Manager may, and not later than the expiry of 12 weeks after the date of practical completion shall, in writing to the Trade Contractor either:

.1 fix a longer Completion Period than that previously fixed if in his opinion the fixing of such longer Completion Period is fair and reasonable having regard to any of the Relevant Events, whether upon reviewing a previous decision or otherwise and whether or not the Relevant Event has been specifically notified by the Trade Contractor under clause 2.2.1; or

.2 fix an earlier Completion Period than that previously fixed under clause 2.3 or stated in a confirmed acceptance of a 3A Quotation if in his opinion the fixing of such earlier Completion Period is fair and reasonable having regard to the omission of any work or obligation instructed or sanctioned by him after the last occasion on which he fixed a new Completion Period; or

.3 confirm to the Trade Contractor the Completion Period previously fixed or stated in a confirmed acceptance of a 3A Quotation.

Provided that no decision under clause 2.3.3.1 or clause 2.3.3.2 shall alter the length of any adjustment to the time required by the Trade Contractor for the completion of the Works in respect of a Variation for which a 3A Quotation has been given and which has been stated in a confirmed acceptance of a 3A Quotation.

2.4 The operation of clauses 2.2 and 2.3 shall be subject to the proviso that:

.1 the Trade Contractor shall constantly use his best endeavours to prevent delay in the progress of the Works, howsoever caused, and to prevent the completion of the Works being delayed beyond the Completion Period;

.2 the Trade Contractor shall do all that may reasonably be required to the satisfaction of the Construction Manager to proceed with the Works;

.3 subject to clauses 3A.2.2 and 3A.3.2.3 no decision of the Construction Manager under clause 2.3 shall fix a Completion Period shorter than the Completion Period stated in the Tender.

Relevant Events

2.5 The following are the Relevant Events:

.1 force majeure;

.2 exceptionally adverse weather conditions;

.3 loss or damage occasioned by any one or more of the Specified Perils;

.4 civil commotion, local combination of workmen, strike or lock-out affecting any of the trades employed upon the Works or any of the trades engaged in the preparation, manufacture or transportation of any of the goods or materials required for the Works;

.5 compliance with the Construction Manager's instructions

.1 under clauses 1.12, 1.13 (except for a confirmed acceptance of a 3A Quotation), 3.21 (except where bills of quantities are included in the

Trade Contract Documents compliance with an instruction for the expenditure of a provisional sum for defined work or of a provisional sum for Performance Specified Work), 2.1.5, or 3.18; or

.2 in regard to the opening up for inspection of any work covered up or the testing of any of the work, materials or goods in accordance with clause 3.11 (including making good in consequence of such opening up or testing) unless the inspection or test showed that the work, materials or goods were not in accordance with this Trade Contract;

.6 .1 where an Information Release Schedule has been provided, failure of the Construction Manager to comply with clause 3.2.3;

.6 .2 failure of the Construction Manager to comply with clause 3.2.4;

.7 Delay on the part of any person other than the Trade Contractor engaged by the Client on or in connection with the Project which the Trade Contractor has taken all practicable steps to reduce;

8. .1 the execution of work not forming part of the Project by the client himself or by persons employed or otherwise engaged by the Client or the failure to execute such work;

.8 .2 the supply by the client of materials and goods which the Client has agreed to provide for the Works or the failure so to supply;

.9 the exercise after the Base Date by the United Kingdom Government of any statutory power which directly affects the execution of the Project by restricting the availability or use of labour which is essential to the proper carrying out of the Project or preventing the Trade Contractor from, or delaying the Trade Contractor in, securing such goods or materials or such fuel or energy as are essential to the proper carrying out of the Project;

.10 where and to the extent that the Appendix states that clause 2.5.10.1 and/or clause 2.5.10.2 shall apply:

.1 the Trade Contractor's inability for reasons beyond his control and which he could not reasonably have foreseen at the Base Date to secure such labour as is essential to the proper carrying out of the Works;

.2 the Trade Contractor's inability for reasons beyond his control and which he could not reasonably have foreseen at the Base Date to secure such goods or materials as are essential to the proper carrying out of the Works;

.11 the carrying out by a local authority or statutory undertaker of work in pursuance of its statutory obligations in relation to the project or the failure to carry out such work;

.12 failure of the Construction Manager to give in due time ingress to or egress from the site of the Project or any part thereof through or over any land, buildings, way or passage adjoining or connected with the site and in the possession and control of the client in accordance with the

Trade Contract Documents, after receipt by the Construction Manager of such notice, if any, as the Trade Contractor is required to give, or failure of the Construction Manager to give such ingress or egress as otherwise agreed between him and the Trade Contractor;

.13 where clause 2.1.2 is stated in the Appendix to apply, the deferment by the Construction Manager of the Date of Commencement;

.14 where bills of quantities are included in the Trade Contract Documents by reason of the execution of work for which an approximate quantity is included in those bills which is not a reasonably accurate forecast of the quantity of work required;

.15 delay which the Trade Contractor has taken all practicable steps to avoid or reduce consequent upon a change in the Statutory Requirements after the Base Date which necessitates some alteration or modification to any Performance Specified Work;

.16 the use or threat of terrorism and/or the activity of the relevant authorities in dealing with such use or threat;

.17 compliance or non-compliance with clause 5.11.1;

.18 delay arising from a suspension by the Trade Contractor of his obligations under the Trade Contract to the Client pursuant to clause 4.11.6;

.19 save as provided for in clauses 2.5.1 to 2.5.18 any impediment, prevention or default, whether by act or omission, by the Client or any person for whom the client is responsible except to the extent that it was caused or contributed to by any default, whether by act or omission, of the Trade Contractor or his servants, agents or sub-contractors.

10.7.2 Significant differences

The structure and wording of this extension of time clause is clearly derived from JCT 98. That is to be expected, because each trade contractor is in direct contract with the employer. The functions carried out by the architect in the JCT 98 clause are carried out by a construction manager whose role is essentially to manage all the consultants and trade contractors. Items to note are:

(1) Under clause 2.2.1, written notice must be given by the trade contractor whenever it becomes reasonably apparent that the 'commencement, progress or completion' of the works is likely to be delayed. It is, therefore, clear that the trade contractor must give this notice even if he has not started on site, provided only that his start is delayed. It is unlikely that the completion will be delayed without a corresponding delay to either commencement or progress and it is difficult to see what the insertion of 'completion' achieves.

(2) Reference is made throughout to the 'Completion Period' rather than to the completion date. That is because, although the period for carrying

out the work is fixed, the commencement and completion dates may change. The result is that this contract makes no express provision for extending the completion date if the commencement date is delayed. Effectively, all the construction manager can do is to extend the period available for carrying out the work. In practice, a delayed commencement may have no effect on the length of the contract period, the whole period being merely moved back in time.

(3) A small inconsistency occurs in clause 2.3.3 where the first sub-clause refers to the fixing of a 'longer' completion period and the second sub-clause refers to fixing an 'earlier' period when one might expect to see reference to a 'shorter' period. This appears to be simply a drafting error.

The relevant events echo those in JCT 98 and the commentary on that clause is generally applicable here.

10.8   Extension of time position under MPF 03

10.8.1   Clause 12

The full text of clause 12 is as follows:

12   Extension of time

12.1   The Contractor shall be entitled to an adjustment to the Completion Date to the extent that, having regard to the principles set out in clause 12.7, completion of the Project or any Section (if applicable) is or is likely to be delayed by:

.1   force majeure;

.2   the occurrence of one or more of the Specified Perils;

.3   the exercise after the Base Date by the United Kingdom Government of any statutory power that directly affects the execution of the Project, other than alterations to Statutory Requirements as referred to by clause 4.5;

.4   the use or threat of terrorism, as defined by the Terrorism Act 2000, and/or the activities of the relevant authorities in dealing with such a threat;

.5   any Change;

.6   interference with the Contractor's regular progress of the Project by Others on the Site;

.7   the valid exercise by the Contractor of its rights under section 112 of the HGCRA 1996; or

.8   any other breach or act of prevention by the Employer or its representative or advisors appointed pursuant to clause 15.2.

Provided always that there shall be no adjustment to the Completion Date in respect of any matter where it is specifically stated by the Contract that such matter will not give rise to a Change.

12.2	Whenever the Contractor becomes aware that the progress of the Project is being or is likely to be delayed due to any cause, it shall forthwith notify the Employer of the cause of the delay and its anticipated effect upon the progress and completion of the Project or any Section (if applicable).

12.3	Where the Contractor considers the cause of the delay is one of those identified in clause 12.1.1 to 12.1.8 it shall also:

.1	provide supporting documentation to demonstrate to the Employer the effect upon the progress and completion of the Project or any Section (if applicable), and

.2	revise the documentation provided so that the Employer is at all times aware of the anticipated or actual effect of the cause upon actual progress and completion of the Project or any Section (if applicable).

12.4	Where the Contractor has notified the Employer under clause 12.2 and the cause of delay is identified as being one of those in clauses 12.1.1 to 12.1.8 the Employer shall, within 42 days of receipt of the notification, either:

.1	notify the Contractor of such adjustment to the Completion Date as it then calculates to be fair and reasonable; or

.2	notify the Contractor why it considers that the Completion Date should not be adjusted.

Any adjustment made to the Completion Date shall be calculated by reference to the documentation provided by the Contractor in accordance with clause 12.3. The Employer may take account of other information available to it.

12.5	Any notification given under clause 12.4 may be reviewed by the Employer at any time in the light of further documentation from the Contractor or when the effects of any identified cause of delay become more apparent.

12.6	No later than 42 days after Practical Completion of the Project the Contractor shall provide documentation to support any further adjustment to the Completion Date that it considers fair and reasonable. Within 42 days of receipt of that documentation the Employer shall undertake a review of its previous adjustments to the Completion Date. The review shall have regard to that documentation and the Employer shall either notify such further adjustment to the Completion Date as is fair and reasonable or confirm the Completion Date previously notified.

12.7	In considering any adjustment to the Completion Date the Employer shall:

.1	implement any agreements about the Completion Date reached in accordance with clauses 13 (*Acceleration*), 19 (*Cost savings and value improvements*) and 20 (*Changes*);

.2	have regard to any breach by the Contractor of clause 9.3;

.3	make a fair and reasonable adjustment to the Completion Date notwithstanding that completion of the Project may also have been delayed due to the concurrent effect of a cause that is not listed in clauses 12.1.1 to 12.1.8.

12.8   Except by agreement with the Contractor (including as set out in clauses 13 (*Acceleration*) and 19 (*Cost savings and value improvements*)), no adjustment to the Completion Date shall give rise to an earlier Completion Date than one that has already been notified.

10.8.2 Significant differences

MP 03 is expressed to be intended for use by employers who have in-house contractual procedures and undertake major projects on a regular basis. The contract is less detailed than some other standard forms, such as JCT 98. Essentially, this is a contract in which the contractor carries out whatever design is required beyond what is contained in the Employer's Requirements. The contractor is also expected to take on more risk than usual. Whatever may be the intention of the draftsman, it is the intentions of the parties as expressed in the contract to which a court must give effect[496].

The extension of time clause bears a passing resemblance to the equivalent clause in WCD 98. However, there are some significant differences which should be noted:

(1) The grounds for extension of time are briefer than under clause 25 of WCD 98. Missing are exceptionally adverse weather conditions, strikes, lockouts and civil commotion, inability on the part of the contractor to obtain labour or materials and delays caused by statutory undertakers.

(2) Although there is a requirement that the contractor must notify the employer of delays due to any cause, therefore including delays which are due to the contractor's own inefficiencies, the method of notification is not specified. That is because clause 38 provides that all communications between the parties must be in writing or by the procedure stated in the Appendix if electronic communication is agreed to be acceptable or by any other agreed method.

(3) The contractor is called upon to form an opinion about whether the delay is one of those in clauses 12.1.1 to 12.1.8. If the opinion is positive, the contractor must supply supporting information. The clause does not specify when this information must be provided, but if it is to be workable, the information must be provided at the same time or very shortly after the notice. Once the contractor has reached a positive conclusion, he has no choice in the matter. Clause 12.3.1 is quite specific that the information must demonstrate to the employer the effect upon progress and completion. Therefore, if the information, viewed objectively, does not do that, the contractor is in breach of his obligations.

(4) Clause 12.3 is important. On receipt of the clause 12.2 notification, the employer has 42 days in which to notify the contractor of an adjustment to the completion date or state why the employer considers there should be no adjustment. Unlike the position under clause 25 of WCD 98, the trigger for action by the employer is not the receipt of sufficient infor-

---

[496] *Schuler A G v. Wickman Machine Tool Sales Ltd* [1974] AC 235.

mation, but receipt of the original notice. However, the employer must calculate the adjustment by reference to the information from the contractor and *may* take other information, such as his own knowledge into account. It, therefore, seems that in stating why the completion date is not to be adjusted, it may be sufficient for the employer to state that the supporting information did not demonstrate any effect upon progress or completion or indeed that he did not receive it. This clause is an improvement upon WCD 98, because the employer must do something on receipt of the notice, even if it is merely to refuse an extension of time, whereas under WCD 98 the employer could theoretically do nothing until after the completion date if the supporting information supplied was not sufficient.

(5) The most significant clause is 12.5. This states that any notification by the employer under clause 12.4 may be reviewed by the employer 'at any time' if further documentation is received from the contractor or if the effects of an identified cause of delay become more apparent. It can be compared with the right to refer a dispute to adjudication 'at any time' under section 108(2)(a) of the Housing Grants, Construction and Regeneration Act 1996 which was held to mean that there was no restriction as to time[497]. Therefore, it appears that if the contractor submits further information, even after the clause 12.6 review, the employer can review his previous decision.

(6) Clause 12.6 divides the review process into two parts. First, the contractor must provide any further information within 42 days after practical completion. The employer has 42 days from receipt of the information to carry out his review of previous extensions of time. The employer must either notify further adjustment or that there is no adjustment.

(7) When considering an adjustment to the completion date, the employer must put into effect any agreements about acceleration, cost savings and value improvements and changes. He must also 'have regard to' any breach of clause 9.3 by the contractor. Clause 9.3 is a 'reasonable endeavours' clause. This is less onerous than the usual JCT 'best endeavours' and the contractor is entitled to have regard to his own financial interests[498].

(8) The contract seems to have adopted one of the principles of the SCL extension of time protocol[499] in clause 12.7.3 by requiring the employer to give the contractor an extension of time even if the project is also delayed concurrently by an unlisted cause or something for which the contractor has agreed under the contract to take the risk, such as exceptionally adverse weather. Contrary to a widely held belief, it is thought that such an approach is not supported by legal authority[500]. Nevertheless, the principle is enshrined in this contract and the employer must comply.

---

[497] *A & D Maintenance & Construction Ltd* v. *Pagehurst Construction Services Ltd* (2000) 17 Const LJ 199.
[498] *UBH (Mechanical Services) Ltd* v. *Standard Life Assurance Co* (1990) BCLC 865; *Phillips Petroleum Co UK Ltd* v. *Enron Europe Ltd* (1997) CLC 329.
[499] See Chapter 2, section 2.7.
[500] The point is considered in Chapter 2, section 2.4.

Chapter 11

Liquidated damages under JCT standard form contracts

11.1 Liquidated damages position under JCT 98

11.1.1 Clause 24

The full text of clause 24 is as follows:

24   Damages for non-completion

24.1   If the Contractor fails to complete the Works by the Completion Date then the Architect shall issue a certificate to that effect. In the event of a new Completion Date being fixed after the issue of such a certificate such fixing shall cancel that certificate and the Architect shall issue such further certificate under clause 24.1 as may be necessary.

24.2   .1   Provided:

— the Architect has issued a certificate under clause 24.1; and
— the Employer has informed the Contractor in writing before the date of the final Certificate that he may require payment of, or may withhold or deduct, liquidated and ascertained damages,

then the Employer may, not later than 5 days before the final date for payment of the debt due under the Final Certificate:

either

.1   require in writing the Contractor to pay to the Employer liquidated and ascertained damages at the rate stated in the Appendix (or at such lesser rate as may be specified in writing by the Employer) for the period between the Completion Date and the date of Practical Completion and the Employer may recover the same as a debt;

or

.2   give a notice pursuant to clause 30.1.1.4 or clause 30.8.3 to the Contractor that he will deduct from monies due to the Contractor liquidated and ascertained damages at the rate stated in the Appendix (or at such lesser rate as may be specified in the notice) for the period between the Completion Date and the date of Practical Completion.

24.2   .2   If, under clause 25.3.3, the Architect fixes a later Completion Date or a later Completion Date is stated in a confirmed acceptance of a 13A Quotation the Employer shall pay or repay to the Contractor any amounts recovered, allowed or paid under clause 24.2.1 for the period up to such later Completion Date.

24.2   .3   Notwithstanding the issue of any further certificate of the Architect under clause 24.1 any requirement of the Employer which has been previously stated in writing in accordance with clause 24.2.1 shall remain effective unless withdrawn by the Employer.

11.1.2 Commentary

There are a number of conditions which must be met before the employer is entitled to liquidated damages. First, there must be a failure by the contractor to complete the works by the date for completion specified or within any extended time. Second, the architect must have performed his duties in deciding upon extensions of time under clause 25. The case of *Token Construction Co Ltd* v. *Charlton Estates Ltd*[501] is instructive. There, an architect sent a somewhat jumbled letter to the employer some 2 years after contract completion in which he said 'with 13 weeks extension of time the adjusted completion date would have been 30.1.68 . . . Details of the 13 weeks' extension of time are being prepared and will be forwarded to you . . . liquidated damages ought to be calculated from 30 January 1968 to 15 July 1968, a period of 24 weeks'. The Court of Appeal held that on the facts there was no valid certificate of delay or extension of time, and that the architect could not certify delay until he had first adjudicated upon all the contractor's applications for extensions of time. This was a decision on a special form of contract in which the wording of the relevant clauses was similar to JCT 63 provisions. It is thought that the decision holds good for JCT 98 also.

The third condition to be observed is that the architect must have issued a certificate to the effect that the contractor has failed to complete by the completion date. He can issue the certificate at any time prior to the issue of the final certificate. In practice, of course, most architects will issue the certificate immediately the completion date has passed in order to allow the employer the maximum possible time and maximum available funds for deduction of liquidated damages. An employer may have a cause of action against an architect who delays the issue of the certificate until just before the issue of the final certificate if by that time it is impossible to recover the liquidated damages. However, once the architect has issued the final certificate under clause 30.8, if no notice of adjudication, arbitration or legal proceedings has been given by either party in accordance with clause 30.9, the architect apparently becomes *functus officio* and is thereby excluded thereafter from issuing any valid certificate under clause 24[502].

The architect's certificate under clause 24 is not, it seems, a condition precedent to arbitration. Once a dispute arises, it can be referred to adjudication or arbitration under clause 41.

---

[501]   (1973) 1 BLR 48.
[502]   *H. Fairweather Ltd* v. *Asden Securities Ltd* (1979) 12 BLR 40.

The architect cannot avoid issuing the certificate of non-completion if the contractor has failed to complete by the due date. It is not something for his discretion. If the architect fixes a new date for completion after the issue of the certificate, the fixing of a new date is said to cancel the existing certificate and the architect must issue a further certificate. The clause refers to 'such further certificate…as may be necessary', because if the architect fixes a new date which is the same as, or later than, the date the contractor actually completes the works, a further certificate is unnecessary.

Clause 24.2.1 introduces a further condition precedent if the employer wishes to deduct them from future payments or if he wishes the contractor to pay him. The employer must give a written notice to the contractor that he may deduct or may require the payment. Clause 24.2.3 provides that the employer need only serve one notice requiring payment. It remains effective, despite the issue of further non-completion certificates, unless the employer withdraws it. Since the decision to deduct liquidated damages rests with the employer, it is unlikely that he would ever, in practice, withdraw the notice. If he decided not to deduct damages, he would simply let the matter rest.

The timing of the written notice sometimes causes difficulty. It seems to suggest that liquidated damages may be deducted provided that the written requirement is served before the date of the final certificate. Thus damages might be deducted from an interim certificate several months before notice is served just before the issue of the final certificate. That, of course, would be a nonsense and the purpose of the clause does not permit such a construction. It is perhaps unfortunate that the wording did not make clear that the date of the final certificate is stated as the deadline for the written requirement and that the requirement must always pre-date the deduction. It should be noted that failure to serve the written requirement will not only prevent deduction, it will also preclude recovery of the liquidated damages as a debt.

Some doubt has been thrown on the precise form to be taken by the employer's written requirement for payment under earlier versions of the standard form. Judge John Newey stated:

> 'There can be no doubt that a certificate of failure to complete given under clause 24.1 and a written requirement of payment or allowance under the middle part of clause 24.2.1 were conditions precedent to the making of deductions on account of liquidated damages or recovery of them under the latter part of clause 24.2.1.'[503]

This seems perfectly clear, but another Official Referee thought:

> '…that there was no condition precedent that the employer's requirement had to be in writing. What was essential was that the contractor should be in no doubt that the employer was exercising its power under 24.2 in reliance on the architect's certificate given under 24.1 and

---

[503] *A. Bell & Son (Paddington) Ltd* v. *CBF Residential Care & Housing Association* (1990) 46 BLR 102.
[504] *Jarvis Brent Ltd* v. *Rowlinson Construction Ltd* (1990) 6 Const LJ 292.

deducting specific sums from monies otherwise due under the contract.'[504]

He went on to hold that the written requirement was satisfied by a letter, written by the quantity surveyor and forwarded to the contractor, which stated the amount which the employer was entitled to deduct; alternatively, that the cheques issued by the employer from which liquidated damages had been deducted constituted such written requirements. In *Holloway Holdings Ltd* v. *Archway Business Centre Ltd*[505] a similar clause in IFC 84 was considered and it was again held:

> 'For (the employer) to be able to deduct liquidated damages there must both be a certificate from the Architect and a written request to (the contractor) from (the employer).'

The matter was finally clarified by a decision of the Court of Appeal in *J. J. Finnegan Ltd* v. *Community Housing Association Ltd*[506] where the court held that the decision in *Bell* was correct and that the employer's written requirement was a condition precedent to the deduction of liquidated damages. Only two things must be specified in the requirement and they are:

- whether the employer is claiming a payment or a deduction of the liquidated damages; and
- whether the requirement relates to the whole or part of the total liquidated damages.

The reworded clause 24.2.1 of JCT 98 leaves the matter in no doubt. Clause 24.2.3 emphasises that a requirement which has been stated in writing remains effective even if the employer issues further non-completion notices. The final condition to be satisfied is that the employer must serve a withholding notice in accordance with clauses 30.1.1.4 and 30.8.3. To summarise, the conditions are:

- The contractor must fail to complete by the contractual completion date or any extended date.
- The architect must have decided all extensions of time.
- The architect must have issued a non-completion certificate.
- The employer must serve a written requirement.
- The employer must serve an effective written withholding notice.

The amount which the employer may deduct is to be calculated by reference to the rate stated in the Appendix. The employer is free to reduce the rate, but not to increase it. Clause 24.2.1 makes clear that the employer need not wait until practical completion before deducting liquidated damages. He may start to deduct them as soon as the clause 24.1 certificate has been issued and the requirement for payment. In

---

[505] 19 August 1991, unreported.
[506] (1995) 65 BLR 103.

practice, such deductions usually commence from the first payment thereafter.

11.2 Liquidated damages position under IFC 98

11.2.1 Clauses 2.6, 2.7 and 2.8

The full text of clauses 2.6, 2.7 and 2.8 is as follows:

Certificate of non-completion

2.6   If the Contractor fails to complete the Works by the Date for Completion or within any extended time fixed under clause 2.3 then the Architect shall issue a certificate to that effect.

In the event of an extension of time being made after the issue of such a certificate such making shall cancel that certificate and the Architect shall issue such further certificate under this clause as may be necessary.

Liquidated damages for non-completion

2.7   Provided:

— the Architect has issued a certificate under clause 2.6; and
— the Employer has informed the Contractor in writing before the date of the final certificate that he may require payment of, or may withhold or deduct, liquidated and ascertained damages,

the Employer may not later than 5 days before the final date for payment of the debt due under the final certificate

either

2.7.1   require in writing the Contractor to pay to the Employer liquidated and ascertained damages at the rate stated in the Appendix for the period during which the Works shall remain or have remained incomplete and may recover the same as a debt

or

2.7.2   give a notice pursuant to clause 4.2.3(b) or clause 4.6.1.3 that he will deduct liquidated damages at the rate stated in the Appendix for the period during which the Works shall remain or have remained incomplete.

Notwithstanding the issue of any further certificate of the Architect under clause 2.6 any written requirement or notice given to the Contractor in accordance with this clause shall remain effective unless withdrawn by the Employer.

Repayment of liquidated damages

2.8   If after the operation of clause 2.7 the relevant certificate under clause 2.6 is cancelled the Employer shall pay or repay to the Contractor any amounts deducted or recovered under clause 2.7 but taking into account the effect of a further certificate, if any, issued under clause 2.6.

11.2.2 Significant differences

This clause is very similar to clause 24 of JCT 98 in wording and in effect. There is no express reference to the employer's right to require payment at a lesser rate than the one stated in the Appendix, but in principle such a right must be implied. In any event, it is unlikely that a dispute would arise on the basis that the contractor insisted on paying the full rate.

11.3 *Liquidated damages position under MW 98*

11.3.1 Clause 2.3

The full text of clause 2.3 is as follows:

> **Damages for non-completion**
> 2.3   If the Works are not completed by the completion date inserted in clause 2.1 hereof or by any later completion date fixed under clause 2.2 hereof the Contractor shall pay or allow to the Employer liquidated damages at the rate of
>
> £ .... per ....
>
> between the aforesaid completion date and the date of practical completion.
>
> The Employer may
>
> either
>
> recover the liquidated damages from the Contractor as a debt
>
> or
>
> deduct the liquidated damages from any monies due to the Contractor under this Contract provided that a notice of deduction pursuant to clause 4.4.2 or clause 4.5.1.3 has been given. If the Employer intends to deduct any such damages from the sum stated as due in the final certificate, he shall additionally inform the Contractor in writing, of that intention not later than the date of issue of the final certificate.

11.3.2 Significant differences

The MW 98 provision marks a very significant departure from the JCT 98 and IFC 98 regime. There is no certificate of non-completion required from the architect. The trigger is simply that the contractor fails to complete the works by the completion date or any extended date. Once that date is passed and the contractor is not finished, the employer may recover the amount of liquidated damages as a debt or he may deduct it from any money due to the contractor under the contract. A written requirement by the employer has been introduced, once again with the date of issue of the final certificate as the deadline. However, as was noted in the commentary to JCT 98, common sense and implication of law would ensure that the notice must predate the deduction. In practice, the architect, complying with his general duty to advise the employer, will usually send him a letter reminding him that the completion date has passed, that the contractor has not completed and that liquidated damages are deductible. The normal withholding notices under the contract must also be served.

11.4   Liquidated damages position under WCD 98

11.4.1 Clause 24

The full text of clause 24 is as follows:

24   Damages for non-completion

24.1   If the Contractor fails to complete the construction of the Works by the Completion Date the Employer shall issue a notice in writing to the Contractor to that effect. In the event of a new Completion Date being fixed after the issue of such notice in writing such fixing shall cancel that notice and the Employer shall issue such further notice in writing under clause 24.1 as may be necessary.

24.2   .1   Provided
    — the Employer has issued a notice under clause 24.1 and
    — the Employer, before the date when the Final Account and Final Statement (or, as the case may be, the Employer's Final Account and Final Statement) become conclusive as to the balance due between the Parties by agreement or by the operation of clause 30.5.5 or clause 30.5.8, has informed the Contractor in writing that he may require payment of, or may withhold or deduct, liquidated and ascertained damages,

then the Employer may not later than 5 days before the final date for payment of the debt due under clause 30.6:

either

.1   require in writing the Contractor to pay to the Employer liquidated and ascertained damages at the rate stated in Appendix 1 (or at such lesser rate as may be specified in writing by the Employer) for the period between the Completion Date and the date of practical Completion and the Employer may recover the same as a debt;

or

.2   give a notice pursuant to clause 30.3.4 or clause 30.6.2 to the Contractor that he deduct from monies due to the Contractor liquidated and ascertained damages at the rate stated in Appendix 1 (or at such lesser rate as may be specified in the notice) for the period between the Completion Date and the date of Practical completion

24.2 .2 If, under clause 25.3.3, the Employer fixes a later Completion Date, the Employer shall pay or repay to the Contractor any amounts recovered allowed or paid under clause 24.2.1 for the period up to such later Completion Date.

24.2 .3 Notwithstanding the issue of any further notice under clause 24.1 any requirement of the Employer which has been previously stated in writing in accordance with clause 24.2.1 shall remain effective unless withdrawn by the Employer.

11.4.2 Significant differences

There is a very distinct family resemblance between this form and JCT 98. The differences spring from the absence of an architect and the design input of the contractor. The main difference in this clause is that it is the employer, or his agent acting on his behalf, who issues a written notice of non-completion to the contractor. Such a notice is intended to be a statement of fact. It is not the expression of an opinion such as would be the case if a certificate was to be issued[507]. The courts have refused to give a notice under this contract the same status as the certificate of an independent architect[508].

11.5 *Liquidated damages position under PCC 98*

11.5.1 Clauses 2.2–2.4

The full text of clauses 2.2–2.4 is as follows:

Certificate of non-completion

2.2 .1 If the Contractor fails to Complete the Works by the Completion date then the Architect shall issue a certificate to that effect.

    .2 In the event of a new Completion Date being fixed after the issue of a certificate under clause 2.2.1 such fixing shall cancel that certificate and the Architect shall issue such further certificate under clause 2.2.1 as may be necessary.

Payment or allowance of liquidated damages

2.3 .1 Provided:

    — the Architect has issued a certificate under clause 2.2.1; and

    — the Employer has informed the Contractor in writing before the date of the Final Certificate that he may require payment of, or may withhold, liquidated and ascertained damages,

    then the Employer may, not later than 5 days before the final date for payment of the debt due under the Final Certificate;

---

[507] *Token Construction* v. *Charlton Estates* (1973) 1 BLR 48.
[508] *J. F. Finnegan Ltd* v. *Ford Seller Morris Developments Ltd* (1991) 53 BLR 38.

either

> .1   require the Contractor to pay to the Employer liquidated and ascertained damages at the rate stated in the Appendix (or at such lesser rate as may be specified in writing by the Employer) for the period between the Completion Date and the date of Practical Completion and the Employer may recover the same as a debt;

or

> .2   give a notice pursuant to clause 4.3.3.2 or clause 4.12.4 to the Contractor that he will deduct from monies due to the Contractor liquidated and ascertained damages at the rate stated in the Appendix (or at such lesser rate as may be specified in the notice) for the period between the Completion Date and the date of Practical Completion.

.2   Notwithstanding the issue of any further certificate of the Architect under clause 2.2.2 any requirement of the Employer which has been previously stated in writing in accordance with clause 2.3.1 shall remain effective unless withdrawn by the Employer.

Later Completion Date: repayment

2.4   If, under clauses 2.5.1 to 2.5.12, the Architect fixes a later Completion Date the Employer shall pay or repay to the Contractor any amounts recovered, allowed or paid under clause 2.3 for the period up to such later Completion Date.

11.5.2 Comments

Clauses 2.2 to 2.4 of PCC 98 are virtually identical to the equivalent clause 24 of JCT 98 and the comments on JCT 98 apply to PCC 98 also.

11.6 *Liquidated damages position under MC 98*

11.6.1 Clauses 2.9–2.11

The full text of clauses 2.9–2.11 is as follows:

Damages for non-completion (2.9 to 2.11)

2.9   If the Management Contractor fails to secure the completion of the Project by the Completion Date then the Architect shall issue a certificate to that effect. In the event of a new Completion Date being fixed after the issue of such a certificate such fixing shall cancel that certificate and the Architect shall issue such further certificate under clause 2.9 as may be necessary.

2.10 .1   Provided the Architect has issued a certificate under clause 2.9 and provided the Employer has informed the Management Contractor in writing before the date of the Final Certificate that he may require payment of, or may withhold or deduct, liquidated and ascertained damages, the Employer may, not later than five days before the final date for payment of the debt due under the final Certificate

either

.1 require in writing the Management Contractor to pay to the Employer liquidated and ascertained damages at the rate stated in the Appendix (or at such lesser rate as may be specified in writing by the Employer) for the period between the Completion Date and the date of Practical Completion and the Employer may recover the same as a debt;

or

.2 give a notice pursuant to clause 4.3.4 or clause 4.12.4 to the Management Contractor that he will deduct from monies due to the Management Contractor liquidated and ascertained damages at the rate stated in the Appendix (or at such lesser rate as may be specified in the notice) for the period between the Completion Date and the date of Practical Completion.

2.11   If after the operation of clause 2.10 the relevant certificate under clause 2.9 is cancelled the Employer shall pay or repay to the Management Contractor any amounts recovered, allowed or paid under clause 2.10 but taking into account the effect of a further certificate, if any, issued under clause 2.9.

11.6.2 Significant differences

(1) The clause is written in terms of the management contractor failing 'to secure the completion of the Project'. This simply reflects the management contractor's obligation to secure the completion of the project by the completion date set out in clause 2.3.1, i.e. his task is to arrange that others complete rather than to physically complete himself.

(2) The employer's right to deduct liquidated damages is made subject to clause 3.21 which essentially provides, in 3.21.2.2, that the employer is to 'keep an account' of liquidated damages which are due, but which have been caused by the breach or non-compliance of the works contractor and which he has not recovered from the management contractor except to the extent that the management contractor can recover the same amounts from the works contractor. These provisions give rise to difficulties in practice. Essentially, the position appears to be that clause 3.21.2.2 gives relief to the management contractor only if the delay is solely caused by the works contractor and the management contractor's obligation is to 'ensure' performance. If the delay is due to the management contractor alone or to a mixture of management contractor and works contractor delays, it seems that the employer is entitled to deduct the liquidated damages from payments due to the management contractor. It is thought the employer may also deduct liquidated damages where the delay is solely due to the works contractor in those instances where the management contractor has failed to comply with his obligations to take all necessary steps under clause 3.21.1[509].

---

[509] See *Copthorne Hotel (Newcastle) Ltd* v. *Arup Associates and Others* (1997) 85 BLR 32 for detailed analysis of the effect of clause 3.21 of JCT 87 which does not differ from MC 98 in any material respect in connection with this issue.

11.7 Liquidated damages position under TC/C 02

11.7.1 Clauses 2.11 and 2.12

There is no liquidated damages provision under this form of contract. Instead there is provision for recovery of unliquidated or actual damages. This is similar to the position under sub-contract forms. Although the client may suffer a loss as a result of delay on the part of a trade contractor, it is impossible to insert a liquidated sum, because several trade contractors may contribute to the loss.

The full text of clauses 2.11 and 2.12 is as follows:

> **Failure of Trade Contractor to complete on time – Client's direct loss and/ or expense**
> *Payment by Trade Contractor – direct loss and/or expense due to failure to complete*

> 2.11 If the Trade Contractor fails to complete the Works within the Completion Period and provided that decisions have been given on all outstanding applications for extension of time the Trade Contractor shall pay or allow to the Client a sum equivalent to any direct loss and/or expense suffered by the Client which has arisen from the failure of the Trade Contractor. Subject to clause 2.12 such sum shall include, but shall not be limited to any damages, costs or expenses which the Client is obliged or may be liable to pay or allow to any other person engaged by the Client on or in connection with the Project as a direct consequence of the Trade Contractor's failure.

> 2.12 If it is stated in the Appendix that clause 2.12 is to apply, the Trade Contractor's maximum financial liability in respect of the matters referred to in clause 2.11 shall not exceed the sum specified in the Appendix.

11.7.2 Key points

Clause 2.11 contains two provisos. First, the trade contractor must have failed to complete within the completion period and, second, the construction manager must have given all decisions on extensions of time. The trade contractor is obliged to 'pay or allow' direct loss and/or expense, but, in *Hermcrest plc* v. *G. Percy Trentham Ltd*[510], that phrase has been held to exclude the right to set off the amount claimed by the party asserting the right to payment. There is an obligation to pay or to allow the sum properly due. Therefore, it can be allowed only insofar as it is agreed and not if it is disputed.

Clause 2.12 provides for a cap on the amount if previously so agreed by the parties and written into the Appendix. This is very useful for the trade contractor when the possible liability might be totally out of proportion to the value of the trade contract. This provision sets out formally what many trade and sub-contractors already include as part of their routine qualification of quotations and tenders.

---

[510] (1991) 53 BLR 104.

11.8 Liquidated damages position under MPF 03

11.8.1 Clause 10

The full text of clause 10 is as follows:

10 Damages for delay

10.1 If the Contractor fails to achieve Practical Completion by the Completion Date, it shall be liable to pay the Employer liquidated damages calculated at the rate stated in the Appendix for the period from the Completion Date to the date of Practical Completion.

10.2 Where liquidated damages have been paid to the employer and the Completion Date is subsequently adjusted in accordance with clause 12 (*Extension of time*), the Employer shall be liable to repay to the Contractor any liquidated damages to which the Employer is no longer entitled.

11.8.2 Significant differences

This is a liquidated damages clause at its simplest. There is no requirement for a non-completion certificate, therefore, no need to provide for its cancellation and re-issue after a further extension of time. The trigger is the contractor's failure to complete by the completion date, the rate is stated in the Appendix and further extensions trigger repayment of any liquidated damages overpaid. There is no express provision for the employer to deduct liquidated damages from payments to the contractor, but that does not appear to preclude the employer from setting off such damages from payments due to the contractor provided that the relevant withholding notices are served.

Chapter 12

Loss and/or expense under JCT standard form contracts

12.1 Introduction

12.1.1 General provisions

It is recognised that there will be many occasions when a contractor is not adequately reimbursed by payment of the contract sum or by valuation of variations in the normal way. Variations, for example, are often disruptive and they may give rise to prolongation of the contract period. As part of the overall scheme of payment, virtually all standard form contracts make provision for the contractor to recover money which he has either lost or expended as an essential part of carrying out the contract. All such provisions place conditions upon the right to recovery and they generally allow additional or alternative claims for damages for breach of contract at common law. So far as the contract machinery is concerned, the parties must take note of and observe the precise wording of the clause if the contractor is to be properly reimbursed.

12.1.2 A practical approach

Many architects, quantity surveyors and contractors are uncertain about the way in which to approach the reimbursement of loss and/or expense. If the contractor does not submit adequate supporting information when he makes his application for reimbursement, it is helpful if the architect and in turn the quantity surveyor, if instructed by the architect, in exercising their powers to request information also take the opportunity to correct any inconsistencies or errors in the contractor's approach. Information should be requested as precisely as circumstances allow. On no account should the contractor simply be requested to provide 'more information' or 'better proof'. The following list, which is not exhaustive, includes some of the points which might be raised with the contractor:

- The architect's and the quantity surveyor's powers are limited by the terms of the contract.
- The contractor must strictly comply with the contract machinery.

- Loss and/or expense does not follow automatically from an extension of time.
- A written application should have been made as soon as it has become, or should reasonably have become, apparent that the regular progress of the works has been or is likely to be affected.
- Cause should be linked to effect.
- Amounts claimed should be linked to the 'matters' in the contract.
- Substantiation in the form of invoices, labour returns, pay slips, detailed schedules of plant and equipment, etc. is required of amounts to which the contractor thinks he is due.
- Where prolongation is involved, the actual dates of the events causing the prolongation are required and the sums claimed must relate to those dates not to the period of prolongation.
- Only actual costs are acceptable. Notional, provisional or formulaic amounts are not acceptable.

12.2　Under JCT 98

12.2.1　Background

There are two provisions in JCT 98 that may give rise to loss and/or expense claims by the contractor. They are clause 26, which deals with loss and expense caused by matters materially affecting regular progress of the works, and clause 34.3, which deals with loss and expense arising from the finding of 'fossils, antiquities and other objects of interest or value' found on the site or when excavating the same.

Both of these clauses impose obligations on the contractor and on the architect and/or quantity surveyor. Each of them confers on the contractor a legally enforceable right to financial reimbursement for 'direct loss and/or expense' suffered or incurred as a result of specified matters provided that he observes the procedures laid down by the provisions. Once the claims machinery has been put in motion by the contractor, the architect and/or quantity surveyor must carry out the duties imposed upon them.

However, the provisions for the giving of notice or the making of applications are not merely procedural. The true position has been well stated by the learned editors of *Building Law Reports* in the following terms:

'Contractors sometimes forget that provisions in the contract requiring them to give notice or to make an application if they wish to recover additional sums are not inserted solely as administrative procedure to be operated (like the handle on a fruit machine) when the right opportunity arrives. They are included in the contract not only to provide the engineer or architect with the opportunity of investigating the claim reasonably close to the time when material events occurred (so that the facts can be established with reasonable certainty), but also to provide the client with the chance to reconsider his budget (so that if necessary additional finance may be made available or savings may be made where it is still possible to do so).'[511]

In addition there is provision in clause 13A which allows the contractor's estimate of the amount of loss and/or expense he will incur in carrying out an instruction to be accepted and in clause 13.4.1.2 to similar effect in connection with the contractor's price statement. These clauses are considered in Chapter 13.

12.2.2 Clause 26

Clause 26 is long and complex but the contractor's entitlement to recovery under the clause depends on the correct operation of its machinery. It is therefore vitally important that all those concerned – contractors, architects and quantity surveyors – should fully understand its intention and the way in which it works.

The greater part of clause 26 deals with the contractor's and nominated sub-contractor's rights to financial reimbursement for events which are breaches of contract by, or which are within the control of, the employer himself, others for whom he is responsible, or the architect acting on the employer's behalf.

It is extremely important to note that a number of the matters to which the clause refers are not breaches of contract by the employer or those for whom he is responsible. For instance, an instruction to open up work for inspection or requiring a variation is one which the architect is specifically empowered to issue under the contract, and its issue, therefore, clearly cannot be a breach of contract which would otherwise entitle the contractor to recover damages. It follows that clause 26 contains the *sole* right of the contractor to compensation for such matters, and if the contractor loses his right to compensation under the clause, for instance by failing to make an application at the proper time, the preservation of his common law rights under clause 26.6 will avail him nothing. This consideration applies to the matters described in clauses 26.2.2, 26.2.5 and 26.2.7 and to deferment of possession of the site where clause 23.1.2 applies.

The clause is divided into six sub-clauses:

26.1 sets out the machinery. It sets out the rights and obligations of the contractor if he wishes to obtain payment and the rights and obligations of the architect and/or quantity surveyor in response thereto.
26.2 sets out the grounds that may entitle a contractor to payment.
26.3 sets out the architect's obligation, in certain circumstances, to notify the contractor of the grounds upon which he has granted extensions of time. This is a highly misleading sub-clause[512].

---

[511] 5 BLR 123 commenting on *J. Crosby & Sons Ltd* v. *Portland Urban District Council* (1967).
[512] This is considered in section 12.2.6.

26.4 provides the machinery whereby the similar provision in the Standard Nominated Sub-Contract Forms NSC/C is operated under the main contract.

26.5 in conjunction with clause 30, provides for certification and payment of amounts found due to the contractor.

26.6 preserves the contractor's common law and other rights if for any reason he cannot or does not wish to claim under clauses 26.1 to 26.5[513].

The full text of clause 26 is as follows:

26   Loss and expense caused by matters materially affecting progress of the Works

26.1   If the Contractor makes written application to the Architect stating that he has incurred or is likely to incur direct loss and/or expense (of which the Contractor may give his quantification) in the execution of this Contract for which he would not be reimbursed by a payment under any other provision in this Contract due to deferment of giving possession of the site under clause 23.1.2 where clause 23.1.2 is stated in the Appendix to be applicable or because the regular progress of the Works or of any part thereof has been or is likely to be materially affected by any one or more of the matters referred to in clause 26.2; and if and as soon as the Architect is of the opinion that the direct loss and/or expense has been incurred or is likely to be incurred due to any such deferment of giving possession or that the regular progress of the Works or of any part thereof has been or is likely to be so materially affected as set out in the application of the Contractor then the Architect from time to time thereafter shall ascertain, or shall instruct the Quantity Surveyor to ascertain, the amount of such loss and/or expense which has been or is being incurred by the Contractor; provided always that:

26.1   .1   the Contractor's application shall be made as soon as it has become, or should reasonably have become, apparent to him that the regular progress of the Works or of any part thereof has been or was likely to be affected as aforesaid; and

26.1   .2   the Contractor shall in support of his application submit to the Architect upon request such information as should reasonably enable the Architect to form an opinion as aforesaid; and

26.1   .3   the Contractor shall submit to the Architect or to the Quantity Surveyor upon request such details of such loss and/or expense as are reasonably necessary for such ascertainment as aforesaid.

26.2   The following are the matters referred to in clause 26.1:

26.2   .1   .1   where an Information Release Schedule has been provided, failure of the Architect to comply with clause 5.4.1;
        .2   failure of the Architect to comply with clause 5.4.2;

26.2   .2   the opening up for inspection of any work covered up or the testing of any of the work, materials of goods in accordance with clause 8.3

---

[513] This is important in view of the Court of Appeal decision in *Lockland Builders Ltd* v. *John Kim Rickwood* (1995) 77 BLR 38, which seems to suggest that in the absence of express provision, contract machinery and common law rights can co-exist only in circumstances where the contractor displays a clear intention not to be bound by the contract. The more recent Court of Appeal decision in *Strachan & Henshaw Ltd* v. *Stein Industrie (UK) Ltd* (1998) 87 BLR 52 takes a different view and, in any event, clause 26.6 puts the matter beyond doubt.

(including making good in consequence of such opening up or testing), unless the inspection or test showed that the work, materials or goods were not in accordance with this Contract;

26.2 .3   any discrepancy in or divergence between the Contract Drawings and/or the Contract Bills and/or the Numbered Documents;

26.2 .4 .1   the execution of work not forming part of this Contract by the Employer himself or by persons employed or otherwise engaged by the Employer as referred to in clause 29 or the failure to execute such work;

      .2   the supply by the Employer of materials and goods which the Employer has agreed to provide for the Works or the failure so to supply;

26.2 .5   Architect's instructions under clause 23.2 issued in regard to the postponement of any work to be executed under the provisions of this Contract;

26.2 .6   failure of the Employer to give in due time ingress to or egress from the site of the Works or any part thereof through or over any land, buildings, way or passage adjoining or connected with the site and in the possession and control of the Employer, in accordance with the Contract Bills and/or the Contract Drawings, after receipt by the Architect of such notice, if any, as the Contractor is required to give, or the failure of the Employer to give such ingress or egress as otherwise agreed between the Architect and the Contractor;

26.2 .7   Architect's instructions issued
      under clause 13.2 or clause 13A.4.1 requiring as a Variation (except for a Variation for which the Architect has given a confirmed acceptance of a 13A Quotation or for a Variation thereto) or
      under clause 13.3 in regard to the expenditure of provisional sums (other than instructions to which clause 13.4.2 refers or an instruction for the expenditure of a provisional sum for Performance Specified Work);

26.2 .8   the execution of work for which an approximate quantity is included in the Contract Bills which is not a reasonably accurate forecast of the quantity of work required;

26.2 .9   compliance or non-compliance by the Employer with clause 6A.1.

26.2 .10   suspension by the Contractor of the performance of his obligations under the Contract to the Employer pursuant to clause 30.1.4 provided the suspension was not frivolous or vexatious;

26.2 .11   save as provided for in clauses 26.2.1 to 26.2.10 any impediment, prevention or default, whether by act or omission, by the Employer or any person for whom the Employer is responsible except to the extent that it was caused or contributed to by any default, whether by act or omission, of the Contractor or his servants, agents or sub-contractors.

26.3   If and to the extent that it is necessary for ascertainment under clause 26.1 of loss and/or expense the Architect shall state in writing to the Contractor what extension of time, if any, has been made under clause 25 in respect of the Relevant Event or Events referred to in clause 25.4.5.1 (so far as that clause refers to clauses 2.3, 13.2, 13.3 and 23.2) and in clauses 25.4.5.2, 25.4.6, 25.4.8 and 25.4.12.

26.4   .1   The Contractor upon receipt of a written application properly made by a Nominated Sub-contractor under clause 4.38.1 of conditions NSC/C shall pass to the Architect a copy of that written application. If and as soon as the Architect is of the opinion that the loss and/or expense to which the said clause 4.38.1 refers has been incurred or is likely to be incurred due to any deferment of the giving of possession where clause 23.1.2 is stated in the Appendix to apply or that the regular progress of the sub-contract works or of any part thereof has been or is likely to be materially affected as referred to in clause 4.38.1 of Conditions NSC/C and as set out in the application of the Nominated Sub-Contractor then the Architect shall himself ascertain, or shall instruct the Quantity Surveyor to ascertain, the amount of loss and/or expense to which the said clause refers.

26.4   .2   If and to the extent that it is necessary for the ascertainment of such loss and/or expense the Architect shall state in writing to the Contractor with a copy to the Nominated Sub-Contractor concerned what was the length of the revision of the period or periods for completion of the sub-contract works or of any part thereof to which he gave consent in respect of the Relevant Event or Events set out in clause 2.6.5.1 (so far as that clause refers to clauses 2.3, 13.2, 13.3 and 23.2 of the Main Contract Conditions), 2.6.5.2, 2.6.6, 2.6.8, 2.6.12 and 2.6.15 of Conditions NSC/C.

26.5   Any amount from time to time ascertained under clause 26 shall be added to the Contract Sum.

26.6   The provisions of clause 26 are without prejudice to any other rights and remedies which the Contractor may possess.

12.2.3 Clause 26.1: The claims machinery

In order to grasp the general meaning of this sub-clause, it is useful to set out a brief and fairly broad paraphrase before discussing the parts in detail:

When the contractor considers that regular progress of any part or the whole of the Works has been or is likely to be substantially affected by any of the matters in clause 26.2 and if he also considers that this or any deferment of possession may result or has resulted in direct loss and/or expense on the contract for which he would not be able to receive payment under any other term in the contract, he may if he so wishes make a written application to the architect, but he must do so as soon as it has become, or should reasonably have become, apparent to him.

The architect must then decide whether or not the contractor is correct and may, to assist him in forming his opinion, request the contractor to submit the information that will reasonably enable him to do so. If the architect agrees with the contractor, he must himself ascertain, or instruct the quantity surveyor to ascertain, the amount of the direct loss and/or expense which the contractor has incurred. The architect or the quantity surveyor may, in order to enable them to carry out the ascertainment, require such details of loss and/or expense as are reasonably necessary for that purpose.

12.2.4 Commentary on the text

Written application essential

The making of a written application by the contractor is clearly a condition precedent. In other words, failure by the contractor to apply in writing as specified in the contract is fatal to his claim for payment under the contractual machinery, as can be seen later.

Nature of application

The contractor's application should be in writing, but no particular form is specified. Clearly it should state that the contractor has incurred or is likely to incur loss and/or expense arising directly from the deferment of giving possession of the site or the material effect upon the regular progress of the works or any part of the works of one or more of the 11 matters listed in clause 26.2. It may be that the application is sufficient if it refers to the general grounds and identifies the occurrence, stating that loss and/or expense is being or is likely to be incurred. A better view is that the application should clearly specify whether it is deferment of possession or which of the matters listed in clause 26.2 is relied upon. It is also advisable, even at this stage, for the contractor to provide a certain amount of detail about the circumstances that have given rise to his application. For example, suppose the architect has failed to supply a necessary piece of information in due time. The contractor's written application should refer to clause 26.2.1, and should state the date on which he specifically applied in writing for the information in question, the date on which it should have been supplied to him, and the effect which the lack of information has had, is having and will have upon contract progress, going into as much relevant detail as possible.

From the wording of clause 26.1 it is plain that the contractor need make only one written application in respect of loss and/or expense arising out of the occurrence of any one event. This will entitle him to recover past, present and future loss and/or expense arising from that event, and there is no need to make a series of applications as was the case under the equivalent provisions of JCT 63. It was to meet the difficulties created by the need for successive applications under JCT 63 that clause 26.1 has been redrafted,

> 'to require applications to be made "as soon as it has become, or should reasonably have become apparent to him [the contractor] that the regular progress of the works or any part thereof has been or is *likely to be affected*" by specified events ... and to state "that he has incurred *or is likely to incur* direct loss and/or expense" ...' (italics in judgment)[514]

It is equally plain that a so-called 'general' or 'protective' notice is not sufficient under clause 26.1. Specific written applications must be made in respect of each event. Further, the issue of an 'automatic' standard letter application every time one of the events listed occurs does not satisfy the requirements of clause 26.1 unless it clearly refers to the appropriate grounds. Where such an application does not satisfy clause 26.1, submitting it is a fruitless exercise.

The contractor's written application under clause 26.1 is related to the degree to which one or more of the matters listed in clause 26.2 has affected or is likely to affect regular progress, and the contractor must have genuine and sustainable grounds for believing this to be the case. Although the point unaccountably seems to be ignored in the construction industry, the making of an application under clause 26 which is not genuine and which the contractor knows not to be genuine is nothing short of attempted fraud.

Timing of application

Clause 26.1.1 requires that the contractor's written application should 'be made as soon as it has become, or should reasonably have become, apparent to him that regular progress of the Works or any part thereof has been or was likely to be' materially affected. In the case of deferred possession, it appears that an application should be made as soon as notification is received from the employer that possession of the site is to be deferred. The application must, therefore, be made at the earliest reasonable time and certainly before regular progress of the works is actually affected, unless there are good reasons why the contractor could not foresee that this would be the case. Although clause 26.1 allows for an application to be made at the time of or after the event, the intention is clearly that the architect should be kept informed at the earliest possible time of all matters likely to affect the progress of the work and likely to result in a claim for loss and/or expense.

---

[514] *F. G. Minter Ltd* v. *Welsh Health Technical Services Organisation* (1980) 13 BLR 7 at 20 per Stephenson LJ.

[515] *Hersent Offshore SA and Amsterdamse Ballast Beton-en-Waterbouw BV* v. *Burmah Oil Tankers Ltd* (1979) 10 BLR 1; *Diploma Constructions Pty Ltd* v. *Rhodgkin Pty Ltd* [1995] 11 BCL 242; *Wormald Engineering Pty Ltd* v. *Resource Conservation Co International* [1992] 8 BCL 158; *Opat Decorating Service (Aust) Pty* v. *Hansen Yuncken (SA) Pty* [1995] BCL vol 11, 360; *City Inn Ltd* v. *Shepherd Construction Ltd*, 17 July 2001, unreported (upheld on appeal).

Application at the right time is clearly a condition precedent to the contractor's entitlement to payment[515]. The architect must not forget that he owes a duty to the employer to reject claims which do not fulfil the time criterion. It is not that he may ignore them; rather that he has no power to consider such claims.

Failure to notify the architect in advance, where it is practicable to do so, will deprive him of the opportunity to take any remedial action open to him and the contractor may therefore be under some difficulty in establishing why he could not give earlier notice. An early, rather than a late, application is therefore essential to enable the contractor to demonstrate that he has taken all reasonable steps to mitigate the effect upon progress and the financial consequences. If the architect fails to take advantage of this, then clearly it is his responsibility and not the contractor's when answering to the employer for the extra cost involved.

The prime objective of the whole machinery of application is to bring the architect's attention to the possibility that regular progress of the works is likely to be affected and will be costly to the employer unless he takes action to avoid it. There should be no question of the contractor's written application being made after the event if it is reasonable for him to anticipate trouble in the future. Even if it is not reasonable for the contractor to anticipate such problems, the application must be made as soon as the trouble occurs, and not just within a 'reasonable time' of it occurring.

There are occasions when the question of whether a written application under clause 26.1 should be made is itself difficult to resolve. Clearly, the making of the application should be the result of a deliberate decision made by the contractor and not an automatic response to, for instance, the issue of every architect's instruction. There are cases in which it will be difficult for the contractor to determine whether progress is likely to be affected. Circumstances may already have affected progress so that the occurrence of the new event may at the time not seem likely to have any further effect. There may also be other factors:

> 'Notice of intention to claim, however, could not well be given until the intention had been formed ... [and] it seems to me that the contractors must at least be allowed a reasonable time in which to make up their minds. Here the contractors are a limited company, and that involves that, in a matter of such importance as that raised by the present case, the relevant intention must be that of the board of management [i.e. directors] ... in determining whether a notice has been given as soon as practicable, all the relevant circumstances must be taken into consideration .... One of the circumstances to be considered in the present case is the fact that it was not easy to determine whether the engineer's orders ... did or did not involve additional work ...'.[516]

---

[515] *Tersons Ltd* v. *Stevenage Development Corporation* (1963) 5 BLR 54 at 68 per Wilmer LJ. It should be noted that the court was there concerned with the ICE Conditions (2nd Edition) and the question of whether certain notices were given 'as soon as practicable'.

Similar circumstances can be envisaged in relation to architect's instructions.

However, such circumstances are likely to be exceptional and it must be emphasised that contractors must make their applications under clause 26.1 at the earliest practicable time. If the architect is in doubt whether the contractor's application has been made in due time, a useful test is for the architect to consider whether the lateness of the application prejudices the employer's interests in any way.

Architect's assumed state of knowledge

Clause 26 now clearly assumes that the architect cannot be expected to have more than a general knowledge of what is happening on site. This view has judicial support:

> 'the architect is not permanently on the site but appears at intervals, it may be of a week or a fortnight …'.[517]

It is the contractor who is responsible for progressing the work in accordance with the requirements of the contract and the architect's instructions. The practical effect of the contractor's obligation to notify as soon as the regular progress is likely to be materially affected is quite significant[518]. The architect is entitled to assume, unless notified to the contrary, that work is progressing smoothly and efficiently and that there are no current or anticipated problems. For instance, if he issues an instruction – even one requiring extra work – and the contractor accepts it without comment, the architect is probably entitled to assume that the effects of that instruction can be absorbed by the contractor into his working programme without any consequential delay or disruption[519]. That is not invariably the case. In *London Borough of Merton* v. *Stanley Hugh Leach Ltd*[520] it was said:

> 'Although I accept that the architect's contact with the site is not on a day to day basis there are many occasions when an event occurs which is sufficiently within the knowledge of the architect for him to form an opinion that the contractor has been involved in loss or expense.'[521]

This is particularly so where there has been some default by the architect himself. It would be unsafe for an architect to rely too heavily on the strict letter of the contractor's obligation as to timing of the notice when he himself has been late in the issue of information which has been requested at the proper time and he must well know that this is bound to have a

[517] *East Ham Corporation* v. *Bernard Sunley & Sons Ltd* [1965] 3 All ER 619 at 636 per Lord Upjohn when speaking of the architect's duty to 'supervise' work.

[518] *Jennings Construction Ltd* v. *Birt* [1987] 8 NSWLR 18.

[519] *Doyle Construction Ltd* v. *Carling O'Keefe Breweries of Canada* (1988) Hudson's Building and Civil Engineering Contracts, 1995, 4.133.

[520] (1985) 32 BLR 51.

[521] *London Borough of Merton* v. *Stanley Hugh Leach Ltd* (1985) 32 BLR 51 at 96 per Vinelott J, quoting the interim award of the arbitrator.

material effect on progress leading to the contractor incurring loss and expense.

Direct loss and/or expense

The loss and/or expense which is the subject of the contractor's claim must be *direct* and not consequential. The listed matter or matters relied upon must be the cause of the loss and/or expense and the phrase may be equated with the common law right to damages. (This matter is fully discussed in Chapters 5 and 7.)

Payment under other contract provisions

The reference to 'would not be reimbursed by a payment under any other provision of this Contract' is to prevent double payment, for instance where increased costs of labour and materials during a period of delay to completion are already being recovered under the fluctuations provisions of the contract. In relation to claims arising from clause 26.2.7 (see below), obviously some care must be taken to distinguish between those costs which are covered by the quantity surveyor's valuation under clause 13 and those for which reimbursement may be obtained under this clause[522]. There is, however, another aspect to this phrase which is often overlooked. Contractors often claim on a 'this or that' basis, hopeful that what they miss under one clause they will recover under the other. This strategy may be successful, but the use of 'would not be' rather than 'has not been' before 'reimbursed' is significant. The effect is that if the contractor is entitled to be reimbursed under any other clause, he is not entitled to be reimbursed under clause 26 whether or not he has actually received reimbursement under any other clause. It seems that if he is entitled to recover under clause 13, he must persevere in his attempts for he cannot recover as loss and/or expense what amounts to a shortfall in clause 13.

Material effect on regular progress

The basis of the contractor's entitlement to claim under clause 26 is that 'the regular progress of the Works or of any part thereof has been or is likely to be materially affected by any one or more' of the 11 matters listed in clause 26.2. In other words, it is the effect of the stated matter upon the regular progress of the Works, i.e. any delay to or disruption of the contract progress.

It is useful to consider the meaning which has been attributed to these important words by various commentators. Some suggest that the effect is to confine the contractor's entitlement to the financial results of *delay* to

---

[522] See the proviso to clause 13.5 and the full discussion of variations under JCT standard form contracts in Chapter 13.

progress only. For example, I. N. Duncan Wallace has said in relation to clause 24 of JCT 63:

> 'Bearing in mind that the most important of the matters listed in this clause (late instructions) is a breach of contract, it seems remarkable that a claim under the clause is limited to heads of damage caused by *delay to progress* and not to other possible consequences of the late instructions. The restriction or qualification seems unnecessary and unfair to the contractor, though no doubt the phrase "regular progress having been materially affected" may be interpreted liberally in the contractor's interest, *but it is clear that it does not cover loss of productivity.*'[523] (emphasis added)

This view may be doubted. It would have been simple to use express words to confine the entitlement to delay to regular progress. The draftsman chose not to do so, preferring the broader expression actually used. Clearly regular progress can be affected other than by delay alone. To exclude other effects pays no attention to the words used and offends against common sense and the straightforward commercial intention of the contract. There can be a disturbance to regular progress, resulting in loss of productivity in working, without there being any delay as such either in the progress or in the completion of the work. The clause cannot be interpreted so as to confine the contractor's right to reimbursement to circumstances that delay progress. It covers circumstances that may give rise, for instance, to reduced efficiency of working without progress as a whole being delayed. Murdoch and Hughes sum up the position like this:

> 'It should at this point be stressed once again that a claim for "loss and/or expense" is based, *not* on delay in completion of the works, but on the fact that the regular progress of those works has been disrupted. It is true that most cases do in fact concern delayed completion, and indeed some types of loss can *only* arise where this is so, but this must be seen as coincidental.'[524]

This seems to be a correct view of the position. It should be noted, however, that this is not the same as saying that merely because the work has proved to cost more or to take longer to complete than was anticipated, the contractor is entitled to additional payment. It must be possible for him to demonstrate that the cause is directly attributable to one or more of the matters set out in clause 26.2 and the effect upon regular progress of the works.

The words 'regular progress' have caused difficulty. They are related to the contractor's obligation under clause 23.1 'regularly and diligently [to] proceed with' the works, which phrase has been the subject of judicial comment:

---

[523] I. N. Duncan Wallace, *Building and Civil Engineering Standard Forms*, 1969 with 1970 and 1973 supps. p. 113 (commenting on clause 24 of JCT 63).
[524] John Murdoch and Will Hughes, *Construction Contracts: Law and Management*, 3rd edition, 2000, Spon Press, p. 218.

'These are elusive words on which the dictionaries help little. The words convey a sense of activity, of orderly progress, of industry and perseverance; but such language provides little help on the question of how much activity, progress and so on is to be expected. They are words used in a standard form of building contract and in those circumstances it may be that there is evidence of usage among architects, builders and building owners or others that would be helpful in construing the words. At present, all I can say is that I remain somewhat uncertain as to the concept enshrined in those words.'[525]

This is not particularly helpful to the contractor. So far as the related phrase 'due diligence' is concerned, a court had this to say:

'If there had been a term as to due diligence, I consider that it would have been, when spelt out in full, an obligation on the contractors to execute the works with such diligence and expedition as were reasonably required in order to meet the key dates and completion date in the contract.'[526]

However, 'regularly and diligently' has been defined more comprehensively recently by the Court of Appeal:

'What particularly is supplied by the word "regularly" is not least a requirement to attend for work on a regular daily basis with sufficient in the way of men, materials and plant to have the physical capacity to progress the works substantially in accordance with the contractual obligations.

What in particular the word "diligently" contributes to the concept is the need to apply that physical capacity industriously and efficiently towards the same end.

Taken together the obligation upon the contractor is essentially to proceed continuously, industriously and efficiently with appropriate physical resources so as to progress the works steadily towards completion substantially in accordance with the contractual requirements as to time, sequence and quality of work.'[527]

Whether or not there is regular progress and whether or not it has been, or is likely to be, materially affected must be a matter of objective judgment in each case. This judgment may be assisted by any programme of works prepared by the contractor, and the requirement for a master programme in clause 5.3.1.2 should be of great assistance in this respect. It is difficult to

---

[525] *London Borough of Hounslow* v. *Twickenham Garden Developments Ltd* (1970) 7 BLR 81 at 120 per Megarry J.
[526] *Greater London Council* v. *Cleveland Bridge & Engineering Co Ltd* (1984) 8 Con LR 30 at 40 per Staughton J. At appeal, the court affirmed the judgment at first instance.
[527] *West Faulkner Associates* v. *London Borough of Newham* (1995) 11 Cost LJ 157 at 161 per Simon Brown LJ.

understand why the requirement for a master programme should be optional: no sensible employer or professional adviser would contemplate deleting this clause. It is not enough, however, to simply request a programme. If it is to be of maximum assistance, the programme should be in the form of, or at least demonstrate, a critical path network, showing all activities, logic links and the associated resources.

The contractor's progress may already not be regular, due to factors within his control or which do not give him entitlement to claim. That is not fatal to his claim under this clause although it will present severe evidential problems. Among other things, he will have to demonstrate what regular progress should have been and further prove that, irrespective of his own failures in this respect, regular progress would have been affected by the matter specified.

The words 'or any part thereof' clearly emphasise the distinction between clauses 26 and 25. Extensions of time under the latter clause must relate to delay in completion of the contract as a whole or, where a sectional completion supplement is used, to any defined section. Reimbursement for the effect on regular progress under clause 26 may relate to circumstances affecting any part of the work, even down to individual operations. The essential difference between the clauses has been neatly summed up in a Scottish case:

> 'Although it took various forms the essential argument presented by counsel for the pursuers, as I understood it, was that a certificate of extension of time issued under clause 25 had no direct bearing on a claim based on disruption under clause 26. While there might, indeed, be many situations on the ground which would result in both clauses being invoked, the purpose of an extension of time certificate was to avoid a claim for liquidated damages rather than found a claim for disruption.
>
> In my opinion one has only to look at the terms of the two clauses to see that this argument is self evidently correct. The operation of clause 25 depends on the occurrence of a "Relevant Event", as there defined whereas the operation of clause 26 depends on whether "*the regular progress of the Works or any part thereof has been or is likely to be materially affected by any one or more of the matters referred to in clause 26.2...*"'[528]

Regular progress must have been, or be likely to be *materially* affected. 'Materially' has been defined as, among other things, 'in a considerable or important degree'[529], and it is suggested that this definition is applicable here. Trivial disruptions such as are bound to occur on even the best-run contract are clearly excluded. The circumstances must be such as to affect regular progress of the works 'in a considerable or important degree'. The affectation must be of some substance. A more recognisable and serviceable

---

[528] *Methodist Homes Housing Association Ltd* v. *Messrs Scott & McIntosh* 2 May 1997, unreported.
[529] *Chambers Twentieth Century Dictionary.*

word is 'substantially', although perhaps less precise. The particular point at which disruption becomes considerable or important is impossible to define in general terms. It must depend upon the circumstances of the particular case.

Architect's opinion

The initiative for taking action under clause 26 comes from the contractor who makes a written application. Provided that the application is properly made in accordance with the terms of the contract, the initiative then passes to the architect. If the architect forms the opinion that the contractor has suffered or is likely to suffer direct loss and/or expense due to deferment of possession, or because regular progress has been substantially affected by matters as stated in the contractor's application, then as soon as he does so, he must initiate the next stage: the ascertainment of the resulting direct loss and/or expense.

Some contractors argue that the architect's opinion must be reasonable, but that is not what the contract says. The only reference to 'reasonable' in this context is in clause 26.1.2, but that clause refers to the contractor's obligation to submit such information as should reasonably enable the architect to form an opinion. It is not the opinion or the information which must be reasonable. It is the enabling which must be reasonable.

It should be noted that the architect's opinion is of prime importance. The mere making of an application does not entitle the contractor to money if, in the architect's opinion, no money is due – though, of course, ultimately it may be the opinion of an arbitrator that will finally determine the matter. The process of ascertainment does not begin unless and until the architect has formed the opinion that deferment of possession has given rise to direct loss and/or expense or that regular progress has been or is likely to be materially affected as *set out in the application of the contractor*.

In *F. G. Minter Ltd* v. *Welsh Health Technical Services Organisation*[530], both the High Court and the Court of Appeal usefully analysed the contractual machinery of the claims provisions of JCT 63 as a temporal sequence in chronological order. The conclusions may be related to JCT 98 and summarised as follows:

(1) the deferment of possession or a matter under clause 26.2;
(2) the incurring of direct loss and/or expense;
(3) a written application by the contractor under clause 26.1;
(4) the forming of an opinion by the architect as to whether the direct loss or expense would or would not have been reimbursed under another provision and whether there has been or is likely to be a material effect on regular progress (to assist him in which he may require further information from the contractor);
(5) the ascertainment;

---

[530] (1980) 13 BLR 7.

(6) certification;
(7) payment.

Where, unusually, the contractor could not reasonably have anticipated that regular progress would be affected, and therefore his application is made at the time of the effect, stages (4) and (5) may overlap or be contemporaneous.

If the contractor's application is not made at the proper time (see section 12.2.4 above) then the architect must reject it (subject to the caveat above), whatever its merits may be, and he is not empowered to form an opinion about it. Further, the architect can deal only with matters that are set out in the contractor's written application; he has no authority to deal with any matters affecting regular progress that are not the subject of a written application from the contractor albeit the architect may be fully aware of them. However, if the contractor has made an application in due time, the architect must make use of any information he has in forming his opinion, because the architect is not a stranger on the works[531].

Further information

Clause 26.1.2 entitles the architect to request the contractor to supply such further information as is reasonably necessary to enable him to form an opinion as to the effect on regular progress or whether the deferment of possession has given or is likely to give rise to direct loss and/or expense. It is in the contractor's own interest to provide as much relevant information as possible at the time of his written application and not to wait until the architect asks for it under this sub-clause. The information which the architect is entitled to request is that which should *reasonably* enable him to form an opinion; an architect is not entitled to delay matters by asking for more information than is reasonably necessary.

It is thought that the architect must attempt to specify the precise information required – for example pages 3 and 4 of the site agent's diary or the time sheets for 27 and 28 July 2004 – rather than simply requiring the contractor to 'prove' his point. The contractor is entitled to know what would satisfy the architect and enable him to form a view. This appears to be the position in law; it certainly should be the aim of the architect, who might otherwise be accused of delaying tactics.

The ascertainment

The word 'ascertainment' is defined as meaning 'find out (for certain), get to know'[532]. It is not therefore simply a matter for the unfettered judgment of the architect or quantity surveyor; the duty upon them is *to find out* the amount of the direct loss and/or expense for certain, not to estimate it. It

---

[531] *London Borough of Merton* v. *Stanley Hugh Leach Ltd* (1985) 32 BLR 51.
[532] *Concise Oxford Dictionary*.
[533] *Alfred McAlpine Homes North Ltd* v. *Property & Land Contractors Ltd* (1995) 76 BLR 65. See the consideration of this point in Chapter 6, section 6.5.

also follows that the loss and/or expense that has to be found out must be that which is being, or has been, actually incurred.[533] References to estimated figures included in the contract bills have no relevance unless, due to the contractor's failure to keep proper records, there are no better data, and therefore an assessment rather than an ascertainment is the best that can be done. The architect cannot refuse to certify payment to the contractor of a reasonable assessment of direct loss and/or expense that he has incurred because proper information is not available, though clearly in such circumstances the architect will be cautious and the contractor may have to accept an assessment figure that is less than his actual loss. It is therefore up to contractors to keep adequate records.

The architect may instruct the quantity surveyor to ascertain the direct loss and/or expense. In these circumstances it will be difficult for the architect to certify anything other than the amount ascertained by the quantity surveyor. However, responsibility for certification of the amount still lies with the architect, who may be held to be negligent if he was to certify without taking reasonable steps to satisfy himself of the correctness of the amount[534]. What such steps may be will depend on all the circumstances, but the architect should, at least, go through the basis of ascertainment with the quantity surveyor to satisfy himself that the correct principles have been put into effect. There is nothing in the contract which suggests that the architect is bound to accept the quantity surveyor's opinion or valuation when he exercises his own function of certifying sums for payment[535]. It is always possible for the architect to ask the quantity surveyor for assistance in ascertaining the amount to be paid without formally instructing him to ascertain, although it seems that there is little to be gained from so doing.

It is essential, however, that the architect's instruction or his request for assistance be precisely set out in writing. The quantity surveyor's agreement to assist must also be in writing so as to establish the quantity surveyor's responsibility to the architect should his advice be given negligently. In any case, whether the architect formally instructs the quantity surveyor or simply asks for his assistance in ascertaining the amount due to the contractor, the employer must be informed of this arrangement, since fees will be involved.

Details of loss and/or expense

As the duty of the architect or the quantity surveyor is to ascertain the amount of the direct loss and/or expense as a matter of fact (so far as this may be attainable) it is necessary for them to look to the contractor as the person in possession of the facts to provide them. Clause 26.1.3 stipulates that the contractor must submit on request such details of the loss and/or expense as are reasonably necessary for the ascertainment. This does not

---

[534] *Sutcliffe* v. *Thackrah* [1974] 1 All ER 859.
[535] *R. B. Burden Ltd* v. *Swansea Corporation* [1957] 3 All ER 243.

necessarily mean the submission of an elaborately formulated and priced claim, although it may well be in the contractor's interest to provide it, particularly if it is expected that the matter may move to adjudication or arbitration, and clause 26.1 expressly allows the contractor to submit one if he so wishes. It is suggested that such details might include comparative programme/progress charts in network form, pin-pointing the effect upon progress, together with the relevant extracts from wage sheets, invoices for plant hire, etc[536].

> 'If [the contractor] makes a claim but fails to do so with sufficient particularity to enable the architect to perform his duty or if he fails to answer a reasonable request for further information he may lose any right to recover loss or expense under those sub-clauses and may not be in a position to complain that the architect was in breach of his duty.'[537]

These are sensible words which highlight not only the contractor's responsibility, but also the consequences if he refuses to help himself. In such a case he is left with only himself to blame. It is important to consider the other side of the coin. The architect's or the quantity surveyor's requests for further information must be reasonably precise. When he receives the request, the contractor should be able to understand with a fair degree of accuracy what he must provide. It is not thought to be sufficient if the architect or the quantity surveyor simply asks for 'proof' or says that the contractor must provide 'more details'. Endless vague requests of this kind, as a delaying tactic, are all too common. Neither the architect nor the quantity surveyor should ask the contractor for information which they already possess. As a basic rule, the contractor should be requested to provide no more than is strictly necessary, indeed clause 26.1.3 states as much, and the necessary information must be particularised by the architect or the quantity surveyor. On receipt of the request the contractor should know that when the information is provided, ascertainment of the whole claim can be completed without delay.

12.2.5 Commentary on the 'matters' giving rise to entitlement

In *Henry Boot Construction Ltd* v. *Central Lancashire New Town Development Corporation* it was said of JCT 63 provisions equivalent to those now found in JCT 98, clauses 25 and 26:

> 'The broad scheme of these provisions is plain. There are cases where the loss should be shared, and there are cases where it should be wholly borne by the employer. There are also those cases which do not fall within either of these conditions and which are the fault of the contractor, where the loss of both parties is wholly borne by the contractor. But in the cases where the fault is not that of the contractor the scheme clearly is that

---

[536] The documentation side of claims is considered in Chapter 9.
[537] *London Borough of Merton* v. *Stanley Hugh Leach Ltd* (1985) 32 BLR 51 at 104 per Vinelott J.

in certain cases the loss is to be shared; the loss lies where it falls. But in other cases the employer has to compensate the contractor in respect of the delay, and that category, where the employer has to compensate the contractor, should, one would think, clearly be composed of cases where there is fault upon the employer or fault for which the employer can be said to bear some responsibility.'[538]

This is sometimes pointed to as a masterly exposition of the position and so it is, but only if the reader accepts the premise that clauses 25 and 26 are linked. Of course they are not linked and it is clear that they are for different purposes. The paragraph is, therefore, not very helpful.

Clause 26.2 deals with those cases where the employer must compensate the contractor and it is only if the contractor can establish that he has incurred direct loss and/or expense not otherwise reimbursable as a direct result of one or more of the 11 matters listed in the clause (or, of course, as a direct result of deferment of possession of the site) that he is entitled to reimbursement. Each of these 11 matters are discussed below.

Late instructions, drawings, details or levels: clause 26.2.1

This ground covers failure by the architect to provide necessary information to the contractor at the proper time to enable the contractor to use it for the purpose of the works. It ties in with the scheme of the contract, which is essentially that the contractor builds only what he is told to build. Consequently, until the contractor has received the appropriate information, he has no obligation and indeed cannot proceed with the contract works. There are two parts to this ground. One covers the situation if there is an information release schedule; a situation so rare as to be virtually non-existent. The other covers the general situation where the architect is obliged to provide information to enable the contractor to complete the works in accordance with the contract. Among the important terms of the contract with which the contractor must comply is the date for completion.

There are two parts to this ground in another sense, because the architect's obligation is not only to provide correct information, but also to provide it at the correct time. The 'matter' simply concerns a failure of the architect to comply with clause 5.4 and, therefore, embraces the failure with regard to time or correct information. Either failure may cause a delay which may affect the completion date.

The commentary on this ground under JCT 98 in Chapter 10, section 10.1 is relevant.

Opening up for inspection and testing: clause 26.2.2

Clause 8.3 entitles the architect to require work to be opened up for inspection and to instruct the contractor to arrange for or to carry out the testing of

---

[538] (1980) 15 BLR 8 at 12 per Judge Edgar Fay.

materials and executed work in order to ensure that they comply with the contract. It is a curious feature of the wording of clause 8.3 that the cost of such opening up and testing is to be reimbursed to the contractor unless the inspection or tests show that the materials or work are not in accordance with the contract. In other words it is for the architect to show that they are not in accordance with the contract and not for the contractor to show that they are. Somewhat similar considerations apply to the question of whether the contractor is entitled to an extension of time under clause 25.4.5.2 and to the question of entitlement to reimbursement under this clause. However, the provision for extension of time is broader, including instructions under clause 8.4.4. Each of these provisions assumes that the contractor will be so entitled unless the inspection or tests demonstrate that the materials or work are not in accordance with the contract.

An interesting situation arises if the specification or bills of quantities direct that work must not be covered up until after inspection by the architect. Failure to observe that provision will clearly place the contractor in breach of contract. However, he will still be entitled to payment under this matter if the work is found to be in accordance with the contract. The solution to this problem lies in the employer's ordinary entitlement to damages for the contractor's breach. The damages are clearly the money that the employer has to pay out under clause 26 and clause 8.3. Although there is no machinery in the contract to enable the employer to recover such money, he can do so, after giving the relevant notices, by setting off the amount he has paid out against the amount payable under the certificate.

Any discrepancy in or divergence between the contract drawings and/or the contract bills and/or the numbered documents: Clause 26.2.3

The contractor's entitlement to direct loss and/or expense is limited to that resulting from discrepancies or divergences in and between the contract drawings and/or the contract bills and/or the numbered documents. The simplicity of the wording of clause 26.2.3 contrasts sharply with the elaborate provisions of clause 2.3, which refers to discrepancies in or divergences between any two or more of the following:

(1) The contract drawings
(2) The contract bills
(3) Any architect's instructions other than variations
(4) Any drawings or documents issued by the architect for the general purposes of the contract
(5) The numbered documents.

Reimbursement of loss and/or expense is confined to discrepancies in or between the contract documents (other than the printed form). The numbered documents are not part of the contract documents, but they are defined in clause 1.3 as being part of the nominated sub-contract documents. Architect's instructions and further drawings and documents are issued during the progress of the works. Therefore, if they differ from each other or from the contract documents, the discrepancy is likely to be discovered virtually on issue and will be promptly corrected by an architect's instruction under clause 2.3 with little or no effect on the progress of the works. The contractor's obligation is not to find discrepancies, but merely to notify the architect if he does find them[539]. Therefore, he may not discover a discrepancy until after that portion of the work has been constructed. That does not prevent the contractor from claiming on this ground, provided he complies with the requirements of clause 26.

Work not forming part of the contract: clause 26.2.4

Clause 26.2.4 covers two situations. The first is where the employer himself carries out work not forming part of the contract or arranges for such work to be done by others while the contract works are being executed by the contractor. It also covers the situation where the employer has undertaken to provide materials or goods for the purposes of the works.

Clause 29 provides that, where the work in question is properly described in the contract bills, the contractor must permit such work to be done. In that situation, it will only be if the work concerned causes an unforeseeable delay or disruption to the contractor's own work that any claim will lie under this sub-clause. That is because the contractor, having had due warning in the contract bills, will have been able to make proper provision in the contract sum and in his programme for the effect on the works of such work. Clause 29 also provides that if the work in question is not adequately described in the contract bills but the employer wishes to have such work executed, the employer may arrange for it to be done with the consent of the contractor, which is not to be unreasonably withheld. In that event, a claim will almost inevitably arise under clause 26.2.4.1, because the contractor will not have been able to make proper allowance in his programme and price.

A claim under clause 26.2.4.1 may, in certain circumstances, relate to work carried out by statutory undertakers, where that work is carried out by them as a matter of contractual obligation, rather than as one of statutory obligation[540]. The general legal position of statutory undertakers is that they are under a statutory obligation to carry out certain work and, when so doing, their obligation if any 'depends on statute and not upon contract'[541]. When acting in that capacity, any interference on their part with the contractor's work does not give rise to any monetary claim against the employer under this or any other provision in the contract, but may give rise to a claim for extension of time under clause 25.4.11. An arbitrator found as a fact in regard to a particular project that the statutory undertakers 'were

---

[539] *London Borough of Merton* v. *Stanley Hugh Leach Ltd* (1985) 32 BLR 51.
[540] This follows from the decision in *Henry Boot Construction Ltd* v. *Central Lancashire New Town Development Corporation* (1980) 15 BLR 8, which is of relevance to JCT 98 regarding the meaning of words in this subclause 'work not forming part of this contract'.
[541] *Clegg Parkinson & Co* v. *Earby Gas Co* (1896) 1 QB 56.

engaged under contract by the respondents to construct the said mains', and it would appear that he also found that most, if not all, of the work being carried out by the statutory undertakers was not covered by their statutory obligation, but was being executed by them voluntarily at the employer's request and expense. At trial it was said:

> 'These statutory undertakers carried out their work in pursuance of a contract with the employers; that is a fact found by the arbitrator and binding on me … In carrying out [their] statutory obligations they no doubt have statutory rights of entry and the like. But here they were not doing the work because statute obliged them to; they were doing it because they had contracted with the [employers] to do it.'[542]

Clause 26.2.4.2 is very odd. It refers to the supply or failure to supply materials and goods which the employer has agreed to supply for the works. But the contract makes no provision for this kind of agreement, although perhaps it should. Therefore, any such agreement would have to be the subject of a separate contract, preferably reduced to writing and by both parties. A further point is that, as a supply only contract, it would not fall under the Housing Grants, Construction and Regeneration Act 1996[543]. The ground on which the contractor may apply is not simply the employer's *failure* to carry out the work or supply materials. He may claim if the employer executes the work or supplies the materials perfectly properly and at the right time. The criteria is whether the execution of work or the supply of materials and goods, whether or not at the correct time, materially affects regular progress of the works. This lays a considerable burden on the employer and makes the employer's decision to employ others upon the works something akin to writing a blank cheque.

Postponement of work: clause 26.2.5

The scope of clause 26.2.5 is often misunderstood. It refers to the power of the architect, under clause 23.2, to 'issue instructions in regard to the postponement of *any work* to be executed under the provisions of (the) Contract' (emphasis added). Clause 23.2 does not empower the architect to issue an instruction postponing the date as stated in the appendix upon which possession of the site is to be given to the contractor, nor does he have power to do so under any other express or implied term of the contract, and this is emphasised by clause 23.1.2 which gives the employer, and not the architect, the right to defer possession of the site for a period not exceeding six weeks or any shorter period stated in the appendix.

Before the introduction of clause 23.1.2, the contractor's right to possession of the site on the date given in the appendix was an absolute one[544].

---

[542] *Henry Boot Construction Ltd* v. *Central Lancashire New Town Development Corporation* (1980) 15 BLR 8 at 20 per Judge Edgar Fay.
[543] See section 105(2)(d) of the Act.
[544] A fuller consideration of this topic is to be found in Chapter 4, section 4.6.

Possession refers to the whole of the site and, in the absence of sectional possession, the employer is not entitled to give possession in parcels[545]. It has been suggested that provided 'the contractor has sufficient possession, in all the circumstances, to enable him to perform, the employer will not be in breach of contract'[546]. This proposition, while it may suggest a way of arriving at the measure of damages, must be treated with caution. His right to possession is an express term of the contract (see clause 23.1) and in any event there is at common law an implied term in any construction contract that the employer will give possession of the site to the contractor in time to enable him to carry out and complete the work by the contractual date[547]. In *London Borough of Hounslow* v. *Twickenham Garden Developments Ltd*[548], Megarry J said, 'The contract necessarily requires the building owner to give the contractor such possession, occupation or use as is necessary to enable him to perform the contract'.

Accordingly, subject to the right to defer possession for a limited time under clause 23.1.2 if that clause is stated in the contract appendix to apply, any failure by the employer to give possession on the due date is a breach of contract, entitling the contractor to bring a claim for damages at common law in respect of any loss that he suffers as a consequence[549]. It can also be addressed in clause 26 as a matter under clause 26.2.11. In an Australian case it was clearly held that, where there was failure to give possession of the building site to a contractor, this constituted a breach of contract, and on the facts the contractor was there to be entitled to treat the contract as repudiated[550]. Under JCT 98, clauses 25.4.19 and 26.2.11 may be able to be used to forestall such action.

Although the clause refers to instructions issued under clause 23.2, such instructions have been held to arise as a matter of fact[551].

Whether a postponement instruction gives rise to any loss and/or expense at all, or to what extent it does so, must be the subject of careful investigation by the architect or, if so instructed, the quantity surveyor. For example, if the instruction is issued relatively early, so that the contractor can use his best endeavours to prevent any delay, if it is of short duration and if, most importantly, it applies to non-critical activities, the effect upon regular progress may be negligible.

---

[545] *Whittal Builders* v. *Chester-Le-Street District Council* (1987) 40 BLR 82.
[546] Stephen Furst and Vivian Ramsey, *Keating on Building Contracts*, 7th edition, 2001, Sweet & Maxwell, p. 711.
[547] *Freeman & Son* v. *Hensler* (1900) 64 JP 260.
[548] (1970) 7 BLR 81.
[549] *London Borough of Hounslow* v. *Twickenham Garden Developments Ltd* (1970) 7 BLR 81. See Chapter 4 for claims at common law.
[550] *Carr* v. *Berriman Pty Ltd* (1953) 27 ALJR 273.
[551] See *M. Harrison & Co (Leeds) Ltd* v. *Leeds City Council* (1980) 14 BLR 118, where an instruction expressed as a variation order was held to be in fact an order for postponement and *Holland Hannen & Cubitts (Northern)* v. *Welsh Health Technical Services Organisation* (1981) 18 BLR 80, where a notice which was apparently intended to notify the contractor of defective work was held to instruct postponement.

Failure to give ingress to or egress from the site: clause 26.2.6

Clause 26.2.6 covers the situation where either there is a provision in the contract bills and/or drawings, or an agreement has been reached between the contractor and the architect, permitting the contractor means of access to the site 'through or over any land, buildings, way or passage adjoining or connected with the site and *in the possession and control of the Employer*'. The contractor must, of course, comply with any requirement as to notice.

The words emphasised above should be noted; the clause does not apply where the employer may have undertaken to obtain a wayleave across land which is not in his possession and control. In such a case failure by the employer to obtain the wayleave would give rise to a claim at common law. It should be further noted that the land etc. must be in both the possession *and* the control of the employer, and the land etc. must also be 'adjoining or connected with the site', which would suggest that there must be contiguity but not necessarily physical contact.

Variations and work against provisional sums: clause 26.2.7

Valuations of the work involved in, and general consequences of, variations and instructions issued by the architect for the expenditure of provisional sums are dealt with in clause 13, which is discussed in Chapter 13. Clause 26.2.7 covers disturbance costs where the introduction of the variation or provisional sum work materially affects the regular progress of the works in general. In considering entitlement to payment under this ground, it must be remembered that clause 26 does not cover a situation where the contractor would be reimbursed under any other clause. Particular care must be taken when considering entitlement under this ground, because clause 13 includes provision for the quantity surveyor to adjust the preliminary items. It is sometimes difficult to decide whether an event should be reimbursed as additional preliminaries or as loss and/or expense. However, the practice which is prevalent among some contractors, of submitting a claim in the alternative (variation or loss and/or expense) is prohibited by the distinction referred to above.

Approximate quantity not a reasonably accurate forecast of the quantity of work: clause 26.2.8

This matter was introduced with the use of the Standard Method of Measurement 7th Edition (SMM7). Quite simply, it is intended to cover the situation where an approximate quantity has been included in the bills of quantity, but the quantity of work actually executed under that item is different, either greater or less. As long as the approximate quantity is reasonably accurate, the contractor has no claim. What is 'reasonably accurate' will depend upon all the circumstances, but as a rule of thumb an approximate quantity which was within 10% of the actual quantity probably would be difficult to demonstrate as unreasonable. The contractor's entitlement will usually be based on the extra time he requires over and above the time he has allowed for doing the quantity of work in the bills. The approximate quantity may be an unreasonable estimate because the actual quantity is substantially less than in the bills. Theoretically, the contractor will also have grounds for a claim under this head, but it will take considerable ingenuity to put together. It should be noted that this clause expressly refers to 'work'. The conclusion is that increases in materials will not entitle the contractor to claim. Generally, it is only an increase in work or labour which will require extra time to execute, but there may be circumstances where increases in the quantity of materials may result in additional off-site and unquantified work, such as in the drawing office or the fabrication shed. It appears that the contractor will have no claim for such matters, at least under this head.

Compliance or non-compliance with duties in relation to the CDM Regulations: clause 26.2.9

Clause 6A provides, among other things, that the employer will ensure that the planning supervisor carries out all his duties under the Construction (Design and Management) Regulations 1994, and where the contractor is not the principal contractor for the purpose of the Regulations, that the principal contractor carries out all his duties under the CDM Regulations. It will be usual for the main contractor to be the principal contractor for the purposes of the Regulations and the employer's duty will just relate to the planning supervisor. The obligation placed upon the employer to 'ensure' is virtually to guarantee that the planning supervisor will carry out his duties[552]. It should be noted that the contractor's entitlement to recover loss and/or expense does not depend on the employer's failure to comply with his obligations under clause 6A. Compliance can also be a ground, provided of course that the other conditions are satisfied.

Suspension by the contractor of performance of his obligations: clause 26.2.10.

This ground is identical to relevant event clause 25.4.18 and the remarks there are equally applicable here. The contractor may suspend performance of his obligations after 7 days written notice if the employer does not pay him sums due.

It is noteworthy that the Housing Grants, Construction and Regeneration Act 1996, from which the power to suspend derives, does not expressly entitle the contractor to recover any losses resulting from the suspension. Nor is such recovery to be easily implied. It is virtually certain that a

---

[552] That has been the view of the court where a party has an obligation to 'ensure' or 'secure' the doing of something: *John Mowlem & Co Ltd* v. *Eagle Star Insurance Co Ltd* (1995) 62 BLR 126 CA, confirming the judgment of the Official Referee. The court made clear their view that 'to ensure' meant exactly what it said and amounted to more than an obligation to use best endeavours.

contractor who resorts to suspension, in an attempt to recover money owed, will be put to expense by the suspension and also by the reorganisation and mobilisation of resources necessary if the suspension comes to an end on payment of the amount due. This matter allows the contractor to be paid such sums which might otherwise be difficult to recover.

Impediment, prevention or default by the employer: clause 26.2.11

This ground is identical to relevant event clause 25.4.19 and the remarks there are generally applicable here. Because this ground is very broadly drafted to include breaches of contract on the part of the employer, it brings within the contractual machinery, and therefore, within the powers of the architect, many occurrences for which the contractor formerly would have had to mount a common law claim. However, such claims are still subject to any provision as to notice, etc., which are imposed by the contract.

12.2.6 Money claims and extensions of time

Clause 26.3 provides that if and to the extent that it is necessary for the purpose of ascertainment of direct loss and/or expense, the architect shall state in writing to the contractor what extension of time, if any, he has granted under clause 25 in respect of those events which are also grounds for reimbursement under clause 26.

There is no logical justification for the inclusion of this provision, which appears to give contractual sanction to the mistaken belief that there is some automatic connection between the granting of an extension of time and the contractor's entitlement to reimbursement. There is not or ought not to be any such connection[553]. An extension of time under clause 25 has only one effect. It fixes a new date for completion and, in so doing, it defers the date from which the contractor becomes liable to pay to the employer liquidated damages. An extension of contract time does not in itself entitle the contractor to any extra money. The correct position has been well stated thus:

> 'JCT 80 clause 25 entitles the contractor to relief from paying liquidated damages at the date named in the contract. It does not in any way entitle him to one penny of monetary compensation for the fact that the architect has extended the contractor's time completion. He is not entitled to claim even items set out in "Preliminaries" for the extended period.'[554]

Moreover, this information is of no interest or relevance to the contractor and cannot and should not have any relevance to any ascertainment of direct loss and/or expense under clause 26. An extension of time is, furthermore, essentially an estimate of what is likely to happen *before* the event. The

---

[553] See H. *Fairweather & Co Ltd* v. *London Borough of Wandsworth* (1987) 39 BLR 106.
[554] John Parris, *The Standard Form of Building Contract*, 2nd edition, 1985, Blackwell Science, in section 10.03: The relationship of JCT 80 clause 25 to clause 26.

ascertainment of direct loss and/or expense under clause 26 is essentially a finding-out of facts *after* the event.

It is for the architect or quantity surveyor (not the contractor) to ascertain the amount of loss and/or expense. It is impracticable to require the architect to provide the contractor with a breakdown of extensions of time between causes of delay, since this is of no relevance to the effect of an extension, as stated previously. If, as so often happens, there are a number of overlapping causes of delay, to apportion the overall extension between those various causes will often be impossible[555]. Clause 26.3 apparently merely requires the architect to specify the relevant events he has taken into account without apportioning. The clue is in the words 'If and to the extent it is *necessary for ascertainment*'. It is difficult to think of any situation where that will apply. *Methodist Homes Housing Association Ltd* v. *Messrs Scott & McIntosh* is relevant[556]. The judgment was very short. The judge said that the action was founded on a claim by the contractors for loss and/or expense due to disruption under clause 26 of the JCT 80 form. He upheld the essential argument of the claimants that a certificate of extension of time under clause 25 had no bearing on a claim based on disruption under clause 26. He went on to consider clause 26.3, which is very similar to clause 26.3 of JCT 98, in these terms:

'It is true that clause 26.3 provides that in certain circumstances the architect shall state in writing to the contractor what extension of time has been made under clause 25 but it is instructive that this provision only operates "if and to the extent that it is necessary for ascertainment under clause 26.1 of loss and/or expense".

In the end, [Counsel] for the pursuers, was, I think constrained to accept that there was no essential link between clause 25 and clause 26, but he nonetheless sought to persuade me that if reference was made to the notification certificates themselves and to the claim document (all of which were lodged in processes and incorporated in the pleadings) it could be seen that the extensions of time granted and the claim proceeded on exactly the same "Architect's Instructions". Having looked at these documents with [Counsel], however, I regret that I am quite unable to take that view. And even if they did, I am not sure that I fully understand the significance bearing in mind the distinct purposes of Clause 25 and Clause 26 respectively.'

12.2.7   Nominated sub-contract claims

Clause 4.38 of JCT Nominated Sub-Contract Form (NSC/C) is a provision corresponding to clause 26 of the main contract form enabling the

---

[555] *H. Fairweather & Co Ltd* v. *London Borough of Wandsworth* (1987) 39 BLR 106.
[556] 2 May 1997, unreported.

nominated sub-contractor to claim against the employer *through* the main contractor in respect of direct loss and/or expense and on similar grounds to those given to the main contractor by clause 26. Clause 26.4 provides the necessary machinery by which the contractor is to pass such claims on to the architect, and the architect's decision is to be passed back to the nominated subcontractor[557].

Clause 26.4.2 corresponds to clause 26.3 in a nominated sub-contractor situation and the same objections are equally sustainable in respect of it.

12.2.8 Certification of direct loss and/or expense

Clause 26.5 provides for amounts ascertained to be added to the contract sum. By clause 3, this also means that as soon as an amount of entitlement is ascertained in whole or in part, that amount is to be taken into account in the next interim certificate. By clause 30.2.2.2, such amounts are not subject to retention. The reference to ascertainment 'in part' in clause 3 means that it is not necessary for the full process of ascertainment to have been completed before an amount must be certified for payment. This is important from the point of view of both the contractor and the employer: in particular, where the direct loss and/or expense is being incurred over a period of time. The proper operation of the contractual machinery should ensure that the matter is dealt with in interim payments from month to month so far as is practicable. This provision for payment of sums ascertained 'in part' permits the inclusion in interim certificates of allowances for direct loss and/or expense where full ascertainment has not been possible in the time available, thereby ensuring proper cash-flow to the contractor and reducing the employer's possible liability for financing charges[558].

12.2.9 Clause 26 claims: a summary

For a claim under clause 26.1 to be sustainable it must meet the following requirements:

(1) The written application must be made at the proper time. If it is made out of time the architect must reject it, irrespective of merits.
(2) It must show that possession of the site has been deferred or that regular progress of the works has been or is likely to be affected by one or more of the 11 matters set out in clause 26.2.
(3) As much relevant detail as possible should be given in the initial application, and the contractor must be prepared on request to give sufficient further information and details as are reasonably necessary.

---

[557] See Chapter 17 for a full discussion of sub-contract claims.
[558] See *F. G. Minter Ltd* v. *Welsh Health Technical Services Organisation* (1980) 13 BLR 7, and the discussion in Chapter 4.

(4) The claim must relate only to direct loss and/or expense not recoverable under any other provision of the contract.

12.2.10 Contractor's other rights and remedies

Clause 26.6 preserves the contractor's common law and other rights[559]. The rights set out in clause 26 confer a specific contractual remedy on the contractor in the circumstances there defined, and subject to the conditions there imposed. But the contractor's other rights are unaffected, in particular his right to claim damages for breach of contract[560].

In other words, the specific contractual right to claim reimbursement for direct loss and/or expense is additional to any rights or remedies which the contractor possesses at law, and notably to damages for breach of contract. It may be that, because of the limitations imposed by the contract machinery, the contractor will be advised to pursue these independent remedies[561]. However, see the earlier consideration of those events which are not breaches of contract[562].

12.2.11 Clause 34.3

Clause 34 generally provides for what is to happen if 'fossils, antiquities and other objects of interest or value' are found on site or during excavation. The contractor is required to use his best endeavours not to disturb the object and to cease work as far as is necessary and to take all steps necessary to preserve the object in its position and condition. He is to inform the architect or the clerk of works. The architect is then required to issue instructions and the contractor may be required to allow a third party, such as an expert archaeologist, to examine, excavate and remove the object. All this will almost undoubtedly involve the contractor in direct loss and/or expense and clause 34.3.1 provides that this is to be ascertained by the architect or quantity surveyor without necessity for further application by the contractor.

The text of clause 34.3 is as follows:

> 34.3.1   If in the opinion of the Architect compliance with the provisions of clause 34.1 or with an instruction issued under clause 34.2 has involved the Contractor in direct loss and/or expense for which he would not be

---

[559] The Court of Appeal decision in *Lockland Builders Ltd* v. *John Kim Rickwood* (1995) 77 BLR 38, suggested that where a party's common law rights were not expressly reserved, they could co-exist with the contractual machinery only where the other party displayed a clear intention not to be bound by the contract. This view seems to ignore earlier contrary authority: *Modern Engineering (Bristol) Ltd* v. *Gilbert Ash (Northern) Ltd* (1974) 1 BLR 73; *Architectural Installation Services Ltd* v. *James Gibbons Windows Ltd* (1989) 46 BLR 91. The more recent Court of Appeal decision in *Strachan & Henshaw Ltd* v. *Stein Industrie (UK) Ltd* (1998) 87 BLR 52 appears to set the matter straight.
[560] *London Borough of Merton* v. *Stanley Hugh Leach Ltd* (1985) 32 BLR 51.
[561] See Chapter 4.
[562] Section 12.2.2.

reimbursed by a payment made under any other provision of this Contract then the Architect shall himself ascertain or shall instruct the Quantity Surveyor to ascertain the amount of such loss and/or expense.

34.3.2   If and to the extent that it is necessary for the ascertainment of such loss and/or expense the Architect shall state in writing to the Contractor what extension of time, if any, has been made under clause 25 in respect of the Relevant Event referred to in clause 25.4.5.1 so far as that clause refers to clause 34.

34.3.3   Any amount from time to time so ascertained shall be added to the Contract Sum.

Rather curiously, it is to be noted that there are no provisions similar to those in clauses 26.1.2 and 26.1.3 requiring the contractor to give further information and details. In practice, of course, an ascertainment can only be made if the relevant information and details are provided by the contractor to the architect or quantity surveyor.

Claims under clause 34.3 are not uncommon and are likely to increase. Modern technology, and especially aerial survey in England and Wales, is constantly revealing new archeological sites. The provisions of the Ancient Monuments and Archaeological Areas Act 1979, which came into force in 1982, affect contractors working in 'areas of archaeological importance', which is a concept introduced by Part II of the Act. Moreover, if an 'ancient monument' was suddenly discovered in the course of contract works, where one was unexpected, the possibility of its being scheduled under the Act (and thus protected) could not be completely discounted. In fact, this has happened only rarely. Instead, the Department of the Environment might make financial help available for the relocation of piling or the rafting-over of the remains, and in such a case the contractor would have a claim under clause 34.3.

Clause 34.3.2 is open to similar objections to those already made in respect of clause 26.3.

12.3 Under IFC 98

12.3.1 Clauses 4.11 and 4.12

The full text of the clauses is as follows:

Disturbance of regular progress

4.11   If, upon written application being made to him by the Contractor within a reasonable time of it becoming apparent, the Architect/the Contract Administrator is of the opinion that the Contractor has incurred or is likely to incur direct loss and/or expense, for which he would not be reimbursed by a payment under any other provision of this Contract, due to

(a)   the deferment of the Employer giving possession of the site under clause 2.2 where that clause is stated in the Appendix to be applicable; or

(b)   the regular progress of the Works or part of the Works being materially affected by any one or more of the matters referred to in clause 4.12, then the Architect/the Contract Administrator shall ascertain, or shall instruct the Quantity Surveyor to ascertain, such loss and expense incurred and the amount thereof shall be added to the Contract Sum provided that the Contractor shall in support of his application submit such information required by the Architect/the Contract Administrator or the Quantity Surveyor as is reasonably necessary for the purposes of this clause 4.11.

The provisions of this clause 4.11 are without prejudice to any other rights or remedies which the Contractor may possess.

Matters referred to in clause 4.11

4.12   The following are the matters referred to in clause 4.11:

4.12.1   .1 where an Information Release Schedule has been provided, failure of the Architect to comply with clause 1.7.1;

.1   failure of the Architect to comply with clause 1.7.1;

4.12.2   the opening up for inspection of any work covered up or the testing of any of the work, materials or goods in accordance with clause 3.12 (including making good in consequence of such opening up or testing), unless the inspection or test showed that the work, materials or goods were not in accordance with this Contract;

4.12.3   the execution of work not forming part of this Contract by the Employer himself or by persons employed or otherwise engaged by the Employer as referred to in clause 3.11 or the failure to execute such work;

4.12.4   the supply by the Employer of materials and goods which the Employer has agreed to supply for the Works or the failure so to supply;

4.12.5   the Architect's/the Contract Administrator's instructions under clause 3.15 issued in regard to the postponement of any work to be executed under the provisions of this Contract;

4.12.6   failure of the Employer to give in due time ingress to or egress from the site of the Works, or any part thereof through or over any land, buildings, way or passage adjoining or connected with the site and in the possession and control of the Employer, in accordance with the Contract Documents after receipt by the Architect/the Contract Administrator of such notice, if any, as the Contractor is required to give, or failure of the Employer to give such ingress or egress as otherwise agreed between the Architect/the Contract Administrator and the Contractor;

4.12.7   the Architect's/the Contract Administrator's instructions issued under clauses

1.4 *(Inconsistencies)* or

3.6 *(Variations)* or

3.8 *(Provisional sums)* except, where the Contract Documents include bills of quantities, for the expenditure of a provisional sum for defined work included in such bills,

or, to the extent provided therein, under clause
3.3 *(Named sub-contractors);*

4.12.8   the execution of work for which an Approximate Quantity is included in the Contract Documents which is not a reasonably accurate forecast of the quantity of work required;

4.12.9   compliance or non-compliance by the Employer with clause 5.7.1.

4.12.10 suspension by the Contractor of the performance of his obligations under this Contract to the Employer pursuant to clause 4.4A provided that the suspension was not frivolous or vexatious;

4.12.11 save as provided for in clauses 4.12.1 to 4.12.10 any impediment, prevention or default, whether by act or omission, by the Employer or any person for whom the Employer is responsible except to the extent that it was contributed to by any default, whether by act or omission, of the Contractor or his servants, agents or sub-contractors.

12.3.2 Main points

It will be seen that these clauses are essentially a version of clause 26 of JCT 98. The basic features remain:

- The requirement for a written application from the contractor as a starting point
- The architect to form an opinion as to whether the contractor has incurred or is likely to incur direct loss and/or expense
- The architect, or the quantity surveyor if so instructed, to ascertain the amount of the loss and/or expense
- The contractor to provide such information 'as is reasonably necessary' to achieve the end envisaged by the clause
- The right of the contractor to reimbursement under the contract is limited to deferment of possession of the site or specified causes materially affecting the regular progress of the works.

IFC 98 contains no equivalent of clause 34.3 of JCT 98 dealing with the discovery of antiquities.

12.3.3 Significant differences

Provision of information

It is not clear whether or not the obligation of the contractor to provide information in support of his application is subject to any request from the architect or quantity surveyor. The words 'required by' at first sight may appear to mean 'needed by', but the words may also mean 'demanded by'. If the contractor is to recover under the clause he must provide whatever information is 'reasonably necessary' to enable the architect and/or the quantity surveyor to carry out their own obligations. In the last edition of this book, the view was expressed that, in this context, 'required' should be interpreted as 'needed'. However, on mature reflection, 'demanded' appears to be the correct interpretation. The best that can be said is that the point is unclear and in the absence of clear words, it behoves the architect or the quantity surveyor to set out the request for information as clearly as possible. The contractor is advised to provide all the information he judges to be necessary without waiting for the request.

Discrepancies

Instead of a reference to 'any discrepancy in or divergence between the Contract Drawings and/or the Contract Bills' as in JCT 98 clause 26.2.3, the reference is to 'the Architect's/Contract Administrator's instructions issued under [clause] 1.4 *(Inconsistencies)*'. That clause requires the architect to issue instructions in regard to the correction of:

- Any inconsistency in or between the contract documents or drawings or details reasonably necessary to enable the contractor to carry out and complete the works (clause 1.7) or the determination of levels and the dimensioned drawings required for setting-out (clause 3.9)
- Any error in description or in quantity or any omission of items in the contract documents
- Any error or omission in the particulars provided of the tender of any named sub-contractor (clause 3.3.1)
- Any departure from the standard method of measurement upon which bills of quantities (if provided as one of the contract documents) have been prepared.

It will be seen surprisingly that this extends considerably beyond the very limited reference in JCT 98 to discrepancies only between or within the contract drawings, the contract bills and the numbered documents, to cover discrepancies between the contract documents and any other details provided by the architect including levels and setting-out drawings, errors in or omissions from the contract documents (which, in addition to drawings, may be bills of quantities, priceable schedules of work or specifications, priceable or otherwise) and errors in the tender documents of named sub-contractors. It is also clear that it is the effect on progress of the architect's instructions for the correction of the error or discrepancy which gives rise to the contractor's entitlement. All this seems so eminently sensible and reasonable that it is difficult to understand why JCT 98 has not been amended similarly.

Named persons as sub-contractors

The effect of certain architect's instructions regarding named sub-contractors upon regular progress is made a ground for claim. The special position of named sub-contractors in IFC 98 is not a subject for this book[563]. Briefly, the contractor is obliged to employ as sub-contractors persons or firms who are named, either in the contract documents or in an instruction from the architect for the expenditure of a provisional sum, to carry out specified parcels of work. The naming process involves the use of a standard form of invitation to tender, tender and articles of agreement (Form NAM/T) and standard conditions of sub-contract incorporated into the sub-contract by reference (Form NAM/SC), together with an optional form of Employer/ Sub-Contractor Agreement (ESA/1) similar in effect to Form NSC/W under JCT 98.

In contrast to the position regarding nominated sub-contractors under JCT 98, named sub-contractors under IFC 98 become effectively 'domestic' sub-contractors to the main contractor. Once the naming procedure has been completed the architect generally speaking has no further involvement with the administration of the sub-contract. He does not direct the main contractor as to the payments to be made to the named sub-contractors and he is not required to certify practical completion of their work. Matters such as extensions of time for the sub-contractor and settlement of his account are matters between the sub-contractor and the main contractor and neither the architect nor the quantity surveyor under the main contract will be involved.

However, the architect may become involved and may be required to issue instructions if things go badly wrong. In certain circumstances the main contractor may be entitled to reimbursement of any direct loss and/or expense arising from the effect of such instructions on the regular progress of the works. Those circumstances are:

(1) If a sub-contractor is named in the contract documents but the main contractor finds that he is unable in the event to enter into a sub-contract with the named firm (perhaps because the firm withdraws or no longer exists or because of some inconsistency between the sub-contract particulars and the main contract) the main contractor is to notify the architect specifying the reason. If the architect is satisfied that the reason is valid he may issue an instruction to

- change the particulars of the proposed sub-contract so as to remove the problem, or
- omit the work from the contract altogether, in which case the employer may have the work carried out under a separate contract, or
- omit the work from the contract documents and substitute a provisional sum, which would then entitle him to name another sub-contractor.

Instructions issued under the first two options are to be regarded as variation instructions, therefore entitling the contractor to an extension of time if the works are delayed and to direct loss and/or expense

---

[563] Readers are referred to Neil F. Jones and Simon Baylis, *JCT Intermediate Form of Building Contract*, 3rd edition, 1999, Blackwell Publishing; David Chappell and Vincent Powell-Smith, *JCT Intermediate Form of Contract – A Practical Guide*, 3rd edition 1998, Blackwell Publishing.

arising from any effect of the instruction (including the effect of the employment of others to carry out the work if the second option is followed) on the regular progress of the works. If the third option is followed, the 'naming' of a replacement sub-contractor becomes a 'naming' against a provisional sum; see below.

(2) If a sub-contractor is named in an instruction for the expenditure of a provisional sum, any direct loss and/or expense caused to the contractor by the effect of the instruction on regular progress of the works is recoverable under the reference to clause 3.8 in clause 4.12.7. For instance, if the named sub-contractor's programme is inconsistent with the main contract programme so that his employment will cause disruption and/or delay, the main contractor may make a 'reasonable objection' to entering into the sub-contract. If he is nevertheless instructed to proceed, he will be entitled to recover the direct loss and/or expense arising from the resulting disruption and/or delay. There may also be an implied term that if the contractor is obliged to accept the sub-contractor, an appropriate extension of time will be granted[564].

(3) If the employment of a named sub-contractor is determined by the main contractor because of the sub-contractor's default or insolvency, the contractor is to notify the architect who may then issue an instruction to

- name another sub-contractor to do the remaining work, or
- instruct the contractor to make his own arrangements for completion of the work, in which case the contractor may sub-let the remaining work to a sub-contractor of his own choice, or
- omit the remainder of the work from the contract, in which case the employer may employ someone else to do it.

If the original sub-contractor was named in the contract documents the exercise of the first option by the architect will entitle the contractor to an extension of time for any resulting delay, but will not entitle him to recover direct loss and/or expense from the employer. Both the other options, and if the original sub-contractor was named against a provisional sum all three options, will entitle the contractor both to an extension of time and to recovery of direct loss and/or expense arising from the resulting effect on progress. If either of the first two options is adopted the contract sum is to be adjusted to take account of the difference in price between what would have been payable to the original sub-contractor and what is now payable to the replacement sub-contractor if one is appointed or to the main contractor if he is instructed to make his own arrangements for completion. But the cost of rectifying the original named sub-contractor's defects is to be excluded. However, subject to certain indemnities from the employer (see clause 3.3.6(b)) the contractor is under an obligation to pursue the original sub-contractor to seek to recover the additional expense which the employer is required to meet.

---

[564] *Fairclough Building Ltd* v. *Rhuddlan Borough Council* (1985) 3 Con LR 38.

All this seems impossibly complex, and is certainly very difficult to follow in the form due to the multiplicity of cross-referencing adopted, but the principle is quite simple. Once a sub-contract has been entered into, if the sub-contractor was named in the main contract documents the contractor is deemed to have accepted that sub-contractor and to have accepted basic responsibility for his performance. If something goes badly wrong and the sub-contractor's employment has to be determined, this is to some extent to be regarded as *force majeure* entitling the contractor to an extension of time for the resulting delay, but he is not entitled to recover any loss or expense in which the determination involves him, unless he himself is required to take on full responsibility for the completion of the work or the employer employs someone else to complete it. If, however, the sub-contractor was named in an instruction for the expenditure of a provisional sum and, therefore, after the main contract was entered into, the employer accepts that greater degree of responsibility and will reimburse to the contractor any direct loss and/or expense resulting from the effect of the determination on progress in any event.

One interesting little point arises if the original sub-contractor is named in the main contract documents, the architect instructs the contractor to employ another sub-contractor and that sub-contractor then delays progress so that the contractor suffers direct loss and/or expense as a result of his default. The instruction naming the replacement sub-contractor is not expressed as being an instruction for the expenditure of a provisional sum and it would, therefore, seem that the contractor would not be entitled to recover the direct loss and/or expense arising from the replacement sub-contractor's delay. Indeed, applying the principles of causation in such a case, there are no grounds for the contractor to make a claim.

12.4 Under MW 98

12.4.1 Clauses 3.6 and 3.7

The text of clauses 3.6 and 3.7 is as follows:

> 3.6   The Architect may, without invalidating this Contract, order an addition to or omission from or other change in the Works or the order or period in which they are to be carried out and any such instruction shall be valued by the Architect on a fair and reasonable basis, using where relevant prices in the priced Specification/schedules/schedule of rates, and such valuation shall include any direct loss and/or expense incurred by the Contractor due to the regular progress of the Works being affected by compliance with such instruction or due to the compliance or non-compliance by the Employer with clause 5.6.
>
> Instead of the valuation referred to above, the price may be agreed between the Architect and the Contractor prior to the Contractor carrying out any such instruction.

Provisional sums

3.7   The Architect shall issue instructions as to the expenditure of any provisional sums and such instructions shall be valued or the price agreed in accordance with clause 3.6 hereof.

12.4.2 Main points

The JCT Agreement for Minor Building Works 1998 (MW 98) does not contain the equivalent of clause 26 of JCT 98 or clauses 4.11 and 4.12 of IFC 98. There is no clause which entitles the contractor to apply to the architect for recovery of loss and/or expense. The reason may be because the simple work content, low value and short contract periods for which the form is intended are unlikely to result in major loss and/or expense claims. That may be the theory, but in practice MW 98 generates claims as frequently if not to the same extent as more complex forms. MW 98 is a very popular form of contract; no doubt due in part at least to its brevity compared with other forms. It is essential, however, that the form should be used as intended. It is not unknown for MW 98 to be employed for projects with contract sums far in excess of the recommended limit[565]. In such cases, the likelihood of substantial claims is greatly increased.

Although the contract does not allow the contractor to instigate a claim for loss and/or expense, it states in clause 3.6 that the architect must include, in the valuation of an instruction regarding an omission, addition or other change in the works, the amount of direct loss and/or expense incurred by the contractor provided that the regular progress of the works has been affected. It cannot be emphasised too much that the loss and/or expense must result from the issue of what might normally be termed a variation instruction or from compliance or non-compliance with the clause requiring the employer's compliance with the CDM Regulations. Instructions about other things do not fall to be valued under this clause. The clause makes no specific reference to 'ascertainment', but the reference to 'incurred' clearly means that it is the actual amount and not some theoretical or formulaic figure which is intended.

The practice of some architects who include the contractor's preliminary costs in certificates following an extension of time, is wrong. The contractor is not entitled under the contract to recover these costs and an architect who certifies them may be negligent. There is no provision for the architect to request information nor is there a stipulation that the contractor must provide it. It is clearly in the contractor's interests to provide whatever information the architect needs. From time to time, it has been stated that, other than the very limited recovery allowed by clause 3.6, the contractor cannot recover any loss and/or expense he incurs due to matters such as failure by the architect to provide information or give instructions at the

---

[565] See *Deciding on the Appropriate Form of JCT Main Contract*, JCT Practice Note 5, Series 2 (2001) RIBA Publications.

right time or postponement of part of the work. Such a view is misconceived. The contractor may always exercise his right to claim damages for breach of contract. The exercise of such right is outside the machinery of the contract and, therefore, the architect has no power to consider a claim made on that basis. The contractor cannot claim in respect of the range of matters included in clause 26 of JCT 98, because some of those matters are not breaches of contract. In practice, architects often do consider common law claims on behalf of their clients, but an architect in this position should be aware that he is exceeding his contractual duties and he must obtain specific authorisation from his client, preferably in writing.

Even if an architect is authorised by the employer to consider common law claims, any amounts which the architect may consider to be due cannot be included in the certification process, because it is not money due under the terms of the contract. Instead, it must be paid directly by the employer to the contractor after correspondence between the parties recording their agreement[566].

12.4.3 The provision summarised

- The architect is the instigator of the process when he values a variation instruction.
- Regular progress of the works must be affected by compliance with the instruction or by compliance or non-compliance by the employer with the CDM Regulations.
- There is no express requirement that the contractor must provide information to enable the architect to form an opinion or ascertain the amount.

12.5 *Under WCD 98*

12.5.1 Clause 26

The text of the clause is as follows:

> 26   **Loss and expense caused by matters affecting regular progress of the Works**
>
> 26.1 If the Contractor makes written application to the Employer stating that he has incurred or is likely to incur direct loss and/or expense (of which the Contractor may give his quantification) in the execution of this Contract for which he would not be reimbursed by a payment under any other provision in this Contract due to deferment of giving possession of the site under clause 23.1.1 where clause 23.1.2 is stated in Appendix 1 to be applicable or because the regular progress of the Works or any part thereof has been or is

---

[566] See Chapter 4 for a full consideration of common law claims.

likely to be materially affected by any one or more of the matters referred to in clause 26.2 the amount of such loss and/or expense which has been or is being incurred by the Contractor shall be added to the Contract Sum; provided always that:

26.1   .1   the Contractor's application shall be made as soon as it has become or, should reasonably have become, apparent to him that the regular progress of the Works or of any part thereof has been or was likely to be affected as aforesaid, and

26.1   .2   the Contractor shall in support of his application and in respect of the amount of the loss and/or expense provide upon request by the Employer such information and details as the Employer may reasonably require.

26.2   The following are the matters referred to in clause 26.1:

26.2   .1   the opening up for inspection of any work covered up or the testing of any of the work materials or goods in accordance with clause 8.3 (including making good in consequence of such opening up or testing), unless the inspection or test showed that the work, materials or goods were not in accordance with this Contract; or

26.2   .2   delay in receipt of any permission or approval for the purposes of Development Control Requirements necessary for the Works to be carried out or proceed, which delay the Contractor has taken all practicable steps to avoid or reduce; or

26.2   .3   .1   the execution of work not forming part of this Contract by the Employer himself or by persons employed or otherwise engaged by the Employer as referred to in clause 29 or the failure to execute such work; or

26.2   .3   .2   the supply by the Employer of materials and goods which the Employer has agreed to provide for the Works or the failure so to supply; or

26.2   .4   Employer's instructions issued under clause 23.2 in regard to the postponement of any work to be executed under the provisions of this Contract; or

26.2   .5   failure of the Employer to give in due time ingress to or egress from the site of the Works or any part thereof through or over any land, buildings, way or passage adjoining or connected with the site and in the possession and control of the Employer, in accordance with the Employer's Requirements after receipt by the Employer of such notice, if any, as the Contractor is required to give, or failure of the Employer to give such ingress or egress as otherwise agreed between the Employer and the Contractor; or

26.2   .6   Employer's instructions issued under clause 12.2 effecting a Change or under clause 12.3 in regard to the expenditure of provisional sums; or

26.2   .7   the Contractor not having received in due time necessary instructions, decisions, information or consents from the Employer which the Employer is obliged to provide or give under the Conditions including a decision under clause 2.4.2 and for which he specifically applied in writing provided that such application was made on a date which having regard to the Completion Date was neither unreasonably distant from nor unreasonably close to the date on which it was necessary for him to receive the same; or

26.2 .8 compliance or non-compliance by the Employer with clause 6A.1; or

26.2 .9 suspension by the Contractor of the performance of his obligations under the Contract to the Employer pursuant to clause 30.3.8 provided the suspension was not frivolous or vexatious; or

26.2 .10 save as provided for in clauses 26.2.1 to 26.2.9 any impediment, prevention or default, whether by act or omission, by the Employer or any person for whom the Employer is responsible except to the extent that it was caused or contributed to by any default, whether by act or omission, of the Contractor or his servants, agents or sub-contractors.

26.3 Any amount from time to time ascertained under clause 26 shall be added to the Contract Sum.

26.4 The provisions of clause 26 are without prejudice to any other rights and remedies which the Contractor may possess.

12.5.2 Main points

The JCT Standard Form of Building Contract With Contractor's Design 1998 (WCD 98) is the JCT design and build contract. It is superficially very like JCT 98. Indeed, it is clearly based on the earlier contract. More correctly, the previous edition, CD 81, was based upon JCT 80 and when JCT 80 was updated to JCT 98, CD 81 was updated in tandem to WCD 98. In any event the superficial resemblance is misleading, because the form is based on a different philosophy. Put shortly, the contract is contractor driven. That means that some of the important actions carried out by the architect or the quantity surveyor under a traditional contract are carried out by the contractor. The architect, or to be precise the employer's agent, simply checks to make sure that the contractor is correct. For example, the contractor is responsible for valuation, determining the amount of payment and ascertainment of loss and/or expense.

This difference of approach is very marked when comparing clause 26.1 in JCT 98 and clause 26.1 in WCD 98. The first part of the clause is identical with the substitution of employer for architect, and the contractor may if he wishes make an application to the employer in writing. It will be seen, however, that the second part of the clause is quite different. There is no requirement for the architect or anyone else to form an opinion or to ascertain the amount of loss and/or expense. There is simply the bald statement that the loss and/or expense incurred must be added to the contract sum. Since it is the contractor's responsibility to make application for interim payment under clause 30.1, he must also calculate the actual loss and/or expense he has suffered. Effectively, the burden of proving that the ascertainment is wrong is laid on the employer under clauses 30.3.3 and 30.3.4. This position is emphasised by the wording of clause 26.1.2 which is not quite an amalgam of clauses 26.1.2 and 26.1.3 of JCT 98, but which simply obliges the contractor to provide 'in support of his application' the information requested by the employer which he reasonably requires. There is no reference to the purpose for which it is required. Contrast with the equivalent clause in JCT 98 wherein it is made very clear that the information and details are required to reasonably enable the architect to form an opinion or as are reasonably necessary for the quantity surveyor to ascertain. The proviso as to timing of the application is identical to JCT 98.

12.5.3 Other significant differences

Development control requirements

A very important clause 26.2.2 is inserted which provides grounds for entitlement if there is any delay in receipt of permission or approval for the purposes of development control requirements. The permission must be necessary for the works to start or to be carried out. In addition, the contractor must have taken all practicable steps to avoid or reduce the delay. What steps a contractor can in practice take are severely limited. Probably all he can do is to make the appropriate development control requirements submission as soon as possible. Development control requirements has a broader meaning than simply planning requirements, but the phrase is not so broad as to encompass all kinds of statutory requirement. The contractor will usually formulate his application under this head when he has been left to make the planning submission and approval has been delayed. Planning approval may be delayed for many reasons and it is always safer for the employer to secure the approval before letting the contract. Attempts to avoid the problem by deleting this ground can be fraught with difficulty, because several clauses refer to this topic and there is a danger that the matter may be dealt with inadequately[567].

Discrepancies

There is no ground which is equivalent to clause 26.2.3 dealing with discrepancies in or between documents. The logic is inescapable: if the contractor is responsible for the whole of the documentation, he cannot blame the employer if there are discrepancies. He may of course be able to seek redress from the consultants he directly employs.

Change instructions

Clause 26.2.6 is apparently the same as the equivalent clause in JCT 98 (26.2.7) except for the terminology: 'variation' in JCT 98, 'change' in WCD

---

[567] *Update Construction Pty Ltd* v. *Rozelle Child Care Ltd* (1992) 9 BLM 2.

98. A change under WCD 98 has a very restricted meaning and refers only to a change in the Employer's Requirements. The employer has no power under the contract to change the design or the construction directly – a point frequently overlooked by employers and their agents. More importantly, this clause is the vehicle by which the contractor can recover if there is a delay in statutory requirements in general and particularly where, under clause 6.3.1, a change in statutory requirements after base date is to be treated as an instruction of the employer requiring a change.

The most significant ground is probably the employer's instruction requiring the expenditure of a provisional sum. Provisional sums must be included in the Employer's Requirements, but they are often provided with little explanation about their purpose. It follows, therefore, that although the contractor must programme the works with reasonable skill and care so that completion by the due date can be achieved, in practice he may be unable to judge the amount of work included in a provisional sum with any confidence. In such circumstances, he is entitled to make only such allowance in the programme as can be justified, which may be nothing at all in many instances. The practical consequence may be that an employer's instruction requiring the expenditure of a provisional sum will have effect on the programme as though it was a simple addition of work and/or materials.

Decisions, information and consent

Clause 26.2.7 is similar in general approach to clause 26.2.1 of JCT 98 and the comments on that clause generally apply here also. Key differences are that under WCD 98, the employer is not obliged to provide drawings, details or levels and of course there is no reference to an information release schedule, but there is reference to decisions, information and consent which he is obliged to provide under the contract. This restricts the contractor's opportunities to claim under the clause, because not only is he rightly prevented from claiming because he has not obtained drawings at the right time, but he cannot claim if the employer is late in giving a decision which he is not strictly obliged to give under the contract.

This clause is identical to clause 25.4.6 except for the addition of the word 'and' after 'clause 2.4.2'. This addition makes clear that the specific written application refers to all decisions and consents and not merely those under clause 2.4.2. It is probably a simple omission from the drafting of clause 25.4.6 which becomes clear when the whole contract is read in context.

Bills of quantities

There is no equivalent clause to clause 26.2.8 of JCT 98, because if there are bills of quantities, normally they will have been produced by the contractor as part of his contract sum analysis. Bills will very rarely be produced by the employer under supplementary provision S5 because an employer who prepares bills of quantities will lose many of the advantages of design and build.

12.5.4 Clause 34.3

The purpose of clause 34 is similar to clause 34 in JCT 98 in that it provides for what is to happen if the contractor finds fossils or other antiquities. The employer may give instructions about the examination, excavation or removal of the object. The contractor must use his best endeavours to avoid disturbing the object and to preserve it in its exact position even if this involves cessation of part or the whole of the work. It is virtually certain that the contractor will incur loss and/or expense and clause 34.3 provides that if the contractor suffers such loss, it must be added to the contract sum. The text is as follows:

> 34.3.1   If compliance with the provisions of clause 34.1 or with an instruction issued under clause 34.2 has involved the Contractor in direct loss and/or expense for which he would not be reimbursed by a payment made under any other provision of this Contract then the amount of such loss and/or expense shall be added to the Contract Sum.
>
> 34.3.2   If and to the extent that it is necessary for the ascertainment of such loss and/or expense the Employer shall state in writing to the Contractor what extension of time, if any, has been made under clause 25 in respect of the Relevant Event referred to in clause 25.4.5.1 so far as that clause refers to clause 34.

The only stipulation is that the contractor would not be reimbursed under any other clause. The comments made in section 12.2.11 generally apply, but it should be noted that, in line with the philosophy of other parts of this contract, whether the contractor has incurred loss and/or expense is not made subject to the opinion of the architect (there is no architect) nor even the opinion of the employer. In practice, it is left for the contractor to calculate the amount and to include it in his application, and for the employer, if he so chooses, to dispute the amount under clause 30.3.4. If the contract states that supplementary provision S7 applies, the contractor must include these amounts in his claims under that provision.

12.5.5 Supplementary provision S7

The text is as follows:

S7   Direct loss and/or expense – submission of estimates by Contractor

S7.1   Clause 26 shall have effect as modified by the provisions of S7.2 to S7.6.

S7.2   Where the Contractor pursuant to clause 26.1 is entitled to an amount in respect of direct loss and/or expense to be added to the Contract Sum, he shall (except in respect of direct loss and/or expense dealt with or being dealt with under S6) on presentation of the next Application for Payment submit to the Employer an estimate of the addition to the Contract Sum which the Contractor requires in respect of such loss and/or expense which he has incurred in the period immediately preceding that for which the Application for Payment has been made.

S7.3 Following the submission of an estimate under S7.2 the Contractor shall for so long as he has incurred direct loss and/or expense to which clause 26.1 refers, on presentation of each Application for Interim Payment submit to the Employer an estimate of the addition to the Contract Sum which the Contractor requires in respect of such loss and/or expense which has been incurred by him in the period immediately preceding that for which each Application for Payment is made.

S7.4 Within 21 days of receipt of any estimate submitted under S7.2 or S7.3 the Employer may request such information and details as he may reasonably require in support of the Contractor's estimate but within the aforesaid 21 days the Employer shall give to the Contractor written notice either

S7.4 .1 that he accepts the estimate; or

S7.4 .2 that he wishes to negotiate on the amount of the addition to the Contract Sum and in default of agreement to refer the issue for decision pursuant to article 6A or 6B whichever is applicable; or

S7.4 .3 that the provisions of clause 26 shall apply in respect of the loss and/or expense to which the estimate relates.

If the Employer elects to negotiate pursuant to clause S7.4.2 and agreement is not reached the provisions of clause 26 shall apply in respect of the loss and/or expense to which the estimate relates.

S7.5 Upon acceptance or agreement under S7.4.1 or S7.4.2 as to the amount of the addition to the Contract Sum such amount shall be added to the Contract Sum and no further additions to the Contract Sum shall be made in respect of the direct loss and/or expense incurred by the Contractor during the period and in respect of the matter set out in clause 26.2 to which that amount related.

S7.6 If the Contractor is in breach of S7.2 and S7.3 direct loss and/or expense incurred by the Contractor shall be dealt with in accordance with clause 26 but any resultant addition to the Contract Sum shall not be included in Interim Payments but shall be included in the adjustment of the Contract Sum under clause 30.5. Provided that such addition shall not include any amount in respect of loss of interest or financing charges in respect of such direct loss and/or expense which have been suffered or incurred by the Contractor prior to the date of issue of the Final Statement and Final Account or of the Employer's Final Statement and the Employer's Final Account as the case may be.

It should be noted that clause 26 is modified, but not superseded by this provision and that any loss and/or expense which is recoverable under clause S6 (valuation of change instructions) is not recoverable under this provision. If the supplementary provisions are stated in the Appendix to apply, the contractor will be expected to initiate this procedure without further prompting. The penalty for failure to proceed in accordance with this clause are severe.

The procedure is triggered as soon as the contractor becomes entitled to have some loss and/or expense added to the contract sum. He must include the amount in the next application for payment. The amount claimed must be referable to the period immediately prior to the application and since the previous application. The contractor must act quickly. He must make application under clause 26.1 and also he must include an estimate of the amount in the next application for payment. Use of the word 'estimate' acknowledges that it may not be a precise figure. With each successive application, the contractor must continue to submit estimates until the loss comes to an end and each estimate must refer only to the period prior to the application. On receipt of each estimate, the employer must give a written notice to the contractor within 21 days. The notice must state either:

- that the employer accepts the estimate; or
- that he wishes to negotiate the amount; or
- that the provisions of clause 26 apply.

The clause does not stipulate a time limit for the negotiations, but there is no reason why a suitable time limit should not be inserted by the parties. Before sending the notice, the employer may request information reasonably required to support the contractor's estimate, but both request and receipt of the information must take place within the 21 days.

When agreement is reached or, failing agreement, the amount of loss and/or expense has been determined by another stipulated method, the sum must be added to the contract sum. No further amount may be added for loss and/or expense in respect of the same time period. If the contractor fails to submit estimates in accordance with the timetable, clause S7 ceases to apply and loss and/or expense is dealt with under clause 26. However, the amounts are not payable until the final account and final statement are agreed. Moreover, the contractor is not entitled to any interest or financing charges incurred before the issue of the final account and final statement. There is much to commend in S7. If the employer is well-organised and sensible, the contractor should have few opportunities to claim. The provision provides a kind of fast track claims procedure which ensures the contractor will receive loss and/or expense quickly if he provides estimates quickly. The parties are encouraged to agree on the amount and it is in their interests to do so. The similarity to claims provisions in the Association of Consultant Architects Form of Building Agreement (ACA 3) form of contract should be noted.

12.6   Under PCC 98

12.6.1   Clauses 4.13–4.16

The text of the clauses is as follows:

Notice by Contractor

4.13 .1 Clauses 4.13 to 4.16 apply where the Contractor gives written notice to the Architect stating that he has incurred or is likely to incur direct loss and/or expense (of which the Contractor may give his quantification) in the execution of this Contract for which he would not be reimbursed by a payment under any other provision in this Contract due to

the deferment of the Employer giving possession of the Site under clause 2.1.2 where that clause is stated in the Appendix to be applicable; or

the regular progress of the Works or part thereof being materially affected by any one or more of the matters referred to in clause 4.14.

Written notice under clause 4.13.1 shall be given as soon as it has become or should have become reasonably apparent to the Contractor that the regular progress of the Works or any part thereof has been so materially affected.

Ascertainment

.2 If and as soon as the Architect is of the opinion that the direct loss and/or expense to which the notice under clause 4.13.1 refers has been incurred or is likely to be incurred, the Architect from time to time thereafter shall ascertain, or shall instruct the Quantity Surveyor to ascertain, the amount of such loss and/or expense.

Submission of information etc. by Contractor

.3 The Contractor shall

submit to the Architect upon request such information as should reasonably enable the Architect to form the opinion to which clause 4.13.2 refers; and

submit to the Architect or to the Quantity Surveyor upon request such details of the loss and/or expense notified by the Contractor under clause 4.13.1 as are reasonably necessary for its ascertainment under clause 4.13.2.

Other rights and remedies

.4 The provisions of clause 4.13 are without prejudice to any other rights and remedies which the Contractor may possess.

The 'matters'

4.14 The following are the matters referred to in clause 4.13.1:

.1 the Contractor not having received in due time necessary instructions, drawings, details or levels from the Architect for which he specifically applied in writing provided that such application was made on a date which having regard to the Completion Date was neither unreasonably distant from nor unreasonably close to the date on which it was necessary for him to receive the same;

.2 the opening up for inspection of any work covered up or the testing of any of the work, materials or goods in accordance with clause 3.11 (including making good in consequence of such opening up or testing). Unless the inspection or test showed that the work, materials or goods were not in accordance with this Contract;

.3   any discrepancy in or divergence between the Specification and/or the Contract Drawings (if any) and/or the Numbered Documents;

.4   the execution of work not forming part of this Contract by the Employer himself or by persons employed or otherwise engaged by the Employer as referred to in clauses 3.13 and 3.14 or the failure to execute such work;

.5   the supply by the Employer of materials and goods which the Employer has agreed to supply for the Works or the failure so to supply;

.6   the Architect's instructions under clause 3.3.4 issued in regard to the postponement of any work to be executed under the provisions of this Contract;

.7   failure of the Employer to give in due time ingress to or egress from the Site or any part thereof through or over any land, buildings, way or passage adjoining or connected with the Site and in the possession and control of the Employer, in accordance with the Contract Documents after receipt by the Architect of such notice, if any, as the Contractor is required to give, or failure of the Employer to give such ingress or egress as otherwise agreed between the Architect and the Contractor;

.8   the Architect's instructions under clause 3.3.2 except to the extent that the instructions are to carry out the items of work described in the Specification and/or as shown upon the Contract Drawings (if any);

.9   compliance with clause 3.16.1 and/or the Architect's instructions under clause 3.16.2;

.10 compliance or non-compliance by the Employer with clause 5.19.

.11 delay arising from a suspension by the Contractor of the performance of his obligations under this Contract to the Employer pursuant to clause 4.3.4;

.12 save as provided for in clauses 4.14.1 to 4.14.11 any impediment, prevention or default, whether by act or omission, by the Employer or any person for whom the Employer is responsible except to the extent that it was caused or contributed to by any default, whether by act or omission, of the Contractor or his servants, agents or sub-contractors.

Statement of extension of time

4.15   If and to the extent that it is necessary for ascertainment under clause 4.13.2 of loss and/or expense the Architect shall state in writing to the Contractor what extension of time, if any, has been made under clause 2.5 in respect of the Relevant Event or events referred to in clauses 2.6.5.1, 2.6.5.2, 2.6.6, 2.6.10, 2.6.11, 2.6.14 and 2.6.15.

Inclusion in Interim Certificates

4.16   Any amount from time to time ascertained under clause 4.13.2 shall be included in the amount stated as due in Interim certificates pursuant to clause 4.5.3.

12.6.2 Differences

Style

The main difference between the loss and/or expense clauses in this contract and in JCT 98 is one of style. The effect is essentially the same and the wording is very similar, but the more spacious layout makes it easier to comprehend, because the information is given in relatively small portions. There may be particular circumstances when the slight differences in wording will have a significant effect, but that will depend on the facts. In general terms, the effect is likely to be the same as under clause 26 of JCT 98.

Instructions

A very marked difference in wording can be observed between clause 26.2.7 of JCT 98 and clause 4.14.8 of PCC 98. In the former, the matter refers to architect's instructions issued in respect of a variation or the expenditure of provisional sums. In the latter, because the contract is structured so that the architect must give instructions for all work to be done including work on drawings and in the specification, the instructions referred to are all instructions issued by the architect for all work to be carried out. Then there is a proviso that the clause does not extend to instructions to carry out work in the specification and/or the drawings. This might be thought to be a clumsy piece of drafting, but nonetheless it is clear in meaning.

Antiquities

Clause 4.14.8 refers to compliance with clause 3.16.1 and with instructions under clause 3.16.2. This allows the contractor to recover loss and/or expense associated with the discovery of antiquities. The effect is broadly the same as clause 34.3 of JCT 98, but its inclusion as part of the major loss and/or expense clause answers the criticisms levelled at clause 34.3 in section 12.2.11 above.

12.7 Under MC 98

12.7.1 Clause 8.5 of MC 98

The text of clause 8.5 is as follows:

> **Loss and expense caused by matters materially affecting regular progress – Works Contracts**
>
> 8.5 Upon receipt of a written application properly made by a Works Contractor under clause 4.45 of the Works Contract Conditions in respect of matters

affecting regular progress of the Works by matters referred to in clause 4.46.1 of the Works Contract Conditions the Management Contractor shall pass to the Architect a copy of that written application together with his comments upon the application. Thereafter, if and as soon as the Architect is of the opinion that the regular progress of the Works Contract or any part thereof has been or is likely to be materially affected as referred to in the aforesaid clause 4.45 and as set out in the application of the Works Contractor then the Architect shall himself ascertain, or shall instruct the Quantity Surveyor to ascertain, the amount of such loss and/or expense in collaboration with the Management Contractor

12.7.2 Comments

The Second Schedule of MC 98 allows the management contractor to recover, in addition to the management fee, reimbursement of all his expenditure in connection with the contract. Therefore he has no need of a claims clause as traditionally understood. Clause 8.5 simply provides for the contractor to pass the works contractor's written application for reimbursement of direct loss and/or expense to the architect. Borrowing something from standard JCT wording, the clause goes on to make clear that it is for the architect alone to form an opinion about the application and then, if appropriate, to ascertain or instruct the quantity surveyor to ascertain the amount. Unlike traditional JCT contracts, the ascertainment must be carried out in collaboration with the contractor. To 'collaborate' is to work in combination with another. It appears, therefore, that although the architect alone will decide in principle if the works contractor is entitled to payment under this clause, he must agree the amount to be paid with the contractor. In general, that is not likely to cause immense problems, but there may be occasions when they cannot agree. The contract is silent on that point.

12.8 Under TC/C 02

12.8.1 Clauses 4.21 to 4.25

The full text of the clauses are as follows:

Application by Trade Contractor – regular progress of the Works materially affected – direct loss and/or expense

4.21   If the Trade Contractor makes written application to the Construction Manager stating that he has incurred or is likely to incur direct loss and/or expense (of which the Trade contractor may give his quantification) in the execution of this Trade Contract for which he would not be reimbursed by a payment under any other provision in this Trade Contract due to the deferment of the Date of Commencement (under clause 2.1.2 where that clause is stated in the Appendix to be applicable) or because the regular progress of the Works or of any part thereof has been or is likely to be materially affected by any one or more of the matters referred to in clause 4.22. If and as soon as the Construction Manager is of the opinion that the direct loss and/or expense has been incurred or is likely to be incurred due to any such deferment of the Date of Commencement or that the regular progress of the Works or of any part thereof has been or is likely to be so materially affected as set out in the application of the Trade Contractor then the Construction Manager from time to time thereafter shall ascertain the amount of such loss and/or expense which has been or is being incurred by the Trade Contractor, provided always that:

.1 the Trade Contractor's application shall be made as soon as it has become, or should reasonably have become, apparent to him that the regular progress of the Works or of any part thereof has been or was likely to be affected as aforesaid; and

.2 the Trade Contractor shall in support of his application submit to the Construction Manager upon request such information as should reasonably enable the Construction Manager to form an opinion as aforesaid; and

.3 the Trade Contractor shall submit to the Construction Manager upon request such details of such loss and/or expense as is reasonably necessary for such ascertainment.

List of matters

4.22 The following are the matters referred to in clause 4.21:

.1 .1 where an Information Release Schedule has been provided, failure of the Construction Manager to comply with clause 3.2.3;
   .2 failure of the Construction Manager to comply with clause 3.2.4;

.2 the opening up for inspection of any work covered up or the testing of any of the work, materials or goods in accordance with clause 3.9 (including making good in consequence of such opening up or testing), unless the inspection or test showed that the work, materials or goods were not in accordance with this Trade Contract;

.3 any discrepancy in or divergence between the Trade Contract Documents;

.4 .1 the execution of work not forming part of this Trade Contract by the client himself or by persons employed or otherwise engaged by him or the failure to execute such work;
.4 .2 the supply by the client of materials and goods which he has agreed to provide for the Works or the failure so to supply;

.5 Construction Manager's instruction under clause 2.1.5 issued in regard to the postponement of any work to be executed under the provisions of this Trade Contract;

.6   failure of the Construction Manager to give in due time ingress to or egress from the site of the Project or any part thereof through or over any land, buildings, way or passage adjoining or connected with the site and in the possession or control of the Client, in accordance with the Trade Contract Documents, after receipt by the Construction Manager of such notice, if any, as the Trade Contractor is required to give, or failure of the Construction Manager to give such ingress or egress;

.7   instructions requiring a Variation (except for a Variation for which the Construction Manager has given a confirmed acceptance of a 3A Quotation) or in regard to the expenditure of provisional sums (other than, where bills of quantities are included in the Trade Contract Documents, an instruction for the expenditure of a provisional sum for defined work or an instruction for the expenditure of a provisional sum for Performance Specified Work);

.8   compliance with clause 3.18.1 and/or the instructions of the Construction Manager under clause 3.18.2;

.9   compliance or non-compliance by the client with clause 5.11.1;

.10 suspension by the Trade Contractor of the performance of his obligations under the Contract to the client pursuant to clause 4.11.6 provided the suspension was not frivolous or vexatious;

.11 save as provided for in clauses 4.22.1 to 4.22.10 any impediment, prevention or default, whether by act or omission, by the client or any person for whom the Client is responsible except to the extent that it was caused by any default, whether by act or omission, of the Trade Contractor or his servants, agents or sub-contractors.

Relevance of certain revisions of the Completion Period

4.23 If and to the extent that it is necessary for ascertainment under clause 4.21 of loss and/or expense the Construction Manager shall state in writing to the Trade Contractor what revision of the Completion Period, if any, has been made under clause 2.3 in respect of the Relevant Event or Events referred to in clause 2.5.5.1 and in clauses 2.5.5.2, 2.5.6, 2.5.8, 2.5.12, 2.5.14, 2.5.17 and 2.5.18.

Amounts ascertained

4.24 Any amount from time to time ascertained under clauses 4.21 to 4.23 shall be added to the Trade Contract Sum or included in the computation of the Ascertained Final Trade Contract Sum as the case may be.

Reservation of rights of Trade Contractor

4.25 The provisions of clauses 4.21 to 4.23 are without prejudice to any other remedies or rights of the Trade Contractor under this Trade Contract.

12.8.2 Significant differences

These provisions are strikingly similar to clause 26 of JCT 98 and the comments on that clause are also applicable here.

12.9 Under MPF 03

12.9.1 Clause 21

The full text of the clause is as follows:

21   Loss and/or expense

21.1 No Change or matter that is required by the Contract to be treated as giving rise to a Change shall, either individually or in conjunction with other Changes, give rise to an entitlement to be reimbursed loss and/or expense under clause 21.

21.2 Subject to clause 21.1 the only matters for which the Employer will be liable to the Contractor in respect of loss and/or expense are:

.1   a breach or act of prevention on the part of the Employer or its representative or advisors appointed pursuant to clause 15.2, other than any matters or actions that are expressly permitted by the Contract and that are stated not to give rise to a Change;

.2   interference with the Contractor's regular progress of the Project by Others on the Site;

.3   the valid exercise by the Contractor of its rights under section 112 of HGCRA 1996.

21.3 As soon as the Contractor becomes aware that the regular progress of the project is or is likely to be materially affected as a consequence of any of the matters set out in clause 21.2 so as to cause loss and/or expense to be incurred it shall notify the Employer. The Contractor shall take all practicable steps to reduce the loss and/or expense to be incurred.

21.4 The Contractor shall provide to the Employer its assessment of the loss and/or expense incurred or to be incurred as a consequence of any matter notified in accordance with clause 21.3 together with such information as is reasonably necessary to enable the Employer to ascertain the loss and/or expense incurred. Such assessment and information shall be updated at monthly intervals until such time as the Contractor has provided all of the information that is reasonably necessary to allow the whole of the loss and/or expense that has been incurred to be ascertained.

21.5 Upon receipt of any information referred to by clause 21.4 regarding loss and/or expense that has been incurred the Employer shall within 14 days notify the Contractor of its ascertainment of the loss and/or expense incurred, that ascertainment being made by reference to the information provided by the Contractor and being in sufficient detail to permit the Contractor to identify any differences between it and the Contractor's assessment of the loss and/or expense incurred.

21.6 No later than 42 days after Practical Completion of the Project the contractor shall provide the documentation in support of any further ascertainment it considers should be made in respect of any matter notified in accordance with clause 21.3. Within 42 days of receipt of such documentation the

Employer shall undertake a review of its previous ascertainment in respect of each matter for which further documentation has now been provided and notify the Contractor of any further ascertainment that it considers appropriate.

21.7   The Employer shall be liable to pay the Contractor any loss and/or expense that has been ascertained in accordance with clause 21.

21.8   No ascertainment of loss and/or expense under clause 21 shall include any element of loss and/or expense if that element was contributed to by a cause other than a Change or a matter set out in clause 21.2. Any loss and/or expense incurred as a consequence of a Change is to be included in a valuation made under clause 20 *(Changes)*.

12.9.2 Main points

This clause is not like JCT 98 clause 26 either in general structure or wording. The first notable thing is that in no less than two clauses (21.1 and 21.8) it is made abundantly plain that no loss and/or expense can be claimed or paid under this clause in connection with a change or variation.

The grounds for loss and/or expense effectively fall under the heading of employer defaults. There are only three grounds and the first deals expressly with breach or act of prevention by the employer, his agents, etc. The second deals with interference with regular progress by others on the site. Such others, by definition, are those for whom the employer, and not the contractor, is responsible. The third ground simply brings a rightful suspension by the contractor due to lack of payment into a category of events giving right to entitlement. Without that inclusion, a contractor who properly suspends performance of his obligations would have great difficulty in obtaining financial recompense. Clause 21.8 expressly excludes from ascertainment of loss and/or expense any element which has been contributed to by a cause other than a change or the three grounds set out. A strict interpretation of that wording leads to the conclusion that if loss and/or expense has been caused by two things, one of which is an acceptable ground under clause 21.2 and another which is not an acceptable ground under clause 21.2, the whole of such loss and/or expense is excluded from calculation of any amount due to the contractor. This is something which contractors must watch carefully as a potentially onerous clause. In such circumstances, it is debatable whether the courts would be prepared to uphold such a clause whose effect is not signalled to the contractor in any particularly noticeable way[568]. It is possible that it could be held to be a penalty.

There is the usual provision for the contractor to notify the employer as soon as he becomes aware that regular progress is or is likely to be affected by one or more of the grounds, but it is significant that it is not made a proviso or condition. Therefore, the contractor's failure to

---

[568] *J. Spurling* v. *Bradshaw* [1956] 2 All ER 121.

notify the employer strictly in accordance with this clause is unlikely to prevent the employer from ascertaining loss and/or expense.

Unlike the position under most JCT forms, the contractor is required to make his own assessment of the loss and/or expense he has incurred and must present this with supporting information to the employer. The supporting information is such as is reasonably necessary to enable the employer to ascertain the loss and/or expense. The contractor is responsible for updating this package on a monthly basis until he has provided information relating to the whole of the loss and/or expense. There appears to be no sanction if the contractor fails to provide the information monthly other than that he will not receive the relevant loss and/or expense. Ascertainment is intended to be a fast-track process. The employer must respond to the contractor within 14 days of receiving the information and notify the contractor of the amount ascertained. It is expressly stated in clause 21.5 that the ascertainment must be made by reference to the contractor's information. The clear message here is that if the contractor fails to provide adequate information, the employer will be unable to properly ascertain the amount due and the contractor may receive less than he expects. However, the ascertainment must be in sufficient detail to allow the contractor to identify the differences.

The extension of time review procedure is echoed in clause 21.6 which requires the contractor to provide any documentation in support of further ascertainment no later than 42 days after practical completion. That does not mean that the contractor who is dissatisfied with an ascertainment has to wait until after practical completion to submit more information. It is simply stating the latest date by which the contractor must have made such submission. The employer must respond within 42 days of receipt of the new material. Thus, the contractor may make his initial submission in week 30 of a 70-week project. The employer must respond within 14 days. If the contractor is dissatisfied, he may submit new material and the employer must now respond within 42 days. The process can be repeated until 42 days after practical completion after which the contractor may not submit further information and the employer must make his final assessment within a further 42 days.

Clause 21.7 makes the employer liable to pay to the contractor any loss and/or expense ascertained in accordance with this clause.

Chapter 13

Variations

13.1 Introduction

Use of the word 'variations' in building contracts usually refers to a change in the works instructed by the architect, contract administrator or the employer as the case may be (but see the quite different 'variation of contract' considered in Chapter 4, section 4.3). There are clauses permitting variations in all standard form contracts. If there was no such clause in lump sum contracts, the contractor could not be compelled to vary the works and he could insist upon completing precisely the work and supplying precisely the materials for which he has contracted. No power to order variations would be implied[569]. Although standard contracts contain a clause permitting variations, the power is not unfettered. A variation may not be ordered if it changes the whole scope and character of the works. To determine whether this has been done in any particular case, reference must be made to the recital which sets out the work to be done. As a broad rule of thumb, if the variation does not invalidate the description in the recital, it is unlikely to be a variation which changes the whole scope and character of the works. To a large extent the point is academic, because a contractor will usually welcome the opportunity to carry out additional work and thereby earn money in the valuation of the variation and possibly in the formation of a claim for disruption or prolongation.

Variations are a fact of life in building contracts. There can only be a minority of contracts of any size in which the subject matter when completed is identical in every respect with what was contemplated at the outset.

Variations are inevitable in even the best-planned contracts simply because, in a matter as complicated as the construction of a building, it is virtually impossible for the building owner and his design team to foresee every eventuality. However, variations often arise quite unnecessarily, simply because the building owner had been unable to make up his mind at the design stage about precisely what he does want or because of lack of foresight on the part of the design team or perhaps, more commonly, because the design team have allowed themselves to be pushed towards a start date which is not realistically achievable. On hospital projects, for instance, medical science advances so rapidly that it is often necessary to make major design

---

[569] *Stockport MBC v. O'Reilly* [1978] 1 Lloyds Rep 595.

changes during the course of construction to accommodate new techniques. Far too often, however, it is simply lack of proper planning, and too much haste to get the project under way, which lead to variations.

It is sometimes said that the bill of quantities system of contracting makes it too easy for a building owner or his team to change their minds. The effects of such changes can be easily seen and readily evaluated, and there is therefore a lack of incentive to make firm design decisions before work starts on the site. This does not allow for the additional loss and/or expense which might be claimed.

13.2 *The baseline*

Where variations are authorised by the terms of the contract, the question will arise: variation from what? In other words, what did the contractor agree to do in the first place? The question may even be whether it is a variation at all, or whether it is something which the contractor agreed, or must be deemed to have agreed, to do as part of the contract[570].

In contracts incorporating bills of quantities the accuracy of which is guaranteed by the employer, these questions will be easy to answer, since anything that involves a change from what is set out in the bills will involve an adjustment of the contract sum based on the bills.

For instance, JCT 98, clause 14.1 states that the quality and quantity of the work included in the contract sum will be deemed to be what is in the contract bills.

Clause 2.2.2.2 says:

'If in the Contract Bills (or in any addendum bill issued as part of the information referred to in clause 13A.1.1 for the purpose of obtaining a 13A Quotation which Quotation has been accepted by the Employer) there is any departure from the method of preparation referred to in clause 2.2.2.1 or any error in description or in quantity or omission of items (including any error in or omission of information in any item which is the subject of a provisional sum for defined work) then such departure or error or omission...shall be corrected...any such correction under this clause 2.2.2.2 shall be treated as if it were a Variation required by an instruction of the Architect under clause 13.2.'

In passing, it should be noted that it is only the correction of departures or errors which is treated as an instructed variation. It seems that if there is *omission*, an architect's instruction is required to add the item.

Clause 2.2.2.2 means that the employer has, in effect, guaranteed the accuracy of the bills of quantities to the extent of agreeing that the contract sum shall be adjusted to take account of any errors of the kind listed. (This does not, of course, mean that the contract sum will be adjusted to take account of any errors in the pricing of the contract bills; the pricing is

---

[570] *Sharpe v. San Paulo Railway* (1873) 8 Ch App 597.

entirely a matter for the contractor, and clause 14.2 says that, subject to clause 2.2.2.2, errors, arithmetic or otherwise in calculating the contract sum will be deemed accepted by the parties. In other words, it is only errors of the kind specified in clause 2.2.2.2 that will lead to an adjustment of the contract sum.) Clause 14.1 means that the contractor has contracted for the contract sum to carry out only what is shown in the contract bills and not necessarily, for instance, what is shown on the contract drawings if they differ from the bills. This is not to say that the contractor can refuse to carry out any work which is different from that shown in the bills if it is shown on the drawings, only that he is entitled to an adjustment of the contract sum if he is directed to follow the drawings rather than the bills.

Clause 2.3 states that if the contractor finds any discrepancy in or divergence between the contract drawings, the contract bills, any architect's instruction (other than those requiring variations, which will, of course, by definition involve a divergence from the contract drawings and/or bills) and any drawings or documents issued by the architect to supplement the contract drawings and bills, he must immediately notify the architect in writing and the architect must then issue instructions. Any such instruction must be of a kind that is specifically authorised by other clauses of the contract conditions and it must have regard to the discrepancy; if it involves a departure from what is set out in the contract bills, it must be considered to be an instruction issued under the authority of clause 13, i.e. a variation.

13.3 Bills of quantities

The characteristic of the bills of quantities contract that has just been discussed, i.e. that the contractor has contracted, for the contract sum, to carry out only the work set out in the bills and no other, has been the subject of criticism by some commentators.

The criticism is on a basic point of principle. Generally, in the absence of specific contract provisions to the contrary, a contractor will be expected to do everything that may be necessary in order to carry the contract works to completion, irrespective of whether it is specifically referred to in the contract documents and even of whether it was possible to foresee the necessity for the work to be carried out beforehand[571].

It is said that the linking of the contract sum to the work set out in the bills transfers a substantial element of risk from the shoulders of the contractor to those of the employer. For example, if the contract works include a substantial element of reinforced concrete this will obviously imply the necessity for formwork; yet, if the quantity surveyor makes a mistake in preparing the bills and fails to measure the necessary formwork, the contractor will be entitled to claim extra payment for it even though, when tendering, he may himself have spotted the mistake. Without the bills, the contractor would be expected to provide the formwork within the contract sum even if not

---

[571] *Thorn* v. *London Corporation* (1876) 1 App Cas 120.

specifically referred to in the contract drawings or specification, and even though he may not himself have appreciated the extent of the formwork required at the time of tender.

The disadvantage of the bills of quantities system can be very much overstated and is, in any case, greatly outweighed by the considerable advantage, to contractor and employer alike, of having a substantial area of possible, or even probable, dispute removed. Had a modern ICE contract been used, the *Sharpe* case[572] would never have reached the courts at all and, even though the employer would have had to pay for the extra 2 million cubic yards of earthworks, he would in fact have been paying the proper price for the job – and would always have had the option of proceeding against the engineer for negligence if real damage had been suffered.

A further consideration, often forgotten by commentators, is that a contractor who is required to take a risk will, unless he wishes to end up insolvent, include an allowance for that risk in his contract price. The employer will then have to pay that price even if the event allowed for never actually happens. If, say, the contractor is expected to allow in his pricing of excavations for the risk of having to excavate in rock if it is encountered, and then no rock is found, the employer will still have to pay whatever price the contractor allowed in his tender for that risk. If, on the other hand, bills of quantities include a provisional quantity for excavation in rock, to be remeasured according to the quantity actually found, the employer will end up paying only for the rock encountered which may, of course, be more, but is just as likely to be substantially less than the provisional quantity allowed in the tender.

Of course it has to be recognised that, for some employers, it is more important that a stated cost for the works will not be exceeded than that they should pay only for what is done, because to exceed the stated cost would render the whole project economically unviable. For example, if a department store will only make a profit if the cost of construction does not exceed £5 million, the owner may be quite happy to pay that sum, even though with remeasurement of risk items the eventual cost might be £4.75 million, rather than run the risk of finding at the end of the contract that he has to pay £5.25 million because the risk items went the other way. In the latter case he would be left with a store which will, at best, only break even or possibly make a thumping loss for every year of its life due to the extra financing costs. For such employers there is no doubt that the standard forms of building contract with quantities are less than satisfactory, although even then an efficient design team and careful cost monitoring during the course of construction will enable design adjustments to be made in order to bring the final cost within the budget.

There is one consideration, perhaps more speculative. Several cases are cited in *Hudson*[573] and deal with the contractor's obligation in certain cir-

---

[572] See Chapter 4, section 4.5.
[573] I. N. Duncan Wallace, *Hudson's Building and Engineering Contracts*, 11th edition, 1995, Sweet & Maxwell.

cumstances to carry out work not shown in documents provided at the time of tender. An example is *Williams* v. *Fitzmaurice*[574] where the specification did not mention floorboards. The specification did say:

> 'the whole of the materials mentioned or otherwise in the foregoing particulars, necessary for the completion of the work, must be provided by the contractor.'

It was held that the provision and laying of the floor boards was included in the contract and the contractor was not entitled to recover for them as an extra.

Most of these cases date from the eighteenth and early years of the nineteenth centuries. The common law has moved rapidly since then, and a number of those cases, including the *Sharpe* case, could well be decided differently were they to come before the courts now. If a document is provided to a contractor in the knowledge that he is bound to rely on it in the preparation of his tender, the courts will require very strong wording indeed to be used in a contract or in the document itself in order to be persuaded that an employer should not be responsible for it[575]. In the *Williams* case, the general clause noted above appears to have been the key factor in the decision. Developments in the law of negligence over the past 30 years or so since *Hedley Byrne & Co Ltd* v. *Heller & Partners Ltd*[576] mean that where, say, a contractor tenders on the basis of quantities provided to him on the principle that errors will not be adjusted and they prove to be negligently inaccurate so that he suffers loss, the contractor may be able to directly sue the person who prepared the quantities[577].

Indeed, bills of quantities enable the allocation of risk between the parties to be very precisely defined, and it is by no means the case that all the risk of unexpected eventualities is placed upon the employer in contracts on a bills of quantities basis. On bills prepared in accordance with the Standard Method of Measurement of Building Works (5th Edition), for instance, the contractor was expected to take the risk of excavating in, and upholding the sides of, any kind of soil encountered except rock and other similar hard material. This has been substantially modified in the 6th and 7th Editions. But bills prepared in accordance with the method of measurement applicable to civil engineering works require much greater degrees of risk to be taken by the contractor.

In *Bowmer & Kirkland Ltd* v. *Wlson Bowden Properties Ltd*[578] the bills contained the following paragraph:

> 'where and to the extent that materials, products and workmanship are not fully specified they are to be:

---

[574] (1858) 3 H & N 844.
[575] *Bacal Construction (Midlands) Ltd* v. *Northampton Development Corporation* (1976) 8 BLR 88.
[576] [1964] AC 465.
[577] *Auto Concrete Curb Ltd* v. *South Nation River Conservation Authority* (1994) Const LJ 39; *Henderson* v. *Merritt Syndicates* (1994) 64 BLR 26.
[578] (1996) 80 BLR 131.

(1) suitable for the purpose of the Works stated in or reasonably to be inferred from the Contract Documents

(2) in accordance with good building practice, including the relevant provisions of current BSI documents.'

The court concluded that the paragraph meant that if the materials and workmanship were fully specified, the contractor had carried out his obligation by doing what was specified. If they were not fully specified, the contractor had the duties set out in sub-paragraphs (1) and (2). This is clearly a valuable paragraph to include in the bills of quantities from the employer's point of view.

13.4 Functions of the architect and the quantity surveyor

The provisions for the valuation of variations now set out in most standard form contracts have, among other things, the advantage of clearly defining the relative responsibilities of architect and quantity surveyor in respect of variations. It is clear that, under the standard form, once the architect has issued an instruction requiring a variation or requiring the contractor to carry out work against a provisional sum, responsibility for defining the financial effect of the work covered by the instruction now passes entirely to the quantity surveyor, subject only to the architect's right and duty to satisfy himself regarding the financial content of his certificates[579]. The quantity surveyor's valuation will be required to cover *all* the effects of the variation up to the point at which it becomes necessary for the contractor to make an application to the architect stating that the introduction of the work in question has affected or is likely to affect the regular progress of the works in some material respect. At that point responsibility passes back to the architect. It is he who bears the responsibility for determining questions concerning the progress of the works and, although the quantity surveyor may be brought into the matter again when ascertainment of the resulting direct loss and/or expense becomes necessary, this will only be at the discretion of the architect who still bears primary responsibility for that aspect. Although some JCT contracts include provision for the contractor to submit his own calculation of the valuation, the quantity surveyor is still responsible for checking that valuation. It has been said, of the quantity surveyor under JCT 63, that his

> 'function and his authority under the contract are confined to measuring and quantifying. The contract gives him authority, at least in certain instances, to decide quantum. It does not in any instance give him authority to determine any liability, or liability to make any payment or allowance.'[580]

---

[579] *R. B. Burden Ltd* v. *Swansea Corporation* [1957] 3 All ER 243.
[580] *County & District Properties Ltd* v. *John Laing Construction Ltd* (1982) 23 BLR 1 at 14 per Webster J.

The judge went on to deal with the words 'the valuation of variations...
unless otherwise agreed shall be made in accordance with the following
rules' in clause 11(4), JCT 63, which counsel had submitted meant 'agreed
by or with the quantity surveyor'. He said:

> 'I reject that submission. In my view the word agreed can only mean
> "agreed between the parties", although it may well be that on occasion a
> quantity surveyor may perhaps be given express authority by the
> employer to make such an agreement.
>
>   But the JCT contract does not give him the authority.
>
>   There are few express references to him in the contract. He is defined in
> Article 4 of the Articles of Agreement. By clause 11(4), he is given the
> express duty of measuring and valuing variations.
>
>   By 11(6), he is given the duty of ascertaining loss and expense involved
> in variation – but only if so instructed by the architect.
>
>   By 24(1), he is given a similar duty in respect of loss or expense caused
> by disturbance of the work etc. – but again only if instructed by the
> architect.'[581]

The position appears to be largely the same under JCT 98.

13.5 Variations position under JCT 98

13.5.1 Clauses 13 and 13A

The full text of clauses 13 and 13A is as follows:

13   Variations and provisional sums
13.1   The term 'variation' as used in the Conditions means:

13.1   .1   the alteration or modification of the design, quality or quantity of the
            Works including

.1   .1   the addition, omission or substitution of any work,
.1   .2   the alteration of the kind or standard of any of the materials or
          goods to be used in the Works,
.1   .3   the removal from the site of any work executed or materials or
          goods brought thereon by the Contractor for the purposes of
          the Works other than work materials or goods which are not in
          accordance with this Contract;

13.1   .2   the imposition by the Employer of any obligations or restrictions in
            regard to the matters set out in clauses 13.1.2.1 to 13.1.2.4 or the addition
            to or alteration or omission of any such obligations or restrictions so
            imposed or imposed by the Employer in the Contract Bills in regard to:

.2   .1   access to the site or use of any specific parts of the site;

---

[581] *County & District Properties Ltd* v. *John Laing Construction Ltd* (1982) 23 BLR 1 at 14 per Webster J.

.2 .2   limitations of working space;

.2 .3   limitations of working hours;

.2 .4   the execution or completion of the work in any specific order; but excludes

13.1 .3   nomination of a sub-contractor to supply and fix materials or goods or to execute work of which the measured quantities have been set out and priced by the Contractor in the Contract Bills for supply and fixing or execution by the Contractor.

13.2 .1   The Architect may issue instructions requiring a Variation.

13.2 .2   Any instruction under clause 13.2.1 shall be subject to the Contractor's right of reasonable objection set out in clause 4.1.1.

13.2 .3   The valuation of a Variation instructed under clause 13.2.1 shall be in accordance with clause 13.4.1 unless the instruction states that the treatment and valuation of the Variation are to be in accordance with clause 13A or unless the Variation is one to which clause 13A.8 applies. Where the instruction so states, clause 13A shall apply unless the Contractor within 7 days (or such other period as may be agreed) of receipt of the instruction states in writing that he disagrees with the application of clause 13A to such instruction. If the Contractor so disagrees, clause 13A shall not apply to such instruction and the Variation shall not be carried out unless and until the Architect instructs that the Variation is to be carried out and is to be valued pursuant to clause 13.4.1.

13.2 .4   The Architect may sanction in writing any Variation made by the Contractor otherwise than pursuant to an instruction of the Architect.

13.2 .5   No Variation required by the Architect or subsequently sanctioned by him shall vitiate this Contract.

13.3   The Architect shall issue instructions in regard to:

13.3 .1   the expenditure of provisional sums included in the Contract Bills; and

13.3 .2   the expenditure of provisional sums included in a Nominated Sub-Contract.

13.4 .1 .1   Subject to clause 13.4.1.3
— all Variations required by the Architect or subsequently sanctioned by him in writing, and
— all work which under the Conditions is to be treated as if it were a Variation required by an instruction of the Architect under clause 13.2, and
— all work executed by the Contractor in accordance with instructions by the Architect as to the expenditure of provisional sums which are included in the Contract Bills, and
— all work executed by the Contractor for which an Approximate Quantity is included in the Contract Bills

shall, unless otherwise agreed by the Employer and the Contractor or unless the Architect has issued to the Contractor, be valued (in the Conditions called 'the Valuation'), under Alternative A in clause

13.4.1.2 or, to the extent that Alternative A is not implemented by the Contractor or, if implemented, to the extent that the Price Statement or amended Price Statement is not accepted, under Alternative B in clause 13.4.1.2. Clause 13.4.1.1 shall not apply in respect of a Variation for which the Architect has issued a confirmed acceptance of a 13A Quotation or it is a Variation to which clause 13A.8 applies.

.1   .2   **Alternative A: Contractor's Price Statement**
Paragraph:

A1   Without prejudice to his obligation to comply with any instruction or to execute any work to which clause 13.1.1 refers, the Contractor may within 21 days from receipt of the instruction or from commencement of work for which an Approximate Quantity is included in the Contract documents or, if later, from receipt of sufficient information to enable the Contractor to prepare his Price Statement, submit to the Quantity Surveyor his price ('Price Statement') for such compliance or for such work.

The Price Statement shall state the Contractor's price for the work which shall be based on the provisions of clause 13.5 *(valuation rules)* and may also separately attach the Contractor's requirements for:

.1   any amount to be paid in lieu of any ascertainment under clause 26.1 of direct loss and/or expense not included in any accepted 13A Quotation or in any previous ascertainment under clause 26;

.2   any adjustment to the time for completion of the Works to the extent that such adjustment is not included in any revision of the Completion Date that has been made by the Architect under clause 25.3 or in his confirmed acceptance of any 13A Quotation. *(See paragraph A7)*

A2   Within 21 days of receipt of a Price Statement the Quantity Surveyor, after consultation with the Architect, shall notify the Contractor in writing

either

.1   that the Price Statement is accepted
or
.2   that the Price Statement, or a part thereof, is not accepted.

A3   Where the Price Statement or a part thereof has been accepted the price in that accepted Price Statement or in that part which has been accepted shall in accordance with clause 13.7 be added to or deducted from the Contract Sum.

A4   Where the Price Statement or a part thereof has not been accepted:

.1   the Quantity Surveyor shall include in his notification to the Contractor the reasons for not having accepted the Price Statement or a part thereof and set out those reasons in similar detail to that given by the Contractor in his Price

Statement and supply an amended Price Statement which is acceptable to the Quantity Surveyor after consultation with the Architect;

.2 within 14 days from receipt of the amended Price Statement the Contractor shall state whether or not he accepts the amended Price Statement or part thereof and if accepted paragraph A3 shall apply to that amended Price Statement or part thereof; if no statement within the 14 days period is made the Contractor shall be deemed not to have accepted, in whole or in part, the amended Price Statement;

.3 to the extent that the amended Price Statement is not accepted by the Contractor, the Contractor's Price Statement and the amended Price Statement may be referred either by the Employer or by the Contractor as a dispute or difference to the Adjudicator in accordance with the provisions of clause 41A.

A5   Where no notification is given pursuant to paragraph A2 the Price Statement is deemed not to have been accepted, and the Contractor may, on or after the expiry of the 21 day period to which paragraph A2 refers, refer his Price Statement as a dispute or difference to the Adjudicator in accordance with the provisions of clause 41A.

A6   Where a Price Statement is not accepted by the Quantity Surveyor after consultation with the Architect or an amended Price Statement has not been accepted by the Contractor and no reference to the Adjudicator under paragraph A4.3 or paragraph A5 has been made, Alternative B shall apply.

A7.1   Where the Contractor pursuant to paragraph A1 has attached his requirements to his Price Statement the Quantity Surveyor after consultation with the Architect shall within 21 days of receipt thereof notify the Contractor

.1 either that the requirement in paragraph A1 in respect of the amount to be paid in lieu of any ascertainment under clause 26.1 is accepted or that the requirement is not accepted and clause 26.1 shall apply in respect of the ascertainment of any direct loss and/or expense; and

.2 either that the requirement in paragraph A1.2 in respect of an adjustment to the time for the completion of the Works is accepted or that the requirement is not accepted and clause 25 shall apply in respect of any such adjustment.

A7.2   If the Quantity Surveyor has not notified the Contractor within the 21 days specified in paragraph A7.1, clause 25 and clause 26 shall apply as if no requirements had been attached to the Price Statement.

Alternative B

The valuation shall be made by the Quantity Surveyor in accordance with the provisions of clauses 13.5.1 to 13.5.7.

.1   .3   The valuation of Variations to the sub-contract works executed by a Nominated sub-Contractor in accordance with instructions of the Architect and of all instructions issued under clause 13.3.2 and all work executed by a Nominated Sub-Contractor for which an Approximate Quantity is included in any bills of quantities included in the Numbered Documents shall (unless otherwise agreed by the Contractor and the Nominated Sub-Contractor concerned with the approval of the Employer) be made in accordance with the relevant provisions of Conditions NSC/C.

13.4   .2   Where under the instruction of the Architect as to the expenditure of a provisional sum a prime cost sum arises and the Contractor under clause 35.2 tenders for the work covered by that prime cost sum and that tender is accepted by or on behalf of the Employer, that work shall be valued in accordance with the accepted tender of the Contractor and shall not be included in the Valuation of the instruction of the Architect in regard to the expenditure of the provisional sum.

13.5   .1   To the extent that the Valuation relates to the execution of additional or substituted work which can properly be valued by measurement or to the execution of work for which an Approximate Quantity is included in the Contract Bills such work shall be measured and shall be valued in accordance with the following rules:

.1   where the additional or substituted work is of similar character to, is executed under similar conditions as, and does not significantly change the quantity of, work set out in the Contract Bills the rates and prices for the work so set out shall determine the Valuation;

.2   where the additional or substituted work is of similar character to work set out in the Contract Bills but is not executed under similar conditions thereto and/or significantly changes the quantity thereof, the rates and prices for the work so set out shall be the basis for determining the valuation and the valuation shall include a fair allowance for such difference in conditions and/or quantity;

.3   where the additional or substituted work is not of similar character to work set out in the Contract Bills the work shall be valued at fair rates and prices;

.4   where the Approximate Quantity is a reasonably accurate forecast of the quantity of work required the rate or price for the Approximate Quantity shall determine the Valuation;

.5   where the Approximate Quantity is not a reasonably accurate forecast of the quantity of work required the rate or price for that Approximate Quantity shall be the basis for determining the Valuation and the Valuation shall include a fair allowance for such difference in quantity.

Provided that clause 13.5.1.4 and clause 13.5.1.5 shall only apply to the extent that the work has not been altered or modified other than in quantity.

13.5 .2 To the extent that the Valuation relates to the omission of work set out in the Contract Bills the rates and prices for such work therein set out shall determine the valuation of the work omitted.

13.5 .3 In any valuation of work under clauses 13.5.1 and 13.5.2:

.3 .1 measurement shall be in accordance with the same principles as those governing the preparation of the Contract Bills as referred to in clause 2.2.2.1;

.3 .2 allowance shall be made for any percentage or lump sum adjustments in the Contract Bills; and

.3 .3 allowance, where appropriate, shall be made for any addition to or reduction of preliminary items of the type referred to in the Standard Method of Measurement; provided that no such allowance shall be made in respect of compliance with an Architect's instruction for the expenditure of a provisional sum for defined work.

13.5 .4 To the extent that the Valuation relates to the execution of additional or substituted work which cannot properly be valued by measurement the Valuation shall comprise:

.4 .1 the prime cost of such work (calculated in accordance with the 'Definition of Prime Cost of Daywork carried out under a Building Contract' issued by the Royal Institution of Chartered Surveyors and the Building Employers Confederation (now Construction Confederation) which was current at the Base Date) together with percentage additions to each section of the prime cost at the rates set out by the Contractor in the Contract Bills; or

.4 .2 where the work is within the province of any specialist trade and the said Institution and the appropriate body representing the employers in that trade have agreed and issued a definition of prime cost of daywork, the prime cost of such work calculated in accordance with that definition which was current at the Base Date together with percentage additions on the prime cost at the rates set out by the Contractor in the Contract Bills.

Provided that in any case vouchers specifying the time daily spent upon the work, the workmen's names, the plant and the materials employed shall be delivered for verification to the Architect or his authorised representative not later than the end of the week following that in which the work has been executed.

13.5 .5 If

compliance with any instruction requiring a Variation or
compliance with any instruction as to the expenditure of a provisional sum for undefined work or
compliance with any instruction as to the expenditure of a provisional sum for defined work to the extent that the instruction for that work differs from the description given for such work in the Contract Bills or

the execution of work for which an Approximate Quantity is included in the Contract Bills to such extent as the quantity is more
or less than the quantity ascribed to that work in the Contract Bills

substantially changes the conditions under which any other work is
executed, then such other work shall be treated as if it had been the
subject of an instruction of the Architect requiring a Variation under
clause 13.2 which shall be valued in accordance with the provisions of
clause 13.

13.5   .6   .1   The Valuation of Performance Specified Work shall include allowance for the addition or omission of any relevant work involved in
the preparation and production of drawings, schedules or other
documents;

13.5   .6   .2   the Valuation of additional or substituted work related to Performance Specified Work shall be consistent with the rates and prices of
work of a similar character set out in the Contract Bills or the
Analysis making due allowance for any changes in the conditions
under which the work is carried out and/or any significant change
in the quantity of the work set out in the Contract Bills or in the
Contractor's Statement. Where there is no work of a similar character set out in the Contract Bills or the Contractor's Statement a fair
valuation shall be made;

13.5   .6   .3   the Valuation of the omission of work relating to Performance
Specified Work shall be in accordance with the rates and prices for
such work set out in the Contract Bills or the Analysis;

13.5   .6   .4   any valuation of work under clauses 13.5.6.2 and 13.5.6.3 shall
include allowance for any necessary addition to or reduction of
preliminary items of the type referred to in the Standard Method
of Measurement;

13.5   .6   .5   where an appropriate basis of a fair valuation of additional or
substituted work relating to Performance Specified Work is daywork the Valuation shall be in accordance with clauses 13.5.4.1 or
13.5.4.2 and the proviso to clause 13.5.4 shall apply;

13.5   .6   .6   if
compliance with any instruction under clause 42.11 requiring a
Variation to Performance Specified Work or
compliance with any instruction as to the expenditure of a provisional sum for Performance Specified Work to the extent that the
instruction for that Work differs from the information provided in
the Contract Bills pursuant to clause 42.7.2 and/or 42.7.3 for such
Performance Specified Work

substantially changes the conditions under which any other work is
executed (including any other Performance Specified Work) then
such other work (including any other Performance Specified Work)
shall be treated as if it had been the subject of an instruction of the
Architect requiring a Variation under clause 13.2 or, if relevant,

under clause 42.11 which shall be valued in accordance with the provisions of clause 13.5.

13.5 .7 To the extent that the Valuation does not relate to the execution of additional or substituted work or the omission of work or to the extent that the valuation of any work or liabilities directly associated with a Variation cannot reasonably be effected in the Valuation by the application of clauses 13.5.1 to .6 a fair valuation thereof shall be made.

Provided that no allowance shall be made under clause 13.5 for any effect upon the regular progress of the Works or for any other direct loss and/or expense for which the Contractor would be reimbursed by payment under any other provision in the Conditions.

13.6 Where it is necessary to measure work for the purpose of the Valuation the Quantity Surveyor shall give to the Contractor an opportunity of being present at the time of such measurement and of taking such notes and measurements as the Contractor may require.

13.7 Effect shall be given to a Valuation under clause 13.4.1.1, to an agreement by the Employer and the Contractor to which clause 13.4.1.1 refers, to a 13A Quotation for which the Architect has issued a confirmed acceptance and to a valuation pursuant to clause 13A.8 by addition to or deduction from the Contract Sum.

13A Variation instruction – Contractor's quotation in compliance with the instruction

13A Clause 13A shall only apply to an instruction where pursuant to clause 13.2.3 the Contractor has not disagreed with the application of clause 13A to such instruction.

13A.1 .1 The instruction to which clause 13A is to apply shall have provided sufficient information to enable the Contractor to provide a quotation, which shall comprise the matters set out in clause 13A.2 (a '13A Quotation'), in compliance with the instruction; and in respect of any part of the Variation which relates to the work of any Nominated Sub-Contractor sufficient information to enable the Contractor to obtain a 3.3A Quotation from the Nominated Sub-Contractor in accordance with clause 3.3A.1.2 of the Conditions NSC/C. If the Contractor reasonably considers that the information provided is not sufficient, then, not later than 7 days from the receipt of the instruction, he shall request the Architect to supply sufficient further information.

13A.1 .2 the Contractor shall submit to the Quantity Surveyor his 13A Quotation in compliance with the instruction and shall include therein 3.3A Quotations in respect of any parts of the Variation which relate to the work of Nominated Sub-Contractors not later than 21 days from
        the date of receipt of the instruction
or   if applicable, the date of receipt by the Contractor of the sufficient further information to which clause 13A.1.1 refers
whichever date is the later and the 13A Quotation shall remain open for acceptance by the Employer for 7 days from its receipt by the Quantity Surveyor.

13A.1   .3   The Variation for which the Contractor has submitted his 13A Quotation shall not be carried out by the Contractor or as relevant by any Nominated Sub-Contractor until receipt by the Contractor of the confirmed acceptance issued by the Architect pursuant to clause 13A.3.2.

13A.2   The 13A Quotation shall separately comprise:

13A.2   .1   the value of the adjustment to the Contract Sum (other than any amount to which clause 13A.2.3 refers) including therein the effect of the instruction on any other work including that of Nominated Sub-Contractors supported by all necessary calculations by reference, where relevant, to the rates and prices in the Contract Bills and including, where appropriate, allowances for any adjustment of preliminary items;

13A.2   .2   any adjustment to the time required for completion of the Works (including where relevant stating an earlier Completion Date than the Date for Completion given in the Appendix) to the extent that such adjustment is not included in any revision of the Completion Date that has been made by the Architect under clause 25.3 or in his confirmed acceptance of any other 13A Quotation;

13A.2   .3   the amount to be paid in lieu of any ascertainment under clause 26.1 of direct loss and/or expense not included in any other accepted 13A Quotation or in any previous ascertainment under clause 26;

13A.2   .4   a fair and reasonable amount in respect of the cost of preparing the 13A Quotation;

and, where specifically required by the instruction, shall provide indicative information in statements on

13A.2   .5   the additional resources (if any) required to carry out the Variation; and

13A.2   .6   the method of carrying out the Variation.

Each part of the 13A Quotation shall contain reasonably sufficient supporting information to enable that part to be evaluated by or on behalf of the Employer.

13A.3   .1   If the Employer wishes to accept a clause 13A Quotation the Employer shall so notify the Contractor in writing not later than the last day of the period for acceptance stated in clause 13A.1.2.

13A.3   .2   If the Employer accepts a clause 13A Quotation the Architect shall, immediately upon that acceptance, confirm such acceptance by stating in writing to the Contractor (in clause 13A and elsewhere in the Conditions called a 'confirmed acceptance'):

.2   .1   that the Contractor is to carry out the Variation;

.2   .2   the adjustment of the Contract Sum, including therein any amounts to which clause 13A.2.3 and clause 13A.2.4 refer, to be made for complying with the instruction requiring a Variation;

.2 .3 any adjustment to the time required by the Contractor for completion of the Works and the revised Completion Date arising therefrom (which, where relevant, may be a date earlier than the Date for Completion given in the Appendix) and, where relevant, any revised period or periods for the completion of the Nominated Sub-Contract work of each Nominated Sub-Contractor; and

.2 .4 that the Contractor, pursuant to clause 3.3A.3 of the Conditions NSC/C, shall accept any 3.3A Quotation included in the 13A Quotation for which the confirmed acceptance has been issued.

13A.4 If the Employer does not accept the 13A Quotation by the expiry of the period for acceptance stated in clause 13A.1.2, the Architect shall, on the expiry of that period

either

13A.4 .1 instruct that the Variation is to be carried out and is to be valued pursuant to clause 13.4.1;

or

13A.4 .2 instruct that the Variation is not to be carried out.

13A.5 If a 13A Quotation is not accepted a fair and reasonable amount shall be added to the Contract Sum in respect of the cost of preparation of the 13A Quotation provided that the 13A Quotation has been prepared on a fair and reasonable basis. The non-acceptance by the Employer of a 13A Quotation shall not of itself be evidence that the Quotation was not prepared on a fair and reasonable basis.

13A.6 If the Architect has not, under clause 13A.3.2, issued a confirmed acceptance of a 13A Quotation neither the Employer nor the Contractor may use that 13A Quotation for any purpose whatsoever.

13A.7 The Employer and the Contractor may agree to increase or reduce the number of days stated in clause 13A.1.1 and/or in clause 13A.1.2 and any such agreement shall be confirmed in writing by the Employer to the Contractor. Where relevant the Contractor shall notify each Nominated Sub-Contractor of any agreed increase or reduction pursuant to this clause 13A.7.

13A.8 If the Architect issues an instruction requiring a Variation to work for which a 13A Quotation has been given and in respect of which the Architect has issued a confirmed acceptance to the Contractor such Variation shall not be valued under clause 13.5; but the Quantity Surveyor shall make a valuation of such Variation on a fair and reasonable basis having regard to the content of such 13A Quotation and shall include in that valuation the direct loss and/or expense, if any, incurred by the Contractor because the regular progress of the Works or any part thereof has been materially affected by compliance with the instruction requiring the Variation.

13.5.2 Commentary

Clause 13.1: Definition

> Clause 13.1 defines two sorts of variation, that is variations in the work as
> set out in the contract documents, and the imposition of any obligations and
> restrictions upon the contractor or any addition, alteration or omission of
> them if already imposed through the medium of the contract bills. It also
> imposes a prohibition (already present at common law) against the nomin-
> ation of a sub-contractor to execute work included in the contract bills for
> execution by the contractor.

Variations in work

> At first sight clause 13.1.1 appears to give the architect almost unlimited
> power to vary the works – see 13.1.1.1 'the addition, omission or substitution
> of any work'. However, this phrase must be read as subject to the overriding
> effect of the preceding reference to 'the alteration or modification of the
> design, quality or quantity of the Works' – i.e. that it is only the Works as set
> out and described in the contract documents that may be altered or modi-
> fied, and the architect's power does not extend to ordering any additions or
> substitutions that would require the contractor to execute work clearly not
> contemplated by the original contract[582]. In other words, the architect
> cannot substantially alter the nature of the Works – for instance by changing
> a steel-framed building into a reinforced concrete-framed building – since
> the contractor could fairly say that he would not have entered into a contract
> had he known that would be the kind of work required, or at least would
> not have done so except on very different terms. Also, the architect cannot,
> by variation instruction, oblige the contractor to build something substan-
> tially in excess of, or substantially less than, what was envisaged under the
> contract. In essence, the works as completed must still be capable of identi-
> fication as the works set out in the first recital.
>
> On a casual reading, the wording of clause 13.1.1.3 could be confused in
> practice with that of clause 8.4.1, which authorises the architect to issue
> instructions 'in regard to the removal from the site of all or any...work,
> materials or goods' which are not in accordance with the contract. Any such
> instruction would not, of course, entitle the contractor to any additional
> payment. However, this clause is at pains to highlight the difference and
> relates to specific instructions for work or materials which are in accordance
> with the contract, but which for some reason the architect wishes to remove
> either temporarily or permanently.

---

[582] *Sir Lindsay Parkinson & Co Ltd* v. *Commissioners of Works* [1950] 1 All ER 208.

Variations in obligations or restrictions

Clause 13.1.2 authorises the imposition of obligations and restrictions upon the contractor and to change obligations and restrictions already imposed through the medium of the contract bills in respect of five specific matters:

(1) access to the site
(2) use to be made of specific parts of the site
(3) limitations of working space available to the contractor
(4) limitations of hours to be worked by the contractor
(5) any requirements that the works be carried out or completed in any specific order.

The curious wording of this clause should be noted. It is effectively varying the contract. It is certainly not varying the works. Perhaps that is why the draftsman refers to the employer. At first sight it appears to be the only variation which the contract expressly authorises the employer to make. Reading clauses 13.1 and 13.1.2 together makes clear that 'The term "Variation" means the imposition by the Employer of any obligations . . . in regard to the matters set out in clauses 13.1.2.1 to 13.1.2.4 . . .'. However, clause 13.2.1 expressly states that the architect can issue instructions requiring a variation. Therefore, the architect may issue instructions in regard to clause 13.1.2 matters. The conclusion to be drawn from this is that if the employer imposes obligations or restrictions or changes them, as set out in clause 13.1.2, it clearly ranks as a variation under the terms of the contract and is to be valued accordingly. However, the architect has power to issue instructions about the same matters, by virtue of clause 13.2.1. In view of the particular wording of clause 13.1.2, however, it is thought that, in giving an instruction requiring a variation under this clause, the architect should always refer to 'the imposition by the Employer'. It is doubtful whether, in practice, employers often, if ever, avail themselves of the powers given to them under this clause; which is fortunate in view of the confusion which could result. What follows assumes that it will be the architect acting.

The architect's powers are limited to those specific obligations and restrictions, and he has no power in respect of obligations or restrictions of any other kind. It is clear that the architect's power is not confined to varying obligations and restrictions already imposed through the medium of the contract bills, but extends to imposing fresh obligations or restrictions – but only of the kinds listed in clauses 13.1.2.1 to 13.1.2.4 and subject to the contractor's right of 'reasonable objection' discussed below.

These five matters appear to attract little attention in practice. Imposing any of them will inevitably give rise to additional costs to the contractor; some of the restrictions may be far reaching and it is surprising that more claims have not been founded on them. The reason is possibly because employers have adopted a sensible approach and declined to impose obligations in any of these categories unless absolutely unavoidable. Restricting access to the site or the times of such access could be quite catastrophic and where there is a need for such restriction, it should be included in the contract documents. It is more likely that architect's instructions issued under this head will be concerned with the relaxation of previously imposed restrictions.

Much the same comment can be made for the other categories. Where the architect issues instructions about the use to be made of various parts of the site, it must not be confused with failure to give possession of the whole of the site on the date of possession in the contract. Possession must be given on that date, but the architect can restrict the use. It is doubtful that this allows the architect to postpone work, because the contract must be read as a whole and he already has that power under clause 23.2. The architect may wish to restrict the contractor from storing certain materials, erecting cranes, siting cabins and the like. Limiting working space probably falls into the same category while limiting hours might be necessary in response to complaints and visits from local authority inspectors when the project is in progress.

The final matter gives rise to most difficulty. The correct vehicle for conveying the employer's wishes in regard to sequence of work and completion is the Sectional Completion Supplement. A strict reading of the clause suggests that although the order of completion and of carrying out the work may be varied, there is no power under this clause to require the contractor to complete parts of the works by any specific dates if there is just one completion date in the contract. In any event, this power must be exercised with great care where a Sectional Completion Supplement is also in use.

The architect cannot alter the content of any of the sections by using this clause, because to do so might invalidate the liquidated damages clause. There is no mechanism in the contract to amend liquidated damages and it could be argued that a substantial change in work content would invalidate the clause.

The powers of the architect to order variations of this kind are, in any case, subject to a special restriction. Clause 4.1.1, JCT 98, reads in part as follows:

> 'The Contractor shall forthwith comply with all instructions issued to him by the Architect in regard to any matter in respect of which the Architect is expressly empowered by the Conditions to issue instructions; save that:
> .1   where such instruction is one requiring a Variation within the meaning of clause 13.1.2 the Contractor need not comply to the extent that he makes reasonable objection in writing to the Architect to such compliance;'

Shorn of its rather convoluted language, what this means quite simply is that, upon receipt of such a variation instruction from the architect, the contractor may object to it, and if that objection proves to be reasonable the contractor will not be obliged to comply with it. The question of whether or not the objection is reasonable is open to immediate adjudication or arbitration. The real difficulty in deciding what constitutes reasonable or unreasonable objection is that the contract provides for the variation to be valued and, therefore, the contractor should be properly recompensed no matter what restrictions are imposed or altered.

By what criteria is 'reasonable objection' to be measured? It is thought that an objection would only be considered reasonable to the extent that the contractor should be relieved of the obligation to comply with it if compliance would make the work impossible or, at least, unduly onerous. Say, for instance, that the architect ordered that noisy machinery should not be used except between certain specified hours, and that the reason for the instruction was that a neighbour had threatened to obtain an injunction against the employer imposing such a restriction upon him; clearly no contractor could 'reasonably' object to compliance with such an instruction since he would be protected as to additional expense and a suitable extension of the time under the contract terms, and non-compliance could well result in severe penalties for the employer. But a contractor could 'reasonably' object to an instruction restricting access to the site if the restriction would mean that essential goods could no longer be brought on to the site; in such a case the contractor's objection would bring the problem to the architect's attention and he could, no doubt, order further variations to the works which could overcome the problem – or withdraw the instruction.

Variation instructions

Clause 13.2 constitutes the architect's authority to issue instructions requiring a variation within the definition set out in clause 13.1, and also authorises him to 'sanction in writing any Variation made by the Contractor otherwise than pursuant to an instruction of the Architect'. It also states that 'No variation . . . shall vitiate this Contract', an injunction the purpose of which is slightly obscure, since the exercise of a power expressly conferred upon the architect by the terms of the contract cannot 'vitiate' it in any circumstances.

Clause 13.3 states that the architect shall issue instructions in regard to the expenditure of provisional sums, both in the contract bills and in a Nominated Sub-Contract.

Clause 4.1.1 (set out in part earlier in this section) requires the contractor 'forthwith' to comply with any instruction issued by the architect which the contract conditions expressly empower him to issue. Clauses 13.2 and 13.3, therefore, set out this express power of the architect to issue instructions requiring variations and in regard to the expenditure of provisional sums, the latter being not only a power but an obligation as well. The contractor must therefore comply with any such instructions 'forthwith', subject to the proviso regarding reasonable objection to instructions about variations in obligations or restrictions imposed upon the contractor. 'Forthwith' in this context clearly does not necessarily mean 'immediately' since the instruction may vary work not yet done, but it imposes an obligation upon the contractor to carry out the work as soon as he reasonably can do so[583].

---

[583] *London Borough of Hillingdon v. Cutler* (1967) 2 All ER 361.

By clause 4.3.1 all architect's instructions, including instructions under clause 13.2 and 13.3, must be issued in writing; any instruction of the architect not issued in writing is therefore not issued in accordance with the contract and is of no effect. By clause 4.3.2, however, any instruction of the architect issued otherwise than in writing is to be confirmed in writing by the contractor to the architect within 7 days, and if not dissented from in writing by the architect within a further 7 days from receipt (not despatch) is to take effect from the expiry of the second 7 days. If the architect himself confirms the instruction in writing within the first 7 days, the contractor need not confirm it himself but it will take effect from the date of the architect's confirmation. If neither the contractor nor the architect confirms the instruction within the time limit, but the contractor nevertheless complies with it, then the architect may confirm it himself at any time prior to the issue of the final certificate, and the instruction will then, strangely, be deemed to have taken effect from the date it was originally given.

What all this means is that the contractor is under no obligation to comply with an architect's instruction issued otherwise than in writing, i.e. orally; indeed, not only is he under no obligation to comply, he is actually at risk if he does so since the architect may withdraw it or deny he ever gave it. The contractor's obligation to comply only arises when he has himself confirmed it within 7 days of issue and has allowed a further 7 days to elapse plus reasonable time for receipt of the confirmation by the architect – say a total period of 15 or 16 days from the actual date upon which the instruction was issued orally[584]. If the contractor inadvertently fails to confirm the instruction but nevertheless complies with it, he will again be at risk since the architect may, but he is not obliged to, confirm it subsequently – he may refuse to do so, and the contractor may find himself unable to recover any cost involved in compliance.

Provisional sums

Provisional sums are sums of money included in the contract bills to cover work which was uncertain in nature and/or scope at the time the bills were prepared. They must not be confused with prime cost (PC) sums which are included to cover work to be executed by nominated sub-contractors or materials to be supplied by nominated suppliers.

Clause 13.3 obliges the architect to issue instructions for the expenditure of such sums whether included in the main contract bills or in a contract with a nominated sub-contractor. There is no express limitation on the instructions that the architect may issue in this regard, and his instruction may, indeed, be simply to omit the sum altogether and to do no work or spend no money against it.

By clause 35.1.2, an architect's instruction for the expenditure of a provisional sum in the main contract bills may include an instruction that the

---

[584] In the absence of any special definition in the contract, a day is a 24 hour period extending from midnight to midnight.

whole or any part of the work covered by the sum be executed by a nominated sub-contractor; by clause 36.1.1.2 it may also require goods to be supplied by a nominated supplier, either specifically, by naming the supplier concerned, or (clause 36.1.1.3) by implication, by requiring the contractor to obtain materials for the work against the sum for which there is only a single source of supply, in which case that source becomes a nominated supplier provided that the materials are, or become, the subject of a prime cost sum.

There is also no express limitation on the instructions that the architect is to issue in regard to the expenditure of provisional sums in nominated sub-contracts. However, as the standard sub-contract form NSC/C contains no provisions similar to those in the main contract for the nomination of sub-sub-contractors and suppliers, any instruction to place orders for work or supply of materials with specified firms will simply result in those firms becoming ordinary sub-sub-contractors or suppliers to the sub-contractor for which no special terms of contract are laid down. However, the sub-contract forms do contain a provision that, if the sub-sub-contractor or supplier concerned insists upon terms of contract which exclude liability in respect of their work or materials to a greater extent than the nominated sub-contractor's liability to the main contractor is limited, the nominated sub-contractor is to notify the main contractor, who will then notify the architect. If they both agree to the limitation in writing then the sub-contractor's liability to the contractor, and the main contractor's liability to the employer, will be limited in the same way. Without such express agreement the nominated sub-contractor is not obliged to enter into the sub-sub-contract or contract of sale containing the limitation.

Variations in nominated sub-contracts

Variations to work executed by nominated sub-contractors are also to be valued by the quantity surveyor appointed under the main contract. In practice, however, variations in specialist sub-contracts such as those for mechanical engineering services are frequently valued by the consulting engineer responsible for their design. Where this is the case it is strongly advised that the situation be clarified with all concerned so that it is absolutely clear with whom the responsibility for valuation actually does lie; if necessary the appropriate clause of the Sub-Contract Form NSC/C should be amended accordingly.

Variations in work done by the main contractor against PC sums

Clause 13.4.2 is to be read in conjunction with clause 35.2.2. If the main contractor tenders for and carries out work for which a PC sum is included in the contract bills, i.e. instead of a nominated sub-contractor, or tenders for and carries out work against a provisional sum for which the architect had otherwise intended to nominate a sub-contractor, any variations in that work are to be valued in accordance with the provisions of the contractor's tender for that work and not in accordance with clause 13. Clause 13.4.2 also provides, for the sake of clarity, that where the contractor tenders for and carries out work against a provisional sum for which the architect had intended to nominate a sub-contractor, that work, whether varied or not, is to be valued in accordance with the accepted tender and not as if it were 'work executed by the Contractor in accordance with instructions by the Architect as to the expenditure of provisional sums' as provided in clause 13.4.1.

Valuation of variations

Clause 13.4.1.1 is not an easy clause to understand. It provides that the first option for valuation of variations is to be under alternative A (the contractor's price statement) if they fall under one of the following categories:

- A variation which has been instructed or sanctioned under clause 13.2.4 by the architect
- Any work which the contract expressly states is to be treated as a variation
- Any work executed by the contractor in accordance with instructions given by the architect about the expenditure of a provisional sum
- Any work carried out for which there is an approximate quantity in the bills of quantities.

However, there are some exceptions. The first is if the price statement is not accepted, when alternative B must be used. Another is nominated sub-contract work which is to be valued under the provisions of the nominated sub-contract NSC/C. Also excluded is the situation where the architect has issued a confirmed acceptance of a clause 13A quotation or if clause 13A.8 applies. Clause 13A.8 refers to the situation if the architect's instruction deals with work which has already been valued under clause 13A. In such a case, the quantity surveyor must value the work on a fair and reasonable basis having regard to the previous clause 13A quotation and he must include any direct loss and/or expense resulting from the effect on the progress of the works.

The contractor's price statement (alternative A)

This is the first option for the valuation of variations. Within 21 days from receiving a instruction or from commencing work for which an approximate quantity is included in the contract, the contractor may submit to the quantity surveyor a price for carrying out the work. If he wishes, he may also include in the quotation the amount of extension of time and loss and/or expense he requires. There is a stipulation that the contractor must calculate the price statement in accordance with the valuation rules in the contract.

Within 21 days of receiving the price statement, the quantity surveyor must give the contractor written notice stating either that the price statement is accepted or that it is not accepted. If it has been accepted, the position is straightforward. The amount is added to or deducted from the contract sum. If there is no acceptance, the quantity surveyor must inform the contractor of the reasons for the rejection in similar detail to that given by the contractor and must include an amended price statement showing the contractor what would be acceptable to the quantity surveyor. The contractor then has 14 days from receipt of the amended price statement within which to state whether he accepts it or rejects it in whole or in part. If the amended price statement is not entirely accepted by the contractor or if the quantity surveyor does not respond within 21 days, either party may refer the price statement and the amended price statement to adjudication.

If the contractor has submitted his requirements for extension of time and loss and/or expense, the quantity surveyor has 21 days in which to notify the contractor that they are acceptable or that clause 25 and/or 26 of the contract as appropriate will apply.

It is notable that it is the quantity surveyor and not the architect who effectively decides whether to accept the contractor's price statement which may include loss and/or expense and the period of extension of time as well as the valuation of the instruction itself. The decision to give the quantity surveyor the final decision on these matters, which appears to fly in the face of the philosophy of the rest of the contract, seems to have been an error. The quantity surveyor is required to come to his decision 'after consultation' with the architect. Although it would be wrong to say that such a requirement is meaningless, it does not appear to impose any serious restrictions on the quantity surveyor. That would be to confuse 'consult' with 'agree'. The requirement to consult does not carry with it the implication to implement the result of the consultation. Therefore, the quantity surveyor must consult with the architect and the architect may state that the price statement should not be accepted and the quantity surveyor is entitled to form his own opinion and to accept it anyway.

It is understandable that the contract may give the quantity surveyor primacy in regard to valuation, albeit the architect is responsible for the certification. It is less comprehensible that the quantity surveyor may in this instance, and this instance alone, decide on the validity of extension of time and loss and/or expense. If the quantity surveyor accepts the contractor's estimate of an extension of time for the instructed work, the architect is not required to confirm with any notice under clause 25 whether he agrees or disagrees. He must not do so, because clause 25 does not expressly provide for such a relevant event and clause 25.3.2 reinforces the position. It does this by providing that no decision of the architect can fix an earlier completion date than the completion date in the Appendix and then separately prohibits the fixing of the date earlier than the date accepted under clause 13.4.1. Presumably, although it is not stated in clause 25, the architect is expected to take into account the quantity surveyor's acceptance of an extension of time against the architect's wishes, when fixing any further completion dates as a result of subsequent delays. Although it is to be hoped that this situation arises but rarely, it poses the question whether the architect can properly estimate a subsequent extension of time in these circumstances.

Function of the quantity surveyor

By clause 13.4.1.1, the default position is that all variations and all work executed by the contractor in accordance with the architect's instructions for the expenditure of provisional sums shall be valued under alternative B by the quantity surveyor whose name appears in the Articles of Agreement.

Unless the contractor submits a price statement under alternative A or the employer and the contractor otherwise agree, therefore, it is solely the responsibility of the quantity surveyor to determine the price to be paid or allowed in respect of a variation. The architect has no authority to determine it or to influence in any way that determination. It follows that if the architect were to include, in an instruction requiring a variation, any purported instruction as to how it should be valued, such as 'the work executed against this instruction is to be valued as daywork', this would be of no effect and the quantity surveyor not only should, but must, ignore it and use his own judgment as to the manner in which the work should be valued under the terms of the contract. Whether such an instruction issued by the architect is rendered void is not clear; it probably remains valid save for the part regarding valuation, which is an instruction which the architect is not empowered to give.

The quantity surveyor's function is to value the work in the manner laid down in the contract. The contract provides that the quantity surveyor is to value the work in accordance with the provisions of clause 13.5 unless otherwise agreed by the employer and the contractor, who – as the parties to the contract – may, of course, agree any variation to its terms they wish. Whether the architect certifies the amount valued by the quantity surveyor is another matter. Financial certification is an onerous burden and the architect must reasonably satisfy himself regarding the quantity surveyor's valuation before certifying. If, perhaps rarely, he believes that the valuation is too high or too low, the architect's duty is to certify what he believes to be the correct amount.

Valuation by measurement

Where the valuation of the work executed against a provisional sum is capable of being valued by measurement, it is to be measured in accordance with the same principles as those governing the preparation of the contract bills, i.e. normally (unless clause 2.2.2.1 is amended to provide otherwise) in accordance with the Standard Method of Measurement of Building Works, 7th Edition, and is to be valued in accordance with the rules set out in clause 13.5.1. So far as variations are concerned, this obviously only applies to

variations in the actual work to be carried out as defined in clause 13.1.1, to
the extent that the work can reasonably be measured and is not of such a
nature that it can only be valued on the basis of 'prime cost', i.e. as daywork.
The rules of valuation are set out in clauses 13.5.1.1 to 13.5.1.5 supplemented
by clauses 13.5.2, 13.5.3, 13.5.5 and 13.5.7. In the basic rules set out in clause
13.5 there are three principal factors to be taken into account: the character
of the work; the Conditions under which the work is carried out; and the
quantity of the work.

In respect of each individual item of measured work:

(1) If all three factors are unchanged from an item of work already set out in
the contract bills – i.e. if the character and conditions are 'similar' and
the quantity is not significantly changed by the variation – then the price
set out in the contract bills against that item must be used for
the valuation of the variation item.

(2) If the character is 'similar' to that of an item of work set out in the
contract bills, but the conditions are not 'similar' and/or the quantity is
significantly changed from the contract bills, then the price set out in
the contract bills against that item must still be used as the basis of the
valuation of the variation item but must be modified so as to make a 'fair
allowance', upwards or downwards, for the changed conditions and/or
quantity.

(3) If all three factors are changed, then the items are to be valued by the
quantity surveyor at 'fair rates and prices'.

It will be seen that, where the character of the variation item is 'similar' to
that of an item in the contract bills these rules leave the quantity surveyor
with little discretion; he must use the rates set out in the bills for their
valuation, and the only modification he can make is in respect of changes
in conditions or significant changes in quantity. It is only if the character is
no longer 'similar' that the quantity surveyor is given complete discretion to
make what, in his opinion, is a fair valuation of the work. Therefore, it is
essential for the proper understanding of the quantity surveyor's function
that the meaning of the phrase 'similar character' should be clearly understood; it is therefore unfortunate that it is this phrase which has given the
most difficulty to commentators so far. One view is this (referring to IFC 98,
but the principle is the same):

'For this provision to operate fairly between the employer and the contractor, this word must in practice be treated as meaning of the same
character. This can be demonstrated by looking at any particular item of
work. The item of work will be covered by a description. If the instruction
requiring a variation alters in any way that description of the work, it
must therefore become different and may well in fairness be deserving of
a different rate. To interpret the words in any other way will be to prevent
the quantity surveyor from applying a fair valuation where the varied
description of the item, although arguably remaining of a similar character to the original, justifies a different rate. Furthermore, the last few lines

of clause 3.7.4 [similar to the last few lines of clause 13.5.1 of JCT 98] make it clear that due allowance for any change in conditions is made only where the work has not been modified other than in quantity so that the character of the work itself must remain unchanged.'[585]

The ordinary meaning of 'similar' would be 'almost but not precisely the same' or 'identical save for some minor particular'[586]. The words 'similar character' when applied to an individual measured item of work probably mean that the item is virtually identical to an item in the contract bills. If the item is of 'similar character' the only grounds upon which the quantity surveyor can vary the price for the item from that which is set out in the bills is that the conditions are not similar or the quantity has significantly changed, otherwise he must use the price in the bills as it stands. 'Similar' must be read in context. It appears that very little change in description would be needed to render the character of work dissimilar for the purpose of this clause. Then, the rules set out in clauses 13.5.1.1 and 13.5.1.2 cannot be applied and the quantity surveyor is given the unfettered discretion under clause 13.5.1.3 to value the item at a 'fair' rate or price.

It is probably necessary to investigate beyond simple matters of description or measurement in determining whether the rates and prices in the bills are to be set aside and a valuation at 'fair rates and prices' substituted.

In respect of 'conditions', of course, it is not necessary to apply the same strict interpretation to the word 'similar'. The 'character' of an item is precisely defined by its verbal description in the bills; the 'conditions' under which it is to be carried out cannot be so precisely defined, and the question of whether the conditions are 'similar' must be judged by vaguer criteria related to the conditions under which the contractor must be deemed to have anticipated that the work in the original bills would be carried out and those under which the varied work actually was carried out. The 'conditions' referred to in the valuation rules are the conditions to be derived from the express provisions of the contract bills, the drawings and other documents. The quantity surveyor is not entitled to take into account the background against which the contract was made[587].

The word 'similar' is not used in connection with quantity, the criterion here being whether or not the quantity had been 'significantly changed' by the variation. This must be a matter for the objective judgment of the quantity surveyor; a small change in quantity may be significant for some items (particularly where the original quantity was small) but a very large change may not be significant for others. No firm rules can be laid down.

No matter how erroneous the rate or price can be demonstrated to be, whether in respect of its being too high or too low, the rate or price in the

[585] Neil F. Jones & Simon Baylis, *Jones & Bergman's JCT Intermediate Form of Contract*, 3rd edition, 1999, Blackwell Science, p.172.
[586] David Chappell, Derek Marshall, Vincent Powell-Smith and Simon Cavender, *Building Contract Dictionary*, 3rd edition, 2001, Blackwell Publishing, p.159.
[587] *Wates Construction (South) Ltd* v. *Bredero Fleet Ltd* (1993) 63 BLR 128.

bills must be used as the basis and only be adjusted to take account of the changed conditions and/or quantity. The contractor has contracted to carry out variations in the work, and the employer has agreed to pay for them on this basis, and neither can avoid the consequences on the grounds that the price in the bills was too high or too low[588]. A contractor will sometimes take a gamble by putting a high rate on an item of which there is a small quantity or a low rate on an item of which there is a large quantity in the expectation that the quantities of the items will be considerably increased or decreased respectively. If the contractor's gamble succeeds, he will make a nice profit. Quantity surveyors checking priced bills at tender stage will be alert to such pricing, but there is little to be done about it. It is not unlawful, but rather part of a contractor's commercial strategy[589].

With respect to what may be considered 'fair rates and prices' for valuation under clause 13.5.1.3, it seems that the word 'fair' must be read in the context of the contract as a whole. A 'fair' price for varied work in a contract where the prices in the bills are 'keen' may be a similarly keen price[590]. The quantity surveyor should determine his 'fair rates and prices' on the basis of a reasonable analysis of the contractor's pricing of the items set out in the bills, including his allowances for head office overheads and profit.

There is a very simple rule for the valuation of omissions from the contract works; they are to be valued at the rates set out in the contract bills. However, the operation of clause 13.5.5 must not be overlooked and if the omissions substantially change, the conditions under which other work is executed must be valued accordingly.

In valuing omissions, additional and substituted work, the quantity surveyor must take into account factors other than the prices set out in the contract bills against individual items or his fair valuation of measured items. He must, by clause 13.5.3.2, make allowance for any percentage or lump sum adjustments in the contract bills; that is, any such percentages or lump sums (usually to be found in the general summary at the end of the bills) must be applied pro rata to all prices for measured work.

By clause 13.5.3.3 the quantity surveyor is also required to make allowance, where appropriate, for any addition to or reduction of preliminary items of the type referred to in the Standard Method of Measurement, 7th Edition, Section A (Preliminaries General Conditions). It is to be noted that the clause does not actually bind the quantity surveyor to use the rates and prices set out in the bills against such items, but simply to make allowance for any addition to or reduction of such items.

In relation to the adjustment of preliminaries, it would be appropriate to draw attention here to the limitation on the duty and power of the quantity surveyor with regard to the valuation of variations set out in the proviso at

---

[588] *Dudley Corporation* v. *Parsons & Morrin Ltd*, 8 April 1959 CA unreported; *Henry Boot Ltd* v. *Alstom Combined Cycles Ltd* [1999] BLR 123.

[589] *Convent Hospital* v. *Eberlin & Partners* (1988) 14 Con LR 1. The case went to appeal, (1989) 23 Con LR 112, but not on this point.

[590] Some support for this view may be extracted from the judgments in *Cotton* v. *Wallis* [1955] 3 All ER 373 and *Phoenix Components* v. *Stanley Krett* (1989) 6-CLD-03-25.

the end of clause 13.5, that no allowance must be made under clause 13.5 for any effect upon the regular progress of the works or for any direct loss and/or expense for which the contractor would be reimbursed by payment under any other provision – the reference being, of course, to clause 26. In making any allowances in respect of preliminary items, therefore, the quantity surveyor must stop short of making allowance for the effect of the variation in question upon regular progress of the works. This is considered further subsequently, when dealing with variations in obligations and restrictions imposed upon the contractor. In practice, the quantity surveyor is often treading a very thin line.

Buried in clause 13.5.7 is an obligation upon the quantity surveyor to make a fair valuation of any liabilities directly associated with a variation, if the valuation cannot reasonably be carried out by the application of clauses 13.5.1 to 13.5.6. This, it seems, will include an obligation to make due allowance for factors such as the loss to the contractor involved where a variation to the work means that materials already properly ordered for the work as originally designed become redundant, which, under the 1963 edition of the JCT Form, had to be the subject of an application by the contractor to the architect for reimbursement of 'direct loss and/or expense' under clause 11(6). It will also include the valuation of the effect of any instruction which does not require the addition, omission or substitution of work, i.e. obligations or restrictions (see section 'Valuation of "obligations and restrictions"' below).

Finally, clause 13.5.5 requires the quantity surveyor, where the introduction of a variation results in other work not itself varied being executed under conditions other than those otherwise deemed to have been envisaged, to revalue that other work as if it had been varied. In practice this will mean that it must be revalued under clause 13.5.1.2 – that is, on the basis of the rates and prices in the contract bills against the appropriate items adjusted in respect of the changed conditions; but it may also be necessary for the quantity surveyor to make allowance for other factors such as consequential changes in preliminary items.

Dayworks

Where additional or substituted work cannot 'properly', i.e. by reference to SMM7, be valued by measurement, it is to be valued as 'daywork', that is on the basis of prime cost plus percentages as set out in clause 13.5.4. Clearly, while this may be a satisfactory method of valuation to the contractor since it ensures that he will, at least, recover the cost to him of the work (subject to the limitations imposed by the relevant 'Definition of Prime Cost' defined in the clause) plus percentages to cover supervision, overheads and profit, it cannot be considered a satisfactory method for the employer since it imposes no incentive on the contractor to work efficiently. It is therefore considered to be used sparingly by the quantity surveyor where measurement is impossible.

The machinery for submission of daywork vouchers, and particularly the timing, is also highly unsatisfactory. The requirement that the vouchers should be 'delivered for verification to the architect or his authorised representative not later than the end of the week following that in which the work has been executed' is quite unworkable. If the architect or his representative is to verify what is set out on the voucher, i.e. to vouch for the truth of it, it is surely wholly unreasonable to expect him to be able to do so when a voucher for work executed on the Monday of one week does not have to be delivered to him for that purpose until Friday of the week following – an interval of 11 days.

There seems also to be a widespread impression that the clerk of works, if there is one, will be the architect's 'authorised representative' in this context – not surprisingly, since this is otherwise the only reference to the possible existence of such an individual as the architect's representative in the entire contract. Reference to clause 12, however, will show that the clerk of works, far from being a representative of the architect, is an 'inspector on behalf of the Employer', and that he therefore has no authority to verify daywork vouchers unless the architect specifically gives him that authority.

The only sensible way to deal with the problem of verification of daywork vouchers seems to be for the contractor to give advance notice to the architect of his intention to keep daywork records of a particular item of work; for the architect himself to attend the site or, if he is unable to do so, to nominate the clerk of works to act as his 'authorised representative' for that purpose and to take his own records of the time spent and materials used; and for the vouchers to be submitted for verification at the end of each day. In that way, at least, the quantity surveyor can be reasonably certain that the vouchers represent an accurate record of time and materials. If this system is to work properly, it requires the quantity surveyor to notify the contractor in advance of his intention to value using daywork. Of course, he is still under no obligation to accept daywork as the method of valuation if, in his opinion, the work can properly be measured.

Verification is normally carried out by signing the sheets. Often the magic formula 'For record purposes only' is added. However, where dayworks is to be the method of valuation in any particular case, the addition of those words has little practical value and certainly does not prevent the contents of the sheets being used for calculation of payment[591]. In these circumstances it appears that the quantity surveyor has no right to substitute his own opinion for the hours and other resources on the sheets[592]. It has been held that where the employer has set out a system of verification by signing, but has neglected to do so, the sheets will stand without further proof as evidence of the work done unless they can be shown to be inaccurate[593].

---

[591] *Inserco* v. *Honeywell* 19 April 1996, unreported.
[592] *Clusky (trading as Damian Construction)* v. *Chamberlain*, Building Law Monthly, April 1995, p.6.
[593] *JDM Accord Ltd* v. *Secretary of State for the Environment, Food & Rural Affairs* (2004) CILL 2067.

Valuation of 'obligations and restrictions'

Clause 13.5.7 (apart from the reference to 'liabilities directly associated with a Variation' already dealt with above) must relate to the valuation of obligations or restrictions imposed by the employer or variations to obligations or restrictions already imposed in the contract bills as defined in clause 13.1.2. That is because the valuation of variations in the work to be executed under the contract or of work to be executed against provisional sums is comprehensively covered by clauses 13.5.1 to 13.5.5.

The clause requires the quantity surveyor to make a fair valuation of such variations; but this is subject to the proviso regarding the exclusion from the quantity surveyor's valuation of any effect of the variation upon the regular progress of the works or of any direct loss and/or expense reimbursed to the contractor under any other provision of the contract. It is difficult to envisage what would remain to be valued in respect of such a variation other than its effect upon the regular progress of the works; what other financial effect could it have? The quantity surveyor's function in respect of such variations would therefore seem to be very limited or even non-existent.

There seems to be one very unfortunate and no doubt unintended effect of the proviso, and that is that, on a strict reading of its wording, it prevents the valuation of the effect of the removal of obligations or restrictions. The clause prevents allowance being made for any effect upon the regular progress of the works – including any possible improvement in progress resulting from the removal of an obligation or restriction except to the extent that it discounts the contractor's entitlement until clause 26. So far as extension of time is concerned, the architect can take account of such removal under clause 25.3.2.

Rights of the contractor in respect of valuation

If the contractor does not avail himself of the opportunity to submit his price statement under alternative A and the architect does not require a quotation under clause 13A then, subject to any agreement to the contrary between the employer and the contractor, the valuation of variations is solely the function of the quantity surveyor. The contractor has no contractual right to be consulted, no more than has the employer or the architect. The contractor has only one right, under clause 13.6, that of being given the opportunity of being present at the time of any measurement and of taking such notes and measurements as he may require. In theory, therefore, the situation is that the quantity surveyor may simply notify the contractor of his intention to take measurements, but if the contractor has a prior appointment the quantity surveyor, having given him the opportunity of being present, may safely proceed without him. When actually valuing the results of such measurement the quantity surveyor has no obligation to consult the contractor at all, but may proceed without him and at the end of the contract, as required by clause 30.6.1.2, may simply, through the architect, present the contractor with the statement of all the adjustments to the contract sum which would include a summary of the variation valuations, possibly without even measurements attached. The contractor would then either have to accept it or, in due course, to refer the matter to adjudication or arbitration when it became enshrined in an architect's certificate.

This is the situation in theory. In practice, of course, quantity surveyors almost invariably measure and value variations in full consultation and, as far as possible, in agreement with the contractor, so that the adjustment of the contract sum will be an agreed document at least insofar as the variations are concerned and a possible area of dispute will be removed. It is in everyone's interests, not least the employer's, that this practice be followed. Nevertheless, it is important to appreciate that if the contractor fails to co-operate, the quantity surveyor has the authority under the contract to proceed unilaterally, as set out earlier. Indeed, the quantity surveyor probably has a duty to do so and certainly has that duty if there is a danger of missing an express contract timetable[594].

Payment in respect of variations

Clause 13.7 provides that a valuation or an agreement under clause 13.4.1.1 or confirmed acceptance under clause 13A must be given effect by adding to or deducting from the contract sum. By clause 3 this form of words should mean that the amount of the valuation must be taken into account in the computation of the next interim certificate. Read strictly, it might be thought that this could pose a difficulty if the amount was taken into account in the next interim certificate after the valuation has been made, but before the work has been carried out. One answer is to ensure that such valuations are not completed until the work is executed. In any event, the position is correctly stated in clause 30.2.1.1, which states that there shall be included in interim certificates 'the total value of the work properly executed by the Contractor including any such work so executed to which Alternative B in clause 13.4.1.2 applies or to which a Price Statement or any part thereof accepted pursuant to clause 13.4.1.2 paragraph A2 or amended Price Statement or any part thereof accepted pursuant to clause 13.4.1.2 paragraph A4.2 applies ...'. If the valuation is made before the work is properly executed, it may be 'taken into account' in the sense of being considered, but it would not be included because not properly executed. If the formal valuation has not been made by the time the work has been properly executed, a reasonably accurate allowance should be made for it in the next interim certificate.

Contractor's quotation under clause 13A

The introduction of clause 13A made necessary a number of small amendments to clause 13 and clause 30 to integrate the procedure into the

---

[594] *Penwith District Council* v. *VP Developments Ltd*, 21 May 1999, unreported.

valuation mechanisms. The result is a very complex set of valuation provisions.

In order to trigger clause 13A, the architect must state in an instruction that the treatment is to be in accordance with 13A. However, nothing in clause 13A expressly empowers the architect to make such a statement. The only clue is to be found in clause 13.2.3 which states the valuation provisions that will apply 'unless the instruction states that the treatment and valuation of the Variation are to be in accordance with clause 13A ...'. If necessary, a term would probably be implied empowering the architect to make such statement to avoid the clause becoming inoperative.

Under clause 13.2.3, the contractor can indicate disagreement within 7 days or such other period as agreed and the instruction is not then to be carried out unless the architect issues a further instruction to that effect. In such a case the instruction will be valued in accordance with clause 13.4.1.1 as usual. Provided that the contractor has received sufficient information with the instruction, he must provide a '13A Quotation', including clause '3.3A Quotations' in respect of variations to nominated sub-contract work where appropriate, not later than 21 days from the date of receipt of the instruction. Although the instruction is to be issued by the architect, the quotation must be sent to the quantity surveyor where it is open for acceptance for 7 days. Unusually, it appears that the contractor cannot withdraw the quotation before acceptance as he could in the course of ordinary negotiations, because in this instance he is bound by the contract terms to keep his offer open. The quotation must contain:

- The value of the adjustment to the contract sum
- Adjustment to the contract period including fixing a new, possibly earlier, completion date
- The amount of loss and/or expense
- The cost of preparation of the quotation.

If the architect specifically states, the contractor must also include:

- Additional resources required
- A method statement.

The employer has an important role to play in clause 13A, probably because the contractor will be asked to quote particularly where it is likely that the instruction will have some significant effect on the contract in terms of additional expenditure or time. It is for the employer to accept the quotation or otherwise and, if he accepts, the architect must confirm the acceptance in writing to the contractor. The purpose of this acceptance is that the architect can formally confirm that the contractor is to proceed, that the adjustment to the contract sum can be made, that a new date for completion (if applicable) can be fixed, including revised sub-contract periods for any nominated sub-contractors, and that the contractor is to accept any clause 3.3A Quotation. The provision that if the employer does not accept, the architect must either instruct that the variation is to be carried out and valued under clause 13.4.1.1 or instruct that the variation is not to be carried out, is remarkable in one particular. There seems to be no provision for the employer or the quantity surveyor on his behalf to negotiate on the quoted price. It is either to be accepted or rejected.

Approximate quantities, defined and undefined provisional sums

The concepts of defined and undefined provisional sums and approximate quantities were introduced into, what was then, JCT 80, in July 1988. Briefly, the situation is this[595]. Where work can be described in items in accordance with SMM7, but accurate quantities cannot be given, an estimate, called 'approximate quantity' must be given. If the work cannot even be described in accordance with SMM7, it must be listed as a provisional sum. Provisional sums are either defined or undefined. If 'defined', the bills must state the nature and construction, how and where the work is fixed to the building and any other work fixed to it, quantities giving an indication of the scope and extent of the work and any specific limitations. If any of this information cannot be given, the work is 'undefined'.

Provisional sums for defined work are deemed to have been taken into account in the contractor's programming and pricing preliminaries. There will be no adjustment unless measured work in the same circumstances would be adjusted. If the work is undefined, the contractor is deemed not to have made any allowance for it in programming and pricing preliminaries.

This addresses a difficulty which contractors have and which is often misunderstood. The fact is that a contractor will be unable to make any sensible attempt to programme or price preliminaries to deal with a provisional sum which may be little more than a title and a figure. For example, 'mechanical installation = £10,000' is almost meaningless. It used to be common, however, for architects to demand that the contractor made some allowance in his programme for provisional sums of this kind. Setting aside the difficulty of complying with such a request, the request itself demonstrates a lack of understanding that the 'bar' on the bar chart is, or should be, simply the result of a series of complicated calculations taking into account the way in which the work will be integrated into other work and the way in which it will be priced. Rules 10.1 to 10.6 usefully set out the minimum information which the contractor must know before he can plan and price the effects of the item in question.

The valuation of work for which an approximate quantity is included in the contract bills is covered by clauses 13.5.1.4 and 13.5.1.5. Where the approximate quantity is a reasonably accurate forecast, the valuation must be in accordance with the rates for the approximate quantity. If it is not a reasonably accurate forecast, the rates form the basis for the valuation, but the quantity surveyor is to make a fair allowance for the difference in quantity.

No allowance for either addition to or reduction of preliminaries can be made under clauses 13.5.1 or 13.5.2 if the valuation relates to an architect's instruction to expend a provisional sum for defined work.

---

[595] A full description is to be found in General Rules 10.1 to 10.6 of SMM7.

Under clause 13.5.5, if the contractor's compliance with an architect's instruction to expend a provisional sum

- for undefined work; or
- for defined work to the extent the instruction is different from the description in the contract bills; or
- the execution of work for which an inaccurate approximate quantity is in the contract bills

substantially changes the conditions under which other work is carried out, the other work is to be treated as though it was the subject of an architect's instruction. This clause simply brings this kind of item under the same rules as variations to measured work. The differences simply highlight the extent of the contractor's knowledge about the kind of work and the amount at the time the contract was made.

Performance specified work

Performance specified work is dealt with under clause 42. It was introduced by JCT Amendment 12 in July 1993. Because, by specifying *performance*, it covers the design and installation or construction of work, the valuation provisions in respect of this work differ from other traditional work. Clause 13.5.6 deals with the position.

The valuation of performance specified work must include addition or omission of the work involved in preparation of drawings and other documents. It seems that the quantity surveyor has discretion to value the design element alone in instances where the work itself is not carried out. Variations to the work are to be valued by using rates consistent with rates in the contract bills for work of a similar character after making allowance for changes in the conditions or the quantity in the bills or in the Contractor's Statement. In the absence of similar rates, the quantity surveyor must make a fair valuation.

Work omitted is valued in accordance with the rates in the contract bills or in the analysis. Addition to or deduction from preliminaries must be taken into account in the valuation. If daywork is the basis of the valuation, the normal daywork clauses 13.5.4.1 and 13.5.4.2 will apply.

Clause 13.5.6.6 provides that if variations to performance specified work or compliance with instructions to carry out such work, to the extent that it differs from information provided in the contract bills, substantially changes the conditions under which other work is carried out, the other work is to be treated as though it was the subject of an architect's instruction requiring a variation.

There is nothing of particular significance here and it is a fairly standard treatment in line with the valuation of measured work and provisional sum work noted above. Although the quantity surveyor must exercise his duties within the parameters set down, he is given considerable scope.

13.6 Variations position under IFC 98

13.6.1 Clauses 3.6–3.8

The full text of clauses 3.6, 3.7 and 3.8 is as follows:

Variations

3.6 The Architect may subject to clause 3.5.1 issue instructions requiring a Variation and sanction in writing any Variation made by the Contractor otherwise than pursuant to such an instruction. No such instruction or sanction shall vitiate the Contract.

Valuation of Variations and provisional sum work – Approximate Quantity, measurement and valuation

3.7 .1 .1 The amount to be added to or deducted from the Contract Sum in respect of
— instructions requiring a Variation and
— any Variation made by the contractor and sanctioned in writing by the Architect and
— all work which under the Conditions is to be treated as if it were a Variation required by an instruction of the Architect under clause 3.6 and
— instructions on the expenditure of a provisional sum and
— work executed by the Contractor for which an Approximate Quantity is included in the Contract Documents
may be agreed between the Employer and the Contractor prior to the Contractor complying with any such instruction or carrying out such work; but if not so agreed there shall be added to or deducted from the Contract Sum an amount determined
— under Option A in clause 3.7.1.2 or
— if Option A is not implemented by the Contractor, under Option B in clause 3.7.1.2.

3.7 .1 .2 **Option A: Contractor's Price Statement**

Paragraph:

.A1 Without prejudice to his obligation to comply with any instruction or to execute any work to which clause 3.7.1.1 refers, the Contractor may within 21 days from receipt of the instruction or from commencement of work for which an Approximate Quantity is included in the Contract Documents or, if later, from receipt of sufficient information to enable the Contractor to prepare his Price Statement, submit to the Quantity Surveyor his price ('Price Statement') for such compliance or for such work.

The Price Statement shall state the contractor's price for the work which shall be based on the provisions of clauses 3.7.2 to 3.7.10 and may also separately attach the Contractor's requirements for:

.1   any amount to be paid in lieu of any ascertainment under clause 4.11 of direct loss and/or expense not included in any previous ascertainment under clause 4.11;

.2   any adjustment to the time for the completion of the Works to the extent that such adjustment is not included in any extension of time that has been made by the Architect under clause 2.3. *(See paragraph A7)*

.A2   Within 21 days of receipt of a Price Statement the Quantity Surveyor, after consultation with the Architect, shall notify the Contractor in writing

either

.A2   .1   that the Price Statement is accepted;

or

.2   that the Price Statement, or a part thereof, is not accepted.

.A3   Where the Price Statement or a part thereof has been accepted the price in that accepted Price Statement or in that part which has been accepted shall in accordance with clause 3.7.1 be added to or deducted from the Contract Sum.

.A4   Where the Price Statement or a part thereof has not been accepted:

.1   the Quantity Surveyor shall include in his notification to the Contractor the reasons for not having accepted the Price Statement or a part thereof and set out those reasons in similar detail to that given by the Contractor in his Price Statement and supply an amended Price Statement which is acceptable to the Quantity Surveyor after consultation with the Architect;

.2   within 14 days from receipt of the amended Price Statement the Contractor shall state whether or not he accepts the amended Price Statement or part thereof and if accepted paragraph A3 shall apply to that amended Price Statement or part thereof; if no statement within the 14 day period is made the Contractor shall be deemed not to have accepted, in whole or in part, the amended Price Statement;

.3   to the extent that the amended Price Statement is not accepted by the Contractor, the Contractor's Price Statement and the amended Price Statement may be referred either by the Employer or by the Contractor as a dispute or difference to the Adjudicator in accordance with the provisions of clause 9A.

.A5   Where no notification has been given pursuant to paragraph A2 the Price Statement is deemed not to have been accepted, and the Contractor may, on or after the expiry of the 21 day period to which paragraph A2 refers, refer his Price Statement as a dispute or difference to the Adjudicator in accordance with the provisions of clause 9A.

.A6   Where the Price Statement is not accepted by the Quantity Surveyor after consultation with the Architect or an amended Price Statement has not been accepted by the Contractor and no reference to the Adjudicator under paragraph A4.3 or paragraph A5 has been made, Option B shall apply.

.A7   .1   Where the Contractor pursuant to paragraph A1 has attached his requirements to his Price Statement the Quantity Surveyor after consultation with the Architect shall within 21 days of receipt thereof notify the Contractor

.1   either that the requirements in paragraph A1.1 in respect of the amount to be paid in lieu of ascertainment under clause 4.11 is accepted or that the requirement is not accepted and clause 4.11 shall apply in respect of the ascertainment of any direct loss and/or expense;

and

.2   either that the requirement in paragraph A1.2 in respect of an adjustment to the time for the completion of the Works is accepted or that the requirement is not accepted and clause 2.3 shall apply in respect of any such adjustment.

.A7   .2   If the Quantity Surveyor has not notified the Contractor within the 21 days specified in paragraph A7.1, clause 4.11 and clause 2.3 shall apply as if no requirements had been attached to the Price Statement.

3.7   .1   .2   **Option B**

The Valuation shall be made by the Quantity Surveyor in accordance with the provisions of clauses 3.7.2 to 3.7.10.

Valuation rules

The valuation rules are:

3.7   .2   'priced document' as referred to in clauses 3.7.2 to 3.7.8 means, where the second recital, alternative A applies, the Specification or the Schedules of Work as priced by the Contractor or the Contract Bills; and, where the second recital, alternative B applies, the Contract Sum Analysis or the Schedule of Rates;

3.7   .3   omissions shall be valued in accordance with the relevant prices in the priced document;

3.7   .4   (a) except for work for which an Approximate Quantity is included in the Contract Documents, for work of a similar character to that set out in the priced document the valuation shall be consistent with the relevant values therein making due allowance for any change in the conditions under which the work is carried out and/or any significant change in quantity of the work so set out;

(b) for work for which an Approximate Quantity is included in the Contract Documents:

— where the Approximate Quantity is a reasonably accurate forecast of the quantity of work required the valuation shall be consistent with the relevant values in the priced document;

— where the Approximate Quantity is not a reasonably accurate forecast of the quantity of work required the relevant values in the priced document shall be the basis for the valuation and that valuation shall include a fair allowance for such difference in quantity;

provided that clauses 3.7.4(a) and 3.7.4(b) shall only apply to the extent that the work has not been modified other than in quantity;

3.7 .5   a fair valuation shall be made

— where there is no work of a similar character to that set out in the priced document, or

— to the extent that the valuation does not relate to the execution of additional, or substituted work or the omission of work, or the execution of work for which an Approximate Quantity is included in the Contract Documents or the omission of work, or

— to the extent that the valuation of any work or liabilities directly associated with the instruction cannot reasonably be effected by a valuation by the application of clause 3.7.3 or clause 3.7.4;

3.7 .6   where the appropriate basis of a fair valuation is Daywork, the valuation shall comprise

— the prime cost of such work (calculated in accordance With the 'Definition of Prime Cost of Daywork carried out under a Building Contract' issued by the Royal Institution of Chartered Surveyors and the Building Employers Confederation (now Construction Confederation), which was current at the Base Date) together with percentage additions to each section of the prime cost at the rates set out by the Contractor in the priced document, or

— where the work is within the province of any specialist trade and the said Institution and the appropriate body representing the employers in that trade have agreed and issued a definition of prime cost of daywork, the prime cost of such work calculated in accordance with that definition which was current at the Base Date together with percentage additions on the prime cost at the rates set out by the Contractor in the priced document;

3.7 .7   the valuation shall include, where appropriate, any addition to or reduction of any relevant items of a preliminary nature; provided that where the Contract Documents include bills of quantities no such addition or reduction shall be made in respect of compliance with an instruction as to the expenditure of a provisional sum for defined work included in such bills;

3.7 .8 no allowance shall be made in the valuation for any effect upon the regular progress of the Works or for any other direct loss and/or expense for which the Contractor would be reimbursed by payment under any other provision in the Conditions;

3.7 .9 if compliance with any such instructions substantially changes the conditions under which any other work is executed, then such other work shall be treated as if it had been the subject of an instruction of the Architect requiring a Variation under clause 3.6 to which clause 3.7 shall apply; Clause 3.7.9 shall apply to the execution of work for which an Approximate Quantity is included in the Contract Documents to such extent as the quantity is more or less than the quantity ascribed to that work in the Contract Documents and, where the Contract Documents include bills of quantities, to compliance with an instruction as to the expenditure of a provisional sum for defined work only to the extent that the instruction for the work differs from the description given for such work in such bills.

3.7 .10 where the priced document is the Contract Sum Analysis or the Schedule of Rates and relevant rates and prices are not set out therein so that the whole or part of clauses 3.7.3 to 3.7.9 cannot apply, a fair valuation shall be made.

Instructions to expend provisional sums

3.8 The Architect shall issue instructions as to the expenditure of any provisional sums.

13.6.2 Commentary

General

It will be seen that these clauses bear a strong resemblance to the equivalent clause 13 in JCT 98, the differences mainly reflecting the much more concise and business-like style of drafting adopted in IFC 98 generally and the absence of an IFC 98 equivalent of JCT 98 clause 13A. A number of interesting points will repay examination.

Instructions

Like clause 4.3.1 of JCT 98, clause 3.5.2 of IFC 98 states that all architect's instructions, including those requiring variations or relating to the expenditure of provisional sums, shall be in writing. However, IFC 98 does not contain any provision similar to clause 4.3.2 of JCT 98 allowing instructions other than in writing to be confirmed by the contractor. It follows that no instruction can be binding on the contractor unless actually issued by the architect in writing. It seems, however, that where a contractor confirms an oral instruction, he is contractually obliged to do the work described in such confirmation and he will be held to have waived his right to rely upon clause 3.5[596]. It is also probable that, where an architect is in the habit of issuing oral instructions without confirmation, the employer will be unable to rely on the absence of a written instruction as an excuse for non-payment[597]. The slight difference in wording between clause 4.3.1 of JCT 98 ('all instructions...shall be *issued* in writing') and clause 3.5.2 of IFC 98 ('all instructions...shall be in writing') is of no real significance[598].

Definition of a variation

The definition of a variation in work in clause 3.6.1 of IFC 98 is virtually identical to that in clause 13.1.1 of JCT 98. The definition of a variation in obligations and restrictions in clause 3.6.2, while not identical with JCT 98 clause 13.1.2, is essentially the same.

Valuation of variations

It will be seen that there is a change in emphasis in clause 3.7 of IFC 98 compared with clause 13.4.1 of JCT 98 so that agreement of a price between the employer and the contractor becomes the preferred option rather than the submission of a price statement or valuation by the quantity surveyor. However, it is still clearly advisable for an employer to seek advice from the quantity surveyor before accepting any price offered by the contractor so that he may be assured that it represents value for money. A more formal process for agreeing a price is by way of option A: the contractor's price statement. This procedure is virtually identical to the equivalent procedure under JCT 98 and the same comments apply.

The rules of valuation reflect the flexibility of contract documentation available under IFC 98. A contract under IFC 98 may be entered into in two ways. It may be on the basis of drawings and a document which has been provided to the contractor by the employer at the time of tendering so that it may be priced by the contractor to form the basis of his tender and ultimately of the contract sum (alternative A of the second recital). Alternatively it may be on the basis of drawings and unpriced specification only (alternative B of the second recital).

A document to be priced by the contractor under alternative A of the second recital may be one of three kinds:

- a full bill of quantities prepared in accordance with a specified method of measurement

---

[596] *Bowmer & Kirkland Ltd* v. *Wilson Bowden Properties Ltd* (1996) 80 BLR 131.
[597] *Redheugh Construction Ltd* v. *Coyne Contracting Ltd and British Columbia Building Corporation* (1997) 29 CLR (2d) 39-46; *Ministry of Defence* v. *Scott Wilson Kirkpatrick and Dean & Dyball Construction* [2000] BLR 20.
[598] Peter Hibberd, *Variations in Construction Contracts*, 1987, Blackwell Science.

- a priceable specification of works, i.e. one set out in such a way that the contractor may attach a price to each item
- a 'schedule of work'; this is defined in clause 8.3 as 'an unpriced schedule referring to the Works which has been provided by the Employer and which if priced by the Contractor (as mentioned in the second recital) for the computation of the Contract Sum is included in the Contract Documents' – i.e. any document which is neither a specification nor a bill of quantities but which in some way describes the works and is set out so that it may be priced by the contractor to form the basis of his tender and of the contract sum.

If the document is a full bill of quantities, clause 1.2 provides that 'the quality and quantity of the work included in the Contract Sum shall be deemed to be that which is set out in the Contract Bills', i.e. they have the same status in defining the work which the contractor has agreed to do for the contract sum as bills of quantities have under JCT 98 with Quantities (see the consideration of this point in section 13.2 above). However, where the document is not a bill of quantities prepared in accordance with a specified method of measurement, clause 1.2 provides that 'where and to the extent that quantities are contained [in the document] the quality and quantity of the work included in the Contract Sum for the relevant items shall be deemed to be that which is set out in the [document]'. For all items to which quantities are not attached, if there is a conflict between such items and the contract drawings, the quality and quantity of work included in the contract sum is deemed to be that which is set out in the drawings. Put simply, it amounts to this. If there are no quantities for a particular item, the contract documents must be read together. If there is conflict between the documents, the drawings prevail. Where quantities are shown, they prevail.

This approach to the priority of documents where a full bill of quantities is not used, but where there is some other priced document which has formed the basis of the contractor's tender, seems illogical. Where the employer has provided a document which is intended by him to be used by the contractor as the basis for the tender, and where that tender becomes the contract sum, then irrespective of the form the document takes, what is set out in that document, whether by way of words or quantities, would usually be expected to be that for which the contractor has priced and would take priority over the contract drawings in that respect. There seems no good reason to separate only those items to which quantities are attached as defining what the contractor has priced to do. This provision is bad in principle and confusing in practice.

Where no document is provided to the contractor for pricing at tender stage, alternative B of the second recital to the form requires the contractor to supply the employer with either a contract sum analysis or a schedule of rates on which the contract sum is based. A contract sum analysis is a type of document first introduced by the JCT as a cost control document in the Standard Form of Contract With Contractor's Design 1981 (CD 81 – the forerunner of WCD 98). It is essentially a breakdown of the contract works into their component parts with a sum stated against each, the whole adding up to the contract sum. The definition in clause 8.3 says that it is to be 'provided by the Contractor in accordance with the stated requirements of the Employer', from which it appears that the employer must specify the form the analysis is to take when inviting tenders. The essential point is that, while the form it is to take is to be specified by the employer, the employer does not actually provide the document himself; if he were to do so it would become a 'priced Schedule of Work' and therefore a contract document.

In contrast to a contract sum analysis, a schedule of rates is simply a list of items with prices attached. It is not added up to show the total tender and contract sum, and there is, therefore and rather dangerously, no means of checking that the prices shown in it actually are the prices used by the contractor in making up his tender. From the employer's point of view, therefore, it is a far less satisfactory document for use as the basis for valuing variations. Clause 3.7.4 requires that the valuation of work of 'similar character' to that set out in the priced document should be 'consistent with the relevant values' set out in that document. This may be contrasted with the equivalent provision in clause 13.5.1.1 of JCT 98 that the rates and prices for the work set out in the contract bills 'shall determine the Valuation'.

There is room for some argument about the significance of this changed wording. It may be that the requirement that the valuation shall be *consistent with* the priced document rather than that the rates and prices in it shall *determine* the valuation gives some scope for greater exercise of discretion by the quantity surveyor. However, there is little doubt that the effect is that, where there are actual rates and prices in the priced document as opposed to lump sums, those rates and prices should be applied to work of 'similar character' making due allowance only for changed conditions or significant changes in quantity. Where the priced document contains lump sums for parcels of work rather than rates and prices for items of measured work, the valuation should take account of the presumed level of pricing underlying the lump sum. Otherwise the general discussion on the valuation of variations by reference to rates and prices in the contract bills under JCT 98 earlier in this chapter can be read as applying equally to the similar exercise under IFC 98.

Daywork

There is an interesting contrast between the words of JCT 98 clause 13.5.4 and IFC 98 clause 3.7.5. The former states that daywork is to be used for the valuation of work 'which cannot properly be valued by measurement'; the latter states that it is to be used where it is 'the appropriate basis of a fair valuation'. It seems quite clear that daywork should only be used where it is the only means of arriving at a fair valuation, so that a fair valuation should be achieved by measurement wherever possible.

It is interesting to note that the provision that vouchers showing the time spent and plant and materials employed on daywork are to be produced to the architect for verification within a stated time limit does not appear in IFC 98. Presumably the appropriate vouchers must, at the latest, be produced by the contractor 'either before or within a reasonable time after Practical Completion of the Works' as required by clause 4.5. Common sense would suggest that any contractor would be well advised to produce the vouchers well before then if he considers daywork to be the appropriate means of arriving at a fair valuation. Any architect asked to verify daywork sheets after practical completion for work carried out much earlier would be unlikely to accept them, because there would then be no means of verification. The daywork sheet might still be relevant as the contractor's record of work done, but the quantity surveyor would doubtless opt for valuing on a different basis.

It must be remembered, however, that if the matter is eventually referred to adjudication or arbitration and dayworks is found to be an appropriate method of valuation, the contractor's own daywork sheets, unverified by the architect, may be held to be the best evidence of what work was done and the resources employed (see the comments on daywork under JCT 98 above). It is desirable to include some provision in the contract documents requiring the contractor to submit daywork sheets within a specified time. The quantity surveyor should give prior notice to the contractor requiring him to keep such records if, in the quantity surveyor's opinion, a particular variation instructed by the architect can only be fairly valued by daywork

13.7 Variations position under MW 98

13.7.1 Clause 3.6

The full text of this clause has been reproduced in Chapter 12, section 12.4.1.

13.7.2 Commentary

The variations provisions in MW 98, like all the other provisions, are very brief. Essentially, they empower the architect to order:

- an addition to the works
- an omission from the works
- a change in the works
- a change in the order in which the works are to be carried out
- a change in the period in which the works are to be carried out.

Unlike the position under other JCT forms, the contractor has no contractual right of objection to changes in the order or timing. The architect is charged with valuing all the types of variation noted in the clause. Although this

contract allows the name of a quantity surveyor to be inserted in the fourth
recital, there are no duties expressly allocated to him. If the name of a
quantity surveyor was not inserted, the architect would not have any
implied authority to appoint a quantity surveyor and in any event it remains
his duty to value[599]. The architect, of course like everyone else, can take
advice from whomsoever he chooses. If a quantity surveyor is appointed,
the architect will no doubt let him carry out the valuation, but the wording
of the clause ('shall be valued by the architect') means that the architect
must understand precisely how the valuation has been carried out and it
will be no defence later to say that he simply adopted the quantity survey-
or's valuation, however eminent the quantity surveyor may be.

A price may be agreed between architect and contractor, but this approach is indicated as an alternative to valuation on a 'fair and reasonable
basis'. It is also clear that, to be effective, the price must be agreed before the
contractor complies with the instruction.

In carrying out his valuation, the architect must use where relevant the
prices in the priced documents. These documents are the priced specification or the priced schedules or the contractor's own schedule of rates. Alone
among the foregoing, the contractor's own schedule of rates is not a contract
document[600]. If the architect waits until the first variation has to be valued
before asking for the contractor's rates, it is an invitation to the contractor to
adjust his rates to suit the circumstances. If the schedule is not provided
before the job starts on site, any later provision will be suspect and the
architect's task of valuation in these circumstances will be very difficult. It is
for the architect to decide whether the prices are 'relevant'. It seems, therefore, that unless the work to be valued is exactly the same and carried out
under the same conditions, he is free to ignore the prices in the priced
document. This is because a very slight change in the conditions under
which work is carried out or in the character of the work may have a
major impact on the contractor's costs.

The valuation must include 'any direct loss and/or expense
incurred by the Contractor due to regular progress of the Works being
affected by compliance with' the variation instruction or compliance or
non-compliance by the employer with clause 5.6 (see Chapter 12, section
12.4).

There is no express requirement that the contractor must submit documentary evidence to help the architect carry out the valuation. No doubt
many contractors will do that as a matter of course in any event. The
valuation must include allowance for profit, overheads and so on, as
usual. A fair and reasonable valuation must include the effect of the instruction on other work not expressly included in the instruction.

Clause 3.7 very simply allows the architect to issue instructions regarding
the expenditure of provisional sums. He must omit the sum and value the
instruction in accordance with the principles in clause 3.6. It is possible to

[599] *Beattie* v. *Gilroy* (1882) 20 Sc LR 162.
[600] See the second recital on the front page of MW 98.

use the provision to nominate a sub-contractor, but it is not particularly wise.

13.8 Variations position under WCD 98

13.8.1 Clauses 12 and S6

The text of clause 12 is as follows:

12   Changes in the Employer's Requirements and provisional sums

12.1   The term 'Change in the Employer's Requirements' or 'Change' means:

12.1   .1   a change in the Employer's Requirements which makes necessary the alteration or modification of the design, quality or quantity of the Works, otherwise than such as may be reasonably necessary for the purposes of rectification pursuant to clause 8 4, including

.1   .1   the addition, omission or substitution of any work,

.1   .2   the alteration of the kind or standard of any of the materials or goods to be used in the Works,

.1   .3   the removal from the site of any work executed or materials or goods brought thereon by the Contractor for the purposes of the Works other than work materials or goods which are not in accordance with this Contract;

.2   the imposition by the Employer of any obligations or restrictions in regard to the matters set out in clause 12.1.2.1 to 12.1.2.4 or the addition to or alteration or omission of any such obligations or restrictions so imposed or imposed by the Employer in the Employer's Requirements in regard to:

.2   .1   access to the site or use of any specific parts of the site,

.2   .2   limitations of working space,

.2   .3   limitations of working hours,

.2   .4   the execution or completion of the work in any specific order.

12.2   .1   The Employer may subject to the proviso hereto and to clause 12.2.2 and to the Contractor's right of reasonable objection set out in clause 4.1.1 issue instructions effecting a Change in the Employer's Requirements. No Change effected by the Employer shall vitiate this Contract. Provided that the Employer may not effect a Change which is, or which makes necessary, an alteration or modification in the design of the Works without the consent of the Contractor which consent shall not be unreasonably delayed or withheld.

12.2   .2   Where the Contractor is and while he remains the Planning Supervisor, he shall, within a reasonable time after receipt of an instruction effecting a Change in the Employer's Requirements or of an instruction in regard to the expenditure of a provisional sum included in the Employer's Requirements, notify the Employer in writing whether, pursuant to his obligations under regulation 14 of the CDM Regulations, he has any objection to such instruction. If he has such objection the Employer shall, to the reasonable satisfaction of the Contractor, vary the terms of the instruction so as to remove that objection; and until the Employer has so varied the terms of the instruction the Contractor shall not be required pursuant to clause 2.1 to comply with such instruction.

12.3   The Employer shall issue instructions to the Contractor in regard to the expenditure of provisional sums (if any) included in the Employer's Requirements.

12.4   .1   The valuation of Changes and of the work executed by the Contractor for which a provisional sum is included in the Employer's Requirements shall, unless otherwise agreed by the Employer and the Contractor, be made under Alternative A in clause 12.4.2 or, if Alternative A is not implemented by the Contractor, under Alternative B in clause 12.4.2. Such valuation shall include allowance for the addition or omission of the relevant design work.

12.4   .2   *Alternative A: Contractor's Price Statement*

Paragraph:

.A1   Without prejudice to his obligation to comply with any instruction or to execute any work to which clause 12.4.1 refers, the Contractor may within 21 days from receipt of the instruction or, if later, from receipt of sufficient information to enable the Contractor to prepare his Price Statement, submit to the Employer his price ('Price Statement') for such compliance or for such work.

The Price Statement shall state the contractor's price for the work which shall be based on the provisions of clause 12.5 (*valuation rules*) and may also separately attach the Contractor's requirements for:

.1   any amount to be paid in lieu of any ascertainment under clause 26.1 of direct loss and/or expense not included in any previous ascertainment under clause 26;

.2   any adjustment to the time for the completion of the Works to the extent that such adjustment is not included in any extension of time that has been made by the Employer under clause 25.3.

.A2   Within 21 days of receipt of a Price Statement the Employer shall notify the Contractor in writing

either

.1   that the Price Statement is accepted;

or

.2   that the Price Statement, or a part thereof, is not accepted.

Variations

.A3 Where the Price Statement or a part thereof has been accepted the price in that accepted Price Statement or in that part which has been accepted shall in accordance with clause 12.6 be added to or deducted from the Contract Sum.

.A4 Where the Price Statement or a part thereof has not been accepted:

.1 the Employer shall include in his notification to the Contractor the reasons for not having accepted the Price Statement or a part thereof and set out those reasons in similar detail to that given by the Contractor in his Price Statement and supply an amended Price Statement which is acceptable to the Employer;

.2 within 14 days from receipt of the amended Price Statement the Contractor shall state whether or not he accepts the amended Price Statement or part thereof and if accepted paragraph A3 shall apply to that amended Price Statement or part thereof; if no statement within the 14 day period is made the Contractor shall be deemed not to have accepted, in whole or in part, the amended Price Statement;

.3 to the extent that the amended Price Statement is not accepted by the Contractor, the Contractor's Price Statement and the amended Price Statement may be referred either by the Employer or by the Contractor as a dispute or difference to the Adjudicator in accordance with the provisions of clause 39A.

.A5 Where no notification has been given pursuant to paragraph A2 the Price Statement is deemed not to have been accepted, and the Contractor may, on or after the expiry of the 21 day period to which paragraph A2 refers, refer his Price Statement as a dispute or difference to the Adjudicator in accordance with the provisions of clause 39A.

.A6 Where the Price Statement is not accepted by the Employer or an amended Price Statement has not been accepted by the Contractor and no reference to the Adjudicator under paragraph A4.3 or paragraph A5 has been made, Alternative B shall apply.

.A7 .1 Where the Contractor pursuant to paragraph A1 has attached his requirements to his Price Statement the Employer shall within 21 days of receipt thereof notify the Contractor:

.1 either that the requirements in paragraph A1.1 in respect of the amount to be paid in lieu of ascertainment under clause 26.1 is accepted or that the requirement is not accepted and clause 26.1 shall apply in respect of the ascertainment of any direct loss and/expense; and

.2 either that the requirement in paragraph A1.2 in respect of an adjustment to the time for the completion of the Works is accepted or that the requirement is not accepted and clause 25 shall apply in respect of any such adjustment.

.A7   .2   If the Employer has not notified the Contractor within the 21 days specified in paragraph A7.1, clause 25 and clause 26 shall apply as if no requirements had been attached to the Price Statement.

3.7.1   .2   *Alternative B*

The Valuation shall be made by the Quantity Surveyor in accordance with the provisions of clauses 12.5.1 to 12.5.6.

12.5   .1   The valuation of additional or substituted work shall be consistent with the values of work of a similar character set out in the Contract Sum Analysis making due allowance for any change in the conditions under which the work is carried out and/or any significant change in the quantity of the work so set out. Where there is no work of a similar character set out in the Contract Sum Analysis a fair valuation shall be made.

.2   The valuation of the omission of work shall be in accordance with the values in the Contract Sum Analysis.

.3   Any valuation of work under clauses 12.5.1 and 12.5.2 shall include allowance for any necessary addition to or reduction of the provision of site administration, site facilities and temporary works.

.4   Where an appropriate basis of a fair valuation of additional or substituted work is Daywork, the valuation shall comprise:

.4   .1   the prime cost of such work (calculated in accordance with the 'Definition of Prime Cost of Daywork carried out under a Building Contract' issued by the Royal Institution of Chartered Surveyors and the Building Employers Confederation (now Construction Confederation) which was current at the Base Date) together with percentage additions to each section of the prime cost at the rates set out by the Contractor in the Contract Sum Analysis; or

.4   .2   where the work is within the province of any specialist trade and the said institution and the appropriate body representing the employers in that trade have agreed and issued a definition of prime cost of daywork, the prime cost of such work calculated in accordance with that definition which was current at the Base Date together with percentage additions on the prime cost at the rates set out by the Contractor in the Contract Sum Analysis.

Provided that in any case vouchers specifying the time daily spent upon the work, the workmen's names, the plant and the materials employed shall be delivered for verification to the Employer not later than the end of the week following that in which the work has been executed.

12.5   .5   If compliance with the instruction effecting a Change or the instruction as to the expenditure of a provisional sum in whole or in part substantially changes the conditions under which any other work is executed, then such other work shall be treated as if it had been the subject of an instruction effecting a Change under clause 12.2 which shall be valued in accordance with the provisions of clause 12.5.

12.5 .6 To the extent that the valuation does not relate to the execution of additional or substituted work or the omission of work or to the extent that the valuation of any work or liabilities directly associated with a Change cannot reasonably be effected in the valuation by the application of clauses 12.5.1 to .5 a fair valuation thereof shall be made.

Provided that no allowance shall be made under clause 12.5 for any effect upon the regular progress of the Works or for any other direct loss and/or expense for which the Contractor would be reimbursed by payment under any other provision in the Conditions.

12.6 Effect shall be given to any agreement by the Employer and the Contractor pursuant to clause 12.4.1 and to the valuation under clause 12.4.1 and 12.4.2 by addition to or deduction from the Contract Sum.

13.8.2 Commentary

General

It will be noted that the wording is very similar to that of JCT 98 clause 13. In administrative terms, however, there is a significant difference – the lack of an independent administrator and certifier. Effectively, in this form of contract the employer and contractor stand facing each other without the benefit of an intervening architect. The employer's agent is exactly that: an agent of the employer and it is probable that he owes no duty to the employer to act fairly between the parties. It seems, however, that he must demonstrate a very high duty of good faith[601]. His notices do not have the status of the certificate of an independent architect[602].

The control documents for the contract works generally which are in place of the contract drawings and the contract bills are the Employer's Requirements and the Contractor's Proposals. A question which often arises is which takes precedence if there is a discrepancy between the Employer's Requirements and the Contractor's Proposals? It is sometimes argued that the employer *accepts* the Contractor's Proposals and that forms the contract and, therefore, the Contractor's Proposals take precedence. That is to view the formation of the contract as a simple matter of offer and acceptance. Although, no doubt, much negotiation may take place, the formation of the contract occurs when the contract documents are executed by both parties. Essentially, those documents consist of the printed form WCD 98, the Employer's Requirements, the Contractor's Proposals and the contract sum analysis.

The whole philosophy of this contract is that the contractor is charged with satisfying the Employer's Requirements. The Employer's Requirements clearly must be the principal document. Clause 8.1 of the contract makes it the prime determinant of the kind and standard of materials and workman-

---

[601] *Balfour Beatty Civil Engineering Ltd* v. *Docklands Light Railway Ltd* (1996) 78 BLR 42.
[602] *J. F. Finnegan Ltd* v. *Ford Seller Morris Developments Ltd* (1991) 53 BLR 38.

ship and only if it does not indicate workmanship or materials does the contractor turn to the Contractor's Proposals. Under clause 12, a 'Change' can refer only to a change in the Employer's Requirements and the employer can only instruct the expenditure of a provisional sum if it is in the employer's requirements. Reference is sometimes made to the third recital which, it is argued, suggests that the employer accepts the Contractor's Proposals. A close reading of the clause indicates that it is of little practical or legal effect. It is a principle of construction of contracts that if words in the main part of the contract are ambiguous, one may turn to the recitals to discover the true meaning of the words. But if the words in the main part of the contract are clear, the recitals cannot change them[603]. In the case of WCD 98, the words in clauses 8.1, 12.1 and 12.3 are unambiguous. In any event, the third recital simply records the employer's general satisfaction with the Contractor's Proposals. The employer, even if in receipt of professional advice, cannot be expected to check the Contractor's Proposals in detail. Any approval or acceptance which the employer gives must be understood in that context[604].

For obvious reasons, there is no provision equivalent to JCT 98 clause 13.6.

Bills of quantities

Perhaps surprisingly, supplementary provision S5 makes provision for the works to be described in the Employer's Requirements by means of bills of quantities. Since the whole idea of a design and build contract is that the contractor not only takes responsibility for building, but also for completing the design (clause 2.1), the scope for describing the Works by means of a bill of quantities might normally be expected to be small, since the bill cannot, or at any rate should not, be prepared until the design is completed. However, it does sometimes happen that the employer requires a virtually complete building design before seeking tenders. This may be partly explained by the current prevalence of 'novation' or 'consultant switch' of the design team[605]. The dangers to the employer of having most of the design completed before tendering may have been lessened following the recent decision dealing with the similar wording in the JCT Designed Portion Supplement which held that the contractor's obligation to complete the design required him to check the adequacy of the preliminary design of others[606]. However, whether that extends to an obligation to check a fully completed design is not at all certain and may be doubted. If, for whatever reason, full bills of quantities are used, clause S5 sets out the following stipulations:

- The Employer's Requirements must state the method of measurement.
- Errors in description or quantity must be corrected and treated as a change instruction.

---

[603] *Leggott* v. *Barrett* (1880) Ch D 306; *Royal Insurance Co Ltd* v. *G & S Assured Investments Co Ltd* [1972] 1 Lloyd's Rep 267.
[604] *Hampshire County Council* v. *Stanley Hugh Leach Ltd* (1991) 8-CLD-07-12.
[605] These practices and other aspects of WCD 98 are considered in David Chappell and Vincent Powell-Smith, *JCT Design and Build Contract*, 2nd edition, 1999, Blackwell Publishing.
[606] *Co-operative Insurance Society Ltd* v. *Henry Boot Scotland Ltd*, 1 July 2002, unreported in this respect.

- Clause 12.5 valuations must use rates and prices in the bills of quantities instead of those in the contract sum analysis.
- Clause 38 (price adjustment formula) must be amended to refer to the bills of quantities instead of the contract sum analysis.

Since the net result of using clause S5 bills of quantities appears to be to place the financial responsibility for errors firmly on the employer, any liability the contractor may have for checking the design appears to be unaccompanied by any financial sanctions.

Clause 12.1 – Definition

The term 'change in the Employer's Requirements' or simply 'Change' is used instead of 'Variation'. The definition of a change is virtually identical to the definition of a variation in clause 13.1 of JCT 98 except in one very important particular. It does not refer to the alteration or modification of the design, quality or quantity of the works, but to a change in the employer's requirements which makes those things necessary. This is fundamental to this contract. The employer has no power to directly alter the design of the works. That cannot be emphasised too much. He can only alter his require- ments on which the contractor has based his design. The contractor is free to respond to that change in any way he wishes and there is no provision for the employer to be consulted. For example, if the employer requires an auditorium to be capable of seating 1500 instead of 1000 people, the contractor may satisfy that change, all other things being equal, by making the auditorium longer, wider, a combination of the two or by introducing a gallery. Where an architect is employed as employer's agent, it is still common for him to issue a change instruction in the form of a detailed drawing showing the precise alteration to the design which the architect believes will satisfy the employer's changed requirements. In such circum- stances, the contractor may not even be given details of such changed requirements. The employer's agent appears to have no power to issue instructions in that form and the contractor may refuse to comply. Alterna- tively, the contractor may choose to consider the drawing to be the employ- er's changed requirements in particularly detailed form. However, the contractor must beware that probably he will be responsible for reviewing such design and making sure that it works.

In any event there is an important restriction on the employer's power to require changes. Clause 12.2.1 states that he cannot instruct the contractor to carry out a change requiring an alteration or modification in the design without the consent of the contractor. Although the contractor cannot with- hold or delay his consent unreasonably, it still leaves plenty of scope for the contractor to refuse consent if he wishes. This provision is entirely consist- ent with the philosophy of the form which assumes that the employer has set out all his requirements and that he is generally satisfied with the contractor's response as embodied in the Contractor's Proposals. Changes must be possible, but they are not encouraged. The word 'design' of course does not simply mean the drawings, but also the written part of the Contractor's Proposals such as specifications and schedules of work[607]. Every change will involve an amendment to some item in one of these documents. This clause, therefore, effectively gives the contractor the right of veto over any change if he cares to use it.

Clause 7 – Site boundaries

Clause 7, with admirable brevity, states that the employer must define the boundaries of the site. The employer has no power to change the definition once it is made, but clause 2.3.1 makes clear that if there is a divergence between the definition and the Employer's Requirements, the employer must issue an instruction to correct it. The instruction is then deemed to be a change issued under clause 12.3 for the purposes of adjustment to the contract sum. Clause 2.3.2 states that if the employer or the contractor finds a divergence, either must give the other a written notice. It may be argued, therefore, that if the contractor fails to spot a divergence which objectively should have been obvious and fails to give notice, he will be unable to recover any resultant expense incurred in correcting the divergence. However, it is clear from the wording of the clause that the contractor has no obligation to look for or find divergencies, merely to give notice if he finds any[608]. In this instance, the contractor's obligation is expressly the same as that of the employer. A better view is that it is for the employer to provide a correct definition.

Clause 12.3 – Provisional sums

The employer may include provisional sums in the Employer's Requirements to cover works for which the contractor is not required to make proposals at tender stage. This may be because it is not practicable to do so, because the employer wishes to keep some significant part of the work under his control, or for some other reason. The issue of instructions regarding the expenditure of such sums is an obligation. It should be noted that there is no provision for the inclusion of provisional sums in the Contractor's Proposals and any such sum must be transferred to the Employer's Requirements before the contract is executed or the employer will be powerless to deal with them.

Clause 12.4 – Valuation of changes

There is no provision in this contract for the valuation of changes by a quantity surveyor. There is provision for the contractor to submit a price statement under clause 12.4.2 (alternative A) and this is the method which

---

[607] *John Mowlem & Co Ltd v. British Insulated Callenders Pension Trust Ltd* (1977) 3 Con LR 63.
[608] *London Borough of Merton v. Stanley Hugh Leach Ltd* (1985) 32 BLR 51.

must be used unless otherwise agreed between employer and contractor or the contractor chooses not to implement it. Except that there is no provision for quantity surveyor or architect and the employer is substituted, this clause is virtually identical to alternative A of JCT 98 (section 13.5). Under clause 30.5, it is for the contractor to submit his final account and final statement at the end of the project setting out the valuation of changes in accordance with the rules set out in clause 12. The provision for valuation of changes by agreement instead of by the strict rules set out in clause 12 remains as in clause 13.4.1.1 of JCT 98. Prior agreement between employer and contractor regarding the valuation of changes is strongly recommended[609]. Clause 12.4.1 could be significant in appropriate circumstances. The last sentence states that such valuation must include allowance for the addition or omission of 'the relevant design work'. It appears that only design work related to a variation can be valued. It emphasises the importance of making sure that there is a clear rate for design work in the contract sum analysis, but also warns the contractor against carrying out design work for changes which may be aborted. The contractor may be unable to claim the cost of such work, because it will not be 'relevant' to any change.

Clause 12.5 – Rules of valuation

The control document for valuation of changes is the contract sum analysis. This is a document which the contractor is required to annex to the Contractor's Proposals. It is referred to in the second and third recitals and in article 4 and there is provision for the identification of the documents which comprise it in Appendix 3.

No practice note has been issued by JCT specifically for WCD 98 at the time of writing, but JCT Practice Note CD/1B (originally issued for CD 81) contains valuable notes about the contract sum analysis and anyone embarking upon a contract under this form, whether as employer, contractor or as adviser to either, would be well advised to read it carefully.

If alternative A is not used, alternative B will govern valuations according to clause 12.5. The rules for valuation of changes in clause 12.5 are very similar to those in IFC 98 clause 3.7, save for the references to 'Approximate Quantity' in the latter.

Clause 12.5 – Dayworks

The provision for valuation of changes by daywork where valuation by measurement is not reasonably practicable is identical to that in JCT 98. The relevant percentage additions to prime cost must be set out in the contract sum analysis.

---

[609] But see the provisions of supplementary provision S6 below.

13.8.3 Supplementary provision S6

The text is as follows:

S6   Valuation of Change instructions – direct loss and/or expense – submission of estimates by the Contractor

S6.1   Clause 12 *(Changes in the Employer's Requirements and provisional sums);* clause 25 *(extension of time)* and clause 26 *(loss and expense caused by matters affecting regular progress of the Works)* shall have effect as modified by the provisions of S6.2 to S6.6.

S6.2   Where compliance with instructions of the Employer under clause 12 will in the opinion of the Contractor or of the Employer entail a valuation under clause 12.4 and/or the making of an extension of time in respect of the Relevant Event in clause 25.2.5.1 and/or the ascertainment of direct loss and/or expense under clause 26 the Contractor, before such compliance, shall submit to the Employer within 14 days of the date of the relevant instruction (or within such other period as may be agreed or, failing agreement, within such other period as may be reasonable in all the circumstances) estimates, or such of those as are relevant, as referred to in S6.3.1 to S6.3.5 unless:

S6.2 .1   the Employer with the instructions or within 14 days thereafter states in writing that such estimates are not required; or

S6.2 .2   the Contractor within 10 days of receipt of the instructions raises for himself or on behalf of any sub-contractor reasonable objection to the provision of all or any of such estimates.

S6.3   The estimates required under S6.2 shall be in substitution for any valuation under clause 12.4 and/or any ascertainment under clause 26 and shall comprise:

S6.3 .1   the value of the adjustment to the Contract Sum, supported by all necessary calculations by reference to the values in the Contract Sum Analysis;

S6.3 .2   the additional resources (if any) required to comply with the instructions;

S6.3 .3   a method statement for compliance with the instructions;

S6.3 .4   the length of any extension of time required and the resultant change in the Completion Date;

S6.3 .5   the amount of any direct loss and/or expense, not included in any other estimate, which results from the regular progress of the Works or any part thereof being materially affected by compliance with the instructions under clause 12.

S6.4   Upon submission of the estimates required under S6.2 the Employer and Contractor shall take all reasonable steps to agree those estimates and upon such agreement those estimates shall be binding on the Employer and Contractor.

S6.5   If within 10 days of receipt of the Contractor's estimates the Contractor and Employer cannot agree on all or any of the matters therein the Employer;

S6.5   .1   may instruct compliance with the instruction and that S6 shall not apply in respect of that instruction; or

S6.5   .2   may withdraw that instruction; or

Where the Employer withdraws the instructions under S6.5.2 such withdrawal shall be at no cost to the Employer except that where the preparation of the estimates involved the Contractor in any additional design work solely and necessarily carried out for the purpose of preparing his estimates such design work shall be treated as if it were in compliance with a Change instruction.

S6.6   If the Contractor is in breach of S6.2 compliance with the instruction shall be dealt with in accordance with clauses 12, 25 and 26 but any resultant addition to the Contract Sum in respect of such compliance shall not be included in Interim Payments but shall be included in the adjustment of the Contract Sum under clause 30.5. Provided that such addition shall not include any amount in respect of loss of interest or any financing charges in respect of the cost to the Contractor of compliance with the instruction which have been suffered or incurred by him prior to the date of issue of the Final Statement and Final Account or the Employer's Final Statement and the Employer's Final Account.

13.8.4 Commentary

Clauses 12, 25 and 26 are modified, but not superseded, by this provision which has some similarities to alternative A of clause 12. If the supplementary provisions are stated in the Appendix to apply, the contractor will be expected to operate these provisions without prompting.

The procedure is triggered when the employer issues an instruction under clause 12. If either the employer or the contractor is of the view that the instruction will involve either valuation or extension of time or loss and/or expense, the contractor must submit certain estimates (noted below) within a particular timescale. The timescale is 14 days from the date of the instruction or within any period agreed or, if no agreement, within such period as may be reasonable in all the circumstances. On the face of it, there is an obvious problem, because the clause calls for the contractor to act on the basis of an opinion which may be held by the employer alone and which the contract provides no mechanism for transmitting to the contractor. Thus if the employer issues an instruction which he believes will give rise to an extension of time, but which the contractor thinks will have no effect, the contractor is obliged to submit estimates within the prescribed period although the employer may not have communicated his opinion. This drafting flaw is made tolerable only by the fact that, in practice, it will be rare that a contractor does not believe an instruction will result in valuation, extension of time or loss and/or expense and even rarer that this lack of belief will be countered by the opposite belief on the part of the employer.

A similar clause (referred to as 'clause 13.8') was considered in the Scottish case *City Inn Ltd* v. *Shepherd Construction Ltd*[610]. There, as in this instance, there were consequences if the contractor failed to operate the provisions of the clause. There it was loss of entitlement to extension of time under clause 13.8.5; here it is loss of entitlement to early payment and to interest or finance charges. The contractor contended that the clause imposed no obligation on him to address his mind to whether the instruction would have the contemplated effects and, therefore, the contractor could not be said to have failed to comply unless he had formed an opinion that the instruction would have those effects, but had not acted accordingly. Lord MacFadyen dismissed this approach:

> 'I am therefore of the opinion that on a sound construction of clause 13.8.1 the contractor, on receipt of an architect's instruction, was obliged to consider whether it would require adjustment of the contract sum and/or an extension of time, so as to place himself in a position (if he formed the opinion that it would have that effect) to comply with his obligations to defer executing the instruction and to provide the requisite details to the architect. The wording of the clause is, it seems to me, less than perfect. It does not expressly address the eventuality of the contractor reasonably and in good faith forming the opinion that the contemplated consequences will not follow from the instruction, and consequently not doing what clause 13.8.21 required, and the need for an extension of time later becoming evident. It is unnecessary, however, for the purposes of this case to decide whether in that event the contractor would have lost his entitlement to an extension of time.'[611]

The question left unanswered by the court in this instance is important. Would the contractor lose entitlement to early payment and to interest or financing charges if he had formed the opinion that the instruction had no consequences, but subsequently that opinion was found to be a wrong conclusion? In such circumstances, it is thought that, upon the contractor making that contention, the burden of proof would switch to the employer to show that no reasonably competent contractor could have formed such an initial conclusion.

The way the provision works is that after every instruction, the contractor considers whether he can produce the necessary estimates within 14 days and, if not, he suggests a longer period. If the employer disagrees, the contractor prepares estimates as quickly as possible and, if the employer feels strongly or, more likely, if the extra time in producing estimates has caused some delay, the employer may, of course, refer the matter to one of the dispute resolution procedures.

---

[610] 17 July 2001, unreported. The case went to appeal to the Scottish Court of Session [2003] BLR 468 but the original decision was upheld.

[611] *City Inn Ltd* v. *Shepherd Construction Ltd*, 17 July 2001, unreported, at paragraph 23.

The employer is empowered to state that he does not want estimates submitting. He may do this either with the instruction or, rather bizarrely, within 14 days, therefore conceivably on the thirteenth day just before, or after, the contractor submits estimates. Whether the draftsman has considered this possibility is not clear. In such circumstances, the contractor should be reimbursed for his costs and, it is thought that such reimbursement would not be restricted to payment for necessary design work as is the case under clause S6.5 if the employer withdraws his instruction. In this instance, reimbursement would be part of the valuation of the instruction under clause 12. The contractor may raise reasonable objections to the provision of estimates either for himself or on behalf of a sub-contractor within 10 days of the issue of an instruction.

Clause S6.3 provides that the estimates replace valuation under clause 12 and ascertainment under clause 26. The estimates required are:

- The value of the instruction with supporting calculations referrable to the contract sum analysis
- Any additional resources required
- A method statement
- Any extension of time and consequent change to the completion date
- The direct loss and/or expense, not included in any other estimate.

Clause S6.4 refers, perhaps optimistically, to the employer and contractor taking all reasonable steps to agree the estimates. If they are successful, the estimates are binding on both parties. Therefore, even if subsequently it is found that the contractor's estimate is wildly wrong, both parties are bound by it. Surprisingly, there is no procedure to record so important an agreement. A brief document setting out the instruction and (possibly revised) estimates with two signatures and the date would appear to be the very minimum requirement if subsequent disputes are to be avoided.

If agreement is not reached within 10 days of receipt of estimates on, effectively, all matters, the employer may do one of two things:

- Instruct compliance with the instruction and that S6 will not apply, thereby reverting to clauses 12, 25 and 26 in full; or
- withdraw the instruction.

Withdrawal of the instruction is not to cost the employer anything other than additional design work which the contractor undertook purely and necessarily to prepare his estimates and for no other reason. Such design work is to be treated as if it was the result of a change instruction.

There is a sting in the tail of this clause. If the contractor does not comply with clause S6.2 and fails to submit estimates or make reasonable objection, clause S6.6 states that clauses 12, 25 and 26 will be applicable, but that any resultant addition to the contract sum will not be included in interim payments, and must wait until the final account and final statement. Moreover, the contractor will not be entitled to any loss of interest or any financing charges for the intervening period. The contractor's obligation to form an opinion has been considered above. The wording of the clause does not preclude any deduction from the contract sum being taken into account in interim payments.

13.9   *Variations position under PCC 98*

13.9.1   Clause 3.3.2, 3.3.3 and 3.4

The text of these clauses is as follows:

Instructions required for all items of work

3.3   .2   The Architect shall issue to the Contractor all instructions reasonably necessary for the carrying out of the Works including instructions on the carrying out of items of work described in the Specification and/or shown on the Contract Drawings (if any). Subject to clauses 3.3.3 and 3.3.5 the Contractor shall forthwith comply with such instructions. Instructions by the Architect in respect of works by Nominated Sub-Contractors to which section 8A applies shall not change those works except by way of variations, as defined in the Conditions NSC/C(PCC), to those works or as a result of instructions on the expenditure of provisional sums included in the Numbered Documents in respect of those works.

Reasonable objection by Contractor

.3   If an instruction of the Architect:

.1   either involves the imposition of obligations or restrictions on access to or use of the Site, on working space, on working hours and on the execution or completion of the work in any specific order or involves changes in such obligations or restrictions set out in the Specification; or

.2   alters the nature and scope of the Works as stated in the First schedule,

the Contractor need not comply with such instruction to the extent that he makes reasonable objection in writing as soon as is reasonably practicable to such compliance

Instruction – alteration in nature and scope of the Works – revision of Contract Fee

3.4   .1   Clause 3.4.2 applies where the Contractor has not made a reasonable objection pursuant to clause 3.3.3 or, if such objection has been made, it has not been accepted.

.2   If the Contractor considers that an instruction of the Architect has altered the nature and scope of the Works as stated in the First Schedule he may within 14 days from the issue of that instruction make a written application to the Employer with a copy to the Architect requesting a revision to the Contract Fee. Upon receipt of such application the Architect shall consider whether it is fair and reasonable to revise the Contract Fee. If the Architect does so consider and the Employer has confirmed in writing to the Contractor his agreement to such revision, the Architect, or, if so instructed by the Architect, the Quantity Surveyor, shall after consultation with the Contractor make an appropriate revision to the Contract Fee; and shall decide the date from which the revised Contract Fee shall be applicable.

13.9.2 Commentary

Since all the work carried out under this kind of contract is uncertain, the architect must issue instructions under clause 3.3.2 for *all* the work required to be carried out even though it is already shown on drawings and specifications. Therefore, there are no variations in the ordinarily understood sense of the word. The architect can change the work only by means of a variation in the nominated sub-contract or by an instruction on the expenditure of provisional sums. Essentially, therefore, the valuation of variations takes place at sub-contractor level, because the employer pays the prime costs of sub-contract work.

Clause 3.3.3 entitles the contractor to refuse to comply with an architect's instruction in two specific situations. The first is similar to clause 13.1.2 of JCT 98. It covers instructions imposing obligations or restrictions on access or use of the site, working space, hours or sequence of work or changes to any such obligations or restrictions which are already in the specification. The second deals with any situation where the architect issues an instruction which alters the nature and scope of the works as stated in the first schedule. The contractor is only entitled to refuse to comply to the extent that he makes reasonable objection in writing and it must be made as soon as reasonably practicable; for example, an instruction restricting the contactor's use of the site might seriously hinder the progress of the works.

Clause 3.4 addresses instructions which alter the nature and scope of the works. The wording is curious. If the contractor has not made reasonable objection or if he has made reasonable objection and it has not been accepted, the second part of the clause takes effect. Presumably it is the architect who accepts or otherwise the contractor's reasonable objection. It is hard to understand this. Clause 3.3.3 gives the contractor the right to withhold compliance with such an instruction 'to the extent that he makes reasonable objection...'. The curious part about clause 3.4 is that it does not refer to the architect's acceptance of the reasonableness or otherwise of the contractor's objection. It would be perfectly sensible to consign that decision to the architect as the professional agreed between the parties as certifier. The clause actually refers to the acceptance or non-acceptance of the contractor's reasonable objection, because it refers to '*such* objection' and the only 'objection' to which that can refer is the 'reasonable objection'. This strange provision immediately raises the question of the grounds on which the architect could refuse to accept an objection which was manifestly reasonable or, at any rate, admitted to be so. One must assume that the draftsman of this contract intended to refer to the architect deciding whether to accept the objection as reasonable. However, that is not what he did say.

The second part of the clause gives the contractor the power to apply to the employer with a copy to the architect requesting a revision to the contract fee. There is nothing remarkable about that, because there is nothing to prevent the contractor from applying to the employer about anything at all at any time. However, the purpose of this clause is to overcome a situation which would ordinarily allow the contractor to refuse performance of the instruction or treat it as a separate contract for which he could either negotiate his own terms or receive reasonable remuneration. Under this clause, the contractor must apply within 14 days of the issue of the instruction in question. Although the application is made to the employer, it is the architect who is charged with considering whether it is fair and reasonable to revise the fee. If the architect does consider a revision to be fair and reasonable and the employer, presumably after receiving the architect's decision, confirms it to the contractor, the architect makes an appropriate revision to the fee and decides the date from which it applies. The architect may delegate this task to the quantity surveyor. It should be noted that the architect is to have a 'consultation' with the contractor; there is no requirement for agreement. Since 'to consult' merely means to seek advice or information, the contractor is in no position to influence the outcome.

13.10 Variations position under MC 98

13.10.1 Clause 3.4

The text of clause 3.4 is as follows:

3.4 .1   Without prejudice to the generality of clause 3.3.1 the Architect may issue to the Management Contractor instructions which may require Project Changes or Works Contract Variations; and shall issue instructions in regard to the expenditure of provisional sums in Works Contracts.

3.4 .2   Where pursuant to clause 3.4.1A.1 of the Works Contract Conditions the Works Contractor has not disagreed with the application of clause 3.14 of the Works Contract Conditions to an instruction which states that the treatment and valuation of a Variation are to be in accordance with that clause 3.14, the Management Contractor shall comply with all the requirements of that clause 3.14 on the Management Contractor in respect of a clause 3.14 Quotation by a Works Contractor.

13.10.2 Commentary

The definition of project change contained in clause 1.3 is very wide; it is 'the alteration or modification of the scope of the Project as shown and described generally in the Project Drawings and the Project Specification'. This definition gives the architect tremendous scope to change the project. In particular, it goes beyond the usual understanding of the architect's power to issue variations. It is difficult to say precisely where the architect's power to change the project ends and one is driven to the conclusion that the overriding consideration may simply be that the project remains capable of description by the entry in the first recital. The definition of works contract variation is in similar terms to the definition of variation found in JCT 98. The valuation of variations is dealt with in Works Contract/2, but there is no provision to value project changes in MC 98. The rationale is presumably that project changes will be reflected either in the individual works packages or in the works contract variations. Although the architect may give instructions about the expenditure of provisional sums in works contracts, such sums are not defined.

The purpose of clause 3.4.2 is not entirely clear. It records that if the works contractor has not disagreed with an instruction being dealt with under clause 3.14 of the works contract, the management contractor must comply with clause 3.14 in relation to a works contractor clause 3.14A quotation.

13.11 *Variation position under TC/C 02*

13.11.1 Clauses 3.19, 3.20, 3.21, 3A, 4.4, 4.5 and 4.6

The text of clauses 3.19, 3.20, 3.21, 3A, 4.4, 4.5 and 4.6 is as follows:

3.19   The term Variation as used in the Conditions means:

.1   the alteration or modification of the design, quality or the quantity of the Works including

.1   the addition, omission or substitution of any work,

.2   the alteration of the kind or standard of any of the materials or goods to be used in the Works,

.3   the removal from the site of any work executed or materials or goods brought thereon by the Trade Contractor for the purposes of the Works other than work materials or goods which are not in accordance with this Trade Contract;

.2   the imposition by the Employer of any obligations or restrictions in regard to the matters set out in clauses 3.19.2.1 to 3.19.2.4 or the addition to or alteration or omission of any such obligations or restrictions so imposed or imposed by the Client in the Trade Contract Documents in regard to:

.1   access to the site or use of any specific parts of the site;

.2   limitations of working space;

.3   limitations of working hours;

.4   the execution or completion of the work in any specific order.

Where Article 2.2 of the Articles of Agreement applies the term 'Variation' has the same meaning but in clause 3.19.1 delete 'design, quality or quantity' and insert 'design or quality'.

3.20 .1 The Construction Manager may issue instructions requiring a Variation.

.2 Any instruction under clause 3.20.1 shall be subject to the Trade Contractor's right of reasonable objection set out in clause 3.2.1.

.3 A Variation instruction under clause 3.20.1 shall be valued pursuant to either clause 4.4 or clause 4.7 as appropriate unless the instruction states that the treatment and valuation of the Variation are to be in accordance with clause 3A or unless the variation is one to which clause 3A.8 applies. Where the instruction so states, clause 3A shall apply unless the Trade Contractor within 7 days (or such other period as may be agreed) of receipt of the instruction states in writing that he disagrees with the application of clause 3A to such instruction. If the Trade Contractor so disagrees, clause 3A shall not apply to such instruction and the variation shall not be carried out unless and until the Construction Manager instructs that the Variation is to be carried out and is to be valued pursuant to clause 4.4 or clause 4.7 as appropriate.

.4 The Construction Manager may sanction in writing any Variation made by the Trade Contractor otherwise than pursuant to an instruction of the Construction Manager.

.5 No Variation required by the Construction Manager or subsequently sanctioned by him shall vitiate this Trade Contract.

3.21 The Construction Manager shall issue instructions in regard to the expenditure of any provisional sums included in the Trade Contract documents.

Variation instruction – Trade Contractor's quotation in compliance with the instruction.
Trade Contractor to submit his quotation ('3A Quotation')

3A.1 Clause 3A shall only apply to an instruction where pursuant to clause 3.20.3 the Trade Contractor has not disagreed with the application of clause 3A thereto.

.1 The instruction to which clause 3A is to apply shall have provided sufficient information to enable the Trade Contractor to provide a quotation, which shall comprise the matters set out in clause 3A.2 (a '3A Quotation'), in compliance with the instruction. If the Trade Contractor reasonably considers that the information provided is not sufficient, then, not later than 7 days from the receipt of the instruction, he shall request the Construction Manager to supply sufficient further information.

.2 The Trade Contractor shall submit to the Construction Manager his 3A Quotation in compliance with the instruction not later than 21 days from

the date of receipt of the instruction

or   if applicable, the date of receipt by the Trade Contractor of the sufficient further information to which clause 3A.1.1 refers

whichever date is the later and the 3A Quotation shall remain open for acceptance by the Construction Manager for 7 days from its receipt by him.

.3 The Variation for which the Trade Contractor has submitted his 3A Quotation shall not be carried out by the Trade Contractor until receipt by the Trade Contractor of the confirmed acceptance issued by the Construction Manager pursuant to clause 3A.3.2.

Content of Trade Contractor's 3A Quotation

3A.2 The 3A Quotation shall separately comprise:

.1 the value of the adjustment to the Trade Contract Sum or the amount to be taken into account in the computation of the Ascertained Final Trade Contract Sum as the case may be (other than any amount to which clause 3A.2.3 refers) including therein the effect of the instruction on any other work and supported by all necessary calculations by reference, where relevant, to the rates and prices in the Trade Contract Documents and including, where appropriate, allowances for any adjustment of preliminary items;

.2 any adjustment to the Completion Period (including if relevant a reduction in the Completion Period stated in the Appendix) to the extent that such adjustment is not included in any revision of the Completion Period that has been made under clause 2.3 (*extension of time*), 2.9 (*acceleration*) or in the Construction Manager's confirmed acceptance of any other 3A Quotation;

.3 the amount to be paid in lieu of any ascertainment under clause 4.21 of direct loss and/or expense not included in any other accepted 3A Quotation or in any previous ascertainment under clause 4.21;

.4 a fair and reasonable amount in respect of the cost of preparing the 3A Quotation;

and, where specifically required by the instruction, shall provide indicative information in statements on

.5 the additional resources (if any) required to carry out the Variation; and

.6 the method of carrying out the Variation.

Each part of the 3A Quotation shall contain reasonably sufficient supporting information to enable that part to be evaluated by the Construction Manager.

Acceptance of 3A Quotation

3A.3 .1 If the Construction Manager wishes to accept a clause 3A Quotation then he shall so notify the Trade Contractor in writing not later than the last day of the period for acceptance stated in clause 3A.1.2.

.2 If the Construction Manager accepts a clause 3A Quotation he shall confirm such acceptance by stating in writing to the Trade Contractor (in clause 3A and elsewhere in the Conditions called a 'confirmed acceptance'):

.1 that the Contractor is to carry out the Variation;
.2 the value of the adjustment to the Trade Contract Sum or to be taken into account in the computation of the Ascertained Final

Trade Contract Sum, as the case may be, including therein any amounts to which clause 3A.2.3 and clause 3A.2.4 refer, to be made for complying with the instruction requiring a Variation;

.3   any adjustment to the time required by the Trade Contractor for completion of the Works and the revised Completion Period arising therefrom (which, where relevant, may be an earlier Completion Period than that stated in the Appendix).

3A Quotation not accepted – payment by Client

3A.4   If the Construction Manager does not accept the 3A Quotation by the expiry of the period for acceptance stated in clause 3A.1.2, he shall, on the expiry of that period

either

.1   instruct that the Variation is to be carried out and is to be valued pursuant to either clause 4.4 or clause 4.7 as appropriate;

or

.2   instruct that the Variation is not to be carried out.

3A.5   If a 3A Quotation is not accepted a fair and reasonable amount shall be added to the Trade Contract Sum or included in the computation of the Ascertained Final Trade Contract Sum as the case may be, in respect of the cost of preparation of the 3A Quotation provided that the 3A Quotation has been prepared on a fair and reasonable basis. The non-acceptance by the Client of a 3A Quotation shall not of itself be evidence that the Quotation was not prepared on a fair and reasonable basis.

Restriction on use of a 3A Quotation

3A.6   If the Construction Manager has not, under clause 3A.3.2, issued a confirmed acceptance of a 3A Quotation neither the Client nor the trade Contractor may use that 3A Quotation for any purpose whatsoever.

Number of days – clauses 3A.1.1 and/or 3A.1.2

3A.7   The Construction Manager and the Trade Contractor may agree to increase or reduce the number of days stated in clause 3A.1.1 and/or in clause 3A.1.2 and any such agreement shall be confirmed in writing by the Construction Manager to the Trade Contractor.

Variations to work for which a confirmed 3A Quotation has been issued – valuation

3A.8   If the Construction Manager issues an instruction requiring a Variation to work for which a 3A Quotation has been given and in respect of which the Construction Manager has issued a confirmed acceptance to the Trade Contractor such Variation shall not be valued under either clause 4.4 or clause 4.7 but the Construction Manager shall make a valuation off such Variation on a fair and reasonable basis having regard to the content of such 3A Quotation and shall include in that valuation the direct loss and/or expense, if any, incurred by the Trade Contractor because the regular progress of the Works or any part thereof has been materially affected by compliance with the instruction requiring the Variation.

Trade Contract Sum Valuation of variations and provisional sum work

4.4   .1   Where Article 2.1 applies then:

— all Variations required by an instruction of the Construction Manager to which clause 3.20 refers or subsequently sanctioned by him in writing, and

— all work which under the Conditions is to be treated as if it were a variation required by an instruction of the Construction Manager to which clause 3.20 refers, and

— all work executed by the Trade Contractor in accordance with instructions by the Construction Manager to which clause 3.21 refers as to the expenditure of provisional sums which are included in the Trade Contract Documents, and

— all work executed by the Trade Contractor for which an Approximate Quantity has been included in any bills of quantities which are included in the Trade Contract Documents shall unless otherwise agreed by the construction manager and the Trade Contractor be valued (called 'the Valuation'), under Alternative A in clause 4.4.2 or, if Alternative A is not implemented by the Trade Contractor, or if implemented, to the extent that the Price Statement or amended Price Statement is not accepted, under Alternative B in clause 4.4.2. Clause 4.4.1 shall not apply in respect of a Variation for which the Construction Manager has issued a confirmed acceptance of a 3A Quotation or is a Variation to which clause 3A.8 applies.

4.4   .2   **Alternative A: Trade Contractor's Price Statement**

Paragraph

4.4   A1   Without prejudice to his obligation to comply with any instruction or to execute any work to which clause 4.4.1 refers, the Trade Contractor may within 17 days from receipt of the instruction or, if later, from receipt of sufficient information to enable the Trade Contractor to prepare his Price Statement, submit to the Construction Manager his price ('Price Statement') for such compliance or for such work.

The Price Statement shall state the Trade Contractor's price for the work which shall be based on the provisions of clause 4.5 and clause 4.6 (*Valuation rules*) and may also separately attach the Trade Contractor's requirements for:

.1   any amount to be paid in lieu of any ascertainment under clause 4.21 of direct loss and/or expense not included in any accepted 3A Quotation or in any previous ascertainment under clause 4.21;

.2   any adjustment to the time for the completion of the Works to the extent that such adjustment is not included in any revision of the period or periods for completion that has been made by the Construction Manager under clause 2.3 or in his confirmed acceptance of any 3A Quotation. (See paragraph A7.)

A2   Within 24 days of receipt of a Price Statement the Construction Manager shall notify the Trade Contractor in writing

either

.1   that the Price Statement is accepted

or

.2   that the Price Statement, or a part thereof, is not accepted.

A3   Where the Price Statement or a part thereof has been accepted the price in that accepted Price Statement or in that part which has been accepted shall be added to or deducted from the Trade Contract Sum.

A4   Where the Price Statement or a part thereof has not been accepted:

.1   the Construction Manager shall include in his notification to the Trade Contractor the reasons for not having accepted the Price Statement, or a part thereof, which reasons shall be set out in similar detail to that given by the Trade Contractor in his Price Statement and supply to the Trade Contractor an amended Price Statement which is acceptable to the Construction Manager;

.2   within 14 days from receipt of the amended Price Statement the Trade Contractor shall state whether or not he accepts the amended Price Statement or part thereof and if accepted paragraph A3 shall apply to that amended Price Statement or part thereof; if no statement within the 14 days period is made the Trade Contractor shall be deemed not to have accepted, in whole or in part, the amended Price Statement;

.3   to the extent that the amended Price Statement is not accepted by the Trade Contractor, the Trade Contractor's Price Statement and the amended Price Statement may be referred either by the Client or by the Trade Contractor as a dispute or difference to the Adjudicator in accordance with the provisions of Section 9 of this Trade Contract.

A5   Where no notification has been given pursuant to paragraph A2 the Price Statement is deemed not to have been accepted, and the Trade Contractor may on or after the expiry of the 24 day period to which paragraph A2 refers, refer his Price Statement as a dispute or difference to the Adjudicator in accordance with the provisions of Section 9 of this Trade Contract.

A6   Where a Price Statement is not accepted by the Construction Manager after consultation with the Architect or an amended Price Statement has not been accepted by the Trade Contractor and no reference to the Adjudicator under paragraph A4.3 or paragraph A5 has been made, Alternative B shall apply.

A7   .1   Where the Trade Contractor pursuant to paragraph A1 has attached his requirements to his Price Statement the Construction Manager shall within 24 days of receipt thereof notify the Trade Contractor:

.1   either that the requirement in paragraph A1.1 in respect of the amount to be paid in lieu of any ascertainment under clause 4.21 is accepted or that the requirement is not accepted and clause 4.21 shall apply in respect of the ascertainment of any direct loss and/or expense; and

.2 either that the requirement in paragraph A1.2 in respect of an adjustment to the time for the completion of the Works is accepted or that the requirement is not accepted and clauses 2.2 to 2.4 shall apply in respect of any such adjustment.

A7 .2 If the Client has not notified the Trade Contractor within the 24 days specified in paragraph A7.1, clauses 2.2 to 2.4 and clause 4.21 shall apply as if no requirements had been attached to the Price Statement.

4.4 .2 **Alternative B**

The valuation shall be made by the Construction Manager in accordance with the provisions of clauses 4.5 and 4.6.

Trade Contractor's schedule of rates or daywork definitions

4.5 Where the Trade Contract Documents include a schedule of rates or prices for measured work and/or a schedule of daywork prices, such rates or prices shall be used in determining the Valuation in substitution for any rates or prices or daywork definitions which would otherwise be applicable under the relevant provisions of clause 4.6.

Valuation Rules

4.6 .1 To the extent that the Valuation relates to the execution of additional or substituted work which can properly be valued by measurement or to the execution of work for which an Approximate Quantity is included in the Contract Bills such work shall be measured and shall be valued in accordance with the following rules:

.1 where the additional or substituted work is of similar character to, is executed under similar conditions as, and does not significantly change the quantity of, work set out in the Trade Contract Documents the rates and prices for the work so set out shall determine the Valuation;

.2 where the additional or substituted work is of similar character to work set out in the Trade Contract Documents but is not executed under similar conditions thereto and/or significantly changes the quantity thereof, the rates and prices for the work so set out shall be the basis for determining the Valuation and the Valuation shall include a fair allowance for such difference in conditions and/or quantity;

.3 where the additional or substituted work is not of similar character to work set out in the Trade Contract Documents the work shall be valued at fair rates and prices;

.4 where any Approximate Quantity is a reasonably accurate forecast of the quantity of work required the rate or price for the Approximate Quantity shall determine the Valuation;

.5 where any Approximate Quantity is not a reasonably accurate forecast of the quantity of work required the rate or price for that

Approximate Quantity shall be the basis for determining the Valuation and the Valuation shall include a fair allowance for such difference in quantity.

Provided that clause 4.6.1.4 and clause 4.6.1.5 shall only apply to the extent that the work has not been altered or modified other than in quantity.

.2   To the extent that the Valuation relates to the omission of work set out in the Trade Contract Documents the rates and prices for such work therein set out shall determine the Valuation of the work omitted.

.3   In any valuation of work under clauses 4.6.1 and 4.6.2:

.1   measurement shall be in accordance with the same principles as those governing the preparation of the Trade Contract Documents;

.2   allowance shall be made for any percentage or lump sum adjustments in the Trade Contract Documents; and

.3   allowance, where appropriate, shall be made for any addition to or reduction of preliminary items of the type referred to in the Standard Method of Measurement; provided that where bills of quantities are included in the Trade Contract Documents no such allowance shall be made in respect of compliance with an instruction for the expenditure of a provisional sum for defined work.

.4   To the extent that the Valuation relates to the execution of additional or substituted work which cannot properly be valued by measurement the Valuation shall comprise:

.1   the prime cost of such work (calculated in accordance with the 'Definition of Prime Cost of Daywork carried out under a Building Contract' issued by the Royal Institution of Chartered Surveyors and the Building Employers Confederation (*now the Construction Confederation*) which was current at the Base Date or the 'Schedule of Dayworks carried incidental ro Contract Work' issued by the Federation of Civil Engineering Contractors which was current at the Base Date) together with percentage additions to each section of the prime cost at the rates set out by the Trade Contractor in the Trade Contract Documents; or

.2   where the work is within the province of any specialist trade and the said Institution and the appropriate body representing the employers in that trade have agreed and issued a definition of prime cost of daywork, the prime cost of such work calculated in accordance with that definition which was current at the Base Date together with percentage additions on the prime cost at the rates set out by the Trade Contractor in the Trade Contract Documents.

Provided that in any case vouchers specifying the time daily spent upon the work, the workmen's names, the plant and the materials employed shall be delivered for verification to the Construction Manager not later than the end of the week following that in which the work has been executed.

.5   If

compliance with any instruction requiring a Variation or

where bills of quantities are included in the Trade Contract Documents compliance with any instruction as to the expenditure of a provisional sum for undefined work or

where bills of quantities are included in the Trade Contract Documents compliance with any instruction as to the expenditure of a provisional sum for defined work to the extent that the instruction for that work differs from the description given for such work in the Trade Contract Documents, or

the execution of work for which an Approximate Quantity is included in the Trade Contract Documents to such extent as the quantity is more or less than the quantity ascribed to that work in the Trade Contract Documents

substantially changes the conditions under which any other work is executed, then such other work shall be treated as if it had been the subject of an instruction of the Construction Manager requiring a Variation which shall be valued in accordance with the provisions of clauses 4.5 and 4.6.

.6   .1   The Valuation of Performance Specified Work shall include allowance for the addition or omission of any relevant work involved in the preparation and production of drawings, schedules or other documents;

.2   the Valuation of additional or substituted work related to Performance Specified Work shall be consistent with the rates and prices of work of a similar character set out in the Trade Contract Documents or the Analysis making due allowance for any changes in the conditions under which the work is carried out and/or any significant change in the quantity of the work set out in the Trade Contract Documents or in the Trade Contractor's Statement. Where there is no work of a similar character set out in the Trade Contract Documents or the Trade Contractor's Statement a fair valuation shall be made;

.3   the Valuation of the omission of work relating to Performance Specified Work shall be in accordance with the rates and prices for such work set out in the Trade Contract Documents or the Analysis;

.4   any valuation of work under clauses 4.6.6.2 and 4.6.6.3 shall include allowance for any necessary addition to or reduction of preliminary items of the type referred to in the Standard Method of Measurement;

.5   where an appropriate basis of a fair valuation of additional or substituted work relating to Performance Specified Work is daywork the Valuation shall be in accordance with clause 4.6.4;

.6   if

compliance with any instruction under Part 11 of the Conditions requiring a Variation to Performance Specified Work or

> compliance with any instruction as to the expenditure of a provisional
> sum for Performance Specified Work to the extent that the instruction
> for that Work differs from the information provided in the Trade
> Contract Documents for such Performance Specified Work

substantially changes the conditions under which any other work is
executed (including any other Performance Specified Work) then such
other work (including any other Performance Specified Work) shall be
treated as if it had been the subject of an instruction requiring a Variation
to be valued in accordance with the provisions of clause 4.6.

.7   To the extent that the Valuation does not relate to the execution of
additional or substituted work or the omission of work or to the extent
that the valuation of any work or liabilities directly associated with a
Variation cannot reasonably be effected in the Valuation by the application of clauses 4.6.1 to 4.6.6 a fair valuation thereof shall be made.

Provided that no allowance shall be made under clause 4.6 for any effect upon
the regular progress of the Works or for any other direct loss and/or expense
for which the Trade Contractor would be reimbursed by payment under any
other provision of this Trade Contract.

13.11.2 Commentary

The construction management form is significantly different from the management contract (MC 98). In the case of MC 98, the management contract
(MC 98) is the contract between employer and contractor. The works contract (Works Contract/2) governs the relationship between management
contractor and works contractor which is sub-contractual. Under construction management, however, the trade contractor contracts directly with the
client under the trade contract (TC/C). The construction manager is also
contracted directly with the client, but more like an organising and adminis-
trating consultant and certainly quite different from a management
contractor.

In general terms, the variation clauses (4.4, 4.5 and 4.6) are very similar to
the equivalent clauses in JCT 98. The definition of 'variation' is almost
identical to the definition in JCT 98. If article 2.2 applies, however, the
definition is modified to omit reference to 'quantity' of the Works. This is
because article 2.2 is used where complete remeasurement of the Works is
required. Clauses 4.7, 4.8 and 4.9 have not been reproduced, because they
refer to complete remeasurement rather than the valuation of variations.

It is the construction manager, not the architect, who may issue instructions
requiring a variation. Valuation is to be in accordance with clause 4.4 or 4.7
unless the instruction states that clause 3A applies or if clause 3A.8 applies.

The construction manager may sanction variations which the trade contractor has made without instruction and there is the usual, unnecessary
proviso that no instruction issued or sanctions given by the construction
manager will vitiate the contract.

Clause 3A provides for the trade contractor to provide a quotation on very similar terms to JCT 98 clause 13A if requested. Clause 3.21 provides that the construction manager must issue instructions regarding the expenditure of provisional sums.

13.12 Variations position under MPF 03

13.12.1 Clause 20

The text of clause 20 is as follows:

20   Changes

20.1   Each party shall immediately notify the other:

.1   whenever it considers that an instruction gives rise to a Change; and/or

.2   of the occurrence of any event that under the Contract is required to be treated as giving rise to a Change.

20.2   Other than in respect of Changes instructed in accordance with the provisions of clauses 13 (*Acceleration*) or 19 (*Cost savings and value improvements*), the consequences of any Change shall be determined in accordance with the provisions of clause 20 so that either:

.1   the value of the Change and any adjustment to the Completion Date is agreed in accordance with clause 20.5 prior to an instruction being issued; or

.2   a fair valuation of the Change is made in accordance with clause 20.6 and any adjustment to the Completion Date is notified in accordance with clause 12 (*Extension of time*).

20.3   Prior to instructing any Change the Employer may provide details of the proposed Change and request the Contractor to submit a quotation in respect of the Change. The Contractor shall provide the quotation within 14 days of the request, or within such longer period as the Employer states in its request.

20.4   The quotation provided by the Contractor shall:

.1   give a valuation of the Change calculated in accordance with the principles set out in clause 20.6;

.2   identify any adjustment to the Completion Date that will be required as a consequence of the Change;

.3   be in sufficient detail for the Employer to assess the amounts and periods required and in particular, shall state separately any amounts included in respect of loss and/or expense;

.4   identify the period, being not less than 14 days, for which the quotation remains open for acceptance.

20.5   The Employer may accept the quotation or request the Contractor to submit a revised quotation. When the Employer accepts a quotation it shall issue an instruction identifying the quotation that is being accepted, the agreed value and any agreed adjustment to the Completion Date.

20.6   Where agreement is not reached under clause 20.5, a fair valuation of any Change shall be made by the Employer. Such valuation shall have regard to the following:

   .1   the nature and timing of the Change;

   .2   the effect of the Change on other parts of the Project;

   .3   the prices and principles set out in the Pricing Document, so far as applicable and,

   .4   any loss and/or expense that will be incurred as a consequence of the Change, provided always that the fair valuation shall not include any element of loss and/or expense if that element was contributed to by a cause other than a Change or a matter set out in clause 21.2.

20.7   Within 14 days of a Change being identified by either party the Contractor shall provide to the Employer details of its proposed valuation of the Change together with such information as is reasonably necessary to permit a fair valuation to be made.

20.8   Within 14 days of receipt of the information referred to by clause 20.7 the Employer shall notify the Contractor of its valuation of the Change, that valuation being calculated by reference to the information provided by the Contractor. The valuation shall be in sufficient detail to permit the Contractor to identify any differences between it and the Contractor's proposed valuation.

20.9   No later than 42 days after Practical Completion of the Project the Contractor shall provide particulars of any further valuation it considers should be made in respect of any Change. Within 42 days of the receipt of these particulars the Employer shall undertake a review of its previous valuations of each Change to which those particulars relate and notify the Contractor of such further valuation as it considers appropriate.

13.12.2 Commentary

Three clauses deal with what this contract refers to as 'Changes', but which are more commonly known as variations. The clauses are 13 (acceleration), 19 (cost savings and value improvements) and 20 (changes). Clause 20 is the main variation clause and the one under consideration here.

There is just one method of valuing changes: by fair valuation. But there are two ways of setting about it. The first is if the employer under clause 20.3 provides the contractor with details of the change before issuing an instruction and asks the contractor to submit a quotation. The contractor has 14 days to do so unless the employer has stated a longer period in his request. Clause 20.4 states that the quotation must value the change in accordance with the principles in clause 20.6. A slight ambiguity is present here, because clause 20.6 refers to making a fair valuation. It provides that the valuation 'shall have regard to' a set of principles. To 'have regard' to something is quite different from calculating 'in accordance with' something. To 'have regard' has the sense that notice must be taken and the principles must be read and considered. However, having read and considered the principles, they need not be strictly observed.

The quotation must also identify any adjustment to the completion date, it must be in enough detail so that the employer can carry out an assessment of amounts and periods, loss and/or expense should be stated separately and, finally, the quotation must state the period of not less than 14 days when it will remain open for acceptance. Unlike the position under the general law when a tender, stated to be open for a period, can be withdrawn without notice at any time if no consideration has been given for keeping it open, once a period has been stated under clause 20.6.4, the quotation cannot be withdrawn, because the procedure of stating a period is part of the contract for which both parties have already provided ample consideration.

Having received the quotation, the employer, under clause 20.5, may either accept it or request a revised quotation. Evidently, no reasons need be given for requesting a revised quotation, but the contractor can probably refuse to provide it. It is clear from clause 20.3 that, on the initial request for a quotation, the contractor 'shall' provide it. However, it is certainly arguable that it does not apply to a request for a revised quotation. If the contractor was not able to refuse, there seems to be no end to the number of revised quotations he could be asked to provide.

To signify acceptance, the employer must issue an instruction noting the quotation, the amount and the adjustment to the completion date (if any). If there is no acceptance, it is for the employer to make a fair valuation under clause 20.6, this time not in accordance with the principles, but having regard to them – a lesser obligation. The principles are the nature and timing of the change, its effect on other parts of the project, prices and principles in the pricing documents, but only to the extent that they are applicable, and any loss and/or expense resulting from the change. However, no loss and/or expense of any kind can be included if anything other than a change contributed to it, because in that case it will be dealt with under clause 21[612].

Clauses 20.7 and 20.8 deal with the situation if the change is instructed without any request for a quotation. In that instance the contractor has 14 days from the date that either party 'identified' the change in which to give the employer details of the contractor's proposal to value the change with supporting information to permit a fair valuation. Use of the word 'identified' is curious. No doubt in most cases a pre-instruction quotation will be requested. Presumably, the contract is attempting to give the contractor the opportunity of submitting a valuation, not only after receipt of an instruction (one way of identifying a change?), but also if he is of the view that a

---

[612] See Chapter 12, section 12.9.

change has occurred in some other way. The contract is silent about the position if the contractor fails to produce proposals within 14 days. Is any valuation after that date invalid or simply a late valuation?

After receipt of the contractor's valuation, the employer has a further 14 days in which to carry out his own valuation. Again the contract is silent about whether the employer can make his own valuation if the contractor does not produce his within the original 14 days or indeed whether the employer's late valuation would be valid. Presumably, business efficacy would require the employer to proceed to make his own valuation. Assuming the employer's valuation proposal is produced in due time, it must be in sufficient detail to allow the contractor to be able to note the differences.

Clause 20.9 is a review clause. Its purpose appears to be to set a timetable for the final consideration of change valuations. The contractor has 42 days from practical completion in which to give particulars to the employer if the contractor considers that a change should have some additional value. The employer has a further 42 days in which to review his previous valuation of the changes notified by the contractor and to notify the contractor of any further valuation considered appropriate.

Although there are one or two loose ends, this clause is relatively simple to understand, but it may be somewhat woolly in application.

PART III

Chapter 14

Claims under GC/Works/1 (1998)

14.1 Introduction

The General Conditions of Government Contracts for Building and Civil Engineering Works, commonly called Form GC/Works/1 (1998), are used by many government departments for construction works. The current version was published in 1998. The conditions can be used as a full bill of quantities form of contract, as a schedule form of contract, or as a lump sum form. They differ basically from the JCT Standard Forms and, unlike the JCT contract, GC/Works/1 (1998) is not a negotiated document. It is drafted on behalf of the relevant government departments and with the employer's interests in mind. Accordingly, it is an employer's 'standard form of contract' for the purposes of the Unfair Contract Terms Act 1977, section 3, which is wide ranging in its effect on contractual responsibilities. The 1977 Act does not apply to the Crown, but the point would be applicable where GC/ Works/1 is adopted by an employer who is not technically an emanation of the Crown. Moreover, any ambiguities in GC/Works/1 will be construed *contra proferentem* the employer.

The principles involved in claims under GC/Works/1 (1998), are, however, identical with those in JCT claims, which have been discussed in earlier chapters. This chapter will examine the particular provisions in GC/Works/1 (1998) that may give rise to claims by the contractor.

14.2 Valuation of instructions

14.2.1 Commentary

For the purpose of these clauses, instructions are considered as variation instructions (VI) and other instructions. Clause 40(1) helpfully makes clear that if the PM issues further drawings, details, instructions, directions and explanations, they must be treated as instructions for the purposes of the contract. The variation or modification of all or any of the specification, drawings or bills of quantities or the design, quality or quantity of the works is dealt with under clause 40(2)(a). Clause 40(5) allows the PM in any VI to require the contractor to submit a quotation for the cost of compliance. Valuation of VIs is dealt with under clause 42 and other instructions are valued under clause 43. Clause 41(2) places an obligation on either the PM or the QS to include the cost of any disruption to or prolongation of both varied and unvaried work in the valuation of an instruction.

Alternative systems of arriving at the value of a VI are prescribed by clause 42(1):

- Acceptance of a lump sum quotation; or
- Valuation of the variation.

Where the PM has required a quotation, the quotation must show how it has been calculated, but the calculation apparently need not be very detailed. All that is required is that it shows the direct cost of compliance and the cost (not described as 'direct') of any disruption or prolongation consequential to compliance. However, although that may simply amount to two lump sum figures, the contractor is obliged to include sufficient other information to enable the QS to 'evaluate' the quotation. Evaluation involves ascertaining an amount[613] while to value is to estimate a value or to appraise[614]. It seems, therefore, that the QS's duty is more onerous when a quotation has been required and that is precisely right of course. Clause 42(3) allows 21 days from receipt for the PM to notify the contractor whether or not a quotation is accepted. If the QS is not prepared to agree the contractor's quotation, the QS may negotiate and agree some other figure. Clause 42(4) makes clear that if either the contractor fails to provide the lump sum quotation or the parties fail to agree, the PM must instruct the QS to proceed to value the VI. It appears, however, that clause 42(1) gives the QS the option of accepting a quotation or valuing the variation as though no quotation had been given.

The principles of valuation of variations are very similar to those under the JCT contracts as explained in Chapter 13. Clause 42(5) states that where the QS is required to value a VI, he must do so in accordance with the following rules:

(1) By measurement and valuation at the rates and prices for similar work in the bills of quantities.
(2) If it is not possible to value by measurement and valuation, at rates and prices deduced therefrom
(3) By measurement and valuation at fair rates and prices if it is not possible to value by the rates and prices for similar work
(4) Where it is not possible to value alterations or additions by any of the foregoing methods, then it must be valued by value of materials used and the plant and labour used following the basis of charge for daywork in the contract. Clause 42(12) deals with the contractor's obligation to produce vouchers, but there is no obligation for the QS formally to verify the vouchers.

Where a VI results in a saving to the contractor, the saving must be passed on to the authority to the amount determined by the QS. It is clear that the QS is given considerable discretion. In the case of a variation which would normally be valued in accordance with options (1) or (2) above, clause 42(11)

---

[613] *Oxford English Dictionary.*
[614] *Oxford English Dictionary.*

permits the QS to ascertain the value by measurement and valuation at fair rates and prices if he is of the opinion that the VI was issued at a time or is of such content that it is unreasonable to value it in the normal way.

The QS must take account of any 'disruptive effect' on work which is not within the direct scope of the VI. He must do this by adjusting the rates of the work which has been disrupted not, strictly, by adding to the value of the VI itself.

There are useful deadlines imposed. No later than 14 days after the QS requests it, the contractor must provide him with any information he requires to enable him to value a VI or to determine any expense in complying with any other instruction. For his part, the QS must respond with his valuation not later than 28 days after receiving the information requested. Cynics may say that the QS can delay his valuation by the simple, and regrettably common, expedient of requesting more and more information. Although no provision can be foolproof, the draftsmen of this contract seem to have devised a quite subtle method of avoiding the common abuses by imposing deadlines on both parties. The contract ensures that if a valuation has not been notified to the contractor within 42 days of a request for information by the QS, either the contractor or the QS must be in breach. It seems that clause 41(4) read in conjunction with clauses 42(7) and (8) does not entitle the QS to request information on more than one occasion in respect of each VI.

If the contractor disagrees with the valuation under clause 42(5) he must give reasons and his own valuation to the QS within 14 days of receipt of notification, otherwise he is to be treated as having accepted the QS valuation.

Clause 43 is a two-edged sword and provides for the adjustment of the contract sum in two sets of circumstances.

Paragraph 43(1)(a) is a claims provision. The contract sum is to be *increased* by the amount of any expense which is more than what is actually provided for in the contract or which the contract reasonably foresees as a result of any instruction other than a VI. The expense must be incurred by the contractor properly and directly.

The scope of this clause is broad. Expense incurred consequent on complying with an instruction under many headings set out in clause 40(2) would be reimbursable. Clause 40(2) authorises the PM to issue instructions in regard to:

- Variation of the specification, drawings or bills of quantities, or the design, quality or quantity of the works
- Discrepancies within the specification, drawings and bills of quantities or between any of them
- Removal of any things intended to be incorporated from the site and the substitution of such things by any other things
- Removal of work and the carrying out of replacement work
- Order in which the works should be carried out
- Hours of working, overtime or night work

- Suspension of carrying out of the works in whole or in part
- Replacement of operatives
- Opening up of work for inspection
- Making good of defects
- Clause 38 cost savings
- The carrying out of emergency work
- Use or disposal of excavated material
- Actions after discovery of antiquities and the like
- Actions for the avoidance of nuisance and pollution
- Contractor's quality control accreditation
- Any other matter which the PM thinks necessary.

It is easy to see that the contractor could incur expense following an instruction to change the hours of working or to suspend part or the whole of the works. Indeed every instruction issued by the PM could potentially involve the contractor in expense. The expense must be directly incurred. In other words it must be akin to damages at common law[615].

The contract mechanism (clause 43(2)) does not require the contractor to make a claim or even an application for reimbursement as would be the case under the JCT contracts. The trigger is the compliance on the part of the contractor with an instruction. The straightforward meaning of the words of the clause are that the operative date is the date on which the instruction has been entirely complied with rather than the date on which compliance begins. The contractor is to submit the information referred to in clause 41(4) within 28 days of compliance. There are two points of note. The first is that nothing prevents the contractor submitting information before compliance is completed except for the second point which is that the information referred to in clause 41(4) is the information 'required' by the QS so that he can value etc. 'Required' is a curious word. It can have an active or passive connotation. If the information has to be actively required by the QS, the provision becomes a nonsense, because the contractor cannot know what information to provide until the QS has informed him. If the QS does not inform him until after 28 days from compliance, the contractor cannot possibly comply with the timetable. The only sensible interpretation to give to 'required' in clause 41(4) is passive. In other words, the contractor must provide the information which viewed objectively the QS will need (require) to determine the expense.

The contractor has 14 days, from receipt of the QS notification of the expense he has determined, to notify the QS of the reasons for any disagreement with the amount and to submit his own estimate of what the amount should be, otherwise the contractor will be taken to have accepted the QS determination. 'Expense' is defined in clause 43(4) as money expended by the contractor, but it does not include interest or finance charges. In the first edition of this book, the view was expressed that 'expense' was limited to money paid out. On that view, loss of profit would clearly be excluded. It is

---

[615] See also Chapter 5, section 5.2 on the meaning of 'direct'.

not thought that this is necessarily the case, because in law the word 'expense' is not necessarily treated as excluding loss: see *Re Stratton's Deed of Disclaimer*[616] where the phrase under consideration was 'at the expense of the deceased'. This, of course, is not the same as the GC/Works/1 (1998) phrase 'incurs any expense', and on different wording in a charterparty the Court of Appeal took a more limited approach where the phrase 'any expense in shifting the cargo' was at issue[617]. There were numerous clauses in the charterparty which drew a distinction between 'expense' or 'expenses' on the one hand and 'time occupied' on the other. In these circumstances, the then Denning LJ was firmly of the opinion that 'expense' in the context meant 'money spent out of pocket and does not include loss of time'[618].

Expenses actually incurred would include any true additional overhead costs to the contractor, for example, the cost of additional supervision, and the cost of keeping men on site, and on the wider interpretation the amount recoverable would extend to fixed overheads such as head office rent and rates and so on[619].

Clause 43(1)(b) deals with the converse case and provides that if the contractor makes any saving in the cost of the execution of the works as a result of complying with an instruction (excluding a VI), the contract sum is to be *decreased* by the amount of that saving. Whether the instruction results in an addition or reduction to the contract sum, the amount is to be calculated by the QS. He would be expected to do this in accordance with the usually recognised principles discussed elsewhere in this book.

14.3 Prolongation and disruption

14.3.1 Introduction

There is no direct equivalent of JCT 98, clause 26, because in this contract, provisions for recovery of disruption and prolongation expenses are scattered over a number of clauses. Very sensibly, recovery of expense associated with instructions is to be recovered, broadly speaking, with the valuation of such instructions. This overcomes a potential grey area familiar to all users of JCT 98. Clauses specifically dealing with expense associated with instructions are: 41(2), 42(2)(b), 42(6) and 43(1)(a). The principal clause for recovery of expense is clause 46. Its objective is to reimburse the contractor for any *expense* in performing the contract as a result of regular progress of the whole or part of the works being materially disrupted or prolonged as an unavoidable result of one or more of the specified matters.

---

[616] [1957] 2 All ER 594.
[617] *Chandris* v. *Union of India* [1956] 1 All ER 358.
[618] *Chandris* v. *Union of India* [1956] 1 All ER 358 at 360 per Denning LJ.
[619] See also the discussion in Chapter 5, section 5.1, in relation to 'loss and/or expense'.

14.3.2 Prolongation and disruption

Clauses 46 and 47 deal with prolongation and disruption, and finance charges respectively.

14.3.3 Commentary

Prolongation and disruption

Although the clause is not long, it is not drafted as simply as might be desired and careful reading is required before its full implications are clear. A point to note is that the clause refers only to 'expense', and it must be an expense which is more than that actually provided for in the contract or which the contract reasonably contemplates. This is an objective test; it is *not* the contemplation of the contractor concerned. It must also be an expense which the contractor would not have incurred in some other way and it must have been *'properly and directly'* incurred by the contractor. In other words, consequential loss is excluded and it would seem that the incurring of the expense must be wholly unexpected. Some slight redrafting has resulted in a proviso being introduced that the expense must not be a result of any default or neglect on the contractor's part or on the part of any of his employees, agents or sub-contractors. That would have been implied in any event.

Although not easy to identify, there are six matters that may give rise to a claim for prolongation and disruption expenses under the clause. Three of them are grouped together. They are:

(1) The carrying out of work under clause 65 (other works).
(2) Delay in being given possession of the site. If this ground was not included, the contractor would be left to claim damages for breach of contract and, if the delay was long enough, he may be able to accept it as a repudiatory breach. The inclusion of the ground in clause 46 does not, of course, preclude the contractor from exercising his common law remedies in any event[620].
(3) Delay in respect of:
— decisions, agreements, drawings, levels, etc. or any other design material to be provided by the PM
— the carrying out of any work or the supplying of any thing by the employer or obtained from anyone except the contractor
— any instruction from the employer or the PM about the issue of a pass to a person or any instruction from the employer or the PM under clause 63(2). A proviso has been added that they must be entitled to a reasonable time for consideration and decision and that

---

[620] *London Borough of Merton* v. *Stanley Hugh Leach Ltd* (1985) 32 BLR 51.

their discretion is not to be fettered. Inevitably, this ground will operate to fetter their discretion to some extent and this proviso is unlikely to change that, because of the nature of discretion.

provided that the employer has failed to supply or do something by a date agreed with the contractor or within any reasonable period in a notice given by the contractor to the employer or to the PM. This is a very restrictive proviso which precludes the contractor from making any claim unless he satisfies one of these conditions, *before* the alleged delay.

(4) Advice, other than required by the CDM Regulations, given by the planning supervisor.

A further stipulation is that one or more of these matters must unavoidably result in the regular progress of the works being disrupted or prolonged. Strictly, this means that there can have been no other outcome no matter what action the contractor took. It is thought that 'unavoidably' should be given the commercial meaning of unavoidably in the context of the ordinary nature of construction operations and not unavoidable in a strict sense. For the contractor to be successful in contending that regular progress has been disrupted or prolonged, he must first be prepared to establish that he was, as a matter of fact, making regular progress[621].

The contractor must satisfy two conditions before he is entitled to payment:

(1) He must give notice to the PM immediately he becomes aware that regular progress of any part of the works has been or is likely to be disrupted or prolonged. 'Immediately' in that context means that the contractor must act with all reasonable speed[622]. Obviously, a commercial interpretation must be given to 'immediately', but it is probably a matter of days rather than weeks. It is likely that the courts will apply notice provisions of this kind strictly[623]. Contrast this with words such as 'as soon thereafter as is practicable', when a broad interpretation can be expected[624]. The only difficulty in this instance is establishing that the contractor has become 'aware' on a particular date. It is important that the contractor does give notice immediately from the purely practical standpoint that the PM may want to give instructions so as to reduce any possible claim. If the notice is given only after the event, the PM will be powerless to influence matters. In addition, clause 25 provides for the contractor to keep such records as may be necessary for the QS, PM or

---

[621] Some guidance on the contractor's obligations in working regularly and diligently has been given in *West Faulkner Associates* v. *London Borough of Newham* (1995) 11 Cost LJ 157. See the discussion in Chapter 12, in section 12.2.4, 'Material effect on regular progress'.

[622] *Hydraulic Engineering Co Ltd* v. *McHaffie, Goslet & Co* (1878) 4 QBD 670 CA.

[623] There is little judicial authority in the English courts other than *Hersent Offshore SA and Amsterdamse Ballast Beton-en-Waterbour BV* v. *Burmah Oil Tankers Ltd* (1979) 10 BLR 1, but there are two Commonwealth cases on the topic: *Jennings Construction* v. *Birt* [1987] 8 NSWLR 18 and *Wormald Engineering Pty Ltd* v. *Resources Conservation Co International* [1992] 8 BCL 158. See also the discussion in Chapter 12, in section 12.2.4, 'Timing of application' and 'Architect's assumed state of knowledge'.

[624] *Tersons* v. *Stevenage Development Corporation* (1963) 5 BLR 54. In *Hersent Offshore* v. *Burmah Oil* (1979), 10 BLR 1, 4 months was held to be outside this period envisaged by those words.

employer to ascertain claims. It seems that the contractor must comply
with the PM's reasonable instructions regarding the particular records
to be kept. If the notice is given late, the PM will be unable to give
instructions in that regard. All notices must be in writing (clause 1(3)).
The notice must specify the causes or likely causes and it must include a
statement that the contractor is entitled to an increase in the contract
sum. The information which the contractor must provide to the PM is
not precisely stated, but it is clear that it must be set out in sufficient
detail so that the basis of his claim is readily identifiable (clause
46(3)(a)).

(2) He must provide to the QS full details of all expenses incurred and
evidence that they *directly* resulted from one of the occurrences in 46(1).
He is to provide the details as soon as reasonably practicable, but in any
event within 56 days of incurring the expense (clause 46(3)(b)).

The respective duties of the PM and the QS can be deduced from this
without too much difficulty.

The PM need take no action until he receives the notice referred to in
clause 46(3)(a) from the contractor. The notice triggers the process and it is
then for the PM to verify that the notice has been correctly served. The
notice must specify the circumstances and it must state that the contractor is
or expects to be entitled to an increase in the contract sum under 46(1).
Applying these criteria, it should be relatively easy to decide whether the
contractor has complied.

If the contractor complies with clause 46(3) in its entirety, it is for the QS to
notify him within 28 days from receipt of all the details and evidence of the
amount due to him. The ascertainment of expense is placed in the hands of
the QS under clause 46(1).

Clause 46(6) is in precisely the same terms as clause 43(4) and the comments on that clause are also applicable here. Although these clauses expressly exclude interest or finance charges however they are described,
clause 47 deals with the circumstances in which the contractor may recover
finance charges.

Finance charges

There are two circumstances in which the employer must pay finance
charges to the contractor. They are that money has been withheld from the
contractor because:

(1) either the employer, the PM or the QS has not complied with a time limit
set out in the contract or any agreed variation to it; or
(2) the QS varies his decision after he has already notified the contractor.

An example of item (1) would be a failure to pay within a prescribed time
period. An example of (2) would be a decision by the QS concerning the
amount of expense by which the contract sum was to be increased. In the
latter case, if it did not result in money withheld, there would be no liability

for finance charges. Therefore, a mistake which gave the contractor too much money and subsequently varied, would not qualify. This position is dealt with in clause 47(4) which expressly requires the QS to take any overpayment into account. Therefore, the QS is entitled to set-off interest earned on overpayment against charges on underpayment. Clause 47(5) provides that the employer is not liable to pay finance charges resulting from act, neglect or default of the contractor or sub-contractors, any failure by the contractor or sub-contractors to supply relevant information or any disagreement about the final account. This is a most important and sensible clause which put the onus on the contractor, among other things, to provide appropriate back-up information and to do it promptly. The finance charges are set in accordance with the percentage to be inserted in the abstract of particulars compounded quarterly on set dates above Bank of England rate to the clearing banks.

Clause 47(3) is unusual. It singles out failure to certify money as a trigger for finance charges provided that it results from one of the circumstances set out in clause 47(1). The applicable period for the charges is between the date on which the certificate should have been issued and the date when, in fact, it was issued under clause 50. Clause 47(6) is noteworthy. It is an attempt to exclude liability for interest and finance charges under the guise of what are usually termed special damages[625]. Whether it would be effective if challenged under the Unfair Contract Terms Act 1977 is uncertain.

14.4 *Extension of time and liquidated damages*

14.4.1 Extension of time

Clause 36 is the provision under which any alteration to the date for completion is made. The 'Date for Completion' is the date calculated from the date of possession of the site. The date of possession must be notified by the authority to the contractor within the period or periods specified in the abstract of particulars.

14.4.2 Clause 36

The full text of clause 36 is as follows:

36   Extensions of time

(1)   Where the PM receives notice requesting an extension of time from the Contractor (which shall include the grounds for his request) or where the PM considers that there has been or is likely to be a delay which will prevent or has prevented completion of the Works or any Section by the relevant Date

---

[625] See the consideration of damages in Chapter 5, section 5.2.

for Completion (in this Condition called delay), he shall, as soon as possible and in any event within 42 Days from the date any notice is received notify the Contractor of his decision regarding an extension of time for completion of the Works or relevant Section.

(2) The PM shall award an extension of time under paragraph (1) only if he is satisfied that the delay, or likely delay, is or will be due to-

(a) the execution of any modified or additional work;

(b) the act, neglect or default of the Employer, the PM or any other person for whom the Employer is responsible (not arising because of any default or neglect by the Contractor or by any employee, agent or subcontractor of his);

(c) any strike or industrial action which prevents or delays the execution of the Works, and which is outside the control of the Contractor or any of his subcontractors;

(d) an Accepted Risk or Unforeseeable Ground Conditions;

(e) any other circumstances (not arising because of any default or neglect by the Contractor or by any employee, agent or subcontractor of his and other than weather conditions) which are outside the control of the Contractor or any of his subcontractors, and which could not have been reasonably contemplated under the Contract;

(f) failure of the Planning Supervisor to carry out his duties under the CDM Regulations properly; or

(g) the exercise by the Contractor of his rights under Condition 52 (suspension for non-payment).

(3) The PM shall indicate whether his decision is interim or final. The PM shall keep all interim decisions under review until he is satisfied from the information available to him that he can give a final decision.

(4) No requests for extensions of time may be submitted after completion of the Works. The PM shall in any event come to a final decision on all outstanding and interim extensions of time within 42 days after completion of the Works. The PM shall not in a final decision withdraw or reduce any interim extension of time already awarded, except to take account of any authorised omission from the Works or any relevant Section that he has not already allowed for in an interim decision.

(5) The Contractor may within 14 days from receipt of a decision, submit a claim to the PM specifying the grounds which in his view entitle him to an extension or further extension of time. The PM shall by notice give his decision on a claim within 28 days of its receipt.

(6) The Contractor must endeavour to prevent delays and to minimise unavoidable delays, and to do all that may be required to proceed with the Works. The Contractor shall not be entitled to an extension of time where the delay or likely delay is, or would be attributable to, the negligence, default, improper conduct or lack of endeavour of the Contractor.

14.4.3 Commentary

The initiative for taking action under this provision lies with either the contractor or the PM. Usually, the contractor will request an extension of time, but he may not do this after completion of the works. Clause 36(1) stipulates that such a request must include grounds for the request. The degree of detail is not specified, but at the very least it must contain sufficient information to enable the PM to understand why the contractor considers he is entitled to an extension of time. In practice, a contractor will be wise to submit a very detailed request. This clause does not expressly empower the PM to require further information but, if such power is needed, it can probably be implied from the wording of clause 25 requiring the contractor to keep records required by the PM. The grounds for extension of time are expressed broadly:

Execution of modified or additional work

For example, following an instruction under clause 40(2)(a). This ground may, and independently, give rise to a money claim under clauses 41(2), 42(2)(b) and 42(6).

Act, neglect or default of the employer or the PM (excluding contractor's fault)

This ground would extend to cover the acts etc. of those for whom the employer or the PM were vicariously responsible in law. A probably unnecessary proviso has been added to make clear that the original cause must not be any default or neglect of the contractor or any of his employees, agents or sub-contractors. At first sight, this kind of proviso might appear designed to get around the ordinary rules of causation[626] and that appears to be the case. On its ordinary meaning, the phrase 'not arising because of' is the same, or nearly so, as 'not arising as a result of'. This is to remove from the scope of this clause any situation where the default or neglect of the contractor has created a situation following which the employer or PM might perform some act, neglect or default which will cause the contractor to be delayed. The phrase 'not arising because of' is much broader than 'not caused by'. The former envisaging a situation which, created by the contractor, permits or probably invites action by the employer or the PM; the latter being a situation which inevitably leads on to the employer's or the PM's actions.

Strike or industrial action outside the control of the contractor or his sub-contractors and which delays or prevents the works

Any strike or any kind of industrial action is included, probably even a work to rule. There are only two stipulations: it must delay or prevent

---

[626] Causation is considered in Chapter 7.

execution of the works and it must be outside the control of the contractor or his sub-contractors.

Accepted risks or unforeseeable ground conditions

The definition of the term 'Accepted Risks' in clause 1(1) means the risks of pressure waves caused by the speed of any aerial machine, ionising radiations or contamination by radioactivity from any nuclear fuel or from its combustion, hazardous properties of any explosive nuclear assembly and war, hostilities, civil war and the like.

Unforeseen ground conditions are the conditions referred to in clause 7.

Any other circumstances outside the control of the contractor or his sub-contractors and which could not have been reasonably contemplated, excluding weather conditions and contractor's fault

The wording is wide, although it is probably aimed at what is usually called Act of God, which is an overwhelming superhuman event, and also at circumstances covered in other forms of contract by the term *force majeure*. This is a French law term which in English law 'is used with reference to all circumstances independent of the will of man, and which it is not in his power to control'[627]. The term *force majeure* has been held to apply to dislocation of business caused by a nationwide coal strike and also accidents to machinery. It did not cover delays caused by bad weather, football matches or a funeral, on the basis that these were quite usual incidents interrupting work and the contractors ought to have taken them into account in making the contract[628]. Delays caused by persons with whom the authority contracts direct under clause 65 fall under this ground. It should be noted that weather conditions are expressly excluded from this ground therefore, long spells of very hot weather or excessively cold weather conditions are at the contractor's risk. Even if a circumstance would otherwise be eligible for consideration under this ground, it will be excluded if it could reasonably have been contemplated. This ground has a similar proviso to the one included in clause 36(2)(b) and the commentary is also applicable.

Failure by the planning supervisor to carry out his duties under the CDM Regulations

This is straightforward, but it should be noted that it only applies if the planning supervisor fails to carry out his duties under the Regulations. There may be many occasions when proper operation of his duties may cause a delay, but the contractor has accepted that risk.

---

[627] *Lebeaupin v. Crispin* [1920] 2 KB 714.
[628] *Matsoukis v. Priestman & Co* [1915] 1 KB 681.

Exercise of the contractor's rights to suspend performance of his obligations

This ground complies with section 112 of the Housing Grants, Construction and Regeneration Act 1996 which provides for extension of the contract period or the fixing of a new date for completion if the suspension provision has been operated correctly.

Clause 36(1) makes clear that the PM can take unilateral action if he considers that there has been or is likely to be a delay which will prevent or has already prevented completion of the works by the date for completion. It is very important that the PM understands that, because failure to make an extension of time in appropriate circumstances may result in time becoming at large[629]. GC/Works/1 (1998) is the only building contract which sets out the PM's power in this respect in such clear terms.

The PM must notify the contractor of his decision as soon as possible, but no later than 42 days from the date he receives notice from the contractor. It is worth giving this provision careful consideration. The PM cannot simply assume that he has 42 days from the date of receipt of notice in which to decide on the extension of time. If it is possible to come to a decision earlier than 42 days, he must do so or be in breach of the provision. It should be noted that the PM has power to extend time for any section under this clause.

In giving an extension, the PM must state whether his decision his final or merely interim. Doubtless most of his decisions taken during the progress of the work will be interim. Where a decision is interim, the PM must review it regularly until he is in a position to give a final decision. The PM has just 42 days from the date when the works are actually completed to come to a final decision about all outstanding (i.e. those he has not yet decided) and interim extensions of time. He cannot withdraw any extension already given nor can he reduce the period unless he does so purely to take account of an omission which he has instructed and which he has not already taken into account in arriving at previous extensions of time. It appears, therefore, that, unlike the position under JCT 98, the PM is not confined to omissions instructed since the last extension was given; he can take into consideration in his final decision an omission which he instructed but failed to consider when giving earlier extensions. It is clear that the final decision under clause 36(4) is intended to be the PM's opportunity to sweep up all outstanding or ill-considered delays.

Clause 36(5) provides what the contractor must do if he believes any decision under clause 36(1) is insufficient. He has 14 days from receipt of the decision within which to submit what the clause refers to as a 'claim' to the PM. The claim must specify the grounds which he thinks entitle him to an extension of time. On receipt, the PM has 28 days in which to respond and it is clear from the wording that the PM must respond whether or not he decides that a further extension is warranted. In the previous edition of this contract, it was unclear whether the contractor's right to submit a claim

---

[629] See Chapter 2, section 2.2.

under this clause extended to a final decision given by the PM after the date of completion under clause 36(4). The view that a decision under clause 36(4) is an opportunity for the PM to review all matters and express his final decision, which the contractor may challenge only by adjudication or arbitration, is now confirmed by the exclusion of a reference to clause 36(1) as suggested in the last edition of this book.

Clause 36(6) requires the contractor to endeavour to prevent delays and to minimise unavoidable delays and to do everything required to proceed with the works. This is very much to the same effect as JCT 'best endeavours' provisions[630]. It goes on to say that the contractor is not entitled to an extension of time if the delay is caused by his negligence, default, improper conduct or lack of endeavour. Read strictly, it appears that if the contractor is responsible for any part of a delay for any of these reasons, he is not entitled to any extension at all for the remaining part of the delay. This is the straightforward meaning of the words used. In practice, of course, an extension of time is often reduced (in the sense of part of the delay being discounted) for these reasons, but rarely is the extension entirely withheld on this ground. Note that the prohibition on extending time in these circumstances is less emphatic under this contract than under MC 98[631].

14.4.4 Liquidated damages

Liquidated damages are dealt with by clause 55 which provides for the payment of such damages by the contractor if he fails to complete the works or a section, if appropriate, before the date for completion or any extension of that date authorised by the PM under clause 36.

14.4.5 Clause 55

The full text of clause 55 reads as follows:

55 Liquidated damages

(1) This Condition applies where a rate of liquidated damages for any delay in the completion of the Works or a Section has been specified in the Abstract of Particulars.

(2) If the Works or a Section are or is not completed by the relevant Date for Completion, the Contractor shall immediately become liable to pay to the Employer liquidated damages at the rate specified in the Abstract of Particulars for the period that the Works or any relevant Section remain or remains uncompleted.

(3) Subject to Condition 50A (withholding payment), the Employer may deduct any amount of liquidated damages to which he may be entitled under this

---

[630] See the comments in Chapter 10, section 10.1.2 on this topic.
[631] See Chapter 10, section 10.6.2 for comparison.

Condition from any advances to which the Contractor may otherwise be entitled under Condition 48 (*Advances on account*).

(4)   If the sum due as liquidated damages exceeds any advance payable to the Contractor under Condition 48 (*Advances on account*), the Contractor shall pay to the Employer the difference. That sum shall be recoverable in accordance with Condition 51 (*Recovery of sums*).

(5)   No payment or concession to the Contractor, or Instruction or VI at any time given to the Contractor (whether before or after the Date or Dates for Completion), or other act or omission by or on behalf of the Employer, shall in any way affect the rights of the Employer to deduct or recover liquidated damages, or shall be deemed to be a waiver of the right of the Employer to recover such damages. The rights of the Employer to deduct or recover liquidated damages may be waived only by notice from the Employer to the Contractor.

14.4.6 Commentary

The liquidated damages clause under this form is very straightforward. There are no certificates of the PM or written requirements of the employer made conditions precedent to recovery of the damages as found in JCT forms of contract. Instead, it is sufficient that the contractor fails to complete by the appointed date for completion or any extension of the contract period. In practice, no doubt the PM will always notify the employer that the time has arrived when liquidated damages can be charged. The rate is to be the rate stated in the abstract of particulars and it should be noted that, under the provisions of clause 55(1), clause 55 only applies if a rate is inserted. This overcomes the uncertainty which may prevail if no figure is inserted and there is a dispute whether the liquidated damages are nothing or whether the clause has no effect. In such an instance, the clause would have effect for the sum or lack of sum inserted as the rate[632]. Thus, the effect of clause 55(1) is that if the employer omits to insert any rate for liquidated damages, the clause will not apply and the employer will be free to revert to unliquidated damages to the extent that they can be proved. However, the proviso in clause 55(1) will not assist an employer who puts any rate at all (even '£nil') in the abstract of particulars[633].

Liquidated damages may be deducted by the employer from any money paid as advance under clause 48. Where there is insufficient money to achieve a deduction of the total amount, the contractor must pay the difference. If the contractor fails to pay, the sum is said to be recoverable under clause 51. Clause 51 purports to allow set-off across contracts and it is doubtful whether it would be effective if challenged[634].

Clause 55(5) is inserted for the avoidance of doubt. It is somewhat expanded since the last edition. It makes clear that the employer will

[632]   *Temloc Ltd* v. *Errill Properties Ltd* (1987) 39 BLR 30.
[633]   *Temloc Ltd* v. *Errill Properties Ltd* (1987) 39 BLR 30.
[634]   See section 10 of the Unfair Contract Terms Act 1977.

waive his rights to recover liquidated damages, if at all, only by formal notice. Therefore, the contractor will be unable to argue that the employer's conduct led him to believe that the employer was waiving his rights in this respect.[635]

However, by going on to state that neither payments, concessions nor instructions to the contractor nor any other act or omission of the employer will operate to affect the employer's rights and they will not be deemed a waiver of rights, the clause seems to go too far. It appears unlikely that what it seems to be saying is what was intended. It seems to be saying that the employer's right to recover liquidated damages will not be affected by acts or omissions of the employer. That is clearly wrong. Liquidated damages will not be recoverable by the employer if the employer is at all responsible for failure to achieve the completion date unless there is power to give an appropriate extension of time and it has been given:

> '[The] cases show that if completion by the specified date was prevented by the fault of the employer, he can recover no liquidated damages unless there is a clause providing for extension of time in the event of delay caused by him.'[636]

In other words, where completion of the contract works or any section (if sectional completion applies) in due time has been prevented by the employer's act or default, the employer cannot recover liquidated damages from the contractor, unless the PM has granted an extension of time on that ground: see clause 36(2)(b). This is the case where there is power to extend time but it has not been properly exercised[637].

GC/Works/1 (1998) is not a negotiated contract, but is one drawn up on behalf of the government departments that use it. Accordingly, as indicated earlier, it is to be construed *contra proferentem* the employer, whose document it is. It has been said:

> 'The liquidated damages and extension of time clauses in printed forms of contract must be construed strictly *contra proferentem*. If the employer wishes to recover liquidated damages for failure by the contractor to complete on time in spite of the fact that some of the delay is due to the employer's own fault or breach of contract, then the extension of time clause should provide, expressly or by necessary inference, for an extension on account of such a fault or breach on the part of the employer . . .'[638]

No liquidated damages are recoverable where the employer has failed to extend time when it should do so.

An earlier illustration of this principle is to be found in the case of *Miller* v. *London County Council* [639]. There, a building contract, not on the same terms

---

[635] *London Borough of Lewisham* v. *Shepherd Hill Civil Engineering* 30 July 2001, unreported.
[636] *Astilleros Canarios SA* v. *Cape Hattera Shipping Co Inc and Hammerton Shipping Co SA* [1982] 1 Lloyds Rep 518 at 526 per Staughton J.
[637] *Peak Construction (Liverpool) Ltd* v. *McKinney Foundations Ltd* (1970) 1 BLR 114.
[638] *Peak Construction (Liverpool) Ltd* v. *McKinney Foundations Ltd* (1970) 1 BLR 114 at 121 per Salmon LJ.
[639] (1934) 50 TLR 479.

as modern JCT or GC/Works/1 contracts, provided that the whole of the work should be completed by 15 November 1931. Clause 31 provided that:

> 'it shall be lawful for the engineer, if he shall think fit, to grant from time to time, and at any time or times, by writing under his hand such extension of time for completion of the work and that either prospectively or retrospectively, and to assign such other time or times for completion as to him may seem reasonable.'

Clause 37 provided that should the contractor fail in due completion of the work, he should pay liquidated damages for delay at a specified rate. The work was not completed until 25 July 1932. Some 4 months after completion of the work the engineer issued a certificate granting an extension of time to 7 February 1932, and subsequently certified a sum of £2625 as payable by the contractor to the building owner under clause 37 for the delay from 7 February to 25 July 1932.

du Parcq J, who construed the relevant clause *contra proferentem* the employer, 'on the ground that it was [they] who prepared and put forward the contract', held that the words 'either prospectively or retrospectively' in clause 31 did not give the engineer a right to fix a new date for completion after the entire work had been completed. They only empowered him, if a delay occurred during the progress of the work, to wait until the cause of the delay had ceased to be operative, and then within a reasonable time after the delay had come to an end 'retrospectively' with regard to the cause of the delay to assign to the contractor a new date for completion. The power to extend the time had not, therefore, been exercised within the time limited by the contract, and so the building owner had lost the benefit of the clause. There was therefore no date from which liquidated damages could run and none were recoverable.

Clause 37 provides for early possession of any part of the works which the PM has certified as completed if it is either a section or some other part of the works agreed by the parties or the subject of the PM's instruction regarding early possession. Surprisingly, and unlike the provisions for 'partial possession' under JCT contracts, there is no provision for any reduction in the rate of liquidated damages where early possession is taken of a part of the works or section. If liquidated damages are expressed as £1000 per week for the whole of the works and early possession is taken of, say, 30% of the works, the employer cannot simply reduce the liquidated damages by 30%[640]. But to deduct the whole £1000 would amount to a penalty, because if £1000 is a genuine pre-estimate of loss for the whole building it cannot be a genuine pre-estimate of loss for 70% of the whole.

---

[640] *Stanor Electric Ltd* v. *Mansell Ltd* (1988) CILL 399.

Chapter 15

Claims Under ACA 3

15.1 Introduction

In October 1982 the Association of Consultant Architects published a Form of Building Agreement (ACA), together with an associated Form of Sub-Contract. It was drafted by one of the original co-authors of this book, Professor Vincent Powell-Smith. A second and substantially revised edition of the ACA form (ACA 2) was published in 1984. This second edition took account of criticisms made of the first edition, and in parts it was substantially redrafted. The ACA form can be used for 'design, develop and construct' contracts, as well as more traditionally where the architect produces all the design and construction information. It can be used with or without bills of quantities but, rather oddly, the use of the Standard Method of Measurement has been ignored. It contains a number of alternative clauses, and it is highly flexible in use. It was revised again in relatively minor aspects in 1990, more significantly in 1995 and in 1998 the third edition (ACA 3) was published to take account of the Housing Grants, Construction and Regeneration Act 1996. Further revisions have been made, the latest at the time of writing was in 2003.

The ACA form has been the target of much criticism from the contracting side of the industry and, unfortunately, it has not become as widely used as the ACA had hoped. In fact, it is a useful contract form and the powers and duties of the parties are clearly set out. Since the ACA form is not a negotiated document (unlike the JCT forms) it will be construed by the courts *contra proferentem*. This means that any ambiguities in it which are not capable of being resolved by any other method of construction may be construed against the employer, whose document it is. The JCT forms are agreed by the whole industry; the ACA form is not. When a contractor enters into a contract in the ACA form, he will be entering into a contract on the employer's 'written standard terms of business' for the purposes of section 3 of the Unfair Contract Terms Act 1977, because any employer who utilises the ACA form makes it his own document in law, just as if he had employed lawyers to draft a special contract for him.

The claims provisions of the ACA form – both as regards money and extensions of time – are different from the corresponding JCT provisions and, indeed, the provisions of other standard forms. In its first alternative, the extension of time provision is narrower than under JCT forms, to the extent that such things as adverse weather conditions, strikes and delays or unforeseeable shortages in the availability of labour and materials are not expressly mentioned as grounds for extension of time. However, it covers 'any act, instruction, default or omission of the Employer, his servants or his agents or of the Architect on his behalf, whether authorised by or in breach of this Agreement'. The second alternative is along more traditional lines and includes *force majeure* as a ground for extension of time, as well as other risks which are the fault of neither party. The contract deals separately with money claims arising in consequence of 'any act, omission or default of the Employer or of the Architect' and those arising as a consequence of architect's instructions, but in both cases the contractor is required to submit an estimate of the money involved.

The principles involved in assessing claims under the ACA provisions, whether for time or for money, are of course identical with those involved in claims under JCT provisions, which have been discussed earlier. This chapter will examine the particular provisions in the ACA form that may give rise to claims by the contractor.

15.2 *Monetary claims under ACA 3*

There are two sets of provisions that enable the contractor to make a money claim. First, clause 7, which is concerned with disruption of the regular progress of the works or of any section of them or which delays their execution in accordance with the dates stated in the time schedule, where the disruption or delay is caused by any act, omission, default or negligence of the employer or of the architect (other than architect's instructions). Second, clauses 8 and 17, which deal with architect's instructions and their valuation. Architect's instructions, whether in respect of variations or otherwise, may have a serious disruptive effect.

15.2.1 Clause 7

The text of this provision reads as follows:

7. EMPLOYER'S LIABILITY

Disturbance to regular progress
7.1 Save in the case of Architect's instructions (to which the provisions of Clause 17 shall apply), if any act, omission, default or negligence of the Employer or of the Architect, or a failure to comply with their duties under the CDM Regulations with all reasonable diligence by the Planning Supervisor, the Principal Contractor (if not the Contractor), or a Designer (not being the Contractor or a Sub-Contractor or Supplier) disrupts the regular progress of the Works or of any Section or delays the execution of them in accordance with the dates stated in the Time Schedule and, in consequence of such disruption or delay, the Contractor suffers or incurs damage, loss and/or expense, he shall be entitled to recover the same in accordance with the provisions of this Clause 7.

Notice of Claim

7.2 Upon it becoming reasonably apparent that any event giving rise to a claim under Clause 7.1 is likely to occur or has occurred, the Contractor shall immediately give notice to the Architect of such event and shall, on presentation of his interim application pursuant to Clause 16.1 next following the giving of such notice, submit to the Architect an estimate of the adjustment to the Contract Sum which the Contractor requires to take account of such damage, loss and/or expense suffered or incurred by him in consequence of such event prior to the date of submission of his estimate.

Submission of estimates

7.3 Following the submission of an estimate under Clause 7.2, the Contractor shall, for so long as the Contractor suffers or incurs loss and/or expense in consequence of such event, on presentation of each interim application pursuant to Clause 16.1, submit to the Architect an estimate of the adjustment to the Contract Sum which the Contractor requires to take account of such damage, loss and/or expense suffered or incurred by him since the submission of his previous estimate.

Agreement of estimates

7.4 Any estimate submitted by the Contractor pursuant to Clause 7.2 or 7.3 shall be supported by such documents, vouchers and receipts as shall be necessary for computing the same or as may be required by the Architect. Within 20* working days of receipt of any such estimate duly supported as aforesaid, the Architect shall give notice of acceptance or rejection of the said estimate. If the estimate is accepted the Contract Sum shall be adjusted accordingly and no further or other additions or payments shall be made in respect of the damage, loss and/or expense suffered or incurred by the Contractor during the period and in consequence of the event in question.

Failure by Contractor to submit estimates

7.5 If the Contractor fails to comply with the provisions of Clause 7.2 or 7.3, then the Architect shall have no power or authority to make, and the Contractor shall not be entitled to any adjustment to the Contract Sum in respect of the damage, loss and/or expense to which such failure relates on any certificate issued under this Agreement prior to the Final Certificate. Such adjustment shall not in such event include an addition in respect of loss of interest or financing charge suffered or incurred by the Contractor between the date of the Contractor's failure so to comply and the date of the Final Certificate.

15.2.2 Commentary

This clause entitles the contractor to claim reimbursement for any damage, loss and/or expense which he suffers or incurs in consequence of delays and disruptions caused by any act, omission or default (save architect's instructions) of the employer or the architect. Delays and disruptions occasioned by architect's instructions – which are by far the commonest ground of claim – are dealt with separately by clauses 8 and 17. The provisions of clause 7 bear some resemblance to the equivalent provisions in the JCT forms (e.g. JCT 98, clause 26) but that resemblance is only superficial. The machinery of this clause is echoed by the very similar procedures in Supplementary Provision S7 added to the JCT Standard Form With Contractor's Design (CD 81) in 1987 and retained in the current form (WCD 98). In issuing the contract form, the ACA stated that under it 'the contractor cannot claim…additional money for a series of things that most clients consider to be normal building risks'. The scope of this clause is in some respects, therefore, limited.

Moreover, the contractor is only entitled to have the contract sum adjusted, i.e. to payment for claims by way of inclusion of the amount in interim certificates, if he complies exactly with the provisions of clauses 7.2 and 7.3. If he fails to comply with those provisions – which lay down procedures for the giving of notice and the submission of estimates – his entitlement to an adjustment to the contract sum is removed until the final certificate. This is the result of clause 7.5, which has the side-heading 'Failure by Contractor to submit estimates'.

Each word of clause 7 is important, and the wording used raises several interesting points. Arguments can be anticipated as to what is properly claimable under the clause, particularly as regards the 'damage, loss and/ or expense' that is recoverable. There is an important difference in the wording used in ACA 3 and in the equivalent provision of JCT forms, since under ACA clause 7 the words 'damage, loss and/or expense' are not limited by the word 'direct'. The qualifying phrase is 'in consequence of such disruption or delay' and on one view, a claim under clause 7 will also extend to consequential loss and cover those heads of claim which, if made as damages at common law, are not excluded if the employer, at the time of entering into the contract, knew or must be taken to have known were liable to result from his or his architect's act, omission, default or negligence[641]. In passing, it should be noted that the word 'damage' is omitted from clause 7.3 although retained in clauses 7.1, 7.2, 7.4 and 7.5. Presumably it is a simple typographical error.

The common law position as regards damages is quite simple in essence, and has been put in these words:

> 'The governing purpose of damage is to put the party whose rights have been violated in the same position, so far as money can do, as if his rights had been observed.'[642]

However, the defaulting party is not liable for all loss actually resulting from a particular breach however improbable or unpredictable it may be. The law, therefore, sets a limit to the loss for which damages are recoverable – sometimes called the general rule as to 'foreseeability'. In both practical and legal terms the contractor's entitlement is to recover only that part of the resulting loss as was reasonably foreseeable as liable to result from the breach relied on. This is to be judged at the time the contract was entered

---

[641] *Hadley* v. *Baxendale* (1854) 9 Ex 341.
[642] *Victoria Laundry (Windsor)* v. *Newman Industries, Coulson & Co* [1949] 1 All ER 997 at 1002 per Asquith LJ.

into[643]. Under clause 7.1, therefore, the amount claimable includes all foreseeable consequential loss. Loss of profits is included[644].

All that is necessary to give rise to a claim under clause 7 is some 'act, omission, default or negligence of the Employer or of the Architect' (other than an architect's instruction). If such an act or omission etc. disrupts regular progress of the works or delays their execution in accordance with the dates stated in the time schedule, and the contractor suffers or incurs damage, loss and/or expense in consequence of the delay or disruption, he is entitled to recover the amount of the damage, loss and/or expense.

The words 'act, omission, default or negligence' are very important because some of the concepts enshrined in them are very difficult to express. Although the reference is only to employer and architect, it is probable that the clause covers others for whom the employer is vicariously responsible in law. It appears that the first three words must be interpreted without reference to the question of whether or not any relevant act or omission could also have provided the foundation for a claim in tort or in contract against the person directly concerned: it is not necessary for there to have been a breach of some legally enforceable duty. Negligence presents very little practical difficulty, because once the facts are established it is usually clear whether a duty of care in the legal sense arises and, if it does, whether it has been broken. In the vast majority of situations covered by clause 7, the contractor would have a right of action against the defaulting party, but the scope of this phrase is not to be limited by considering whether the contractor could sue employer or architect direct.

The legal meaning of 'default' has been considered, albeit in the very different context of a contract for the sale of real property:

'Default must, I think, involve either not doing what you ought or doing what you ought not, having regard to your relations with the other parties concerned in the transaction; in other words, it involves the breach of some duty you owe to another or others. It refers to personal conduct and is not the same thing as breach of contract.'[645]

In the context of a construction industry contract, in construing similar words to the present in an indemnity clause and having cited with approval the above passage, it has been said that:

'default would be established if one of the persons covered by the clause either did not do what he ought to have done, or did what he ought not to have done in all the circumstances, provided of course...that the conduct in question involves something in the nature of a breach of duty so as to be properly describable as a default'.[646]

---

[643] *H. Parsons (Livestock) Ltd* v. *Uttley Ingham & Co Ltd* [1978] 1 All ER 525. See the more comprehensive consideration of this in Chapter 5, section 5.2.
[644] *Wraight Ltd* v. *P. H. & T. Holdings Ltd* (1968) 13 BLR 27.
[645] *In Re Bayley Worthington & Cohen's Contract* [1909] 1 Ch 648 at 656 per Parker J.
[646] *City of Manchester* v. *Fram Gerrard Ltd* (1974) 6 BLR 70 at 90 per Kerr J.

This appears to be the correct interpretation to be put upon 'default' in clause 7, and this was reinforced by *Greater London Council* v. *The Cleveland Bridge & Engineering Co. Ltd*[647], which emphasised that 'default' is a narrower term than breach of contract. However, when considering the wording of a bond, the Court of Appeal in *Perar BV* v. *General Surety & Guarantee Co Ltd*[648] held that 'default' did not mean anything other than a breach which was its common meaning and the meaning to be derived from the context of the bond and the reference to damages. Therefore, in this contract, the precise meaning remains open.

This is not to suggest, of course, that it is easy to establish a claim under the clause, and it is important to note that the contractor must be able to show that he 'suffers or incurs damage, loss and/or expense' *in consequence* of the disruption or delay occasioned by the employer's (or architect's) act, omission, etc. The damage, loss and/or expense must have been *caused* by the breach and not merely be the occasion for it[649].

In other words, the loss, etc. must follow directly and in the natural course of things from the event that gives rise to it; and the event giving rise to the claim is the 'act, omission, default or negligence of the employer or of the architect' resulting in delay or disruption.

There is a further limitation, because the disruption or delay must be referable to the dates stated in the time schedule. The time schedule is thus the yardstick against which delays and disruption are to be measured. The time schedule is a feature peculiar to ACA 3, although similar in some respects to the abstract of particulars under Form GC/Works/1(1998). It is described by the ACA as 'an essential part of the ACA Form of Agreement'. The time schedule is a contract document and is not the same as the contractor's programme.

The time schedule is printed at the back of the form of agreement, and is in two alternative versions, one providing for normal single completion and the other for sectional completion. For ordinary completion, it contains the following information:

- date for possession (clause 11.1);
- taking-over and commencement of maintenance period ('Taking-Over' is described by the ACA as 'similar to "practical completion" which has been described as "when the building is reasonably safe and not unreasonably inconvenient"'. This, of course, is not the definition of practical completion favoured by the courts.);
- weekly rate of liquidated damages (clause 11.3). If a daily rate is required the necessary amendments must be made to the wording. (This item is, however, to be deleted where the employer wishes to exercise his option

[647] (1987) 8 Con LR 30.
[648] (1994) 66 BLR 72. In this case, the court declined to follow *Northwood Development Company Ltd* v. *Aegon Insurance Company (UK) Ltd* (1994) 10 Const LJ 157, which held that 'default' was wider in meaning than 'breach' and included non-fulfilment of the contractor's obligation under the contract, whether or not the contractor was in breach.
[649] *Weld-Blundell* v. *Stephens* [1920] AC 956.

of recovering unliquidated damages for late completion and based on
actual loss, and is one alternative open to him under the contract, as the
alternative clause 11.3 so provides and entitles the employer to deduct
such unliquidated damages from amounts otherwise payable to the contractor.),

- the maintenance period (clause 12.2).

The same information is required as regards the alternative version for
sectional completion, with appropriate adjustments. The final part of the
time schedule is headed 'Issue of Information'. Under the normal contract
procedure, where the architect is to be responsible for the preparation and
issue of all drawings etc., the architect may set out under this heading the
items he proposes to issue and when, to the intent that the contractor can
then prepare his construction programme taking this into consideration.
(An alternative heading is for use where the contractor is to supply drawings etc., in which case he completes it appropriately.)

The dates stated in the time schedule are those against which 'regular
progress of the Works' is to be judged and, under clause 7, failure by the
employer or the architect to adhere to those dates may give rise to a claim
and will do so if delay or disruption results.

Clause 7.2 and clause 7.3 lay down the actual claims procedure, and exact
adherence to it is essential if the contractor is to be reimbursed as the
contract proceeds. The procedural steps may be summarised in this way:

(1) The contractor must give written notice to the architect of any event
giving rise to a claim, i.e. any act, omission, default or negligence of the
employer or his architect. The notice must be specific and identify the
event(s) relied on.

(2) This notice must be given *immediately* 'upon it becoming reasonably
apparent that any event giving rise to a claim under clause 7.1 *is likely
to occur or has occurred'*. This means that the notice must be given as soon
as regular progress is likely to be disrupted or delayed.

(3) When he makes his next interim application for payment after the issue
of his notice, the contractor must submit to the architect a written
estimate of the adjustment to the contract sum that he requires to take
account of the claim, in respect of the damage etc. which he has suffered
prior to the date of submission of the estimate. The use of the word
'estimate', although often taken to mean 'an approximate judgment' is
here used in its strictly accurate sense[650]. If the contractor's estimate is
accepted by the architect then, as will be seen, the contract sum is to be
adjusted accordingly 'and no further or other additions or payments
shall be made in respect of such claim'. The contractor's estimate is to be
supported 'by such documents, vouchers and receipts as shall be necessary for computing the same *or as may be required by the architect'*: see

---

[650] See *Crowshaw* v. *Pritchard & Renwick* (1899) 16 TLR 45, where the court considered that a
contractor's estimate, dependent on its terms, may amount to a firm offer and then acceptance by
the employer will result in a binding contract.

clause 7.4. In practical terms the clause requires moneyed-out claims supported by substantiating documentation. The italicised phrase is not qualified in any way by the word 'reasonably' or otherwise and there-fore it seems that the architect can call for whatever documents, vouchers and receipts as he may in his absolute discretion require. Clause 7.3 is important since it covers continuing losses. The contractor is to submit further estimates with each subsequent interim application for payment in respect of damage, loss and/or expense which he has suffered or incurred since the submission of his previous estimate.

(4) The onus then passes to the architect. Within 20 working days from receipt of the contractor's estimate and supporting documentation, the architect must give written notice to the contractor that he either accepts or rejects the estimate of quantum. The second edition of this contract (ACA 2) gave the option to negotiate, but not to reject. This edition has dispensed with the negotiation option which may well have the unfor-tunate effect of pushing a contractor whose estimate is rejected into adjudication as the only real remedy.

(5) The contract sum is to be adjusted – and the amount of the claim included in the next interim certificate – as soon as the contractor's estimate is accepted. Clause 7.4 says that 'the Contract Sum shall be adjusted accordingly *and no further or other additions or payments shall be made in respect of the damage, loss and/or expense suffered or incurred by the Contractor during the period'*.

(6) If the contractor fails to submit estimates as required by clauses 7.2 and 7.3 then, by virtue of clause 7.5, he loses any entitlement to reimburse-ment in respect of his claim until the final certificate. Furthermore, he is disentitled to any interest or financing charges element in his claim which he suffers or incurs in the period between the date of his failure to submit the estimate and the date of the final certificate.

It would seem that, under clause 7, the architect also has power to settle what would be considered to be common law claims provided that such arise from 'any act, omission, default or negligence of the Employer or of the Architect'.

The effect of clause 7.5 is straightforward. If the contractor fails to comply with the procedures of clauses 7.2 and 7.3 as to the giving of notice and the submission of his estimate, he loses his entitlement to an adjustment of the contract sum until the final certificate, as well as any interest or financing charges which might otherwise form part of his claim. Failure to so comply with these provisions means that 'the Architect shall have no power or authority to make, and the Contractor shall not be entitled to, any adjust-ment to the Contract Sum in respect of [his claim] on any certificate issued ... prior to the Final Certificate'.

If properly operated, these claims provisions should prove of benefit to contractors and employers alike. Clearly there is room for argument whether it is an 'act, omission, default or negligence' of employer or archi-tect that is the cause of the delay; but the existence of the time schedule as a measure of progress does away with some of the practical difficulties encountered under the money claims clauses of other contracts.

15.3 Architect's instructions

Clause 8 deals with the issue of architect's instructions (which are normally to be in writing) and gives the extent of the architect's authority in this respect. The provisions for the confirmation of oral instructions are more satisfactory than those under JCT forms since they must be confirmed within 5 working days of being issued. Very sensibly, oral instructions can only be issued in an emergency, the primary meaning of which is a 'sudden juncture demanding immediate action'[651]. The cost of complying with architect's instructions is to be settled under clause 17.

15.3.1 Clauses 8 and 17

The text of clauses 8 and 17 is as follows:

Architect's instructions

8.1   The Architect shall have the authority to issue instructions at any time up to the Taking-Over of the Works and the Contractor shall (subject to Clause 17.1) immediately comply with all instructions issued to him by the Architect in regard to any of the following matters:

    (a)   the removal from the Site of any work, materials or goods which are not in accordance with this Agreement or the CDM Regulations;

    (b)   the dismissal from the Works of any person employed on them if, in the opinion of the Architect, such person misconducts himself or is incompetent or negligent in the performance of his duties;

    (c)   the opening up for inspection of any work covered up or the carrying out of any test of any materials or goods or of any executed work;

    (d)   the addition, alteration or omission of any obligations or restrictions in regard to any limitations of working space or working hours, access to the Site or use of any parts of the Site;

    (e)   the alteration or modification of the design, quality or quantity of the Works as described in the Contract Documents, including the addition, omission or substitution of any work, the alteration of any kind or standard of any materials or goods to be used in the Works and the removal from the Site of any materials or goods brought on to it by the Contractor for the Works;

    (f)   any matter connected with the Works; and

    (g)   pursuant to Clauses 1.5, 1.6, 1.7, 2.6 (if applicable), 3.5, 9.4, 9.5, 10.2, 11.8, 12.2 and 14.

After Taking-Over of the Works the Architect shall have authority to issue instructions at any time up to completion by the Contractor of his obligations under

---

[651] *Concise Oxford Dictionary.*

Clause 12.2 and the Contractor shall (subject to Clause 17.1) immediately comply with all instructions issued to him by the Architect in regard to any of the matters referred to in Clauses 8.1(a), (b), (c) and (d).

Valuation of Architect's instruction

8.2   If any instruction issued under Clauses 1.6, 3.5, 8.1(c), (d), (e) or (f) or 14 or any instruction issued under Clause 1.5 or 2.6 (if applicable) to which this Clause 8.2 shall apply, shall require the Contractor to undertake work or do any other thing not provided for in, or to be reasonably inferred from, the Contract Documents, or shall require the omission of any work or of any obligation or restriction, and provided the same shall not have arisen out of or in connection with, or shall not reveal, any negligence omission or default of the Contractor or of any sub-contractor or supplier or his or their respective servants or agents, the provisions of Clause 17 shall apply. Otherwise, no adjustment shall be made to the Contract Sum in respect of compliance by the Contractor with any such instruction.

Oral instructions

8.3   Notwithstanding the provisions of Clauses 17 and 23, in an emergency the Architect may issue an oral instruction under Clause 8.1. Such oral instruction shall be confirmed in writing by the Architect within 5 working days of being issued. The Contractor shall immediately comply with such an oral instruction.

17.   VALUATION OF ARCHITECT'S INSTRUCTIONS

Submission of estimates by Contractor

17.1   Where, in the opinion of the Contractor or of the Architect, any instructions issued by the Architect to the Contractor under Clause 1.6, 3.5, 8.1(c), (d), (e) or (f) or 14 or any instructions issued under Clause 1.5 or 2.6 (if applicable) to which Clause 8.2 shall apply, will require an adjustment to the Contract Sum and/or affect the Time Schedule, the Contractor shall not comply with them (subject to Clauses 8.3 and 17.5) but shall first furnish the Architect within 10 working days (or within such other period as may be agreed between the Contractor and the Architect) of receipt of the instruction with estimates of:

(a)   the value of the adjustment (providing him with all necessary supporting calculations by reference to the Schedule Rates or otherwise); and

(b)   the length of any extension of time to which he may be entitled under Clause 11.5; and

(c)   the amount of any loss and/or expense which may be suffered or incurred by him arising out of or in connection with such instruction.

Agreement of Contractor's estimates

17.2   The Contractor and the Architect shall then take reasonable steps to agree the Contractor's estimates and any agreement so reached shall be binding upon the Contractor and the Employer. The Contractor shall immediately thereafter comply with the instruction and the Architect shall grant an extension of time under Clause 11.6 of the agreed length (if any) and the agreed adjustments (if any) in relation to the Contractor's estimates under Clauses 17.1(a) and/or 17.1(c) shall be made to the Contract Sum.

Failure to agree Contractor's estimates

17.3 If agreement cannot be reached within 5 working days of receipt by the Architect of the Contractor's estimates on all or any of the matters set out in them, then

(a) the Architect may nevertheless instruct the Contractor to comply with the instruction in which case the provisions of Clause 17.5 shall apply as if the Architect had dispensed with the Contractor's obligation under Clause 17.1; or

(b) the Architect may instruct the Contractor not to comply with the instruction.

Instruction not to comply

17.4 If the Architect instructs the Contractor not to comply under Clause 17.3(b), the Contractor shall have no claim arising out of or in connection with such instruction or with any failure to reach agreement.

Valuation if no agreement of estimates

17.5 The Architect may, by notice to the Contractor before or after the issue of any instruction, dispense with the Contractor's obligation under Clause 17.1, in which case the Architect shall, within a reasonable time after the issue of such instruction, ascertain and certify a fair and reasonable adjustment to the Contract Sum based on (where appropriate) the Schedule of Rates in respect of compliance by the Contractor with such instruction and any loss and/or expense suffered or incurred by the Contractor arising out of or in connection with it and a fair and reasonable extension of time shall be granted under Clause 11.6.

Non-compliance by Contractor

17.6 If the Contractor fails to comply with any one or more of the provisions of Clause 17.1 where the Architect has not dispensed with compliance under Clause 8.3 or Clause 17.5, the Architect shall have no power or authority to make, and the Contractor shall not be entitled to, any addition to the Contract Sum in respect of any instructions issued by the Architect to which this Clause 17 relates on any certificate issued under this Agreement prior to the Final Certificate. Such addition shall not in any such event include any adjustment in respect of loss of interest or financing charges suffered or incurred by the Contractor prior to the issue of the Final Certificate.

15.3.2 Commentary

Only compliance with architect's instructions relating to the following matters may give rise to payment:

(1) Opening up of work for inspection or testing (clause 8.1(c)), where such inspection and testing is in the contractor's favour.

(2) Additions, alterations or omission of any obligations or restrictions to working space or working hours, access to the site or parts of the site (clause 8.1(d)).

(3) The alteration or modification of the design, quality or quantity of the works as described in the Contract Documents, including the addition, omission or substitution of any work, alteration of any kind or standard, or the removal or bringing on to site of any materials or goods brought thereon by the contractor for the works (clause 8.1(e)).

(4) On any matter connected with the works (clause 8.1(f)).

(5) Relating to any ambiguity or discrepancy falling within clause 1.5. This provision relates to ambiguities and discrepancies in the contract documents which could not reasonably have been foreseen or found at the date of the making of the contract by a contractor of the prescribed degree of competence (clause 8.1(g)).

(6) Compliance with statutory requirements under clause 1.6 (clause 8.1(g)).

(7) Where applicable, in respect of adverse ground conditions or artificial obstructions at the site as referred to in the optional clause 2.6 (clause 8.1(g)).

(8) Provision of samples of the quality of any goods, materials or standards of workmanship under clause 3.5 (clause 8.1(g)).

(9) Relating to the execution of work or installation of materials, etc. not forming part of the contract by the employer's direct contractors in accordance with clause 10.2 (clause 8.1(g)).

(10) Acceleration or postponement instructions under clause 11.8 (clause 8.1(g)).

(11) Relating to the discovery of antiquities etc. under clause 14.2 (clause 8.1(g)).

However, there is a very important limitation on the contractor's entitlement to payment for compliance with these architect's instructions and it is found in clause 8.2. First, the instruction must require 'the Contractor to undertake work or do any other thing not provided for in, *or to be reasonably inferred from*, the Contract Documents' or 'shall require the omission of any work or of any obligation or restriction'. Second, the need for the instruction 'shall not have arisen out of or in connection with, or shall not reveal, any negligence, omission or default of the Contractor or of any *sub-contractor or supplier of his or their respective servants or agents*'. Only if these conditions are satisfied will compliance with an architect's instruction rank for payment.

The position under the valuation provisions in clause 17 should be carefully noted[652]. Clause 17.1 provides that if the architect issues an instruction ranking for payment (i.e. one related to those matters listed above) which, in the opinion of the architect or the contractor, requires an adjustment to the contract sum under clause 8.2 and/or will affect the time schedule, the contractor must provide to the architect the following estimates before complying with the instruction:

(1) The value of the adjustment with all necessary supporting calculations; and

---

[652] Compare this provision with the Supplementary Provision S6 of WCD 98 (Chapter 13, Section 13.8.3) and the clause 13A quotation procedure of JCT 98 (Chapter 13, Section 13.5.2), both drafted subsequently.

(2) The length of any extension of time to which he may be entitled under clause 11.5; and

(3) The amount of any loss and or expense which he may have suffered or incurred arising out of or in connection with the instruction (note that 'damage' appears to have been deliberately omitted from this clause).

In principle these estimates must be provided with 10 working days of receipt of the relevant instruction, or such other period as the architect and the contractor agree.

Moreover, there is a saving provision in clause 17.5 (below), which provides that the architect may waive the contractor's obligation to provide estimates and himself make a fair and reasonable adjustment to the contract sum and a fair and reasonable extension of time and, in any case, an emergency oral instruction under clause 8.3 is to be excluded from this estimate procedure.

Assuming that the contractor is required to submit these estimates – which it is obviously practicable to do where a reasonable time is allowed for compliance with the instruction – clause 17.2 imposes on architect and contractor a duty to take *reasonable steps* to agree the contractor's estimate. Any agreement so reached is binding upon both employer and contractor, and thereupon the contractor is to comply with the instruction immediately.

Agreement is to be reached within 5 working days of receipt of the contractor's estimates by the architect – an unusually short time period, although of course this and all other time provisions can be amended when the contract is made. If there is failure to reach agreement about any of the matters set out in the estimates then:

- The architect may nevertheless instruct the contractor to comply with the instruction, in which case it is to be valued by the architect under clause 17.5; or
- The architect may instruct the contractor not to comply.

Where the architect withdraws his instruction, the contractor has no claim for reimbursement of any loss or expense incurred in the preparation of the estimate and, of course, this could be considerable: clause 17.4.

Clause 17.5 applies where the architect waives the contractor's obligation to submit estimates of cost etc. and also where he decides, failing agreement as to price etc., to instruct compliance in any event. The architect's duty then is to ascertain and certify 'a fair and reasonable adjustment to the Contract Sum based on (where appropriate) the Schedule of Rates' (and presumably in other cases on the basis of fair valuation). This ascertainment is to include any loss or expense suffered or incurred by the contractor arising out of or in connection with the instruction. The architect is also bound to grant a fair and reasonable extension of time under clause 11.6.

The terms of clause 17.6 are important. If the contractor fails to submit estimates under clause 17.1, and the architect has not waived compliance under clause 17.5 or 8.3 (emergency oral instructions), the contractor is deprived of any entitlement to payment until the final certificate. In other words, he loses his right to payment for compliance with the architect's instructions on an interim certificate basis. The contractor's non-compliance also means that he forfeits any right to interest or finance charges for the period prior to the issue of the final certificate.

These provisions seem extremely complex but are, in fact, based on similar provisions in certain specialist forms of contract and sub-contract. If it is practicable for the contractor to submit estimates and these can be agreed, this is to his immediate benefit. If not, then provided the provisions of clause 17.5 are operated fairly and properly, there should be no problem, and it is believed that few difficulties have been encountered in practice. The long-stop is payment at final certificate stage.

Under ACA terms, there is a provision (clause 10) whereby the contractor must permit work to be done by others on the site: it has the heading 'Employer's Licensees' and covers work etc. to be done on site 'by the Employer or his employees, agents and contractors as provided in the Contract Documents' as well as that done by statutory undertakers such as water, gas and electricity suppliers. Clause 10.4 then provides that if the regular progress of the works or any section is disrupted or delayed by the employer's licensees, the contractor is entitled to recover any loss, damage or expense under the provisions of clause 7.4. It is important to note that there is no claim in respect of statutory undertaker's work carried out under clause 10.3. If the statutory undertakers are acting under their statutory powers they fall under clause 10.3. If, on the other hand, they are acting under contract with the employer, they fall under clause 10.2, in which case the contractor will have a disturbance claim[653]. The point is both important and significant.

A monetary claim may also arise as a result of the operation of clause 11.8, which in broad terms entitles the architect to issue an instruction to bring forward or postpone dates shown on the time-schedule for the taking over of any part of the works. It is then provided that the architect must ascertain and certify a fair and reasonable adjustment to the contract sum, if appropriate, to cover compliance by the contractor with the instruction together with any damage, loss and/or expense suffered by the contractor arising out of the instruction. There is a proviso that, if before giving the instruction the architect requires the contractor to give an estimate of the adjustment to the contract sum then, except for the provisions relating to extensions of time, clause 17 applies just as if the instruction was included in clause 17.1.

15.4 Extensions of time and liquidated damages

The important related matters of extensions of time and liquidated damages are dealt with in the lengthy clause 11, and it should be noted that the provision for damages is in the alternative. Alternative 1 of clause 11.3 is an ordinary liquidated damages clause; alternative 2 unusually provides for

---

[653] *Henry Boot Construction Ltd* v. *Central Lancashire New Town Development Corporation* (1980) 15 BLR 8, decided under JCT terms, is relevant.

unliquidated damages and gives the employer power to deduct such sums from moneys otherwise due to the contractor. There are also alternative provisions as to extensions of time.

15.4.1 Clause 11

The full text of clause 11 is as follows:

11. **COMMENCEMENT AND DELAYS IN THE EXECUTION OF THE WORKS**

Commencement and Taking-Over of the Works

11.1 Subject to Clauses 11.6 and 22.4 and any provisions to the contrary referred to in Clause 1.3, the Employer shall give to the Contractor possession of the Site, or such part or parts of it as may be specified, on the date or dates stated in the Time Schedule. The Contractor shall then immediately commence the execution of the Works and shall proceed with the same regularly and diligently and in accordance with the Time Schedule so that the Works and each Section are fit and ready for Taking-Over by the Employer in accordance with the provisions of Clause 12.1 on the date or dates for the Taking-Over of the same stated in the Time Schedule, subject to the provisions of Clauses 11.6 and 11.7, and to the Architect's powers pursuant to Clause 11.8.

Certificate that Works not fit and ready

11.2 If any Section is, or the Works are, not fit and ready for Taking-Over by the Employer in accordance with Clause 11.1, the Architect shall issue a certificate to that effect.

Alternative 1
Damages for delay

11.3 If the Architect issues a certificate under Clause 11.2 in respect of any Section or of the Works, the Contractor shall pay or allow to the Employer liquidated and ascertained damages at the rate or rates stated in the Time Schedule for the period between the date stated in the Time Schedule for the Taking-Over of such Section or of the Works, or such other date as may have been granted or adjusted under Clauses 11.6 or 11.8, and the Taking-Over of the same and, if having given due notice the Employer so requires, the Employer may, subject to the prior issue of the certificate under Clause 11.2, deduct such damages from the amount which would otherwise be payable to the Contractor on any certificate or the Employer may recover them from the Contractor as a debt. If a deduction is to be made from any amount due, the Employer shall give the prior written notice referred to in Clause 16.6.

Alternative 2
Damages for delay

11.3 If the Architect issues a certificate under Clause 11.2 in respect of any Section or of the Works, the Employer shall be entitled to recover from the Contractor such damage, loss and/or expense as may be suffered or incurred by him arising out of or in connection with the Contractor's breach of his obligations under Clause 11.1. The Employer may deduct such damage, loss and/or expense from the amount which would otherwise be payable to the Contractor on any certificate. If a deduction is to be made from any amount due, the Employer shall give the prior written notice referred to in Clause 16.6.

Adjustment of damages for delay

11.4   If, after the issue of a certificate under Clause 11.2, the Architect fixes a later date for the Taking-Over of any Section or of the Works under Clause 11.6 or 11.7, the Employer shall pay or repay to the Contractor any amounts paid, allowed or recovered by the Employer under Clause 11.3.

Alternative 1

Grounds for extension of time

11.5   Subject to Clause 11.8, no extension of time shall be granted to the Contractor except in the case of any failure to comply with their duties under the CDM Regulations with all reasonable diligence by the Planning Supervisor or the Principal Contractor (in either case if not the Contractor), or a Designer (not being a Sub-Contractor or Supplier), or any act, instruction, default or omission of the Employer, his servants or agents or of the Architect on his behalf, whether authorised by or in breach of this Agreement, which in the reasonable opinion of the Architect causes the Taking-Over of the Works or of any Section by the date or dates for Taking-Over stated in the Time Schedule to be prevented. The Contractor shall immediately upon it becoming reasonably apparent that the Taking-Over of the Works or of any Section is being or is likely to be so prevented submit to the Architect a notice specifying such act, instruction, default or omission and, as soon as possible thereafter, submit full and detailed particulars of the extension of time to which the Contractor may consider himself entitled and the Contractor shall keep such particulars up-to-date by submitting such further particulars which may be necessary or may be requested from time to time by the Architect, so as to enable the Architect fully and properly to discharge his duties under this Clause 11.6 at the times specified for the discharge of the same.

Alternative 2

Grounds for extension of time

11.5   Subject to Clause 11.8, no extension of time shall be granted to the Contractor except in the case of:

(a)   *force majeure;*

(b)   loss or damage to the Works by fire, lightning, explosion, storm, tempest, flood, bursting or over-flowing of water tanks, apparatus or pipes, earthquake, aircraft and other aerial devices or articles dropped therefrom, riot and civil commotion;

(c)   war, hostilities (whether war be declared or not), invasion, act of foreign enemies, rebellion, revolution, insurrection, military or usurped power, civil war, riot, commotion or disorder;

(d)   delay or default by a governmental agency, local authority or statutory undertaker or private utility in carrying out work in pursuance of its statutory obligations in relation to the Works or the exercise after the date of this agreement of any statutory power which restricts the availability or use of labour, or prevents or delays the Contractor obtaining goods, materials, fuels or energy;

(e) any act, instruction, default or omission of the Employer, or of the Architect or of their servants or their agents on his behalf, whether authorised by or in breach of this Agreement;

(f) failure to comply with their duties under the CDM Regulations with all reasonable diligence by the Planning Supervisor or the Principal Contractor (in either case if not the Contractor), or a Designer (not being a Sub-Contractor or Supplier);

(g) discovery of antiquities. Any instruction issued to the Contractor under clause 14.2

(h) determination or discharge of any sub-contract of the kind referred to in Clause 9.7 as a result of any breach or repudiation of such sub-contract by the sub-contractor.

and only then to the extent that the Contractor shall prove to the satisfaction of the Architect that the taking-over of the Works or of any Section by the date or dates stated in the Time Schedule is prevented.

Provided that no account shall be taken of any of these circumstances (except in the case of the circumstances referred to in Clause 11.5 (e)) unless the Contractor:

immediately upon it becoming reasonably apparent that the Taking-Over of the Works or of any Section is being or is likely to be so prevented, shall have submitted to the Architect a notice specifying the circumstance or circumstances; and

as soon as possible thereafter, shall have submitted full and detailed particulars of the extension of time to which the Contractor may consider himself entitled; and

shall have kept such particulars up-to-date by submitting such further particulars which may be necessary or may be requested from time to time by the Architect;

so as to enable the Architect fully and properly to discharge his duties under Clause 11.6 at the times specified for the discharge of the same.

Extensions of time

11.6 Either:

(a) so soon as may be practicable, but in any event not later than 60 working days after receipt of all the particular referred to in Clause 11.5; or

(b) in the case of any act, instruction, default or omission of the Employer, or of the Architect on his behalf or their servants or their agents and where the Contractor has failed to provide such particulars, at any time,

the Architect shall grant to the Contractor such extension of time for the Taking-Over of the Works and/or of any Section as he then estimates to be fair and reasonable: Provided always that an extension of time for the Taking-Over of one Section shall not as a necessary consequence entitle the Contractor to an extension of time for the Taking-Over of any other Section and/or of the Works and provided further that the Contractor shall not be entitled to any extension of time in respect of any delay attributable to any negligence, default or improper conduct by him or by his sub-contractors or suppliers at any tier or his or their servants or agents. The Architect shall be entitled to take into account at any time before the Taking-Over of any Section of the Works, the effects of any omission from such Section or from the Works.

Review of extensions of time granted

11.7   Within a reasonable time after the Taking-Over of the Works, the Architect shall confirm the dates for the Taking-over of the Works or any Section previously stated, adjusted or fixed or may fix a date for the taking-Over of the Works or of any Section which is later than previously stated, adjusted or fixed, whether as a result of reviewing all or any previous decisions under Clause 11.6 given prior to the date stated in the certificate under Clause 11.2 or as a result of any act, instruction, default or omission of the Employer, or of the Architect or their servants or their agents on his behalf, whether authorised by or in breach of this Agreement, having occurred after the date stated in the certificate issued under clause 11.2. The Architect shall notify the Contractor of his final decision under this Clause 11.7.

Acceleration and postponement

11.8   The Architect may at any time, but not unreasonably, issue an instruction to the Contractor to bring forward or postpone the dates shown on the Time Schedule for the Taking-Over of the Works, any Section or any part of the Works and the Contractor shall immediately take such measures as are necessary to comply with such instruction and the provisions of Clause 11.3 shall apply to the adjusted date. The Architect shall ascertain and certify a fair and reasonable adjustment (if appropriate) to the Contract Sum in respect of compliance by the Contractor with such instructions and any damage, loss and/or expense suffered or incurred by the Contractor arising out of or in connection with it: Provided that if prior to giving any such instruction the Architect requires the Contractor to give an estimate of the adjustment to the Contract Sum, the provisions of Clause 17 (other than the provision relating to extensions of time therein contained) shall apply as if an instruction given under this Clause 11.8 were included in clause 17.1.

Revisions to the Time Schedule

11.9   If the Architect shall adjust the date for the Taking-Over of the Works or any Section under Clause 11.6 or shall issue an instruction under Clause 11.8, the Contractor shall submit to the Architect a time schedule revised to take account of such adjustment or instruction for the Architect's consent within 10 working days of the date of the Architect's notice or instruction or the Adjudicator's decision (as the case may be) and, upon the Architect giving his consent to the same, such revised time schedule shall be the Time Schedule.

15.4.2   Commentary

General

Clauses 11.1 and 11.2 call for little comment. Subject to any provision to the contrary, the employer must give the contractor possession of the site as set out in the time schedule. Breach of this requirement may give rise to a claim at common law or under clause 11.5 in either alternative. The contractor is then to commence and proceed with the works 'regularly and diligently and in accordance with the Time Schedule' until they are ready for taking-over. If the works (or a section of them) are not ready by the date shown in the time schedule (or as extended under clause 11 or by the adjudicator under clause 25.2) the architect is to certify to that effect in writing.

Liquidated damages

Clause 11.3, alternative 1, is a common-form provision for liquidated and ascertained damages. Where alternative 1 is used, the figure stated will be exhaustive of the employer's rights arising out of delay in completion, even if third-party claims are made against the employer[654]. Deduction of liquidated damages is subject to the condition precedent of the issue by the architect of a certificate that the works (or a section) are not fit and ready for taking-over. Subject to the issue of the certificate and of course to a written withholding notice in accordance with clause 16.6, the clause permits the employer to deduct liquidated damages at the rate or rates stated in the time schedule. The period for deduction of liquidated damages runs from the date of taking over stated in the time schedule, or any adjusted date, to the date of taking-over (clause 12) or under the provisions of clause 13. If the employer so requires, he may deduct liquidated damages from the amount that would otherwise be certified as payable to the contractor, or he may recover them from the contractor as an ordinary debt by legal action.

Unliquidated damages

Alternative 2 of clause 11.3 is a curious provision. It says that, subject to the issue of a clause 11.2 certificate, the employer is entitled to recover from the contractor such damage, loss and/or expense as he may suffer arising out of the contractor's failure to meet his obligations under clause 11.1. At first sight, it would appear that if the contractor is in breach of *any* of his obligations under clause 11.1, so that he is liable to the employer for damages, these are recoverable by deduction under this clause and without resort to arbitration. However, the side-note refers to 'damages for delay', and it is clearly the draftsman's intention that the employer's right to set off unliquidated damages is limited to the breach of late completion. How such damages are to be assessed is not stated; but they are the equivalent of unliquidated damages at common law, which are recoverable *on proof of actual loss*. What this version of clause 11.3 does, however, is to confer on the employer a right to deduct such damage, loss and/or expense from the amount which would otherwise be payable to the contractor in any certificate, i.e. it confers an express right of set-off on the employer who, presumably, will quantify the amount himself or through his professional advisers. The wording of the clause is very broad and the use of this optional provision appears to be contrary to the interests of contractors, who might be well advised to resist it. It represents a startling departure from the accepted

---

[654] *Temloc Ltd v. Errill Properties Ltd* (1987) 39 BLR 30.

procedures in the construction industry and it is an open invitation to adjudication or arbitration.

Adjustment of damages for delay

Clause 11.4 applies whichever of the foregoing alternative versions of clause 11.3 is used. If the architect grants extensions of time and adjusts the taking-over date to a later date (or the adjudicator does so), and damages have been deducted from sums due to the contractor, the employer is to repay any excess damages so deducted for the period between the original take-over date and the later date as adjusted. The previous edition of this contract contained an entitlement for the contractor to have interest added to the retention repaid. That provision has now been removed and, without such express provision, it is not thought the contractor would have any right to interest in these circumstances.

Grounds for extension of time

Clause 11.5 is given in alternative forms.
Alternative 1 In this form, clause 11.5 provides that an extension of time shall be granted to the contractor only where there is 'any act, instruction, default or omission of the Employer, his servants or his agents, or of the Architect on his behalf, whether authorised by or in breach of this Agreement'. Moreover, the contractor's entitlement to an extension of time on these grounds is qualified: the act etc. relied on must in the reasonable opinion of the architect prevent the taking-over of the works by the date stated in the time schedule. Obviously this is a limiting factor since the contractor must not only establish that there is a cause of delay, i.e. some act etc., but also that this is causing delay in the architect's reasonable opinion. However, the reasonableness or otherwise of the architect's opinion is open to review on adjudication or arbitration as appropriate and, indeed, in litigation. 'Prevented' means 'hindered or stopped'; it has no special meaning in this context.

This clause also contains provisions for notices by the contractor. Immediately it is reasonably apparent that the taking-over is being or is likely to be prevented by the act etc. of the employer or his architect, the contractor must serve written notice on the architect. The notice must specify the particular act, instruction, default or omission relied on. As soon as possible after that notice, the contractor must submit to the architect full and detailed particulars of the extension of time to which he thinks he is entitled and it is suggested that this is most usefully accomplished by reference to the time schedule and the contractor's own programme[655]. The contractor is under a further duty; he must keep the particulars up to date, by submitting 'such

---

[655] See the very useful comments of the court in *John Barker Construction Ltd* v. *London Portman Hotels Ltd* (1996) 12 Const LJ 277.

further particulars which may be necessary or may be requested from time to time by the Architect, so as to enable the Architect fully and properly to discharge his duties' regarding extension of time at the times specified by the contract.

Under this version of clause 11.5 the grounds on which extension of time can be claimed are, broadly speaking, all defaults of the employer or his agent, the architect. Normally, failure by the architect properly to exercise his duty to extend time will undoubtedly mean that time under the contract will be at large; in other words, the date fixed for completion will cease to be applicable and the contractor's obligation will then be to complete within a reasonable time[656]. This is so under this form also, save to the extent that clause 11.6(b) enables the architect to grant an extension at any time where the contractor has failed to provide the necessary information. In such cases, where no notice is given, he can delay his decision right up until his powers come to an end with the issue of the final certificate. The architect must act if the contractor fails to give the necessary particulars in order to preserve the employer's right to liquidated damages and prevent time becoming at large.

This version of clause 11.5 is limited in its extent since it does not cover any events outside the control of the employer, the architect or the contractor, and it has been criticised as being much narrower than is felt to be reasonable. Under it, no extensions of time are available for such things as exceptionally adverse weather conditions, strikes and delays or unforeseeable shortages in the availability of labour or materials, etc. These eventualities are beyond the control of the contractor but no doubt contractors invited to tender under ACA terms with this version of clause 11.5 in force will adjust their tender price accordingly. Contracts, after all, are about the distribution of risk and provided that the risk is known, it can usually be priced.

It used to be possible to contend that the clause did not cover default etc. by anyone (other than the architect) for whom the employer is vicariously responsible in law. However, that possible defect has now been corrected.
Alternative 2   The second version of clause 11.5 is more traditional in its drafting. It is wider than the first alternative and is akin to the JCT provisions for extension of time.

There are eight matters for which the contractor is entitled to claim an extension of time and, as in the case of alternative 1, the architect is expected to adjudicate upon some matters which may be his own fault. The eight causes or matters giving rise to an extension of time are:

(1)   *Force majeure.*
(2)   The list of events set out in the insurance clause (clause 6.4), that is, those contingencies 'covered by the Contractor's policy' referred to (or by the employer's policy, as appropriate).
(3)   War, hostilities (whether war be declared or not), invasion, act of foreign enemies, rebellion, revolution, insurrection, military or usurped power, civil war, riot, commotion or disorder.

---

[656] *Percy Bilton Ltd* v. *Greater London Council* (1982) 20 BLR 1.

(4)   Delay or default by governmental agency, local authority or statutory undertaker in carrying out work in pursuance of its statutory obligations in relation to the works, i.e. there is no extension of time where a statutory undertaker, such as an electricity supplier, is carrying out work pursuant to contract[657]. Delay that is caused by a statutory undertaker carrying out work in pursuance of its obligations in relation to the works will come under this heading, but not otherwise.

(5)   Any act, instruction, default or omission of the employer or of the architect on his behalf, whether authorised by or in breach of the contract. Thus, failure to give possession of the site (which is a breach of contract at common law) is within this provision.

(6)   Failure of the planning supervisor, the principal contractor (if not the contractor) or a designer to comply with their duties under the CDM Regulations. A sub-contractor or supplier carrying out design functions is expressly excluded.

(7)   Any instruction issued following the discovery of antiquities.

(8)   Determination of a named sub-contract as a result of its breach or repudiation by the sub-contractor. This is a valuation right for the contractor which is not applicable to ordinary domestic sub-contracts.

From the contractor's point of view, this alternative is to be preferred to alternative 1. As regards *force majeure,* it must be some matter not within the control of the person relying on it and it is certainly wider than the English law term 'Act of God'. It has been said that:

> 'This term is used with reference to all circumstances independent of the will of man, and which it is not in his power to control...Thus, war, inundations and epidemics are cases of *force majeure*; it has even been decided that a strike of workmen constitutes a case of *force majeure*.[658]

It seems that any direct legislative or administrative interference would, of course, come within the term: for example, an embargo.

Force majeure in the context of ACA contracts has a rather more restricted meaning than might seem to be the case at first sight because many matters – such as war, governmental delays, etc., which are otherwise dealt with in the term – are covered expressly by the provisions of this clause. It seems that severe weather conditions are not within the term *force majeure* although major strikes probably are[659].

As in the former alternative, there is an important limitation; the onus of proof is on the contractor. He is stated to be entitled to an extension of time on the foregoing grounds (or any of them) only 'to the extent that [he] shall prove to the satisfaction of the Architect that the taking-over of the Works...by the date...stated in the Time Schedule is prevented'.

---

[657] *Henry Boot Construction Ltd* v. *Central Lancashire New Town Development Corporation* (1980) 15 BLR 8.
[658] *Lebeaupin* v. *Crispin* [1920] 2 KB 714, in which McCardie J quoted with approval the definition of the French writer Goirand.
[659] *Matsoukis* v. *Priestman & Co* [1915] 1 KB 681, where it was held to include the general coal strike and the breakdown of machinery.

The change in wording is noteworthy: 'to the satisfaction of the Architect', and not in his 'reasonable opinion', and the standard of proof required is that of the balance of probabilities. On the face of it, it is arguable that where the second alternative is used the architect's satisfaction need not be reasonable.

The proviso is important and governs all those events listed above except clause 11.5(e), i.e. it does not apply where the delay is caused by acts or omissions of the employer or his architect acting on his behalf. (The reason for this is clear: without this provision the employer would be liable to forfeit his right to liquidated damages if the contractor failed to give notice[660].) Except in that special case, no account is to be taken of any of the other listed circumstances unless the contractor shall have given written notice to the architect as soon as it becomes reasonably apparent that the taking-over is being or going to be prevented. *No notice is required where the delaying cause falls within clause 11.5(e).* For the other causes, the notice must be given 'immediately upon it becoming apparent that the Taking-Over ... is being or is likely to be so prevented'; and the notice must be specific. It must specify the circumstance or circumstances relied on and must be followed up as soon as possible by 'full and detailed particulars of the extension of time to which the Contractor may consider himself entitled'. The contractor must keep the particulars up to date as necessary and also supply, on request, any further particulars that the architect may require to enable him fully and properly to discharge his duties under this clause. In seven cases, therefore, the giving of notice etc. appears to be a condition precedent to the contractor's entitlement to extension of time during the running of the contract. Only in the case of delay occasioned by any act, default or omission of the employer, or of the architect on his behalf, is notice not essential. Therefore, the contractor would be well advised to submit early notices as a matter of course and good practice.

There are two situations to be considered as to when the architect must exercise his duty to grant an extension of time. The first and normal case is dealt with in clause 11.6(a). When the architect has received the contractor's written notice and he has received all the particulars that he requires to enable him to form an opinion, he must make a decision about an extension of time 'so soon as may be practicable, but in any event not later than 60 working days after *receipt*' of the contractor's particulars. The wording is unhappy; but it appears that the period of 60 working days does not begin to run until the receipt of the later particulars referred to in clause 11.5, i.e. those submitted on the architect's request, or which are necessary to update the position.

As regards the second case, where act, instruction, omission, etc is concerned and the contractor has failed to provide the particulars referred in clause 11.5, it is significant that the reference is only to 'particulars' and not to 'notice'. It is suggested that notice should be given in any event, but the contractor's failure to submit the required particulars is not fatal to his claim

---

[660] *Dodd v. Churton* [1897] 1 QB 562.

for an extension. In this case, the architect may exercise his duties as regards extending the contract period at any time, i.e. so long as he is not *functus officio*, which is the case after the issue of the final certificate under the contract.

The architect's duty is to grant to the contractor 'such extension of time ... as he *then* estimates to be fair and reasonable'. The first proviso is of great importance. Where there is a sectional completion, an extension of time for the taking-over of one section does not necessarily entitle the contractor to an extension for the taking over of another section or of the works as a whole.

There is a common misconception that an extension of time for one section has a knock-on effect on other sections. That is particularly the case where the employer has unwisely specified specific dates for possession of subsequent sections which are actually dependent on the first section. In order to achieve a practical outcome, dependent sections should have their dates for possession expressed as related to the relevant take-over dates.

There is a second proviso to the effect that the contractor cannot rely on his own breach to claim an extension of time. That would be the case in any event under the general law, but it is useful to have the position set out clearly to put the matter beyond doubt.

The contractor is not entitled to any extension of time for delays caused by sub-contractors or suppliers (named or domestic) even to the limited extent laid down by the House of Lords in *Westminster Corporation* v. *J. Jarvis & Sons Ltd*[661], as regards nominated sub-contractors under the JCT 63 form. It is doubtful whether the words 'his or their respective servants or agents' would extend to cover sub-sub-contractors and a very limited meaning must be given to these words[662]. However, delays caused by sub-sub-contractors might equally well not rank for an extension of time unless they could be proved to fall within the meaning of the term *force majeure* (clause 11.5(a)), which is unlikely, or alternatively be shown to fall within clause 11.5(e).

The final sentence of clause 11.6 establishes that the architect is entitled to take into account any omission instructions which he has issued. An omission instruction may, of course, and usually will if issued early enough, effect a saving of time; and he can do this 'at any time before the Taking-Over of any Section or of the Works'. It is the *effect* on time of the omission instruction that is important.

Clause 11.7 – with the side-note 'Review of extensions of time granted' – imposes on the architect a mandatory duty to 'confirm the dates for the Taking-Over of the Works or any Section previously stated, adjusted or fixed ...'. He must exercise this duty 'within a reasonable time after the taking over of the Works' as a whole; what is a reasonable time is a question of fact, but clearly the power must be exercised before the architect is *functus*

---

[661] (1970) 7 BLR 64.
[662] *City of Manchester* v. *Fram Gerrard Ltd* (1974) 6 BLR 70.

officio. Alternatively, the architect may fix a later date than previously fixed, whether as a result of reviewing any clause 11.6 decisions or because there has been some act, instruction, default or omission of the employer or the architect on his behalf which has 'occurred after the date stated in the' clause 11.2 certificate. He must then notify the contractor of his final decision. This provision differs from the analogous provision under JCT 98 in so far as under it the architect cannot reduce an extension of time already granted; he may only 'fix a date...which is later than previously stated, adjusted or fixed'; but he has powers under the next sub-clause to order acceleration.

Clause 11.8 is an unusual provision to find in a construction contract, although its rationale is easy to understand. Generally, under other forms, acceleration is only possible outside the terms of the contract by agreement between the employer and the contractor. Under clause 11.8 the architect may, at any time, issue an instruction to bring forward or postpone dates shown on the time schedule for the taking-over of the works or any section or part. Although it does not so state and, therefore, the contractor is not concerned about it, it is implied, if not express, in the architect's terms of engagement that he must obtain the employer's agreement before giving such instruction. The architect must not issue the instruction unreasonably. Therefore, any instructed acceleration must not only be possible, it must be a practical proposition. If the architect issues such an acceleration or postponement instruction, the contractor must comply with it immediately; and the clause 11.3 provision as to liquidated or unliquidated damages applies as appropriate from the adjusted date.

Any acceleration or postponement is *prima facie* at the employer's cost. The architect must ascertain and certify a fair and reasonable adjustment to the contract sum in respect of the contractor's compliance with such an instruction, along with any damage, loss and/or expense which the contractor suffers or incurs as a direct result of the instruction. In theory, at any rate, the adjustment may be up or down. In fact, the proviso states that the architect may require the contractor to give an estimate of the cost of complying with the instruction before he issues it, and if he exercises this option, the relevant provisions of clause 17 apply, i.e. contractor's written estimate of cost of compliance, agreement or negotiation, etc.

Clause 11.9 deals with revisions to the time schedule where the architect adjusts the completion date by granting an extension of time under clause 11.6 or issues an instruction for postponement or acceleration under clause 11.8 or where the adjudicator adjusts the date under clause 25.2. The contractor must then submit a revised time schedule to the architect within 10 working days of the architect's notice or instruction or the adjudicator's decision for the architect's consent. The revised time schedule then takes the place of the original time schedule. This is a sensible and workmanlike provision. It does not state what is to happen if the architect cannot consent for some reason. The contract appears to rely upon the common sense of the parties to come to an agreement.

15.5 *Claims on the final certificate*

If the contractor fails to comply with the procedural provisions of clauses 7.2 and 7.3 then he loses his entitlement to settlement of his claim until the final certificate (clause 7.5).

The issue of the final certificate is provided for by clause 19.2 and is purely financial in its effect. It is to be issued by the architect within 30 working days after completion by the contractor of his obligations under the agreement.

Before the final certificate is issued, the contractor must submit to the architect within 30 working days of the expiry of the maintenance period his final account for the works, and this will include any outstanding claims. The final account must be fully documented and, in fact, the wording is 'all documents, vouchers and receipts as shall be necessary for computing the Final Contract Sum or as may be required by the Architect'. Any outstanding claim must be supported by the necessary vouchers etc. and what is required is a claim which is properly calculated and backed up by such supporting evidence as the architect may require. No interest or finance charges element is allowable in respect of the period between the date of the contractor's failure to submit the clause 7.2 and 7.3 notice and estimate and the date of the final certificate.

If the contractor is not satisfied with the amount stated in the final certificate, he may refer the matter to adjudication for a rapid decision or he may choose whichever of the arbitration or litigation options applies. The parties can agree to use clause 25A and attempt to settle the matter by conciliation. However, the agreement of both parties is required for this route.

Chapter 16

Claims under NEC

16.1 Introduction

The first edition of the New Engineering Contract was published in 1993. Although from its title, one might be excused for believing that it was not suitable for use for building works, its authors maintained that it could be used for either engineering or building. The second edition was published in 1995 and its title had been changed to the Engineering and Construction Contract although it is still known by the initials NEC. Amendments have been issued since then to deal with the Housing Grants, Construction and Regeneration Act 1996. The philosophy of this form is intended to be different from that of the more common JCT and other contracts.

Although the form was praised by Sir Michael Latham as containing the kind of provisions advocated in his report *Constructing the Team*[663], it has been the subject of much criticism by legal commentators[664]. Some perceived shortcomings are that the syntax and grammar are not good. This is partly because the present tense is used almost exclusively even where one would normally expect to the see past or future tense. The form is supposed to be a model of simple English, but sticking almost exclusively to one tense does not assist comprehension and certainty is a prime requisite for a legally binding contract. It is difficult to know whether something is being expressed as a power or a duty or just as a fact. There are other difficulties. Clause numbering is quite strange; for example, clause 21.5 is a sub-clause of clause 2, not clause 21, 93.2 is a sub-clause of 9, etc. One would have expected to see reference to clauses 2.1.5 and 9.3.2 respectively. The most confusing, for some reason, is reference to sub-clauses of clause 1 – 11.1, 11.2, 14.3, etc. Defined terms have capitals and terms identified in the contract data (something like the JCT Appendix) are in italics. It is difficult to get used to this, particularly in a phrase such as: 'The Completion Date is the *completion date* unless . . .'. Perhaps the strongest criticism is that the form appears to go out of its way to avoid using words and phrases which are in common use in other construction contracts. The effect of that is that it becomes very difficult to interpret the meaning of such words and phrases by reference to decisions in the courts and other legal authorities. For example, are 'delay damages' the same as 'liquidated damages'? Interpret-

---

[663] HMSO July 1994.
[664] D. G. Valentine, *The New Engineering Contract* (1996) 12 Const LJ 305 is well worth reading.

ation of this contract is, therefore, to some extent a venture into the unknown.

Guidance notes are published. Although such notes are certainly useful in
operating the contract, it must be remembered always that they have no
legal significance. In law, what the draftsman intended to mean when the
contract was drafted is irrelevant. It is the meaning in the words of the
contract which is important. Obviously, a court has to interpret a contract in
line with the intentions of the parties, but it is the intentions of the parties as
revealed by the words of the contract, and not some extraneous document
such as guidance notes, which matters.

The form does have its good points. It has a set of nine core clauses which
are intended to be present in every version of the contract. There are then six
options (A, B, C, D, E and F), one of which must be included in every
contract. The purpose of these is to convert the contract for use with
different procurement systems:

A – priced contract with activity schedule
B – priced contract with bill of quantities
C – target contract with activity schedule
D – target contract with bill of quantities
E – cost reimbursable contracts
F – management contract.

In addition, there is a selection of clauses which may, but need not, be
included in the contract. These cover such matters as performance bonds,
advance payment to the contractor, sectional completion, delay damages.

Before turning to a consideration of the relevant clauses, it is worth noting
that there is no architect or contract administrator mentioned in the contract.
The parties to the contract are the employer and the contractor, but a key
player is the project manager whose function it is to administer the contract
by issuing instructions, certifying payment, assessing compensation events
and the like. However, there is also the supervisor, who broadly deals with
standards and quality of the work.

16.2 Compensation events

16.2 .1 Clause 6

The text of this clause, together with clauses imported by the various
options, is as follows:

6   Compensation events

60
60.1   The following are compensation events.

(1) The *Project Manager* gives an instruction changing the Works Information except

- a change made in order to accept a Defect or
- a change to the Works Information provided by the *Contractor* for his design which is made at his request or to comply with other Works Information provided by the *Employer*.

(2)   The *Employer* does not give possession of a part of the Site by the later of its *possession date* and the date required by the Accepted Programme.

(3)   The *Employer* does not provide something which he is to provide by the date for providing it required by the Accepted Programme.

(4)   The *Project Manager* gives an instruction to stop or not to start any work.

(5)   The *Employer* or Others do not work within the times shown on the Accepted Programme or do not work within the conditions stated in the Works Information.

(6)   The *Project Manager* or the *Supervisor* does not reply to a communication from the *Contractor* within the period required by this contract.

(7)   The *Project Manager* gives an instruction for dealing with an object of value or of historical or other interest found within the Site.

(8)   The *Project Manager* or the *Supervisor* changes a decision which he has previously communicated to the *Contractor*.

(9)   The *Project Manager* withholds an acceptance (other than acceptance of a quotation for acceleration or for not correcting a defect).

(10) The *Supervisor* instructs the *Contractor* to search and no Defect is found unless the search is needed only because the *Contractor* gave insufficient notice of doing work obstructing a required test or inspection.

(11) A test or inspection done by the *Supervisor* causes unnecessary delay.

(12) The *Contractor* encounters physical conditions which

- are within the Site,
- are not weather conditions
- an experienced contractor would have judged at the Completion Date to have such a small chance of occurring that it would have been unreasonable for him to have allowed for them.

(13) A *weather measurement* is recorded

- within a calendar month,
- before the Completion Date for the whole of the *works* and
- at the place stated in the Contract Data

the value of which, by comparison with the *weather data*, is shown to occur on average less frequently than once in ten years.

(14) An *Employer*'s risk event occurs.

(15) The *Project Manager* certifies takeover of a part of the *works* before both Completion and the Completion Date.

(16) The *Employer* does not provide materials, facilities and samples for tests as stated in the Works Information.

(17) The *Project Manager* notifies a correction to an assumption about the nature of a compensation event.

(18) A breach of contract by the *Employer* which is not one of the other compensation events in this contract.

60.2   In judging the physical conditions, the *Contractor* is assumed to have taken into account

- the Site Information,
- publicly available information referred to in the Site Information,
- information obtainable from a visual inspection of the Site and
- other information which an experienced contractor could reasonably be expected to have or to obtain.

60.3   If there is an inconsistency within the Site Information (including the information referred to in it), the *Contractor* is assumed to have taken into account the physical conditions more favourable to doing the work.

60.4   A difference between the final total quantity of work done and the quantity stated for an item in the *bill of quantities* at the Contract Date is a compensation event if
- the difference causes the Actual Cost per unit of quantity to change and
- the rate in the *bill of quantities* for the item at the Contract Date multiplied by the final total quantity of work done is more than 0.1% of the total of the Prices at the Contract Date.

If the Actual Cost per unit of quantity is reduced, the affected rate is reduced. (OPTIONS B & D)

60.5   A difference between the final total quantity of work done and the quantity for an item stated in the *bill of quantities* at the Contract Date which delays the Completion is a compensation event. (OPTIONS B & D)

60.6   The *Project Manager* corrects mistakes in the *bill of quantities* which are departures from the *method of measurement* or are due to ambiguities or inconsistencies. Each such correction is a compensation event which may lead to reduced Prices. (OPTIONS B & D)

60.7   Suspension of performance is a compensation event if the *Contractor* exercises his right to suspend performance under the Act.

61   Notifying compensation events
61.1   For compensation events which arise from the *Project Manager* or the *Supervisor* giving an instruction or changing an earlier decision, the *Project Manager* notifies the *Contractor* of the compensation event at the time of the event. He also instructs the *Contractor* to submit quotations, unless the event arises from a fault of the *Contractor* or quotations have already been submitted. The *Contractor* puts the instruction or changed decision into effect.

61.2   The *Project Manager* may instruct the *Contractor* to submit quotations for a proposed instruction or a proposed changed decision. The *Contractor* does not put a proposed instruction or a proposed changed decision into effect.

61.3   The *Contractor* notifies an event which has happened or which he expects to happen to the *Project Manager* as a compensation event if

- the *Contractor* believes that the event is a compensation event,
- it is less than two weeks since he became aware of the event and
- the *Project Manager* has not notified the event to the *Contractor*.

61.4   The Prices and the Completion Date are not changed if the *Project Manager* decides that an event notified by the *Contractor*

- arises from a fault of the *Contractor*,
- has not happened and is not expected to happen,

- has no effect upon Actual Cost or Completion or is not one of the compensation events stated in this contract.

If the *Project Manager* decides otherwise, he instructs the *Contractor* to submit quotations for the events. Within either

- one week of the *Contractor*'s notification or
- a longer period to which the *Contractor* has agreed

the *Project Manager* notifies his decision to the *Contractor* or instructs him to submit quotations.

61.5   If the *Project Manager* decides that the *Contractor* did not give an early warning of the event which an experienced contractor would have given, he notifies this decision to the *Contractor* when he instructs him to submit quotations.

61.6   If the *Project Manager* decides that the effects of a compensation event are too uncertain to be forecast reasonably, he states assumptions about the event in his instruction to the *Contractor* to submit quotations. Assessment of the event is based on these assumptions. If any of them is later found to have been wrong, the *Project Manager* notifies a correction.

61.7   A compensation event is not notified after the *defects date*.

62   Quotations for compensation events

62.1   The *Project Manager* may instruct the *Contractor* to submit alternative quotations based upon different ways of dealing with the compensation event which are practicable. The *Contractor* submits the required quotations to the *Project Manager* and may submit quotations for other methods of dealing with the compensation event which he considers practicable.

62.2   Quotations for compensation events comprise proposed changes to the Prices and any delay to the Completion Date assessed by the *Contractor*. The *Contractor* submits details of his assessment with each quotation. If the programme for remaining work is affected by the compensation event, the *Contractor* includes a revised programme in his quotation showing the effect.

62.3   The *Contractor* submits quotations within three weeks of being instructed to do so by the *Project Manager*. The *Project Manager* replies within two weeks of the submission. His reply is

- an instruction to submit a revised quotation,
- an acceptance of a quotation,
- a notification that a proposed instruction or a proposed changed decision will not be given or
- a notification that he will be making his own assessment.

62.4   The *Project Manager* instructs the *Contractor* to submit a revised quotation only after explaining his reasons for doing so to the *Contractor*. The *Contractor* submits the revised quotation within three weeks of being instructed to do so.

62.5   The *Project Manager* extends the time allowed for

- the *Contractor* to submit quotations for a compensation event and
- the *Project Manager* to reply to a quotation

if the *Project Manager* and the *Contractor* agree to the extension before the submission or reply is due. The *Project Manager* notifies the extension that has been agreed to the *Contractor*.

63   Assessing compensation events

63.1   The changes to the Prices are assessed as the effect of the compensation event upon

- the Actual Cost of the work already done,
- the forecast Actual Cost of the work not yet done and
- the resulting Fee.

63.2   If the effect of a compensation event is to reduce the total Actual Cost, the Prices are not reduced except as stated in this contract. If the effect of a compensation event is to reduce the total Actual Cost and the event is

- a change to the Works Information or
- a correction of an assumption stated by the *Project Manager* for assessing an earlier compensation event,

the Prices are reduced.

63.3   A delay to the Completion Date is assessed as the length of time that, due to the compensation event, planned Completion is later than planned Completion as shown on the Accepted Programme.

63.4   If the *Project Manager* has notified the *Contractor* of his decision that the *Contractor* did not give an early warning of a compensation event which an experienced contractor could have given, the event is assessed as if the *Contractor* had given early warning.

63.5   Assessment of the effect of a compensation event includes cost and time risk allowances for matters which have a significant chance of occurring and are at the *Contractor*'s risk under this contract.

63.6   Assessments are based upon the assumptions that the *Contractor* reacts competently and promptly to the compensation event, that the additional Actual Cost and time due to the event are reasonably incurred and that the Accepted Programme can be changed.

63.7   A compensation event which is an instruction to change the Works Information in order to resolve an ambiguity or inconsistency is assessed as follows. If Works Information provided by the Employer is changed, the effect of the compensation event is assessed as if the Prices and the Completion Date were for the interpretation most favourable to the Contractor. If Works Information provided by the Contractor is changed, the effect of the compensation event is assessed as if the Prices and the Completion Date were for the interpretation most favourable to the Employer.

63.8   Assessments for changed Prices for compensation events are in the form of changes to the *activity schedule*. (OPTIONS A & C)

63.9   Assessments for changed Prices for compensation events are in the form of changes to the *bill of quantities*. If the *Project Manager* and the *Contractor* agree, rates and lump sums in the bill of quantities may be used as a basis for assessment instead of Actual Cost and the resulting Fee. (OPTIONS B & D)

63.10 The assessment of a compensation event which is or includes subcontracted work has the *Contractor's fee percentage* added to Actual Cost but fees paid or to be paid by the *Contractor* to the Subcontractors are not added. (OPTIONS A & B)

63.11 If the *Project Manager* and the *Contractor* agree, the *Contractor* assesses a compensation event using the Shorter Schedule of Cost Components. The *Project Manager* may make his own assessments using the Shorter Schedule of Cost Components. (OPTIONS A, B, C, D & E)

64   The Project Manager's assessments

64.1   The Project Manager assesses a compensation event

- if the *Contractor* has not submitted a required quotation and details of his assessment within the time allowed,
- if the *Project Manager* decides the *Contractor* has not assessed the compensation event correctly in a quotation and he does not instruct the *Contractor* to submit a revised quotation,
- if, when the *Contractor* submits a quotation for a compensation event, he has not submitted a programme which this contract requires him to submit or
- if when the *Contractor* submits quotations for a compensation event the *Project Manager* has not accepted the *Contractor*'s latest programme for one of the reasons stated in this contract.

64.2   The *Project Manager* assesses a compensation event using his own assessment of the programme for the remaining work if

- there is no Accepted Programme or
- the *Contractor* has not submitted a revised programme for acceptance as required by this contract.

64.3   The *Project Manager* notifies the *Contractor* of his assessment of a compensation event and gives him details of it within the period allowed for the *Contractor*'s submission of his quotation for the same event. This period starts when the need for the *Project Manager's* assessment becomes apparent.

65   Implementing compensation events

65.1   The Project Manager implements each compensation event by notifying the Contractor of the quotation which he has accepted or of his own assessment. He implements the compensation event when he accepts a quotation or completes his own assessment or when the compensation event occurs, whichever is latest.

65.2   The assessment of a compensation event is not revised if a forecast upon which it is based is shown by later recorded information to have been wrong.

65.3   The *Project Manager* includes the changes to the forecast amount of the Prices and the Completion Date in his notification to the *Contractor* implementing a compensation event. (OPTIONS E & F)

65.4   The *Project Manager* includes the changes to the Prices and the Completion Date from the quotation which he has accepted or from his own assessment in his notification implementing a compensation event. (OPTIONS B, C & D)

65.5 The *Contractor* does not implement a subcontract compensation event until it has been agreed by the *Project Manager*. (OPTIONS E & F)

Other compensation events:

J1.2 The advanced payment is made either within four weeks of the Contract Date or, if an advanced payment bond is required, within four weeks of the later of

- the Contract Date and
- the date when the *Employer* receives the advanced payment bond.

The advanced payment bond is issued by a bank or insurer which the *Project Manager* has accepted. A reason for not accepting the proposed bank or insurer is that its commercial position is not strong enough to carry the bond. The bond is for the amount of the advanced payment and in the form set out in the Works information. Delay in making the advanced payment is a compensation event (SECONDARY OPTION J).

T1.1 A change in the law of the country in which the Site is located is a compensation event if it occurs after the Contract Date. The *Project Manager* may notify the Contractor of a compensation event for a change in the law and instruct him to submit quotations. If the effect of a compensation event which is a change in the law is to reduce the total Actual Cost, the prices are reduced (SECONDARY OPTION T).

U1.1 A delay to the work or additional or changed work caused by the application of The Construction (Design and Management) Regulations 1994 is a compensation event if an experienced contractor could not reasonably be expected to have foreseen it (SECONDARY OPTION U).

16.2.2 Commentary

Put simply, compensation events are events which entitle the contractor to be compensated in terms of time and/or adjustment of the prices of the activities in the activity schedule. The idea is to deal with extensions of time, loss and/or expense and the valuation of variations at the same time. Clause 6 is very complex. The clause lists the events and sets out the procedure. The adjustment of the prices varies dependent on which of the main options is incorporated into the contract.

16.2.3 The compensation events

Project manager's instructions changing works information: clause 60.1(1)

Works information is defined in clause 11.2(5) as information specifying the works or stating constraints on carrying out the works. Essentially, it appears that it will comprise drawings and specifications and similar restrictions to those possible under JCT 98 clause 13.1.2. Where the contractor is to carry out some of the design, he will provide that part of the works information. The project manager may issue instructions to change the works information under clause 14.3. A change for any reason except:

- if it is made under clause 44.2 which is merely accepting a defect to avoid correcting it; or
- if it is made to the works information provided by the contractor at his request or in order to make it comply with the employer's works information

will fall under this ground. The reasons for the exceptions are so that the contractor benefits neither from his own defects, nor from his failure to make his design suit the employer's works information, nor from his decision to change his design later.

The employer's failure to give possession by whichever is the later of the possession date and the date in the accepted programme: clause 60.1(2)

The accepted programme is the programme identified in the contract data unless the project manager has accepted a later programme. At any time, the accepted programme is the latest programme accepted by the project manager. The programme is dealt with in clause 31 and revisions to the programme under clause 32. This is a perfectly straightforward provision. The possession date is to be inserted by the employer into the contract data. There is provision for possession to be given in parts. In normal circumstances, that is the date on which the contractor must have possession of the site or the part indicated and the contract data would also show that a programme had been produced which reflected that date. However, if the contractor prepares a programme, accepted by the project manager, which indicates a later date for possession, it is the later date which applies. Every time the contractor produces another programme with an adjusted date which is accepted by the project manager, the adjusted date applies. It is clear from clause 31.2 that the contractor cannot submit a programme showing a date earlier than the date in the contract data.

The extent of the contractor's possession is not exclusive possession as under JCT contracts. That is clear from clause 25.1 which states that he 'shares' (presumably meaning that he *must* share) the working areas with others. Curiously, clause 33.2 refers to the employer giving the contractor 'access to and use of' the site while the contractor has possession of it. In the ordinary meaning of possession, the contractor would certainly have access, use and control of the site in order to enable him to carry out the contract works. The employer's obligation to give possession of the site is governed by clause 33.1[665]. Failure to give possession on the due date is a serious

---

[665] For a full consideration of the position regarding possession of the site by the contractor, see Chapter 4, section 4.6.

breach of contract. Depending on the circumstances, it may amount to repudiation on the part of the employer.

The employer's failure to provide something by the date in the accepted programme: clause 60.1(3)

Clause 31.2 contains what appears to be a mandatory list of the contents of any programme submitted for acceptance. It is impossible to be definite about the matter, due to the curious use of the present tense mentioned earlier. In such circumstances, the normal rules for construing a contract should be employed. It is noted that the word 'may' is used in some clauses. That clearly denotes a power, but not an obligation to do something. In other words, where 'may' is used, it means that the action can be taken if desired. On the basis that the draftsman of the contract proceeded in a logical manner, the absence of the word 'may' must mean something different. It seems logical that it means that the particular action is obligatory. In other words, it is a duty.

The list seems to be fairly comprehensive regarding what the contractor *must* include in his programme, but it does not include reference to something being provided by the employer or the date for so providing. However, that does not preclude the contractor from incorporating other things. The conclusion is that this ground refers to the situation where it has been agreed between the parties that the employer will provide something to the contractor, perhaps paint, bricks or other building material which the employer can obtain cheaply. Having come to an agreement, the contractor has quite reasonably included it and the date by which he must have it if the works are not to be delayed, in his programme for acceptance by the project manager.

It is noteworthy that, under clause 31.3, the project manager has very limited scope to refuse to accept a programme. If the contractor has inserted the wrong date for requiring the information, the only grounds for rejection appear to be that either the contractor's plans are not practicable or it does not represent the contractor's plans realistically.

The project manager's instruction stopping work or preventing it from starting: clause 60.1(4)

This appears to be the equivalent to a postponement instruction under a JCT contract.

The failure of the employer or others to work within the times on the accepted programme or the conditions stated in the Works information: clause 60.1(5)

'Others' are defined in clause 11.2(2) as people or organisations not being the employer, the project manager, the supervisor, the adjudicator, the contractor or any employee, sub-contractor or supplier of the contractor.

Therefore, such people can be anyone else. The limiting factor is that they must be referred to in the accepted programme. Express mention of such reference is made in clause 31.2 (sixth bullet point).

   This is a sensible provision, but the employer, having agreed appropriate dates with the contractor, would be wise to keep careful watch on a future revised programme to ensure that the date is not changed to something which the employer cannot easily achieve. The project manager also has a duty to discuss with the employer any change to the original date. The project manager can probably reject a programme showing a date which the employer cannot meet as 'not practicable' (clause 31.2).

The project manager or the supervisor fails to reply to the contractor within the period required in this contract: clause 60.1(6)

This is the sanction if the project manager or the supervisor fails to comply with the provisions of clause 13.3. This requires them, and the contractor, to reply to communications within the 'period for reply' stated in the contract data. The NEC is full of good ideas, but this is not one of them. The period during which a reply must be made depends on all the circumstances. For example, some communications do not need a reply at all, some need an immediate reply, while others can safely be left some weeks before a response is necessary. It is also notable that although this is the sanction if the project manager or the supervisor fails to respond by the due time, there is no express sanction if the contractor fails. In any event, subsequent clauses regulate the effects so that the compensation event is only relevant if the failure has some detrimental effect on the other party.

The project manager gives instructions about an object of value, or historical or other interest found within the site: clause 60.1(7)

This is similar to the JCT provisions regarding what are termed 'antiquities' in those contracts. However, it is clear that the NEC definition in clause 73 is very much broader than under JCT contracts, and embraces not only items of historic interest but also anything valuable and anything which could objectively be classified as interesting. The only constraint seems to be that it must be an 'object'. That is to say, it must be something separate from the site itself. Therefore, curious or even significant rock strata would not be covered by this clause, but pieces of jewellery, old weapons and the foundation of a Roman villa would be covered, as would a piece of modern sculpture.

The project manager changes a decision he has previously communicated to the contractor: clause 60.1(8)

Contrary to the position regarding 'possession', nothing in this contract suggests that the word 'decision' has any special, broad or restricted meaning and its ordinary meaning can be assumed. Therefore, this ground applies whenever the project manager gives the contractor a decision and then subsequently changes it.

The project manager withholds an acceptance for a reason not stated in the contract: clause 60.1(9)

The exception to this is if the acceptance in question is in regard to a quotation for acceleration or for not correcting a defect. Clause 13.8 gives the project manager the power ('may withhold') to withhold acceptance of any submission by the contractor. There are two things to note about this power.

The first is that it must be implied that the submission concerns something which the contractor is entitled to submit. Examples are to be found throughout the contract and its options together with contractually valid reasons for non-acceptance by the project manager.

The second is that although the project manager has a very broad power, the exercise of it may result in a compensation event if the reason for acceptance does not fall within one of the reasons stated in the contract. Therefore, whenever the contractor makes a proper submission under the contract, it is imperative that the project manager carefully considers the contractually valid reasons for rejection before coming to a decision. Obviously, there may be instances when the project manager does not wish to accept a programme, particulars of a design or a quotation, etc. for reasons not included in the contract. The potential effect of the resultant compensation event requires a careful weighing of the balance between the benefits to the employer compared to the likely cost in terms of time and money.

No defect is found after the supervisor instructs the contractor to search: clause 60.1(10)

There is a proviso that excludes the situation where the contractor gave insufficient notice before doing work which obstructed a required test or inspection. Such notice is required under the provisions of clause 40.3.

Clause 4 deals with testing and defects. The works information may require some tests and inspections to be carried out, but this ground deals with the situation under clause 42.1 where the supervisor may instruct the contractor to search, provided that he gives reasons. Searching may include what is commonly understood by the expression 'opening up and testing' under JCT 98 clause 8.3, but not the kind of opening up and testing following the finding of defective work as set out in JCT 98 clause 8.4.4. In a similar way to that employed under JCT 98, this ground protects the contractor, if the search reveals that there is no defect, by providing the contractor with a right to additional time and/or money.

A test or inspection done by the supervisor causes unnecessary delay: clause 60.1(11)

This ground is a reference to clause 40.5 which requires the supervisor to carry out his tests and inspections without causing unnecessary delay. 'Unnecessary delay' is a difficult concept. It presupposes that some delay is necessary. The difficulty with that is the absence of any criteria which can be used to separate necessary from unnecessary delay. A tentative view can be advanced that the ground excludes any delay which one might expect to be caused as a result of the opening up or testing. That would be 'necessary' delay in the sense of being 'indispensable' or 'inevitably resulting from the nature of things'[666]. On that reading of the clause, 'unnecessary delay' probably refers to delays which are not inevitable, but which are nonetheless caused by the opening up and testing. If that view is correct, it appears that the contractor is expected to have allowed in his price and timetable for carrying out opening up and testing, because he should be aware of the tests required from the works information and what the contract refers to as the 'applicable law' (clause 40.1). Unfortunately 'applicable law' is neither defined nor is it part of the contract data (although the law of the contract is in the data). This seems to be a case of sloppy drafting and the meaning of 'applicable law' and its effect on this ground will depend upon all the circumstances. On a strict reading, 'applicable law' means the law which applies to the particular provision of the contract. It is not easy to see how the general law will 'require' tests and inspections.

In due course a court will no doubt have the task of explaining this particular compensation event.

Physical conditions encountered by the contractor in the site which are not weather conditions and which an experienced contractor would have judged the chance of occurring to be so small that no allowance need by made for them: clause 60.1(12)

This ground is similar, but not the same as clause 12(1) of the ICE 6th Edition form of contract. Both use the expression 'physical conditions'. Had the reference been to 'site', 'ground' or 'soil' conditions, the expression would have a limited rather than a wide effect. So far as the soil conditions are concerned, the expression has been held to apply to both transient and intransient combinations of stresses[667]. In other words, pre-existing permanent conditions are included, but so are conditions which may change for one reason or another. It is often overlooked that the expression also refers to above ground conditions.

Weather conditions are excepted and that means exactly what it says. Therefore, snow, ice, rain or excessive heat are not included under this ground. It is thought that if the weather conditions caused the site to be

---

[666] *Concise Oxford Dictionary.*
[667] *Humber Oil Terminals Trustee* v. *Harbour & General* (1991) 59 BLR 1.

flooded, that would also be excluded, as it would if the site became snow or ice-bound.

The final proviso could cause difficulties. It must be read in conjunction with clause 60.2 and 60.3. These clauses stipulate that, in judging the physical conditions, the contractor is assumed to have taken into account the site information which describes the site and surroundings and any publicly available information noted therein, what the contractor can obtain from a visual inspection and any other information he could reasonably be expected to have or to acquire. Moreover, if there is an inconsistency within the site information, the contractor is assumed to have taken account of the conditions most favourable to doing the work. It should be noted that 'assumed' used in this context appears to be similar to 'deemed' as used in many contracts[668]. Therefore, there are three possible situations:

- Physical conditions which will not occur
- Physical conditions which have such a small chance of occurring that it would be unreasonable for the contractor to allow for them
- Physical conditions which are very likely to occur.

Effectively, the contractor is assumed to have made his decision based on the information in the site information (this might well include a soil investigation report), any other available information, his own experience and what he can actually see. Therefore the job of deciding whether the first or last situations apply should be fairly easy.

The task of deciding whether physical conditions have so small a chance of occurring that it would be unreasonable to allow for them requires the contractor to calculate the risk. Whether it would be unreasonable is clearly to be decided based on what the contractor would view as unreasonable taking all factors into account. In practice, it is likely to be difficult to argue that a contractor was being unreasonable unless the matter was glaring. Obviously, in preparing his tender, a contractor is unlikely as a matter of principle to take unreasonable risks.

A weather measurement is recorded: within a calendar month, before completion date for the whole works and at the place stated in the contract data, the value of which, compared to weather data, occurs on average less frequently than once in 10 years: clause 60.1(13)

The idea is that a record of past weather over the last 10 years is provided in the contract data. Weather conditions only qualify to rank as a compensation event if they exceed what is in the contract data. The weather data containing the past weather measurements are divided into cumulative rainfall, the number of days when rainfall exceeds 5 mm or when minimum air temperature is less than 0°C or when snow is lying at a specific

[668] Meaning that circumstances are to be treated as existing even if manifestly they are not: *Re Cosslett (Contractors) Ltd, Clark, Administrator of Cosslett (Contractors) Ltd (in Administration)* v. *Mid Glamorgan County Council* [1997] 4 All ER 115.

time – usually, no doubt, at some time early in the working day. If no recorded data are available, assumed values are to be inserted in the contract data. It appears that the intention is to provide a simpler system of deciding when the contractor is entitled to further time and or money without having to make a judgment about whether the weather conditions are exceptional. The result has a high chance of being accurate, but it will give rise to problems of fairness if the weather is exceptional and pushes the boundaries of one or more of the criteria at once, but keeps within the parameters set down.

An employer's risk event occurs: clause 60(14)

This is simple. Employer's risks are set out in clause 80.1 and an event presumably occurs when one of the risks manifests itself. In such a case, it becomes a compensation event. The risks are broadly:

- Claims, proceedings and so on due to use or occupation of the site by the works, negligence or breach of statutory duty by the employer or a fault of the employer or in his design.
- Loss or damage to plant and materials supplied by the employer up to the day the contractor has accepted them.
- Loss or damage to plant and materials due to war and the like, strikes and the like and radioactive contamination.
- Loss or damage to parts of the works taken over by the employer unless the loss or damage is due to existing defects, an event which was not an employer's risk or the contractor's on-site actions after takeover occurred before the defects certificate.
- Loss or damage to the works and equipment, materials, etc. kept on site by the employer after termination unless damaged by the contractor on site after termination.
- Additional risks stated in the contract data.

Takeover of part of the works is certified by the project manager before both completion and the completion date: clause 60.1(15)

Takeover is dealt with in clause 35. Under clause 35.4, the project manager is to certify the date and extent when the employer takes over any part of the works. It should be noted, however, that the employer may use any part of the works before completion has been certified and in that case he has taken over that part of the works when he begins to use it unless there is a contrary reason in the works information or the use is to suit the contractor's method of working (clause 35.3). Takeover is not quite the same as practical completion under JCT contracts, although in practice it may amount to the same. Takeover depends on some action by the employer, while practical completion under JCT contracts depends on whether practical completion has taken place in the opinion of the architect, such opinion being exercised according to law.

Completion is defined in clause 11.2(13) as being when the contractor has carried out the work in the works information stated to be carried out by the completion date and he has corrected the defects which would have stopped the contractor using the works.

The completion date is defined in clause 11.2(12) as the completion date stated in the contract data unless it has been changed in accordance with the contract. If so, it is presumably the date as changed although the definition does not expressly so state.

Therefore, this compensation event depends upon completion not having taken place and the completion date being still in the future so that the contractor is not in culpable delay. Provided those two criteria are met, all that is required is for the project manager to certify takeover of part of the works. It has been seen that this happens when the employer begins to use the part, provided that the works information does not say otherwise and that the employer is not simply using the part to suit the contractor's way of working. But if the project manager fails to certify, there is no compensation event.

The employer, contrary to the works information, fails to provide materials, facilities and samples for tests: clause 60.1(16)

This is a straightforward breach of the obligation stated in clause 40.2 which requires both employer and contractor to provide these things. Of course, it is only the employer's breach which is of consequence here.

The project manager notifies an assumption regarding the nature of a compensation event: clause 60.1(17)

This refers to the procedure noted above in section 16.2.2. If the project manager has stated an assumption under clause 61.6 which is later found to be wrong, he must notify a correction. The importance of this event is related to clause 65.2 which provides that the assessment of a compensation event is not revised if the forecast is found to be wrong. Clearly, the necessary adjustment must be carried out here.

The employer's breach of contract which is not otherwise a compensation event: clause 60.1(18)

This is clearly an attempt at a catch-all clause. The idea is to allow all breaches of contract to be dealt with under the contractual mechanism rather than having to be dealt with at common law. It also prevents time becoming at large if an employer's action or default is responsible for delay, but for which the contract does not otherwise expressly provide. This contract does not expressly reserve the contractor's common law rights

and remedies, but it is likely that the contractor would still have this option and clear words would be required to displace them[669].

A difference between the final quantity of work done and the quantity for an item in the bill of quantities at the contract date which changes the actual cost per unit if the affect is more than 0.1% of the total of the prices at the contract date: clause 60.4 (Options B & D)

It is easy to be confused by the use in this contract of 'Actual Cost' as opposed to the bill of quantities rate. Actual cost is the cost of the components in the schedule of cost components. It is used for the calculation of compensation, and bill of quantities rates are irrelevant for that purpose. Nevertheless, one of the two criteria which must be satisfied before a compensation event can be said to have occurred concerns bill of quantities rates. The item will not qualify for a compensation event unless its bill of quantities rate, when multiplied by the final quantity of work carried out under that item, exceeds 0.1% of the total of the 'Prices at the Contract Date' (a somewhat awkward way of describing what other contracts would call the 'contract sum'). This is apparently to avoid trivial effects on items and leaving only important items for consideration.

The main criterion is that there is a difference between the quantity of work actually done and the quantity in the bill of quantities for any important item, and that this difference causes the actual cost to change. It should be noted that the end result may be a reduction in actual cost.

A difference between the final quantity of work done and the quantity for an item in the bill of quantities at the contract date which delays completion: clause 60.5 (Options B & D)

This is similar to the last event save that there is no limit on the items to be considered. That is clearly sensible, because even items which are relatively small in quantity, may have a significant effect if they are delayed. The key criterion is that the difference, of whatever amount, must delay completion.

Mistakes in the bill of quantities which are corrected by the project manager: clause 60.6 (Options B & D)

The mistakes must be either departures from the method of measurement stipulated in the contract data or due to ambiguities or inconsistencies. The departure from the method of measurement is clear enough. Ambiguity and

---

[669] The Court of Appeal decision in *Lockland Builders Ltd* v. *John Kim Rickwood* (1995), 77 BLR 38, suggested the contrary, but seems to ignore *Modern Engineering (Bristol) Ltd* v. *Gilbert Ash (Northern) Ltd* (1974) 1 BLR 73; *Architectural Installation Services Ltd* v. *James Gibbons Windows Ltd* (1989) 46 BLR 91 and the more recent Court of Appeal decision in *Strachan & Henshaw Ltd* v. *Stein Industrie (UK) Ltd* (1998) 87 BLR 52.

inconsistency are ordinary English words. In order to correct an ambiguity, the project manager must clarify it. In order to correct an inconsistency, he must choose one of the inconsistent elements over the others. There appears to be no restriction on whether his correction results in more or less or whether higher or lower quality. In any event, his correction is dealt with as a compensation event. Obviously, it can lead to reduced prices.

It is not thought that there is any justification for the view of some commentators that there is an inconsistency between this clause and clause 63.7. Clause 63.7 does not deal with bills of quantities. Clearly, the project manager must be allowed to deal with 'mistakes' in the bills of quantities as he sees fit. There may or may not be a price for the employer to pay.

Suspension of performance by the contractor under the Housing Grants, Construction and Regeneration Act 1996: clause 60.7

This is straightforward. The Act entitles the contractor to suspend performance of all his obligations if the employer has not made payment of all money properly due by the final date for payment and has not served any effective withholding notice. This clause makes such suspension into a compensation event. By doing so, it goes further than the Act which only entitles the contractor to an extension of the contract period. Because it is a compensation event, the contractor may be entitled also to additional payment.

Delay in making the advanced payment: clause J1.2

Where secondary option J applies, the employer must make an advanced payment to the contractor of the amount stated in the contract data. It must be made within the time specified; that is within 4 weeks of the contract date. If an advanced payment bond is required, the payment may be delayed until not later than 4 weeks from the date the employer receives the bond, if that is later than the contract date. If the employer is late in making the payment, it ranks as a compensation event. Late receipt of the bond may be because it does not conform to the requirement in the works information, or because the project manager cannot accept the bank or insurer proposed by the contractor on account of its poor commercial position relative to the value of the bond.

Clearly, delay in receiving the advanced payment may result in serious financial and time consequences for a contractor.

A change, after the contract date, in the law of the country in which the site is located: clause T1.1

Where this secondary option applies, 'law' would be given its ordinary meaning and in this instance, the ordinary meaning is not restricted in any way. Therefore, the clause is very broad and any change in any aspect of the applicable law is a compensation event. It is quite conceivable that there may be dozens of compensation events on this ground during the life of any contract. It should be noted that, although the contract date marks the beginning of such events, there is no concluding date indicated. In practice, of course, it is only those changes in the law which have an effect on the cost of the project which will be worth notifying by the contractor. The effect of the change in the law may be to reduce the actual cost, in which case the prices are also to be reduced. Perhaps, for that reason, the clause makes provision for the project manager, *if he wishes to do so*, to give notice to the contractor and to instruct him to submit quotations.

Some commentators make reference to the guidance notes issued in support of this contract. However, care should be taken in using the notes. It should be remembered that, although they may be helpful, they do not have the force of law nor do they bind the parties. An adjudicator, arbitrator or judge trying to decide the meaning of the clauses would not be able refer to the notes.

A delay to the work, additional or changed work, caused by the application of the CDM Regulations 1994 if an experienced contractor could not reasonably be expected to have foreseen it: clause U1.1

There is no contractual obligation to comply with the CDM Regulations and, therefore, failure by any party to so comply is not a breach of contract as it would be under the JCT contracts. Therefore, the parties do not have any general remedy for such failure. The obligation lies under statute. This secondary optional clause provides the only contractual remedy and that is for the benefit of the contractor. The remedy is qualified to the extent that an experienced contractor might reasonably be expected to foresee the delay. If so, there is no compensation event. In practice, such a qualification may not be easy to agree and recourse to an adjudicator may be indicated.

16.2.4 Procedure

The procedure is set out in clauses 61 to 65 inclusive. These are very complex in operation.

Clause 61: notifying compensation events

If the event is due to either the project manager or the supervisor giving an instruction or changing a decision, clause 61.1 states that it is for the project manager to notify the contractor of the event. He must do this 'at the time of the event'. Of course, he may not then know that it is a compensation event, but there is no provision for the project manager to notify an event later. The notification must be separate from the instruction (clause 13.7). It is suggested that such communications are put into separate envelopes, even if sent on the same day. That may be overcautious and it is difficult to envisage any tribunal coming to the conclusion that a notice is invalid because it was in the same envelope. What is clear is that separate pieces of paper are required. There will be a need for close liaison between project manager and supervisor if the supervisor's actions are not to be overlooked.

The compensation events which fall into this category are not listed in the contract, but an inspection of clause 60 suggests that they are 60.1(1), (4), (7), (8), (10), (17) and 60.6. Instructions or changed decisions must be put into effect by the contractor immediately. If the project manager simply wishes to discover what it may cost to issue an instruction or change a decision, clause 63.2 permits him to instruct the contractor to submit a quotation for 'proposed' instructions or changed decisions. In this instance, of course, the contractor 'does not put' them into effect.

The contractor's duty to notify a compensation event arises under clause 61.3 if he believes it is a compensation event which has happened or which he expects to happen and if the project manager has not already notified the event to the contractor. The contractor must act no later than 2 weeks after he became aware of the event. The project manager's obligation to notify the contractor arises under clause 61.1 only in connection with a minority of the events. Therefore, the onus is on the contractor to notify in most instances. No doubt the contractor will do this in all instances, just to be sure. Where the project manager has a duty to notify, but has failed, it is arguable that it becomes the employer's breach of contract and thus creates another compensation event under clause 60.1(18). It will be a matter of fact, although perhaps difficult to prove, just when the contractor 'became aware' of the event. The only restriction on such notification taking place at any time, provided only that it is no later than 2 weeks after the contractor 'became aware', is set out in clause 61.7. That clause states that a compensation event 'is not notified' (presumably this means 'is not to be notified' in NEC parlance) after the defects date. The defects date is found in the contract data supplied by the employer. It appears to be roughly the equivalent of the end of the defects liability period in JCT contracts. Therefore, no notification by either contractor or project manager can take place after this date. In practice, it will be difficult for a contractor to demonstrate that he did not become aware of an event very soon after it occurred except in wholly unusual circumstances.

If it can be shown that the contractor became aware on a certain date and did not notify within 2 weeks of that date, it appears that he is not entitled to the benefit of additional time and/or money. This is a fairly onerous condition hidden away in clause 61.3 and it is to be doubted whether the courts would be prepared to apply it strictly[670]. It is perhaps more likely that the condition would be applied only to the extent that the employer could show that he had suffered prejudice by the late notification.

---

[670] *J. Spurling Ltd* v. *Bradshaw* [1956] 2 All ER 121.

Under clause 61.1, the project manager must request a quotation from the contractor unless it has been already submitted or the event is due to the fault of the contractor. Clause 61.4 deals with the project manager's duty if he receives the contractor's notification. He must decide, within a week of the notification (whether the date of issue or receipt is not clear) or such longer period as agreed by the contractor, whether the event:

- arises from the contractor's fault; or
- has not and is not expected to happen; or
- has no effect on actual cost or completion; or
- is not a compensation event.

If he decides that the event falls into one or more of these categories, it seems that the project manager need do no more than notify the decision to the contractor. However, if he decides that it does not so fall, in other words that it is valid, he must instruct the contractor, as in clause 61.1, to submit a quotation and the contractor has only a week in which to respond unless the contractor has agreed to a longer period. This seems to imply that the initiative for suggesting a longer period lies with the project manager and that the contractor can only agree or disagree. There is no mechanism to allow the contractor to request a longer period.

Under clauses 16.1 to 16.4 inclusive, the contractor and the project manager have a duty to give early warning of various matters which obviously include compensation events. Clause 61.5 provides that if the project manager comes to the decision that the contractor did not give early warning of the event which an experienced contractor would have given, he must so notify the contractor at the time he instructs him to submit a quotation.

This becomes important when the project manager comes to assess the events, because clause 63.4 states that where such notification has been given under clause 61.5, the event is assessed as if the contractor had given early warning. Obviously, if the contractor had given early warning, the project manager might have taken various steps to reduce the impact of the event. On the other hand, it is conceivable that to treat it as if early warning had been given, when in fact it had not, could produce a result in the contractor's favour. For example, if early warning was given in a particular instance, the project manager might have issued further instructions to reduce the impact of the event. In fact, the early warning was not given and the instruction remained unissued. However, if it is treated as if early warning had been given, it might be argued that the lack of instructions on the part of the project manager was a matter solely for the project manager and could not be the cause of any reduction in time and/or money for the contractor. The precise intention behind this provision is difficult to discover.

In submitting a quotation, the contractor will be obliged to incorporate some forecasts of the effects of the event. Clause 65.2 provides that an assessment will not later be revised just because the forecast is found to be wrong. Due to the short timescale when most events must be notified, the effects of the event will be unknown or only partly known. The contractor will, therefore, take some care about his forecasts, usually erring on the generous side. If the project manager comes to the decision that the future effects of a compensation event are too uncertain to be reasonably forecast, he 'states assumptions' about the event. Once again, the strange use of the present tense does not make the task easy, but it is presumed that the clause means that the project manager *must* state the assumptions, i.e. he has a duty to do so. From the contractor's point of view, the good thing about the project manager's assumptions is that, under clause 61.6, unlike forecasts, if any of them is later found to be wrong, the project manager notifies (again presumably *must* notify) a correction. The added advantage of this is that the notification of a correction by the project manager itself ranks as a compensation event (clause 60.1(17)).

Clause 62: quotations for compensation events

It has been seen that quotations usually cannot be amended to suit changed circumstances which come to light as the project proceeds. With the exception noted in clause 61.6, the quotations are fixed. They are to comprise the Contractor's Proposals about changes to the prices and his assessment of the amount of delay to the completion date resulting from the event. It is clear that the contractor has a duty to submit the details or calculations of his assessment. He must also submit a revised programme with his quotation if he believes that the existing programme will be affected (clause 62.2). Clause 62.1 provides that either the project manager may instruct the contractor to submit alternative quotations based on different ways of dealing with the event or the contractor may submit additional quotations on his own initiative, provided the contractor's suggestions are thought to be practicable. There may be alternative ways of dealing with the event. One may be cheaper, but the other more effective. The contractor may well think of an entirely different way of resolving what is essentially a problem in each case. It is for the employer, advised by the project manager, to decide.

Under clause 62.3 quotations must be submitted within 3 weeks of the project manager's request. It is at least arguable that the time limit does not apply to a contractor submitting an alternative quotation under clause 62.1. The project manager has very little time to respond to a quotation; just 2 weeks from the submission of the quotations. These time periods are subject to clause 62.5 which allows them to be relaxed if the project manager and the contractor agree to an extension of the time periods before the quotation or the project manager's response is due. It is the duty of the project manager to notify the agreed extensions to the contractor. But clause 61.4 gives the contractor only 1 week after his notification.

The project manager may reply in one of four ways. First, he may give the contractor an instruction to submit a revised quotation under clause 62.4, but before doing so, he must explain his reasons to the contractor. These reasons should be in writing. These may simply be that the quotations submitted triggered the idea of an entirely new approach. The project manager must give his reasons, it appears, for the purpose of assuring the contractor that the revised quotation is not requested on a whim. However, there is nothing which seems to prevent the project manager so requesting and, indeed, giving that as his reason. The contractor has a further 3 weeks to submit his revised quotation. Presumably the 3 weeks commence on receipt of the instruction. The second mode of reply may take the form of a simple acceptance of the contractor's quotation. The third reply results from a request under clause 61.2 for a quotation and it may simply be that a proposed instruction or changed decision will not be given. Finally, if the project manager is not minded to accept any quotation and does not believe that the position can be rectified by asking for a revised quotation, he may notify the contractor that he will make his own assessment.

Clause 63: assessing compensation events

Although the word 'assessed' is used to state how the effects of a compensation event are to be determined, rather than such words as 'calculated' or 'ascertained', the word is nowhere defined in the contract. The ordinary everyday meaning of assess is to fix an amount or to estimate[671]. It therefore seems that something less than complete accuracy is required.

The key figures are the contractor's actual costs. It is the effect of the compensation events on such actual costs which determines the amount of compensation. But it is not actual costs in the commonly accepted meaning of that term, i.e. the actual (real) costs to the contractor. Clause 52.1 deals with actual costs. They are to be at open market or competitively tendered prices allowing for the deduction of discounts, rebates and taxes which can be recovered. In this respect, the term 'Actual Cost' is not quite accurate. That is because, if the contractor's real cost of any item is greater than what might be termed 'normal costs', perhaps due to special rates being charged, the excess is disregarded. Actual costs are the costs of the items in the schedule of cost components attached to the contract. It should be noted that the contractor's costs which are not included in actual cost are deemed to be included in the fee percentage. Therefore, in the event that costs incurred by the contractor are not in the schedule of cost components, it seems that the contractor will be deprived of any reimbursement for those items.

The assessment, which must be carried out in the first instance by the contractor in his quotations and, possibly later, by the project manager if necessary, is a complex affair. Partly, that is due to the structure of the main options and partly due to the NEC approach to the subject. It has to be said again that many clauses are less than crystal clear.

Clause 63.1 makes reasonably clear that the effect of the compensation event upon actual cost of work done, the forecast cost of work not yet done

---

[671] *Concise Oxford Dictionary.*

and the resulting fee, is to be expressed as a change to the prices. Therefore, broadly it is the effect on the actual cost and forecast actual cost which is translated to the prices. Since the prices will be the original tendered rates, the notional actual cost (that is the actual cost before the effect of the compensation event is taken into account) should not be very different, if at all, from the original rates. Nevertheless, it is the notional actual cost which should form the baseline for the effect of the compensation event. That is to avoid the contractor being able to recover for the financial effects of his own inefficient working.

Except in the situations where the contract expressly so states (e.g. clause 60.6) or the event is a change to the works information or the correction of an assumption under clause 61.6, the prices are not reduced if the effect of the event is to reduce the total final cost (clause 63.2).

The difference between 'planned Completion' and 'Completion Date' is important in clause 63.3. The delay to the completion date is to be assessed as the length of time that the compensation event has caused planned completion to be later than planned completion shown on the accepted programme. The exercise will necessitate careful examination of the pro-grammes in the normal way, but with regard to the planned completion. It is only when a decision has been reached about the delay to the planned completion that the length of that delay is transferred to the completion date. The contractor should be able to demonstrate his reasoning by means of the revised programme that he is obliged, under clause 62.2, to provide to the project manager.

Clauses 63.5, 63.6 and 63.7 set out various terms of general application. Clause 63.6 reasonably states that assessments are to be made on the as-sumption that the contractor has reacted both competently and promptly to the event, that additional actual costs are incurred reasonably and that the accepted programme can be changed. The general law would imply these assumptions in any event.

Clause 63.7 deals specifically with compensation events which arise from instructions to change the works information in order to resolve an ambigu-ity or inconsistency. The way that it is to be treated depends, again quite reasonably, on whether it is the information provided by the employer or by the contractor which is changed. If it is the employer's information, the effect of the event is assessed in the way most favourable to the contractor. If it is the contractor's information, the assessment is to be done in the way most favourable to the employer. It is unlikely that the general law would imply this procedure.

The most difficulty is caused by clause 63.5. It states: 'Assessment of the effect of a compensation event includes cost and time risk allowances . . .'. Matters which are at the contractor's risk are set out in clause 81.1. Some commentators seem to believe that the effect of this clause is that the contractor must bear the risk of such things as ground conditions which an experienced engineering contractor should have anticipated. It is reason-ably clear that certain risks under clause 81.1 must be borne by the con-tractor. Clause 81.1 states that those risks are the risks not carried by the employer. The contractor must bear such risks from the starting date of the contract until the defects certificate has been issued. It is perfectly sensible and in accordance with law that the parties must stand by their bargain and that, in respect of the works for which the contractor originally contracted, he will bear such risks. However, in this instance, the clause is dealing with the effects of a compensation event. By its very nature, a compensation event is one over which the contractor has little or no control. Many of the compensation events rank as breaches of contract on the part of the employer for which the contractor could expect to recover damages sufficient to put himself in the position he would have occupied if the breach had not occurred. It is obvious that the effect of a compensation event could well include what would normally be considered as contractor's risk items under clause 81.1. The contractor's costs for dealing with such items would also be part of the recoverable damages. Indeed, it appears that the contractor can raise a claim at common law for the whole of his damages where a compensation event is also a breach of contract on the part of the employer. This is not a matter that the contractor must include for his risk items as part of a quotation to carry out additional work. Here, the contract is dealing with the assessment (which may be the contractor's quotation or it may be the project manager's assessment) consequent upon a compensation event.

Although the wording of clause 63.5 could be much clearer (hence the difficulty) the common-sense interpretation of it is that the assessment of the effect of a compensation event must include the cost to the contractor of what would otherwise be cost and time risk allowances for matters which have a significant chance of occurring and are at the contractor's risk under the contract. That also appears to be the legal interpretation. If the contractor is preparing the assessment, he will include his costs and, if the effects are in the future, he will have to make a forecast. Such a forecast will no doubt be on the generous side.

Clauses 63.8, 63.9, 63.10 and 63.11 deal with variations concerning the main options. Assessments under options A and C will be in the form of changes to the activity schedule, but assessments under options B and D will be in the form of changes to the bills of quantities. However, under options B and D, the project manager and the contractor may agree to use the lump sums and rates in the bill of quantities instead of the actual cost and resulting fee. No doubt this will appeal to many. So far as A and B are concerned, the assessment which includes sub-contracted work has the contractor's fee percentage added to the actual cost, but obviously any fees payable by the contractor to the sub-contractors are not to be added. In the case of main contract options A, B, C, D and E, if the project manager and the contractor agree, the contractor may carry out the assessment by using the shorter schedule of cost components. In any event, where the project manager is to assess, he may use the shorter schedule. The recovery of the fee will usually amount to the difference between the fee percentage applied to actual cost before and after the effect of the compensation event is taken into account.

Clause 64: the project manager's assessments

This clause sets out stipulations regarding the assessment of a compensation event by the project manager. This should be the exception, because usually the contractor will have been requested to provide a quotation using the principles set out in clause 63. Only if the contractor fails to satisfy all the criteria may the project manager act. Of course, it is initially a matter for the project manager to decide whether such failure has occurred so as to justify his intervention. There are six instances where the project manager may carry out the assessment himself. They are set out in clauses 64.1 and 64.2. They are if:

- the contractor fails to supply the quotation and details within the time allowed. (This is straightforward. If the contractor does not comply with the 3 weeks set out in clause 62.3 or such extension of that time as agreed under clause 62.5);
- the contractor, in the opinion of the project manager, fails to assess the event correctly in his quotation and the project manager does not request a revised quotation. (Reference to 'correctly' can only mean in accordance with the rules set out in the contract. The contractor's assessment is not incorrect simply because it does not accord with the project manager's view);
- the contractor fails to submit a programme required by the contract when he submits his quotation. (Again, this is straightforward. Obviously, if the programme for the remaining work is unaffected by the compensation event, the requirement to provide a revised programme under clause 62.2 falls away);
- the contractor has submitted a quotation, but the project manager has not accepted the accompanying programme for one of the reasons stated in the contract. (Clause 31.3 sets out these reasons as being if the programme is not practicable, if it does not show the information required under clause 31.2, if it is not a realistic representation of the contractor's plans or if it does not comply with works information);
- there is no accepted programme; or
- the contractor has failed to submit a revised programme for acceptance.

It is not quite clear why the last two reasons are separated from the others in the contract. In front of the first four, the contract states: 'The Project Manager (sic) assesses a compensation event.' In front of the last two, the contract states: 'The *Project Manager* assesses a compensation event using his own assessment of the programme for the remaining work...' The only difference is that the project manager seemingly has a free hand in assessing the compensation event in the first instance, but in the second instance he must use his own assessment of the programme. Since the second instance deals with those circumstances where there effectively is no programme, it seems he has little choice in any event. It is difficult to see why he should not do that if he wishes so far as the first instance is concerned also.

Clause 64.3 could be drafted more clearly. The first phrase: 'The *Project Manager* notifies the *Contractor* of his assessment of a compensation event...' is clear if 'notifies' is taken to mean 'must notify'. However, the remainder of the first sentence: '...and gives him details of it within the period allowed for the *Contractor's* submission of his quotation for the same event' is less clear. Presumably, the project manager is to notify the assessment and give details at the same time. But this time is the period allowed for the contractor to submit his quotation. Clause 62.3 states that such period is 3 weeks of being instructed to do so by the project manager (assuming for simplicity that no extension has been agreed under clause 62.5). But it is usually not until the end of the 3 weeks period that the project manager will know that he has to carry out the assessment. The last sentence caters for this by stating that the '...period starts when the need for the *Project Manager's* assessment becomes apparent'. The clause does not state to whom it should become apparent, presumably the project manager.

However, the clause could conceivably be interpreted as meaning that the project manager must notify the contractor of the assessment and it is only the details which are subject to the time constraints.

Clause 65: implementing compensation events

Implementing the compensation events is not straightforward. Until they are implemented, the contractor is entitled to neither additional money nor additional time. Clause 65.1 purports to state how and when implementation occurs. The first sentence is very clear. The project manager implements each event by notifying the contractor, either that he accepts his quotation, or of the project manager's own assessment. The clause proceeds to give a definition of when implementation takes place by stating that it is the *latest* of:

- his acceptance of a quotation; or
- his completion of his own assessment; or
- when the compensation event occurs.

The wording of this clause could be greatly improved in order to make it easily intelligible.

Following from the provisions of clause 65.1, it appears that, under the provisions of clause 61.1, the contractor is required to carry out the instruction or changed decision before the event is implemented.

Clauses 65.3, 65.4 and 65.5 set out certain variations applicable to the main options. Under options A, B, C and D, the project manager is to include in his notification the changes to the prices and the completion date resulting from an accepted quotation or his own assessment. However, under options E and F, cost reimbursable and management contracts respectively, he is to include the changes to the forecast amount of prices and completion date and the contractor must not implement a sub-contract compensation event until it has been agreed by the project manager. Both options E and F produce what are cost reimbursable contracts.

16.3 Delay damages

16.3.1 Clause R1

Delay damages is not part of the core clauses of NEC. It is only a secondary option: option R. If delay damages is not chosen, the employer must resort to ordinary unliquidated damages as a remedy for the contractor's late completion[672].

The full text of clause R1 is as follows:

Delay Damages R1

R1.1   The *Contractor* pays delay damages at the rate stated in the Contract Data from the Completion Date for each day until the earlier of

- Completion and
- the date on which the *Employer* takes over the works.

R1.2   If the Completion Date is changed to a later date after delay damages have been paid, the *Employer* repays the overpayment of damages with interest. Interest is assessed from the date of payment to the date of repayment and the date of repayment is an assessment date.

16.3.2 Commentary

Delay damages appear to be the same as liquidated damages under other contracts and at law. There is no good reason for adopting a new name, because the courts will look to the substance rather than the form to decide whether a provision is, in fact, liquidated damages[673]. The general comments on liquidated damages in Chapter 3 are equally applicable to delay damages under NEC.

The contractor is to pay delay damages from the completion date until either completion or when the employer takes over the works. There is no requirement for the equivalent of a certificate of non-completion from the project manager. The rate of delay damages is to be set out in the contract data. Where options L (sectional completion) and R are used together, the individual damages are to be set down for each section. However, the contract data also provide for the situation where 'Option R is used (whether or not Option L is also used)' and proceed to state: 'Delay damages for the whole of the *works* are . . .'. The purpose of stipulating delay damages for the whole of the works as well as for the individual sections is not clear. It seems that it can only lead to problems. For example, if each section has its own delay damages totalling £5000, the delay damages for the whole of the works is presumably £nil. Otherwise, the employer might be able to argue

---

[672] See Chapter 3, section 3.1.
[673] *Kemble* v. *Farren* [1829] All ER 641.

that he is entitled to £10,000 if he does not take over any section until he takes over the whole of the works after the completion date.

The delay damages are set at a rate per day. This might lead to the contractor successfully arguing that the amount is really a penalty unless, for example, the employer can show that the daily rate for Saturday and Sunday, which is the same as for the other days, was a genuine pre-estimate of the loss likely to be suffered by the employer in the case of an overrun. Of equal concern is the fact that, although clause 35 provides for the employer to be able to take over a part of the works, there is no provision under clause R for the delay damages to be proportionately reduced. Therefore, in the event that partial takeover is carried out, it is likely that the delay damages clause will become inoperable and the employer will be thrown back on his common law remedies for breach of contract.

Clause R1.2 provides that if the completion date is changed after delay damages have been paid, the employer must pay the overpayment with interest. Interest is to run from the date of payment to the date of repayment, which is an assessment date. This is more advantageous for contractors than JCT contracts which make no provision for interest in these circumstances.

Chapter 17
Sub-contract claims

17.1 Introduction

This chapter considers sub-contract claims arising under six forms of sub-contract:

- The JCT Standard Form of Nominated Sub-Contract, NSC/C
- The JCT Standard Form of Sub-Contract for Domestic Sub-Contractors DSC/C
- The JCT conditions of sub-contract for sub-contractors 'named' under IFC 98, NAM/SC
- Domestic Sub-Contract Form IN/SC
- The JCT Works Contract/2 for use with MC 98
- The ACA Form of Sub-Contract.

It is important to remember that the legal principles which apply to claims under the main contract forms equally hold good for sub-contract forms. There is nothing to prevent a sub-contractor bringing his claims as claims for damages under the common law and reference should be made to Part I of this book. Therefore, only claims made under the express provisions of the sub-contracts are considered here.

17.2 Sub-contract form NSC/C

17.2.1 Introduction to the form

The Joint Contracts Tribunal issued the JCT Standard Form of (Nominated) Sub-Contract at the same time as JCT 1980 and it originally existed in two versions. NSC/4 for nominated sub-contractors who had tendered on Tender NSC/1 and executed Agreement NSC/2 and been nominated by Nomination NSC/3 under clause 35.10.2, i.e. what the JCT termed the 'basic method', and NSC/4a for use where the architect had adopted some other method of selecting a nominated sub-contractor and where Tender NSC/1 was not used. A new version of the nominated sub-contract form was introduced in 1991. Clause 35 of JCT 80 was amended and NSC/4 and NSC/4a were combined into one form NSC/C, which was given a modern structure like JCT Works Contract/2, IFC 84 and PCC 92. The current form of nominated sub-contract has been changed to suit JCT 98. The most recent version incorporates JCT amendments 1–4. The complete set of forms are:

NSC/T	1	Invitation to tender
	2	Tender document
	3	Particular conditions
NSC/A	Agreement	
NSC/C	Conditions	
NSC/W	Warranty	
NSC/N	Nomination form	

17.2.2 Extensions of time under NSC/C

Extensions of sub-contract time are dealt with by clauses 2.2 to 2.6, which provide as follows:

2.2 .1 If and whenever it becomes reasonably apparent that the commencement, progress or completion of the Sub-Contract Works or any part thereof is being or is likely to be delayed, the Sub-Contractor shall forthwith give written notice to the Contractor of the material circumstances including the cause or causes of the delay and identify in such notice any matter which in his opinion comes within clause 2.3.1. The Contractor shall forthwith inform the Architect of any written notice by the Sub-Contractor and submit to the Architect any written representations made to him by the Sub-Contractor as to such cause as aforesaid.

.2 In respect of each and every matter which comes within clause 2.3.1, and identified in the notice given in accordance with clause 2.2.1, the Sub-Contractor shall, if practicable in such notice, or otherwise in writing as soon as possible after such notice:

.1 give particulars of the expected effects thereof; and

.2 estimate the extent, if any, of the expected delay in the completion of the Sub-Contract Works or any part thereof beyond the expiry of the period or periods stated in the agreed programme details in NSC/T Part 3 item 1, or beyond the expiry of any extended period or periods previously fixed under clauses 2.2 to 2.7 which results therefrom whether or not concurrently with delay resulting from any other matter which comes within clause 2.3.1; and

.3 give such further written notices to the Contractor as may be reasonably necessary or as the Contractor may reasonably require for keeping up-to-date the particulars and estimate referred to in clauses 2.2.2.1 and 2.2.2.2 including any material change in such particulars or estimate.

.3 The Contractor shall submit to the Architect the particulars and estimate referred to in clauses 2.2.2.1 and 2.2.2.2 and the further notices referred to in clause 2.2.2.3 to the extent that such particulars and estimate have not been included in the notice given in accordance with clause 2.2.1 and shall, if so requested by the Sub-Contractor, join with the Sub-Contractor in requesting the consent of the Architect under clause 35.14 of the Main Contract Conditions.

2.3 If on receipt of any notice, particulars and estimate under clause 2.2 and of a request by the Contractor and the Sub-Contractor for his consent under clause 35.14 of the Main Contract Conditions the Architect is of the opinion that:

.1   any of the matters which are stated by the Sub-Contractor to be the cause of the delay is the occurrence of a Relevant Event or is an act, omission or default of the Contractor (including, where the Contractor is the Principal Contractor, any omission or default in the discharge of his obligations as the Principal Contractor) or of any person for whom the Contractor is responsible (see clause 6.1); and

.2   the completion of the Sub-Contract Works is likely to be or has been delayed thereby beyond the period or periods stated in the agreed pro-gramme details in NSC/T Part 3 item 1, or any revision of such period or periods

then the Contractor shall, with the written consent of the Architect, give in writing to the Sub-Contractor an extension of time by fixing such revised or further revised period or periods for the completion of the Sub-Contract Works as the Architect in his written consent then estimates to be fair and reasonable. The Contractor shall, in agreement with the Architect, when fixing such revised period or periods state:

.3   which of the matters, including any of the Relevant Events, referred to in clause 2.3.1 they have taken into account; and

.4   the extent, if any, to which the Architect, in giving his written consent, has had regard to any instructions under clause 13 of the Main Contract Conditions requiring the omission of any work or obligation or restriction,

and shall, if reasonably practicable having regard to the sufficiency of the aforesaid notice, particulars and estimate, fix such revised period or periods not later than 12 weeks from the receipt by the Contractor of the notice and of reasonably sufficient particulars and estimates, or, where the time between receipt thereof and the expiry of the period or periods for the completion of the Sub-Contract Works or the applicable part thereof is less than 12 weeks, not later than the expiry of the aforesaid period or periods.

If, upon receipt of the aforesaid notice, particulars and estimate and request of the Contractor and the Sub-Contractor, the Architect is of the opinion that it is not fair and reasonable to give his written consent to any revision or further revision of the period or periods for completion of the Sub-Contract Works or any part thereof, the Architect shall if reasonably practicable having regard to the sufficiency of such notice, particulars and estimate, so inform the Contractor who shall inform the Sub-Contractor of the opinion of the Architect not later than 12 weeks from the receipt by the Contractor of the aforesaid notice, particulars and estimate and request by the Sub-Con-tractor or, where the period of time between such receipt and the expiry of the period or periods for the completion of the Sub-Contract Works or the applicable part thereof is less than 12 weeks, not later than the expiry of the aforesaid period or periods.

2.4 After the first exercise by the Contractor of the duty under clause 2.3, the Contractor, or after any revision to the period or periods for the completion of the Sub-Contract Works stated by the Contractor in his acceptance of a 3.3A Quotation in respect of a Variation, the Contractor with the written consent of the Architect, may in writing to the Sub-Contractor fix a period or periods for completion of the Sub-Contract Works or the applicable part thereof shorter than that previously fixed under clause 2.3 or stated by the Contractor in his acceptance of a 3.3A Quotation if, in the opinion of the Architect, the fixing of such shorter period or periods is fair and reasonable having regard to any instructions issued under clause 13 of the Main Contract Conditions requiring the omission of any work or obligation or restriction where such issue is after the last occasion on which the Contractor with the consent of the Architect made a revision of the aforesaid period or periods. Provided that no decision under clause 2.4 shall alter the length of any revision to the period or periods for the completion of the Sub-Contract Works in respect of a Variation for which a 3.3A Quotation has been given and which has been stated by the Contractor in his acceptance of the 3.3A Quotation.

2.5 If the expiry of the period or periods when the Sub-Contract Works should have been completed in accordance with the agreed programme details in the NSC/T Part 3 item 1, as revised by any operation of the provisions of clause 2.3 or 2.4 or stated by the Contractor in his acceptance of a 3.3A Quotation, occurs before the date of practical completion of the Sub-Contract Works certified under clause 35.16 of the Main Contract Conditions, the Contractor, with the consent of the Architect, may

and

not later than the expiry of 12 weeks from the aforesaid date of practical completion of the Sub-Contract Works, the Contractor, with the consent of the Architect, shall

either:

.1 fix such a period or periods for completion of the Sub-Contract Works longer than that previously fixed under clause 2.3 or 2.4 or stated by the Contractor in his acceptance of a 3.3A Quotation as the Architect in his written consent considers to be fair and reasonable having regard to any of the matters referred to in clause 2.3.1 whether upon reviewing a previous decision or otherwise and whether or not such matters have been specifically notified by the Sub-Contractor under clause 2.2.1; or

.2 fix such a period or periods for completion of the Sub-Contract Works shorter than that previously fixed under clause 2.3 or 2.4 or stated by the Contractor in his acceptance of a 3.3A Quotation as the Architect in his written consent considers to be fair and reasonable having regard to any instruction issued under clause 13 of the Main Contract Conditions requiring the omission of any work or obligation or restriction where such issue is after the last occasion on which the Contractor, with the consent of the Architect, made a revision of the aforesaid period or periods; or

.3   confirm to the Sub-Contractor the period or periods for the completion of the Sub-Contract Works previously fixed.

Provided always the Sub-Contractor shall use constantly his best endeavours to prevent delay in the progress of the Sub-Contract Works or any part thereof, howsoever caused, and to prevent any such delay resulting in the completion of the Sub-Contract Works or any part thereof being delayed or further delayed beyond the period or periods for completion stated in the agreed programme details in NSC/T Part 3 item 1; and the Sub-Contractor shall do all that may reasonably be required to the satisfaction of the Architect and the Contractor to proceed with the Sub-Contract Works or any part thereof.

Provided that no decision under clause 2.5.1 or clause 2.5.2 shall alter the length of any revision to the period or periods for the completion of the Sub-Contract Works in respect of a Variation for which a 3.3A Quotation has been given and which has been stated by the Contractor in his acceptance of the 3.3A Quotation.

2.6   The following are the Relevant Events referred to in clause 2.3.1:

.1   force majeure;

.2   exceptionally adverse weather conditions;

.3   loss or damage occasioned by any one or more of the Specified Perils;

.4   civil commotion, local combination of workmen, strike or lock-out affecting any of the trades employed upon the Works or any of the trades engaged in the preparation, manufacture or transportation of any of the goods or materials required for the Works;

.5   compliance by the Contractor and/or Sub-Contractor with the Architect's instructions:

   .1   under clauses 2.3, 13.2 (except for a confirmed acceptance of a 13A Quotation), 13.3 (except where bills of quantities are included in the Numbered Documents, compliance with an Architect's instruction for the expenditure of a provisional sum for defined work), 13A.4.1, 23.2, 34, 35 or 36 of the Main Contract Conditions, or

   .2   in regard to the opening up for inspection of any work covered up or the testing of any of the work, materials or goods in accordance with clause 8.3 of the Main Contract Conditions (including making good in consequence of such opening up or testing) unless the inspection or test showed that the work, materials or goods were not in accordance with the Main Contract or the Sub-Contract as the case may be;

.6   .1   where an Information Release Schedule has been provided, failure of the Architect to comply with clause 5.4.1 of the Main Contract Conditions.

   .2   failure of the Architect to comply with clause 5.4.2 of the Main Contract Conditions except to the extent that the failure referred to in clause 2.6.6.1 or clause 2.6.6.2 results from a breach by the Sub-Contractor of his obligations to the Employer under clause 3.2 of Agreement NSC/W.

.7 delay on the part of nominated sub-contractors (other than the Sub-Contractor) or of nominated suppliers in respect of the Works which the Contractor has taken all practicable steps to avoid or reduce;

.8 .1 the execution of work not forming part of the Main Contract by the Employer himself or by persons employed or otherwise engaged by the Employer as referred to in clause 29 of the Main Contract Conditions or the failure to execute such work;

.2 the supply by the Employer of materials and goods which the Employer has agreed to provide for the Works or the failure so to supply;

.9 the exercise after the Base Date by the United Kingdom Government of any statutory power which directly affects the execution of the Works by restricting the availability or use of labour which is essential to the proper carrying out of the Works, or preventing the Contractor or Sub-Contractor from, or delaying the Contractor or Sub-Contractor in, securing such goods or materials or such fuel or energy as are essential to the proper carrying out of the Works;

.10 .1 the Contractor's or Sub-Contractor's inability for reasons beyond his control and which he could not reasonably have foreseen at the Base Date for the purposes of the Main Contract or the Sub-Contract as the case may be to secure such labour as is essential to the proper carrying out of the Works; or

.2 the Contractor's or Sub-Contractor's inability for reasons beyond his control and which he could not reasonably have foreseen at the Base Date for the purposes of the Main Contract or the Sub-Contract as the case may be to secure such goods or materials as are essential to the proper carrying out of the Works:

.11 the carrying out by a local authority or statutory undertaker of work in pursuance of its statutory obligations in relation to the Works, or the failure to carry out such work;

.12 failure of the Employer to give in due time ingress to or egress from the site of the Works or any part thereof through or over any land, buildings, way or passage adjoining or connected with the site and in the possession and control of the Employer, in accordance with the Contract Bills and/or Contract Drawings and/or Specification/Schedules of Work and/or Numbered Documents, after receipt by the Architect of such notice, if any, as the Contractor is required to give, or failure of the Employer to give such ingress or egress as otherwise agreed between the Architect and the Contractor;

.13 delay arising from

.13 .1 a suspension by the Contractor of the performance of his obligations under the Main Contract to the Employer pursuant to clause 30.1.4 of the main Contract Conditions; and/or

.2 the valid exercise by the Sub-Contractor of the right pursuant to clause 4.21.1 to suspend the performance of his obligations under the Sub-Contract to the Contractor;

.14 where it is stated in the completed Appendix of the Main Contract Conditions (attached to NSC/T Part 1 or, if different, in the completed Appendix of the Main Contract Conditions enclosed with the copy of Nomination NSC/N sent to the Sub-Contractor by the Architect) that clause 23.1.2 of the Main Contract Conditions applies to the Main Contract, any deferment by the Employer in giving possession of the site of the Works to the Contractor;

.15 where bills of quantities are included in the Numbered Documents, by reason of the execution of work for which an Approximate Quantity is included in those bills which is not a reasonably accurate forecast of the quantity of work required;

.16 the use or threat of terrorism and/or the activity of the relevant authorities in dealing with such use or threat;

.17 compliance or non-compliance by the Employer with clause 6A.1 of the Main Contract Conditions (Employer's obligation – Planning Supervisor – Principle Contractor where not the Contractor);

.18 save as provided for in clauses 2.6.1 to 2.6.17 any impediment, prevention or default, whether by act or omission, by the Employer or any person for whom the Employer is responsible except to the extent that it was caused or contributed to by any default, whether by act or omission, of the Sub-Contractor or his servants, agents or sub-sub-contractors.

17.2.3 Commentary

The text should be compared with clause 25 of JCT 98, which it closely parallels, and the comments in this chapter will be confined to the differences between claims under the two forms. Largely, these arise from the nature of the contractual relationship. The only parties to the sub-contract are the main contractor and the sub-contractor. There is no contractual relationship between sub-contractor and employer. The practical consequence for present purposes is that the sub-contractor makes claims against the main contractor who passes them up the contractual chain, if the 'claim' is one for which the employer may be responsible to the contractor in law. Despite the Contracts (Rights of Third Parties) Act 1999, third party rights are expressly excluded from JCT 98 by clause 1.12 and from NSC/C by clause 1.6

The first sentence of clause 2.2.1 is akin to JCT 98, clause 25.2.1.1, and it is for the sub-contractor to give written notice of delay to the main contractor. He must do this 'forthwith' (i.e. as soon as he reasonably can[674]) 'if and whenever it becomes reasonably apparent that the commencement,

---

[674] *Hudson* v. *Hill* (1874) 43 LJCP 273; *London Borough of Hillingdon* v. *Cutler* [1967] 2 All ER 361.

progress or completion' of the whole or part of the sub-contract works 'is being or is likely to be delayed'. The notice must identify any matter which in his opinion comes within clause 2.3.1. Clause 2.3.1 refers to 'the occurrence of a Relevant Event or . . . an act, omission or default of the Contractor or any person for whom the Contractor is responsible'. The sub-contractor's notice to the contractor must, therefore, identify not only a relevant event, i.e. one of the grounds listed in clause 2.6, but also refer to any alleged act, default or omission of the main contractor or those for whom he is responsible in law, if the sub-contractor alleges that this is a cause of delay.

The contractor is to notify the architect forthwith of the sub-contractor's written notice; he is also to let the architect have any written representations made by the sub-contractor regarding the causes of delay, which may include his own default, of course. The sub-contractor's notice is itself to state the 'material circumstances' including the cause or causes of the delay, and it must set out how and why the delay is likely to occur, in some detail.

The sub-contractor is under further obligations. *If practicable* in the notice itself, or otherwise as soon as possible after the notice, he must 'give particulars of the expected effects' of each delaying cause identified in the notice, and it seems that each cause is to be considered in isolation. He must also give his own estimate of the expected delay in completion of the sub-contract works resulting from each delaying cause he has identified in his notice – again considered in isolation whether or not it is concurrent with a delay resulting from any other matter which comes within clause 2.3.1, i.e. acts, omissions or defaults of the contractor or those for whom he is responsible in law and 'Relevant Events'.

The sub-contractor's duty does not stop there; he must keep each notice of delay under review and revise his statement of particulars and estimates and/or give whatever further notices may reasonably be necessary to keep the contractor up to date with developments as they occur.

By clause 2.2.3 the contractor must submit the particulars, estimates, etc. to the architect, and undertakes, if the sub-contractor requests it, to join with the sub-contractor in requesting the architect's consent to an extension under JCT 98, clause 35.14.1.

When, and only when, he has received the notice, particulars, estimate, and request for his consent from the contractor, the architect must consider and decide (1) whether any of the causes of delay specified in the notice is a relevant event or default etc. of the contractor, and (2) whether completion of the sub-contract works or any part of them is, in fact, likely to be delayed thereby.

If his conclusions are positive, but only if it is reasonably practicable for him to do so having regard to the sufficiency of the information supplied, the architect must give his written consent to the contractor, granting an extension of time. It is the contractor who fixes the new period for completion, but he can only do this with the architect's written consent. The position regarding the architect's consent is somewhat ambiguous. The contractor is to fix such new period for completion as the architect 'then estimates to be fair and reasonable'. It is sometimes thought that this clause gives the architect power to specify the period of extension. Clause 35.14.2 of JCT 98, however, gives no such power and imposes no such duty. Indeed, it clearly confines the architect's powers to the giving or withholding of consent:

> 35.14   .1   The Contractor shall not grant to any Nominated Sub-Contractor any extension of the period or periods within which the sub-contract works (or where the sub-contract works are to be completed in parts any part thereof) are to be completed except in accordance with the relevant provisions of Conditions NSC/C which require the written consent of the Architect to any such grant.
>
> .2   The Architect shall operate the relevant provisions of Sub-Contract NSC/C upon receiving any notice, particulars and estimate and a request from the Contractor and any Nominated Sub-Contractor for his written consent to an extension of the period or periods for the completion of the sub-contract works or any part thereof as referred to in clause 2.3 of Conditions NSC/C.

That also appears to be the position under NSC/C, which makes no mention of the architect 'fixing' or 'specifying' a period, but only that the contractor's fixing is subject to the architect's consent. Of course, the architect must estimate whether he believes the contractor's proposed extension is fair and reasonable. Common sense alone dictates that the architect cannot specify the period. Compared with the contractor, and even when supplied with information, he is not in the best position to determine the appropriate extension of time for the sub-contract.

It is stated that 'in agreement with the architect' the contractor is to state in his extension (1) which matters they have taken into account, and (2) the extent, if any, to which the architect, in giving consent to the extension, has taken into account any omission instruction under clause 13.2 of JCT 98. The last part of this clause, which used to restrict the consideration to omission instructions or restrictions since the first fixing of a revised period or periods, is omitted in this edition, freeing the architect to consider any such omissions or restrictions from the beginning. This change is to be welcomed, because it puts this provision on a similar basis to the equivalent provision in JCT 98.

The contractor is given a time limit of 12 weeks from receipt of a notice of delay and of 'reasonably sufficient particulars and estimate' from the sub-contractor in which to give an extension of time. If there are fewer than 12 weeks left between receipt of the notice, particulars and estimate and the currently fixed completion date, the architect must reach his decision, and give or withhold his written consent, and the contractor must grant any extension no later than that date. However, as under JCT 98, that is provided it is reasonably practicable to do so.

As in JCT 98, if he does not consider any extension of time to be due, the architect must withhold consent and inform the contractor of that decision, which must then be passed down to the sub-contractor not later than 12 weeks after receipt of the requisite notice, particulars or estimate or not later than the expiry of the sub-contract period or periods, whichever is the earlier.

Clause 2.4 is the provision empowering the contractor (and the architect) to take into account any main contract omission instructions issued, omitting work, obligations or restrictions since the completion date was last fixed, so that the time necessary for completion has in their opinion been thereby reduced. This clause appears to give the contractor power to act unilaterally, i.e. without any notice from the sub-contractor. In other words, as soon as the contractor becomes aware that the omission of work has resulted in the sub-contractor requiring less time to complete the sub-contract works, he can act. There is no express provision equivalent to clause 25.3.6 JCT 98, but on general principle the sub-contractor cannot be deprived of his original completion period without his own express consent.

There are two points to note. First, while extensions previously granted can be reduced, or even extinguished completely so as to return to the original date, no earlier date than that can be fixed by the contractor unilaterally no matter how much work is omitted. Second, extensions can only be reduced on account of omissions of work instructed since an extension was last granted. Each extension is deemed to take into account omissions of work instructed up to the date of the extension.

Clause 2.5, in common with JCT 98 clause 25.3.3, makes it clear that the architect in making his final decision on extensions of time, can take into account any events entitling the sub-contractor to an extension of time which occur between the expiry of the period or periods previously fixed for completion of the sub-contract works and actual practical completion, as well as reviewing the events of the whole contract period.

As under clause 25.3.4 of JCT 98, clause 2.5 of NSC/C requires the sub-contractor to 'use constantly his best endeavours to prevent delay in the progress of the Sub-Contract Works...*howsoever caused*'. This is an important sub-contractual obligation. He is also required to use his best endeavours to prevent any such delay resulting in the completion of the sub-contract works being delayed or further delayed. The sub-contractor is also required to do all reasonably required to the satisfaction of the architect and the contractor to proceed with the sub-contract works. The requirement is to *prevent* delay and it is suggested that as under JCT 98, the requirements in this proviso do not require the sub-contractor actually to spend substantial sums of money[675].

The contractor is to review the completion date in light of any relevant events whether or not they have been notified by the sub-contractor. He must do this not more than 12 weeks after the issue of the certificate of practical completion issued under clause 35.16 of the main contract. The obligation is quite plain: the contractor is to review the extensions of time granted and make a final decision as to the sub-contractor completion date, which he must convey to the sub-contractor (with the architect's written

---

[675] See the consideration of 'best endeavours' in Chapter 10, section 10.1.2.

consent) in writing. In doing this he must review all extensions previously granted and take into account any causes of delay of which he is aware and which he considers should fairly entitle the sub-contractor to an extension even though the sub-contractor has not notified him of them.

This final review is only to enable the contractor (and the architect) either to confirm the completion date previously fixed or to grant a further extension of time. The contractor may not reduce extensions previously granted unless, *since an extension was last granted*, the architect has issued an omission instruction under main contract clause 13.2, thereby reducing the time needed to complete what is left.

17.2.4 Relevant events

Clause 2.6 lists the relevant events, the occurrence of which, in principle, gives rise to an extension of time. Grounds 2.6.1 to 2.6.18 parallel those listed in JCT 98, clause 25.4, and reference should be made to the commentary thereon[676].

'An act, omission or default of the Contractor or any person for whom the Contractor is responsible' is an additional ground for extension of time: see clause 2.3.1. This phrase now extends to sub-sub-contractors (or 'tertiary contractors' as they are sometimes called)[677].

Finally, if the sub-contractor is aggrieved by the architect's failure to give his written consent (clause 2.3) to an extension in due time or at all and/or by the terms of that consent, e.g. the extension granted, he has a right to require the main contractor to join in dispute resolution procedures, subject to safeguards as to costs: clause 2.7.

17.2.5 Direct loss and/or expense claims under NSC/C

Clauses 4.38 to 4.41 provide:

4.38 .1 If the Sub-Contractor makes written application to the Contractor stating that he has incurred or is likely to incur direct loss and/or expense in the execution of the Sub-Contract for which he would not be reimbursed by a payment under any other provision in the Sub-Contract due to deferment by the Employer of giving to the Contractor possession of the site of the Works where it is stated in the completed Appendix of the Main Contract Conditions (attached to NSC/T Part 1 or, if different, in the completed Appendix of the Main Contract Conditions enclosed with the copy of Nomination NSC/N sent to the Sub-Contractor by the Architect) that clause 23.1.2 of the Main Contract Conditions applies to the Main Contract by reason of the regular progress of

---

[676] See Chapter 10, section 10.1.3.

[677] The previous phrasing 'his servants or agents or his sub-contractors, their servants or agents' was held not to be sufficient to include sub-sub-contractors: *City of Manchester* v. *Fram Gerrard Ltd* (1974) 6 BLR 70.

the Sub-Contract Works or of any part thereof having been or being likely to be materially affected by any one or more of the matters referred to in clause 4.38.2, the Contractor shall require the Architect to operate clause 26.4 of the Main Contract Conditions so that the amount of that direct loss and/or expense, if any, may be ascertained. Provided always that:

.1 the Sub-Contractor's application shall be made as soon as it has become, or should reasonably have become, apparent to him that the regular progress of the Sub-Contract Works or of any part thereof has been or was likely to be affected as aforesaid; and

.2 the Sub-Contractor shall submit to the Contractor such information in support of his application as the Contractor is requested by the Architect to obtain from the Sub-Contractor in order reasonably to enable the Architect to operate clause 26.4 of the Main Contract Conditions; and

.3 the Sub-Contractor shall submit to the Contractor such details of such loss and/or expense as the Contractor is requested by the Architect or the Quantity Surveyor to obtain from the Sub-Contractor in order reasonably to enable the ascertainment of that loss and/or expense under clause 26.4 of the Main Contract Conditions.

.2 The matters to which clause 4.38.1 applies are:

.1 .1 where an Information Release Schedule has been provided failure of the Architect to comply with clause 5.4.1 of the Main Contract Conditions;

.2 failure of the Architect to comply with clause 5.4.2 of the Main Contract Conditions.

except to the extent that the failure referred to in clause 4.38.2.1.2 results from a breach by the Sub-Contractor of his obligations to the Employer under clause 3.2 of the Agreement NSC/W;

.2 the opening up for inspection of any work covered up or the testing of any of the work, materials or goods in accordance with clause 8.3 of the Main Contract Conditions (including making good in consequence of such opening up or testing), unless the inspection or test showed that the work, materials or goods were not in accordance with the Main Contract or the Sub-Contract as the case may be; or

.3 any discrepancy in or divergence between the Contract Drawings and/or the Contract Bills and/or the Numbered Documents; or

.4 the execution of work not forming part of the Main Contract by the Employer himself or by persons employed or otherwise engaged by the Employer as referred to in clause 29 of the Main Contract Conditions or the failure to execute such work or the supply by the Employer of materials and goods which the Employer has agreed to provide for the Works or the failure so to supply;

.5 Architect's instructions issued in regard to the postponement of any work to be executed under the provisions of the Main Contract or the Sub-Contract; or

.6   failure of the Employer to give in due time ingress to or egress from the site of the Works, or any part thereof through or over any land, buildings, way or passage adjoining or connected with the site and in the possession and control of the Employer, in accordance with the Contract Bills and/or the Contract Drawings and/or the Numbered documents, after receipt by the Architect of such notice, if any, as the Contractor is required to give or failure of the Employer to give such ingress or egress as otherwise agreed between the Architect and the Contractor; or

.7   Architect's instructions issued

under clause 13.2 or under clause 13A.4.1 of the Main Contract Conditions requiring a Variation except for a Variation for which the Architect has issued to the Contractor a confirmed acceptance of a 13A Quotation pursuant to clause 13A.3.2 of the Main Contract Conditions or for a Variation to such work or

under clause 13.3 of the Main Contract Conditions in regard to the expenditure of provisional sums (other than an instruction to which clause 13.4.2 of the Main Contract Conditions refers or, where bills of quantities are included in the Numbered Documents, an instruction for the expenditure of a provisional sum for defined work); or

.8   where bills of quantities are included in the Numbered Documents the execution of work for which an Approximate Quantity is included in those bills which is not a reasonably accurate forecast of the quantity of work required; or

.9   compliance or non-compliance by the Employer with clause 6A.1 of the Main Contract Conditions (Employer's obligation – Planning Supervisor – Principal Contractor where not the Contractor); or

.10 save as provided for in clauses 4.38.1 to 4.38.9 any impediment, prevention or default, whether by act or omission, by the Employer or any person for whom the Employer is responsible except to the extent that it was caused or contributed to by any default, whether by act or omission, of the Sub-Contractor or his servants, agents or sub-sub-contractors.

.3   Any amount from time to time ascertained as a result of the operation of clause 4.38.1 shall be added to the Sub-Contract Sum or included in the calculation of the Ascertained Final Sub-Contract Sum.

.4   The Sub-Contractor shall comply with all directions of the Contractor which are reasonably necessary to enable the ascertainment which results from the operation of clause 4.38.1 to be carried out.

4.39 If the regular progress of the Sub-Contract Works (including any part thereof which is sub-sub-contracted) is materially affected by any act, omission or default of the Contractor (including, where the Contractor is the Principal Contractor, any omission or default in the discharge of his obligations as the Principle Contractor), or any person for whom the Contractor is responsible (see clause 6.3.1), the Sub-Contractor shall within a reasonable time of such material effect becoming apparent give written notice thereof to the Contractor and the agreed amount of any direct loss and/or expense thereby caused to the Sub-Contractor shall be recoverable by the Sub-Contractor from the Contractor as a debt. Provided always that:

.1   the Sub-Contractor's application shall be made as soon as it has become, or should reasonably have become, apparent to him that the regular progress of the Sub-Contractor Works or of any part thereof has been or was likely to be affected as aforesaid; and

.2   the Sub-Contractor, in order to enable the direct loss and/or expense to be ascertained, shall submit to the Contractor such information in support of his application including details of the loss and/or expense as the Contractor may reasonably require from the Sub-Contractor.

4.40   If the regular progress of the Works (including any part thereof which is sub-contracted) is materially affected by any act, omission or default of the Sub-Contractor or any person for whom the Sub-Contractor is responsible (see clause 6.3.1), the Contractor shall within a reasonable time of such material effect becoming apparent give written notice thereof to the Sub-Contractor and the agreed amount of any direct loss and/or expense thereby caused to the Contractor (whether suffered or incurred by the Contractor or by sub-contractors employed by the Contractor on the Works from whom claims under similar provisions in the relevant sub-contracts have been agreed by the Contractor, sub-contractor and the Sub-Contractor) may be deducted from any monies due or to become due to the Sub-Contractor or may be recoverable from the Sub-Contractor as a debt. Provided always that:

.1   the Contractor's application shall be made as soon as it has become, or should reasonably have become, apparent to him that the regular progress of the Works, (including any part which is sub-contracted) has been or was likely to be affected as aforesaid; and

.2   the Contractor, in order to enable the direct loss and/or expense to be ascertained, shall submit to the Sub-Contractor such information in support of his application including details of the loss and/or expense as the Sub-Contractor may reasonably request from the Contractor.

4.41   The provisions of clause 4.38 to 4.40 are without prejudice to any other rights or remedies which the Contractor or Sub-Contractor may possess.

17.2.6 Commentary

Sub-contractor's claims

Clause 4.38 gives the sub-contractor a right to claim *through* the main contractor for direct loss and/or expense not covered by a payment under any other provision in the sub-contract. Other than clauses 4.39 and 4.40 it almost parallels JCT 98, clause 26 and, so far as claims made by the sub-contractor are concerned, for the most part the situation is exactly the same as with claims by the contractor under JCT 98, clause 26[678]. However and unlike other sub-contracts, there appears to be no provision which entitles the sub-contractor to any loss and/or expense if he properly suspends performance of his obligations under clause 4.21. Causation is an obstacle to bringing the claim under clause 4.39 as being due to a default on the part of the contractor. The plain fact is that, on a failure to pay by the contractor, the sub-contractor is not obliged to suspend performance of his obligations. He may do so if he so wishes. Therefore, the suspension breaks the chain of causation. It is the suspension and not the failure to pay which causes the sub-contractor loss and/or expense.

Specific comment is, however, required on the terms of clause 4.39, of which the margin note states 'Disturbance of regular progress of Sub-Contract Works – Sub-Contractor's claims'. It deals with disturbance claims that arise as a result of the fault of the main contractor or of those for whom he is responsible in law. For a claim under this head to be successful, the *regular progress* of the sub-contract works including any part which is sub-sub-contracted must be *materially* affected by any act, omission or default of the contractor or any person for whom the contractor is responsible. It should be noted that this clause does not refer to occurrences which, put bluntly, the contractor can blame on the employer or his architect.

The onus is on the sub-contractor to give written notice of the claim to the main contractor. This he must do as soon as it has become, or reasonably should have become, apparent to him of the material effect on progress. The clause envisages that the amount of such a disturbance claim will be agreed between the parties by negotiation: *'the agreed amount* of any direct loss and/or expense thereby caused … shall be recoverable … from the Contractor as a debt'. If the parties are unable to agree, then the dispute is referable to adjudication or arbitration. In practice, of course, it is very unlikely that the contractor will agree any amount due to the contractor, but failing agreement, there can be no automatic recovery of the sum suggested by the sub-contractor[679].

Main contractor's claims

Clause 4.40 is a matching provision dealing with main contractor's claims that arise as a result of any act, omission or default of the sub-contractor or any person for whom he is responsible, and the procedure is similar to that in clause 4.39. It must be initiated by the contractor's application and the sub-contractor is entitled to request supporting information provided only that the request is reasonable. Under clause 4.39, the contractor can *require* details, the sub-contractor may only *request*. It is not clear why different words have been used and it is always possible that it is a mistake. In

---

[678] See Chapter 12, section 12.2.2.
[679] *Hermcrest plc* v. *G. Percy Trentham Ltd* (1991) 53 BLR 104.

practice it will make little difference in that a failure on the part of either contractor or sub-contractor will not encourage the other to *agree* even if there was the will to do so. It should be noted that such claims extend to claims by other sub-contractors of the main contractor, provided they are agreed between all three parties. When claims made under clause 4.40 are agreed, the main contractor may deduct the amount agreed from monies due or to become due to the sub-contractor or, if necessary, recover the sums due as a debt. It is clear that agreement is essential under this clause, as under clause 4.39.

Clause 4.41 preserves to both parties their other rights and remedies[680].

17.2.7 Delayed completion by the sub-contractor

The main contractor's right to claim against a nominated sub-contractor for delay in completion of the sub-contract works is set out in clauses 2.8 and 2.9 of form NSC/C as follows:

2.8 If the Sub-Contractor fails to complete the Sub-contract Works or any part thereof within the period or periods for completion stated in the agreed programme details in NSC/T Part 3 item 1 or any revised period or periods fixed under clauses 2.2 to 2.7 or any revised period stated by the Contractor in his acceptance of a 3.3A Quotation, the Contractor shall so notify the Architect and give the Sub-Contractor a copy of such notification.

2.9 The Sub-Contractor shall pay or, subject to clauses 4.26 to 4.29, allow to the Contractor a sum equivalent to any loss or damage suffered or incurred by the Contractor and caused by the failure of the Sub-Contractor as aforesaid. Provided that the Contractor shall not be entitled to such sum unless the Architect in accordance with clause 35.15 of the Main Contract Conditions shall have issued to the Contractor (with a copy to the Sub-Contractor) a certificate in writing certifying any failure notified under clause 2.8.

Clause 35.15 of the Main Contract Conditions runs as follows:

35.15.1 If any Nominated Sub-Contractor fails to complete the sub-contract works (or where the sub-contract works are to be completed in parts any part thereof) within the period specified in the Nominated Sub-Contract or within any extended time granted by the Contractor with the written consent of the Architect, and the Contractor so notifies the Architect with a copy to the Nominated Sub-Contractor, then, provided that the Architect is satisfied that clause 35.14 has been properly applied, the Architect shall so certify in writing to the Contractor. Immediately upon the issue of such a certificate the Architect shall send a duplicate copy thereof to the Nominated Sub-Contractor.

35.15.2 The certificate of the Architect under clause 35.15.1 shall be issued not later than 2 months from the date of notification to the Architect that the Nominated Sub-Contractor has failed to complete the sub-contract works or any part thereof.

---

[680] i.e., claims at common law, see Chapter 4.

17.2.8 Commentary

The main contractor's right to claim for delayed completion of the sub-contract works is therefore dealt with on a different basis from his right to claim for the effect of any act, omission or default of the sub-contractor upon regular progress of the main contract works[681]. It is a condition precedent to the contractor's right to claim that he should have the architect's certificate issued under main contract clause 35.15.1. That certificate, as with the non-completion certificate under the main contract clause 24.1[682] is to be a simple statement of the fact that the sub-contractor has not completed within the specified period or any extended period yet granted. There is, however, the important proviso that the architect is only to issue the certificate if he is satisfied that the provisions of clause 35.14 of the main contract (also embodied in clause 2.3 of the sub-contract) regarding extensions of time have been properly applied, and the architect is therefore given a period of 2 months before he may become obliged to issue the certificate. The architect is thus clearly placed under an obligation to make proper investigation to satisfy himself that the main contractor has passed to him all notices and applications from the sub-contractor regarding delays to the sub-contract works (including those alleging delay caused by the main contractor's own default) and that all extensions of time for which he has given consent have been properly granted.

Indeed, this has been emphasised by the case of *Brightside Kilpatrick Engineering Services* v. *Mitchell Construction (1973) Ltd*[683] where the court decided that the refusal of the architect to grant a 27(d)(ii) certificate[684] under JCT 63 could not be a matter to be arbitrated upon in an arbitration under the sub-contract, since neither the main contractor nor the sub-contractor had any responsibility for the architect's decisions.

It is likely that an architect would be failing in his duty to a sub-contractor, and it is possible that he may even be liable to him in damages if he issues a 35.15 certificate simply at the request of the contractor and without making proper investigation and enquiry, including seeking the views of the sub-contractor, in order to satisfy himself that he is in possession of sufficient facts to form a proper opinion[685]. It may be, for instance, that a simple enquiry to the sub-contractor will reveal that the main contractor has failed to pass on notices of delay sent to him by the sub-contractor, perhaps because they allege that the delays are due to the main contractor's default. On the other hand the architect would be failing in his duty to the main contractor if he flatly refused to issue a 35.15 certificate; it is

---

[681] See the consideration of clause 4.40 above.
[682] See Chapter 11, section 11.1.2.
[683] (1973) 1 BLR 64.
[684] Generally equivalent to the architect's certificate under clause 35.15 of JCT 98.
[685] See the developments in the law relating to assumption of responsibility: *Henderson* v. *Merritt Syndicates* (1994) 69 BLR 26 and Cartwright, John: *Liability in Negligence: New Directions or Old*, (1997) 13 Const LJ 157.

up to the architect to make the proper enquiries and, if he is satisfied following such enquiries that there is no legitimate reason for the sub-contractor's failure to complete the work within the period established under the sub-contract, to issue a certificate to that effect so that the contractor may claim his proper entitlement against the sub-contractor.

The importance of the issue of the architect's certificate under clause 35.15.1 of JCT 98 as a condition precedent to the main contractor's right to claim against the sub-contractor was emphasised by the case of *Tubeworkers Ltd* v. *Tilbury Construction Ltd*[686]. The case was under the old 'Green Form' of sub-contract under JCT 63, but the principles are the same.

The main contractor applied to the architect for a certificate stating that the nominated sub-contractor was in delay. The architect purported to issue such a certificate on 14 July 1983, but it was generally agreed that it was defective and the main contractor did not at that time take any steps to rely upon it. On 13 September 1983, the architect issued an interim certificate for payment under the main contract including a direction for the payment of a substantial sum to the sub-contractor. On 31 October 1983, the main contractor asked the architect to issue a new, proper, certificate of delay and this was issued by the architect on 9 November 1983. On 16 November 1983, having withheld payment to the sub-contractor under the interim certificate in the meantime, the main contractor put forward a counterclaim to the sub-contractor for a considerably larger sum as a purported 'set-off' under clause 13A of the Green Form (equivalent to clauses 4.26 to 4.29 of NSC/C).

The Official Referee held that the main contractor, under the terms of the sub-contract relating to set-off, was not entitled to withhold payment since, at the time that payment became due, he did not have a valid architect's certificate of delay, and this was not cured by the subsequent issue of a valid certificate of delay.

While it might be argued that the absence of the word 'direct' would give rise to a wider entitlement, it is not thought that in practice there is any real distinction between the 'loss or damage suffered or incurred' which the main contractor can recover under this clause and the 'direct loss and/or expense' which he can recover under clause 4.40, although it is possible that he can claim consequential loss under this provision in the correct circumstances. The main contractor's right to deduct the amount of any loss or damage from payments certified as due to the sub-contractor will be subject to the conditions governing set-off set out in clauses 4.26 to 4.29 of Form NSC/C.

17.3 Domestic sub-contract form DSC/C

17.3.1 Introduction

Domestic sub-contractors are work and material sub-contractors who are not nominated by the architect under clause 35 of JCT 98. The full title of this

---

[686] (1985) 30 BLR 67.

document, which is issued by the Joint Contracts Tribunal, is The JCT Standard Form of Domestic Sub-Contract 2002 Edition It is known as DSC/C. It supersedes the DOM/1 which was introduced in 1980[687].

A version of DOM/1 is still retained (DOM/2) for use in conjunction with the JCT Standard Form With Contractor's Design (WCD 98). DOM/2 is essentially a series of amendments to DOM/1 in order to import a sub-contract design responsibility reflecting the contractor's own design responsibility under WCD 98.

17.3.2 Extensions of time under DSC/C

Extensions of time are dealt with by clauses 2.1 to 2.10, which read as follows:

> **Sub-Contractor's obligation – carrying out and completion of Sub-Contract Works – extension of Sub-Contract time (2.1 to 2.10)**

2.1   The Sub-Contractor shall carry out and complete the Sub-Contract Works in accordance with the programme details in item 1 of the Particular Conditions in Agreement DSC/A and reasonably in accordance with the progress of the Works but subject to receipt of the notice to commence work on site as stated in item 1 of the Particular Conditions in Agreement DSC/A, and to the operation of clauses 2.2 to 2.10 and any revision to the period or periods for the completion of the Sub-Contract Works in respect of a Variation for which a 3.4 Quotation has been given and which has been stated by the Contractor in his acceptance of the 3.4 Quotation.

2.2   .1   If and whenever it becomes reasonably apparent that the commencement, progress or completion of the Sub-Contract Works or any part thereof is being or is likely to be delayed, the Sub-Contractor shall forthwith give written notice to the Contractor of the material circumstances including, insofar as the Sub-Contractor is able, the cause or causes of the delay and identify in such notice any matter which in his opinion comes within clause 12.3.1.

      .2   In respect of each and every matter which comes within clause 2.3.1, and identified in the notice given in accordance with clause 2.2.1, the Sub-Contractor shall, if practicable in such notice, or otherwise in writing as soon as possible after such notice:

            .1   give particulars of the expected effects thereof; and

            .2   estimate the extent, if any, of the expected delay in the completion of the Sub-Contract Works or any part thereof beyond the expiry of the period or periods stated in item 1 of the Particular Conditions in Agreement DSC/A or beyond the expiry of any extended period or periods previously fixed under clauses 2.3, 2.6 or 2.7 which results

---

[687] The commentary on DOM/1 which was contained in the previous edition of this book is reproduced in Appendix B.

therefrom whether or not concurrently with delay resulting from any other matter which comes within clause 2.3.1; and

.3 the Sub-Contractor shall give such further written notices to the Contractor as may be reasonably necessary or as the Contractor may reasonably require for keeping up to date the particulars and estimate referred to in clauses 2.2.2.1 and .2 including any material change in such particulars or estimate.

2.3 If on receipt of any notice, particulars and estimate under clause 2.2 the Contractor properly considers that:

.1 any of the causes of the delay is an act, omission or default of the Contractor, (including where the Contractor is the Principle Contractor, any act, omission or default in the discharge of his obligations as the Principal Contractor) his servants or agents or his sub-contractors, their servants or agents (other than the Sub-Contractor, his servants or agents) or is the occurrence of a Relevant Event; and

.2 the completion of the Sub-Contract Works is likely to be delayed thereby beyond the period or periods stated in item 1 of the Particular-Conditions in Agreement DSC/A or any revised such period or periods.

then the Contractor shall, in writing, give an extension of time to the Sub-Contractor by fixing such revised or further revised period or periods for the completion of the Sub-Contract Works as the Contractor then estimates to be reasonable. Provided that

.3 where the Contractor has stated a revision to the period or periods for the completion of the Sub-Contract Works in his acceptance of a 3.4 Quotation in respect of any Variation; or

.4 where bills of quantities are included in the Numbered Documents compliance with a direction for the expenditure of a Provisional Sum for defined work or of a provisional Sum for Performance Specified Work,

such revision shall be made and no further or other revisions of time shall be given in respect of the Variation to which the 3.4 Quotation or such direction relates.

2.4 .1 When fixing such revised period or periods, the Contractor shall, if reasonably practicable having regard to the sufficiency of the notice, particulars and estimate, fix such revised period or periods within the following time limit:

.1 not later than 16 weeks from the receipt by the Contractor of the notice and of reasonably sufficient particulars and estimates, or

.2 where the time between receipt thereof and the expiry of the period or periods for the completion of the Sub-Contract Works is less than 16 weeks, not later than the expiry of the aforesaid period or periods.

2.4 .2 The Contractor, when fixing such revised period or periods, shall state:

.1   which of the matters, including any of the Relevant Events, referred to in clause 2.3.1 he has taken into account; and

.2   the extent, if any, to which the Contractor has regard to any direction requiring as a Variation the omission of any work or obligation or restriction issued since the previous fixing of any such revised period or periods for the completion of the Sub-Contract Works.

2.5   If, upon receipt of any notice, particulars and estimate under clause 2.2, the Contractor properly considers that he is unable to give, in writing, an extension of time to the Sub-Contractor, the Contractor shall, if reasonably practicable having regard to the aforesaid notice, particulars and estimate, so notify the Sub-Contractor in writing not later than 16 weeks from receipt of the notice particulars and estimate, or, where the time between such receipt and the expiry of the period or periods for the completion of the Sub-Contract Works is less than 16 weeks, not later than the expiry of the aforesaid period or periods.

2.6   After the first exercise by the Contractor of the duty under clause 2.3, or after any revision to the period or periods for the completion of the Sub-Contract Works stated by the Contractor in his acceptance of a 3.4 Quotation in respect of a Variation, the Contractor may in writing fix a period or periods for completion of the Sub-Contract Works shorter than that previously fixed under clause 2.3 or stated by the Contractor in his acceptance of a 3.4 Quotation if, in the opinion of the Contractor, the fixing of such shorter period or periods is fair and reasonable having regard to any direction issued requiring as a Variation the omission of any work or obligation or restriction where such issue is after the last occasion on which the Contractor made a revision of the aforesaid period or periods. Provided that no decision under clause 2.6 shall alter the length of any revision to the period or periods for the completion of the Sub-Contract Works in respect of a Variation for which a 3.4 Quotation has been given and which has been stated by the Contractor in his acceptance of the 3.4 Quotation, or any decision of the Adjudicator under clause 9A or of an arbitrator or the courts under clause 9B or 9C, as the case may be.

2.7   If the expiry of the period when the Sub-Contract Works should have been completed in accordance with clause 2.1 occurs before the date of practical completion of the Sub-Contract Works established under clause 2.13 or 2.14, the Contractor may and not later than the expiry of 16 weeks from the aforesaid date of practical completion of the Sub-Contract Works, the Contractor shall either:

.1   fix such a period or periods for completion of the Sub-Contract Works longer than that previously fixed under clauses 2.3 or 2.6 or stated by the Contractor in his acceptance of a 3.4 Quotation as the Contractor properly considers to be fair and reasonable having regard to any of the matters referred to in clause 2.3.1 whether upon reviewing a previous decision or otherwise and whether or not the matters referred to in clause 2.3.1 have been specifically notified by the Sub-Contractor under clause 2.2; or

.2   fix such a period or periods for completion of the Sub-Contract Works shorter than that previously fixed under clauses 2.3 or 2.6 as the Contractor properly considers to be fair and reasonable having regard to any direction issued requiring as a Variation the omission of any work where such issue is after the last occasion on which the Contractor made a revision of the aforesaid period or periods; or

.3 confirm to the Sub-Contractor the period or periods for the completion of the Sub-Contract Works previously fixed.

Provided that no decision under clause 2.7.2 or 2.7.3 shall alter the length of any revision to the period or periods for the completion of the Sub-Contract Works in respect of a Variation for which a clause 3.4 Quotation has been given and which has been stated by the Contractor in his acceptance of a clause 3.4 Quotation, or any decision of the Adjudicator under clause 9A or of an arbitrator or the courts under clause 9B or 9C, as the case may be.

2.8 The operation of clauses 2.1 to 2.10 shall be subject to the proviso that the Sub-Contractor shall use constantly his best endeavours to prevent delay in the progress of the Sub-Contract Works or any part thereof, however caused, and to prevent any such delay resulting in the completion of the Sub-Contract Works being delayed or further delayed beyond the period or periods for completion, and the Sub-Contractor shall do all that may reasonably be required to the satisfaction of the Architect and the Contractor to proceed with the Sub-Contract Works.

2.9 No decision of the Contractor under clauses 2.2 to 2.7 inclusive shall fix a period or periods for completion of the Sub-Contract Works which will be shorter than the period or periods stated in item 1 of the Particular Conditions in Agreement DSC/A.

2.10 The following are the Relevant Events referred to in clause 2.3.1:

.1 force majeure;

.2 exceptionally adverse weather conditions;

.3 loss or damage occasioned by any one or more of the Specified Perils;

.4 civil commotion, local combination of workmen, strike or lockout affecting any of the trades employed upon the Works or any of the trades engaged in the preparation, manufacture or transportation of any of the goods or materials required for the Works;

.5 .1 compliance by the Contractor with the Architect's instructions (which shall be deemed to include compliance by the Sub-Contractor with the Contractor's directions which pass on such instructions):

.1 under clause 2.3, 2.4.1, 13.2 (except for a confirmed acceptance of a 13A Quotation), 13.3 (except compliance with an Architect's instruction for the expenditure of a provisional sum for defined work or of a provisional sum for Performance Specified Work) 13A.4.1, 23.2, 34, 35 or 36; or

.2 in regard to the opening up for inspection of any work covered up or the testing of any of the work, materials or goods in accordance with clause 8.3 (including making good in consequence of such opening up or testing) unless the inspection or test showed that the work, materials or goods were not in accordance with the Main Contract;

.2 compliance by the Sub-Contractor with the Contractor's directions in regard to the opening up for inspection of any work covered up or the testing of any of the work, materials or goods in accordance with clause 3.5 (including making good in consequence of such opening up or testing) unless the inspection or test showed that the work, materials or goods were not in accordance with this Sub-Contract;

.6 .1 where an Information Release Schedule has been provided to the main Contractor failure of the Architect to comply with clause 5.4.1 of the Main Contract Conditions;

.2 failure of the Architect to comply with clause 5.4.2 of the Main Contract Conditions;

.7 delay on the part of Nominated Sub-Contractors which the Contractor has taken all practicable steps to avoid or reduce;

.8 .1 the execution of work not forming part of the Main Contract by the Employer himself or by persons employed or otherwise engaged by the Employer as referred to in clause 29 of the Main Contract Conditions or the failure to execute such work;

.8 .2 the supply by the Employer of materials and goods which the Employer has agreed to provide for the Works or the failure so to supply;

.9 the exercise after the Main Contract Base Date by the United Kingdom Government of any statutory power which directly affects the execution of the Works by restricting the availability or use of labour which is essential to the proper carrying out of the Works, or preventing the Contractor or Sub-Contractor from, or delaying the Contractor or Sub-Contractor in, securing such goods or materials or such fuel or energy as are essential to the proper carrying out of the Works;

.10 .1 the Contractor's or Sub-Contractor's inability for reasons beyond his control and which he could not reasonably have foreseen at the Main Contract Base Date:

.1 to secure such labour as is essential to the proper carrying out of the Works; or
.2 to secure such goods or materials as are essential to the proper carrying out of the Works;

.11 the carrying out by a local authority or statutory undertaker of work in pursuance of its statutory obligations in relation to the Works, or the failure to carry out such work;

.12 failure of the Employer to give in due time ingress to or egress from the site of the Works or any part thereof through or over any land, buildings, way or passage adjoining or connected with the site and in the possession and control of the Employer, in accordance with the Contract Bills/Specifications/Schedule of Works and/or the Contract Drawings, after receipt by the Architect of such notice, if any, as the Contractor is required to give, or failure of the Employer to give such ingress or egress as otherwise agreed between the Architect and the Contractor;

.13 where it is stated in the completed Appendix of the Main Contract Conditions annexed to Agreement DSC/A that clause 23.1.2 of the Main Contract Conditions applies to the Main Contract, any deferment by the Employer in giving possession of the site of the Works to the Contractor under clause 23.1.2 of the Main Contract Conditions.

.14 .1 by reason of the execution of work for which an Approximate Quantity is included in the Contract Bills which is not a reasonably accurate forecast of the quantity of work required;

.2 where bills of quantities are included in the Numbered Documents, by reason of the execution of work for which an Approximate Quantity is included in those bills which is not a reasonably accurate forecast of the quantity of work required;

.15 .1 delay which the Contractor has taken all practicable steps to avoid or reduce consequent upon a change in the Statutory Requirements after the Main Contract Base Date which necessitates some alteration or modification to any Performance Specified Work in the Main Contract;

.2 delay which the Sub-Contractor has taken all practicable steps to avoid or reduce consequent upon a change in the Statutory Requirements after the Sub-Contract Base Date which necessitates some alteration or modification to any Performance Specified Work;

.16 the use or threat of terrorism and/or the activity of the relevant authorities in dealing with such use or threat;

.17 compliance or non-compliance by the Employer with clause 6A.1 of the Main Contract Conditions;

.18 .1 delay arising from a suspension by the Contractor of the performance of his obligations under the Main Contract to the Employer pursuant to clause 30.1.4 of the Main Contract Conditions;

.2 delay arising from a suspension by the Sub-Contractor of the performance of his obligations under the Sub-Contract to the Contractor pursuant to clause 4.20;

.19 save as provided for in clauses 2.10.1 to 2.10.18 any impediment, prevention or default, whether by act or omission, by the Employer or any person for whom the Employer is responsible except to the extent that it was caused or contributed to by any default, whether by act or omission, of the Sub-Contractor or his servants, agents or sub-sub-contractors.

17.3.3 Commentary

This contract is the successor to DOM/1 to which it bears a strong resemblance. It is also generally similar to NSC/C. The principal difference is that the architect plays no part in the process. The commentary above on NSC/C is, therefore, relevant. There are differences in wording and arrangement of clauses and the major points are as follows.

Progress

The words of clause 2.1 have been construed very broadly by contractors over the years as obliging the sub-contractor to work in accordance with the contractor's progress on the works and that if the contractor's progress slowed or quickened, the sub-contractor was obliged to follow suit. It has been held, under clause 11.1 of DOM/1, the relevant words of which have been reproduced in clause 2.1 of DSC/C, that the sub-contractor may plan and perform the work as he pleases if there is no indication to the contrary, provided that he finishes it by the time fixed in the contract. The sub-contractor's only obligation so far as programming requirements are concerned are those requirements expressly contained in the sub-contract itself[688]. It is likely that the same principle applies in the case of other similarly worded sub-contracts such as NSC/C.

Notice

In clause 2.2.1 the sub-contractor's obligation to give notice of the material circumstances is a qualified one. The material circumstances are to include the cause or causes of the delay, *insofar as the sub-contractor* is able to identify them. In practice, this qualification probably will make little difference, but it does offer the sub-contractor some limited protection against allegations that he has not included all the required information in his notice.

Extension of time

The architect is not involved at all in giving extensions of time. The duty is the contractor's alone. This apart, the procedure is the same as under NSC/C.

Time period

The time period within which the contractor must respond to the sub-contractor's notice is 16 weeks, as opposed to 12 weeks in NSC/C; this time limit runs from receipt of notice and what the contractor considers to be 'reasonably sufficient' particulars and estimate.

Omissions

Omission directions given by the contractor can be taken into account.

---

[688] *Pigott Foundations* v. *Shepherd Construction* (1996) 67 BLR 48.

Best endeavours

In clause 2.8, the operation of the extension of time provisions is made subject to the proviso that the sub-contractor shall 'use constantly his best endeavours to prevent delay ... and the Sub-Contractor shall do all that may reasonably be required to the satisfaction of the *Contractor to proceed with* the Sub-Contract Works' (emphasis added). In the equivalent clause 11.8 of DOM/1, the architect was coupled with the contractor and the removal of the obligation of the contractor to satisfy the architect on this matter is an improvement. In any event the dual obligation was of no practical significance, because the architect had, and has, no right to require anything directly from a domestic sub-contractor.

Completion date

Clause 2.9 is vital; it parallels clause 25.3.6 of JCT 98 and means that, no matter how much work is omitted, the sub-contractor is always entitled to his original sub-contract completion date. These points apart, the provision operates in a similar way as the corresponding clause in NSC/C.

Clause 11 of DOM/2 is the corresponding provision under that sub-contract form and follows the wording of clause 11, DOM/1, with the necessary drafting changes to take account of the differing circumstances in which it is used.

17.3.4 Direct loss and/or expense claims under DSC/C

This matter is dealt with in clauses 4.24 to 4.28, and is the matching provision to NSC/C clauses 4.38 to 4.41. The text reads as follows:

> **Matters affecting regular progress – direct loss and/or expense – Contractor's and Sub-Contractor's rights (4.24 to 4.28)**

> 4.24   If the regular progress of the Sub-Contract Works is materially affected due to deferment of giving to the Contractor possession of the site of the Works where it is stated in the Appendix to the Main Contract Conditions annexed to Agreement DSC/A that clause 23.1.2 of the Main Contract Conditions applies to the Main Contract or the regular progress of the Sub-Contract Works is materially affected by any act, omission or default of the Contractor, (including, where the Contractor is the Principal Contractor, any act, omission or default in the discharge of his obligations as the Principal Contractor) his servants or agents, or any sub-contractor, his servants or agents (other than the Sub-Contractor, his servants or agents), or is materially affected by any one or more of the Relevant Matters referred to in clause 4.26 and if the Sub-Contractor shall within a reasonable time of such material effect becoming apparent make written application to the Contractor, the amount agreed by the Parties of any direct loss and/or expense thereby caused to the Sub-Contractor shall either be included in the adjustment of the Sub-Contract Sum or in the computation of the Ascertained Final

Sub-Contract Sum, whichever is applicable, or be recoverable from the Contractor as a debt. Provided always that:

.1   the Sub-contractor's application shall be made as soon as it has become, or should reasonably have become, apparent to him that the regular progress of the Sub-Contract Works or of any part thereof has been or is likely to be affected as aforesaid; and

.2   the Sub-Contractor shall submit to the Contractor such information in support of his application as is reasonably necessary to show that the regular progress of the Sub-Contract Works or any part thereof has been or is likely to be affected as aforesaid; and

.3   the Sub-contractor shall submit to the Contractor such details of such loss and/or expense as the Contractor requests in order reasonably to enable that direct loss and/or expense as aforesaid to be agreed; and

.4   where the amount of loss and/or expense in respect of a Variation for which a 3.4 Quotation has been accepted by the Contractor, such amount shall be paid by the Contractor to the Sub-Contractor and clause 4.24 shall not apply.

4.25   If, and to the extent that, it is necessary for the agreement of any direct loss and/or expense applied for under clause 4.24, the Contractor shall state in writing to the Sub-contractor what extension of time, if any, has been made under clauses 2.3, 2.6 and 2.7 in respect of the Relevant Events referred to in clause 2.10.5.1 (so far as that clause refers to clauses 2.3, 13.2, 13.3 and 23.2 of the Main Contract Conditions) and in clauses 2.10.5.2, 2.10.6, 2.10.8, 2.10.12 and 2.10.15.

4.26   The following are the Relevant Matters referred to in clause 4.24:

.1   .1   where an Information Release Schedule has been provided to the Main Contractor failure of the Architect to comply with clause 5.4.1 of the Main Contract conditions;

     .2   failure of the Architect to comply with clause 5.4.2 of the Main Contract Conditions;

.2   .1   the opening up for inspection of any work covered up or the testing of any work, materials or goods in accordance with clause 8.3 of the Main Contract Conditions (including making good in consequence of such opening up or testing) unless the inspection or test showed that the work, materials or goods were not in accordance with the Main Contract;

     .2   the opening up for inspection of any work covered up or the testing of any work, materials or goods in accordance with clause 3.5 (including making good in consequence of such opening up or testing) unless the inspection or test showed that the work, materials or goods were not in accordance with this Sub-Contract;

.3   any discrepancy in or divergence between the Contract Drawings and/or the Contract Bills/Specification/Schedules of Work and/or the Numbered Documents as applicable to the Main Contract;

.4 .1 the execution of work not forming part of the Main Contract by the Employer himself or by persons employed or otherwise engaged by the Employer as referred to in clause 29 of the Main Contract Conditions or the failure to execute such work;

.2 the supply by the Employer of materials and goods which the Employer has agreed to provide for the Works or the failure so to supply;

.5 Architect's instructions under clause 23.2 of the Main Contract Conditions issued in regard to the postponement of any work to be executed under the provisions of the Main Contract which shall be deemed to include Contractor's directions which pass on such instructions;

.6 failure of the Employer to give in due time ingress to or egress from the site of the Works, or any part thereof through or over any land, buildings, way or passage adjoining or connected with the site and in the possession and control of the Employer, in accordance with the Contract Bills/Specification/Schedules of Work and/or the Contract drawings, after receipt by the Architect of such notice, if any, as the Contractor is required to give or failure of the Employer to give such ingress or egress as otherwise agreed between the Architect and the Contractor;

.7 Architect's instructions issued:

under clause 13.2 or clause 13A.4.1 of the Main Contract Conditions requiring a Variation (except for a Variation for which the Architect has given a confirmed acceptance of a 13A Quotation or for a Variation thereto) or under clause 13.3 of the Main Contract Conditions in regard to the expenditure of provisional sums (other than work to which clause 13.4.2 of the Main Contract Conditions refers or an instruction for the expenditure of a provisional sum for defined work or of a provisional sum for Performance specified Work);

which shall be deemed to include Contractor's directions which pass on such instructions.

.8 .1 the execution of work for which an Approximate Quantity is included in the Contract Bills which is not a reasonably accurate forecast of the quantity of work required;

.2 where bills of quantities are included in the Numbered Documents the execution of work for which an Approximate Quantity is included in those bills which is not a reasonably accurate forecast of the quantity of work required;

.9 compliance or non-compliance by the Employer with clause 6A.1 of the Main Contract Conditions;

.10 .1 suspension by the Contractor of the performance of his obligations under the main Contract to the Employer pursuant to clause 30.1.4 of the main Contract conditions provided the suspension was not frivolous or vexatious;

.2   suspension by the Sub-Contractor of the performance of his obligations under the Sub-Contract to the Contractor pursuant to clause
4.20 provided the suspension was not frivolous or vexatious;

.11   save as provided for in clauses 4.26.1 to 4.26.10 any impediment,
prevention or default, whether by act or omission, by the Employer
or any person for whom the Employer is responsible except to the
extent that it was caused or contributed to by any default, whether
by act or omission, of the Sub-Contractor or his servants, agents or sub-
sub-contractors.

4.27   If the regular progress of the Works is materially affected by any act,
omission or default of the Sub-Contractor, his servants or agents, and if
the Contractor shall within a reasonable time of such material effect becoming apparent make written application to the Sub-Contractor, the amount
agreed by the Parties of any direct loss and/or expense thereby caused to
the Contractor shall either be deducted from any monies due or to become
due to the Sub-Contractor or be recoverable by the Contractor from the Sub-
Contractor as a debt. Provided always that:

.1   the Contractor's application shall be made as soon as it has become, or
should reasonably have become, apparent to him that the regular progress of the Works or of any part thereof has been or is likely to be
affected as aforesaid; and

.2   the Contractor shall submit to the Sub-Contractor such information in
support of his application as is reasonably necessary to show that the
regular progress of the Works or of any part thereof has been or is likely
to be affected as aforesaid; and

.3   the Contractor shall submit to the Sub-Contractor such details of such
loss and/or expense as the Sub-Contractor requests in order reasonably
to enable the ascertainment and agreement of that direct loss and/or
expense as aforesaid.

4.28   The provisions of clauses 4.24 and 4.27 are without prejudice to any other
rights or remedies which the Contractor or Sub-Contractor may possess.

17.3.5 Commentary

Sub-contractor's claims

Again, the commentary on NSC/C clauses 4.38 to 4.41 is relevant, as is the
consideration of the position under clause 26 of JCT 98[689]. However, the
architect is not involved in the process. The text should be compared with
NSC/C and the following points should be noted:

(1)   Clause 4.24: The wording indicates that compliance with the provisions
as to written notice is a condition precedent. The sub-contractor must
make written application to the main contractor within a reasonable

---

[689] See Chapter 12, section 12.2.4.

time of the material effect on progress becoming apparent. Proviso .1 adjusts the requirement so that the sub-contractor must make application 'as soon as it has become, or should reasonably have become apparent'. The timescale in the proviso is stricter than that in the introductory part of clause 4.24 and there seems no reason why it should not be enforceable.

(2)   The sub-contractor's obligation is to 'submit ... such information in support of his application as is reasonably necessary to show that ... regular progress ... has been or is likely to be affected ...'.

(3)   He is also to submit 'such details of such loss and/or expense as the Contractor requests in order reasonably to enable that direct loss and/or expense ... to be agreed', i.e. claims backed up by supporting evidence.

(4)   Clause 4.25, like JCT 98, clause 26.3, appears confusingly to link extensions of time to direct loss and/or expense claims and the list of relevant events should be noted[690].

(5)   Clause 4.2.4 refers to 'Relevant Matters', i.e. grounds giving rise to a potential claim. Apart from acts, omissions, or defaults of the main contractor or those for whom he is responsible in law, the 'Relevant Matters' are listed in clause 4.26.

(6)   Unlike the provision under NSC/C, the contractor's acts, omissions or defaults are to be considered together with the relevant matters in 4.24. They are separated under NSC/C, because although the architect has a key role in deciding the loss and/or expense due to the nominated sub-contractor in regard to the matters, he has no role in deciding whether the contractor has committed any act, omission or default which has materially affected the sub-contractor's progress. Effectively, the contractor decides that question himself, at least in the first instance.

(7)   The direct loss and/or expense must be agreed between the contractor and the sub-contractor and, if not so agreed, no sum is payable and the sub-contractor's recourse is to adjudication or to arbitration proceedings as the case may be.

Main contractor's claims

Clause 4.27 deals with claims by the main contractor against the sub-contractor in respect of disturbance of regular progress of the main contract works. Unlike the corresponding provision in NSC/C (clause 4.40) sub-sub-contractors are not included. The contractor is required to make written application to the sub-contractor (and not merely to give written notice): the main contractor's right to claim is subject to virtually the same three conditions as is the sub-contractor's right to claim against him. Clause 4.28 preserves the common law rights of both contractor and sub-contractor.

---

[690] See the comments on such linkage in Chapter 12, section 12.2.6.

17.4 Sub-contract conditions for sub-contractors named under IFC 98 (NAM/SC)

17.4.1 Extension of time and direct loss and/or expense claims under NAM/SC

The provisions for extensions of time and claims for and against sub-contractors named under clause 3.3 of the JCT Intermediate Form (IFC 98) are contained in clause 12 and 14 of Form NAM/SC and run as follows:

12. **Sub-Contractor's obligation – carrying out and completion of Sub-Contract Works – extension of Sub-Contract period**

12.1 The Sub-Contractor shall carry out and complete the Sub-Contract Works in accordance with NAM/T, Section I, item 15, and Section II, item 1, and reasonably in accordance with the progress of the Works subject to receipt of the notice to commence work on site as stated in NAM/T, Section II, item 1, and to the operation of clause 12.

12.2 Upon it becoming reasonably apparent that the commencement, progress or completion of the Sub-Contract Works is being or is likely to be delayed, the Sub-Contractor shall forthwith give written notice of the delay, specifying the cause in so far as he is able, to the Contractor, and if the completion of the Sub-Contract Works is likely to be or has been delayed beyond the period or periods referred to in NAM/T, Section II, item 1, or any extended period or periods previously fixed under this clause, by:

.1 any act, omission or default of the Contractor (including where the Contractor is the Principal Contractor, any omission or default in the discharge of his obligations as the Principal Contractor) his servants or agents or his sub-contractors, their servants or agents (other than the Sub-Contractor, his servants or agents), or

.2 by any of the events in clause 12.7,

then the Contractor shall so soon as he is able to estimate the length of delay beyond that period or periods make in writing a fair and reasonable extension of the period or periods for completion of the Sub-Contract Works.

12.3 If any act, omission or default as referred to in clause 12.2.1 or an event referred to in 12.7.5 to 12.7.8, 12.7.11 or 12.7.13 occurs after the expiry of the period or periods stated in NAM/T, Section II, item 1, (or after the expiry of any extended period or periods previously fixed under this clause) but before practical completion of the Sub-Contract Works is achieved the Contractor shall so soon as he is able to estimate the length of the resulting delay, if any, beyond that period or periods make in writing a fair and reasonable extension of the period or periods for completion of the Sub-Contract Works.

12.4 At any time the Contractor may make an extension of the period or periods for completion of the Sub-Contract Works in accordance with the provisions of clause 12.2 or 12.3, whether upon reviewing a previous decision or otherwise, whether or not the Sub-Contractor has given notice as referred to in clause 12.2. Such an extension of time shall not reduce any previously made.

12.5 Provided always that the Sub-Contractor shall use constantly his best endeavours to prevent delay and shall do all that may be reasonably required to the satisfaction of the Contractor to proceed with the Sub-Contract Works.

12.6 The Sub-Contractor shall provide such information required by the Contractor as is reasonably necessary for the purposes of clauses 12.2, 12.3 and 12.4.

12.7 The following are the events referred to in clause 12.2.2:

12.7 .1 force majeure;

12.7 .2 exceptionally adverse weather conditions;

12.7 .3 loss or damage caused by any one or more of the Specified Perils;

12.7 .4 civil commotion, local combination of workmen, strike or lock-out affecting any of the trades employed upon the Works or any trade engaged in the preparation, manufacture or transportation of any of the goods or materials required for the Works;

12.7 .5 compliance by the Contractor with the Architect's instructions (which shall be deemed to include compliance by the Sub-Contractor with the Contractor's directions which pass on such instructions):

    .1 under clause 1.4, 3.3.1, 3.3.3, 3.6, 3.8 (except, where bills of quantities are included in the Numbered Documents, compliance with an Architect's instruction for the expenditure of a provisional sum for defined work) or 3.15 of the Main Contract Conditions; or

    .2 in regard to the opening up or the testing of any of the work, materials or goods in accordance with clause 5.5 or with clause 3.12 or 3.13.1 of the Main Contract Conditions (including making good in consequence of such opening up or testing) unless the inspection or test showed that the work, materials or goods were not in accordance with the Main Contract or the Sub-Contract as the case may be;

12.7 .6 .1 where an Information Release Schedule has been provided failure of the Architect to comply with clause 1.7.1 of the Main Contract Conditions;

    .2 failure of the Architect to comply with clause 1.7.2 of the Main Contract Conditions;

12.7 .7 the execution of work not forming part of the Main Contract by the Employer himself or by persons employed or otherwise engaged by the Employer as referred to in clause 3.11 of the Main Contract Conditions or the failure to execute such work;

12.7 .8 the supply by the Employer of materials and goods which the Employer has agreed to provide for the Works or the failure so to supply;

12.7 .9   where it is stated in NAM/T, Section I, item 6 that clause 2.4.10 of the Main Contract Conditions applies, the Contractor's or the Sub-Contractor's inability for reasons beyond his control and which he could not reasonably have foreseen at the Base Date to secure such labour as is essential to the proper carrying out of the Works;

12.7 .10 where it is stated in NAM/T, Section I, item 6 that clause 2.4.11 of the Main Contract Conditions applies, the Contractor's or the Sub-Contractor's inability for reasons beyond his control and which he could not reasonably have foreseen at the Base Date to secure such goods or materials as are essential to the proper carrying out of the Works;

12.7 .11 failure of the Employer to give in due time ingress to or egress from the site of the Works or any part thereof through or over any land, buildings, way or passage adjoining or connected with the site and in the possession and control of the Employer, in accordance with the Contract Documents, after receipt by the Architect of such notice, if any, as the Contractor is required to give, or failure of the Employer to give such ingress or egress as otherwise agreed between the Architect and the Contractor;

12.7 .12 the carrying out by a local authority or statutory undertaker of work in pursuance of its statutory obligations in relation to the Works, or the failure to carry out such work;

12.7 .13 the deferment of the Employer giving possession of the site under clause 2.2 of the Main Contract Conditions;

12.7 .14 delay arising from a suspension by the Contractor of the performance of his obligations under the Main Contract to the Employer pursuant to clause 4.4A of the Main Contract Conditions and/or from the valid exercise by the Sub-Contractor of the right pursuant to clause 19.6 to suspend the performance of his obligations under this Sub-Contract to the Contractor;

12.7 .15 by reason of the execution of work for which an Approximate Quantity is included in the Numbered Documents which is not a reasonably accurate forecast of the quantity of work required;

.16 the use or threat of terrorism and/or the activity of the relevant authorities in dealing with such use or threat;

.17 compliance or non-compliance by the Employer with clause 5.7.1 of the Main Contract Conditions;

.18 save as provided for in clauses 12.7.1 to 12.7.17 any impediment, prevention or default, whether by act or omission, by the Employer or any person for whom the Employer is responsible except to the extent that it was caused or contributed to by any default, whether by act or omission, of the Sub-Contractor or his servants, agents or sub-sub-contractors.

12.8 In clauses 12 and 13 any references to delay, notice or extension of period or periods include further delay, further notice or further extension of period or periods as appropriate.

14 *Disturbance of regular progress*

14.1   If the Sub-Contractor shall make written application to the Contractor within a reasonable time of it becoming apparent that the Sub-Contractor has incurred or is likely to incur direct loss and/or expense, for which he would not be reimbursed under any other provision of this Sub-Contract, due to the commencement or regular progress of the Sub-Contract Works or any part thereof having been materially affected by any act, omission or default of the Contractor (including where the Contractor is the Principal Contractor, any omission or default in the discharge of his obligations as the Principal Contractor), his servants or agents, or any sub-contractor, his servants or agents (other than the Sub-Contractor, his servants or agents) or is materially affected by any one or more of the matters referred to in clause 14.2, the agreed amount of any direct loss and/or expense caused to the Sub-Contractor shall be added to the Sub-Contract Sum; provided that the Sub-Contractor shall in support of his application submit such information required by the Contractor as is reasonably necessary for the purposes of this clause.

14.2   The following are matters referred to in clause 14.1:

14.2   .1   .1   where an Information Release Schedule has been provided failure of the Architect to comply with clause 1.7.1 of the Main Contract Conditions;
         .2   failure of the Architect to comply with clause 1.7.2 of the Main Contract Conditions;

14.2   .2   the opening up for inspection of any work covered up or the testing of any work, materials or goods in accordance with clause 5.5 or with clause 3.12 of the Main Contract Conditions (including making good in consequence of such opening up or testing), unless the inspection or test showed that the work, materials or goods were not in accordance with the Main Contract or the Sub-contract as the case may be;

14.2   .3   the execution of work not forming part of the Main Contract by the Employer himself or by persons employed or otherwise engaged by the Employer as referred to in clause 3.11 of the Main Contract Conditions or the failure to execute such work;

14.2   .4   the supply by the Employer of materials and goods which the Employer has agreed to provide for the Works or the failure so to supply;

14.2   .5   failure of the Employer to give in due time ingress to or egress from the site of the Works, or any part thereof through or over any land, buildings, way or passage adjoining or connected with the site and in the possession and control of the Employer in accordance with the Contract Documents after receipt by the Architect of such notice, if any, as the Contractor is required to give or failure of the Employer to give such ingress or egress as otherwise agreed between the Architect and the Contractor;

14.2   .6   compliance by the Contractor with the Architect's instructions (which shall be deemed to include compliance by the Sub-Contractor with the

Contractor's directions which pass on such instructions) issued under clause 1.4, 3.3.1, 3.3.3, 3.6, 3.8 (except, where bills of quantities are included in the Numbered Documents, an Architect's instruction for the expenditure of a provisional sum for defined work) or 3.15 of the Main Contract Conditions and the Contractor's directions consequent thereon issued under clause 5;

14.2   .7   the deferment of the Employer giving possession of the site under clause 2.2 of Main Contract Conditions:

14.2   .8   the valid exercise by the Sub-Contractor of the right in clause 19.6 to suspend the further execution of the Sub-Contract Works;

14.2   .9   the execution of work for which an Approximate Quantity is included in the Numbered Documents which is not a reasonably accurate forecast of the quantity of work required;

.10 compliance or non-compliance by the Employer with clause 5.7.1 of the Main Contract Conditions;

.11 .1   a suspension by the Contractor of the performance of his obligations under the Main Contract to the Employer pursuant to clause 4.4A of the Main Contract Conditions;

.2   the valid exercise by the Sub-Contractor of the right pursuant to clause 19.6 to suspend the performance of his obligations under this Sub-Contract to the Contractor provided the suspension was not frivolous or vexatious;

.12 save as provided for in clauses 14.2.1 to 14.2.11 any impediment, prevention or default, whether by act or omission, by the Employer or any person for whom the Employer is responsible except to the extent that it was caused or contributed to by any default, whether by act or omission, of the Sub-Contractor or his servants, agents or sub-sub-contractors.

14.3   If the regular progress of the Works is materially affected by any act, omission or default of the Sub-Contractor, his servants or agents, and if the Contractor shall within a reasonable time of such material effect becoming apparent make written application to the Sub-Contractor, the agreed amount of any direct loss and/or expense thereby caused to the Contractor may be deducted from any monies due or to become due to the Sub-Contractor or may be recoverable from the Sub-Contractor as a debt; provided that the Contractor shall, in support of his application submit such information required by the Sub-Contractor as is reasonably necessary for the purposes of this clause.

14.4   The provisions of clause 14 are without prejudice to any other rights or remedies which the Contractor or Sub-Contractor may possess.

17.4.2 Commentary

The references to NAM/T are to the Form of Tender and Agreement. This is divided into three parts:

I Invitation to tender
II Tender document
III Articles of Agreement.

It is this last document which constitutes the sub-contract as actually executed by the main contractor and the sub-contractor, form NAM/SC containing the conditions of sub-contract being issued separately and
incorporated by reference into the agreement.

Clauses 12 and 14 of NAM/SC are essentially simplified versions of the
equivalent clauses 2.1 to 2.10 and 4.24 to 4.28 of the domestic sub-contract
Form DSC/C for use with JCT 98. This reflects the position of 'named' sub-
contractors under IFC 98. In contrast to nominated sub-contractors under
JCT 98, for instance, the architect is not required to certify amounts for
payment to the sub-contractor, there is no provision for direct payment
by the employer if the main contractor defaults on payment and the architect is not required separately to certify practical completion of the sub-
contractor's work. The architect will only be involved if either the sub-contractor or the main contractor is so seriously in default under the
sub-contractor that either becomes entitled to determine the employment
of the other, when the architect is required to step in to deal with the
situation that then arises.

The simplification in comparison with DSC/C is similar in nature to the
simplification that has taken place in the extension of time and 'claim'
provisions in IFC 98 as compared with JCT 98. There are two changes to
which attention should be drawn:

• In parallel with clause 2.3 of IFC 98, clause 12.3 makes specific provision
  for delaying events occurring after the sub-contract works ought to have
  been completed and before actual practical completion.
• Clause 12.4 empowers the contractor to grant an extension of time of his
  own volition and without notice from the sub-contractor at any time
  during the progress or after completion of the sub-contract works. In
  DSC/C (clause 2.7) this power can only be exercised within the period
  of 16 weeks after practical completion of either the sub-contract or main
  contract works, whichever is the earlier. It would appear that, under
  NAM/SC, this power could be exercised by the main contractor at any
  time up to the issue of the final certificate under the main contract[691].

17.5 *The IN/SC form of domestic sub-contract for use with IFC 98*

The equivalent clauses 12 and 14 in Domestic Sub-Contract Form IN/SC
produced by the Building Employers' Confederation and the sub-
contractors' organisations for use with IFC 98 are virtually identical to those
in Form NAM/SC and therefore do not require separate consideration.

---

[691] The clauses are covered in Peter R. Hibberd, *Sub-Contracts under the JCT Intermediate Form*, 1987,
Blackwell Science, Chapter 10. That commentary is still useful although now somewhat out of date.

17.6 The Works Contract/2 for use with MC 98

17.6.1 The clauses dealing with extension of time

The text of clauses 2.1–2.10 is as follows:

Works Contractor's obligation – carrying out and completion of the Works – extension of Works Contract time (2.1 to 2.10)

2.1   The Works Contractor shall carry out and complete the Works in accordance with the details in Works Contract/1, Section 2, item 1, and reasonably in accordance with the progress of the Project but subject to receipt of the notice to commence work on site as stated in Works Contract/1, Section 2, item 1 and to the operation of clauses 2.1 to 2.10 and any revision of the period or periods for the completion of the Works in respect of a Variation for which a 3.14 Quotation has been given and which has been stated by the Management Contractor in his acceptance of a 3.14 Quotation. The Management Contractor shall give to the Works Contractor sufficient information on the progress of the Project to enable him to fulfil his obligations under clauses 2.1 and 2.2.

2.2   .1   If and whenever it becomes reasonably apparent that the commencement, progress or completion of the Works or any part thereof is being or is likely to be delayed, the Works Contractor shall forthwith give written notice to the Management Contractor of the material circumstances including, insofar as the Works Contractor is able, the cause or causes of the delay and identify in such notice any matter which in his opinion comes within clause 2.3.1.1.

.2   In respect of each and every matter which comes within clause 2.3.1.1, and identified in the notice given in accordance with clause 2.2.1, the Works Contractor shall, if practicable in such notice or otherwise in writing as soon as possible after such notice:

.1   give particulars of the expected effects thereof; and

.2   estimate the extent, if any, of the expected delay in the completion of the Works or any part thereof beyond the expiry of the period or periods stated in Works Contract/1, Section 2, item 1, or any revision of such period or periods which results therefrom whether or not concurrently with delay resulting from any other matter which comes within clause 2.3.1.1; and

.3   give such further written notices to the Management Contractor as may be reasonably necessary or as the Management Contractor may reasonably require for keeping up-to-date the particulars and estimate referred to in clause 2.2.2.1 and 2.2.2.2 including any material change in such particulars or estimate.

2.3   .1   If on receipt of any notice, particulars and estimate under clause 2.2 the Management Contractor properly considers that:

.1   any of the causes of the delay is the occurrence of a Relevant Event, or is an act, omission or default of the Management Contractor (including, where the Management Contractor is the Principal Contractor, any omissions or default in the discharge of his obligations as the Principal Contractor) or any person for whom the Management Contractor is responsible as defined in clause 6.1 (other than the Works Contractor or of any person for whom the Works Contractor is responsible as defined in clause 6.1); and

.2   the completion of the Works is likely to be or has been delayed thereby beyond the period or periods stated in Works Contract/1, Section 2, item 1, or any revision of such period or periods,

then the Management Contractor shall, in writing, having first notified the Architect, give an extension of time to the Works Contractor by fixing such revised or further revised period or periods for the completion of the Works as the Management Contractor then estimates to be reasonable. If the Management Contractor on receipt of the notice, particulars or estimate under clause 2.2 properly considers that any extension of time should not be given to the Works Contractor then the Management Contractor shall, in writing, having first notified the Architect, so inform the Works Contractor. If under clause 2.14 of the Management Contract Conditions the Architect has expressed dissent from the decision of the Management Contractor on giving or not giving an extension of time or the length thereof the Management Contractor shall notify in writing the terms of such dissent to the Works Contractor.

.2   If the Works Contractor suspends the performance of his obligations under the Works Contract to the Management Contractor under clause 4.28 and delay to the progress or completion of the Works results from such suspension, such delay shall for the purposes of clause 2.3.1.1 be deemed to be a delay caused by an act, omission or default of the Management Contractor.

2.4   When fixing such revised period or periods, the Management Contractor shall, if reasonably practicable having regard to the sufficiency of the notice, particulars and estimate, fix such revised period or periods within the following time limit:

.1   not later than 12 weeks from the receipt by the Management Contractor of the notice and of reasonably sufficient particulars and estimates; or

.2   where the time between receipt thereof and the expiry of the period or periods for the completion of the Works is less than 12 weeks, not later than the expiry of the aforesaid period or periods.

2.5   The Management Contractor, when fixing such revised period or periods, shall state:

.1   which of the matters, including any of the Relevant Events, referred to in clause 2.3.1.1 he has taken into account; and

.2   the extent, if any, to which the Management Contractor has had regard to any Instruction requiring the omission of any work or obligation or restriction issued since the previous fixing of any such revised period or periods for the completion of the Works.

2.6   After the first exercise by the Management Contractor of the duty under clause 2.3.1 or after any revision of the period or periods for the completion of the Works stated by the Management Contractor in his acceptance of a 3.14 Quotation in respect of a Variation, the Management Contractor may in writing, having notified the Architect, fix a period or periods for completion of the Works or the applicable part thereof shorter than that previously fixed under clause 2.3.1 or stated by the Management Contractor in his acceptance of a 3.14 Quotation if, in the opinion of the Management Contractor, the fixing of such shorter period or periods is fair and reasonable having regard to any Instruction requiring the omission of any work or obligation or restriction which has been issued after the last occasion on which the Management Contractor made a revision of the aforesaid period or periods. Provided that no decision under clause 2.6 shall alter the length of any revision to the period or periods for the completion of the Works in respect of a Variation for which a 3.14 Quotation has been given and which has been stated by the Management Contractor in his acceptance of a 3.14 Quotation.

2.7   If the expiry of the period or periods when the Works should have been completed in accordance with the details in Works Contract/1, Section 2, Item 1 as revised by the operation of the provisions of clauses 2.2 to 2.10 or stated by the Management Contractor in his acceptance of a 3.14 Quotation, occurs before the date of practical completion of the Works certified by the Management Contractor under clause 2.14 the Management Contractor, having first notified the Architect in writing, may

and

not later than the expiry of 12 weeks from the aforesaid date of practical completion of the Works, the Management Contractor, having first notified the Architect in writing, shall

either

.1   fix such a period or periods for completion of the Works longer than that previously fixed under clauses 2.3 and 2.6 or stated by the Management Contractor in his acceptance of a 3.14 Quotation as the Management Contractor properly considers to be fair and reasonable having regard to any of the matters referred to in clause 2.3.1.1 whether upon reviewing a previous decision or otherwise and whether or not such matters have been specifically notified by the Works Contractor under clause 2.2; or

.2   fix such a period or periods for completion of the Works shorter than that previously fixed under clauses 2.2 to 2.10 or stated by the Management Contractor in his acceptance of a 3.14 Quotation as the Management Contractor properly considers to be fair and reasonable having regard to any Instruction requiring the omission of any work or obligation or restriction which has been issued after the last occasion on which the Management Contractor made a revision of the aforesaid period or periods; or

.3   confirm to the Works Contractor the period or periods for the completion of the Works previously fixed.

Provided that no decision under clause 2.7.1 or clause 2.7.2 shall alter the length of any revision to the period or periods for the completion of the Works in respect of a Variation for which a 3.14 Quotation has been given and which has been stated by the Management Contractor in his acceptance of a 3.14 Quotation.

2.8   The operation of clauses 2.3, 2.6 and 2.7 shall be subject to the proviso that the Works Contractor shall use constantly his best endeavours to prevent delay in the progress of the Works or any part thereof, however caused, and to prevent any such delay resulting in the completion of the Works being delayed or further delayed beyond the period or periods for completion, and the Works Contractor shall do all that may reasonably be required to the satisfaction of the Management Contractor to proceed with the Works.

2.9   Except where clauses 3.4.2 to 3.4.7 apply no decision of the Management Contractor under clauses 2.3, 2.6 or 2.7 inclusive or any revised period stated by the Management Contractor in his acceptance of a 3.14 Quotation shall fix a period or periods for completion of the Works which will be shorter than the period or periods stated in Works Contract/1, Section 2, item 1.

2.10   The following are the Relevant Events referred to in clause 2.3.1.1:

.1   force majeure;

.2   exceptionally adverse weather conditions;

.3   loss or damage occasioned by any one or more of the Specified Perils;

.4   civil commotion, local combination of workmen, strike or lock-out affecting any of the trades employed upon the Works or any of the trades engaged in the preparation, manufacture or transportation of any of the goods or materials required for the Works;

.5   compliance by the Management Contractor with Instructions (which shall be deemed to include compliance by the Works Contractor with Instructions):

.1   under clause 1.5A.1, under clauses 3.3 to 3.5 (except for a conformed acceptance of a 3.14 Quotation and except, where bills of quantities are included in the Numbered Documents, compliance with an instruction for the expenditure of a provisional sum for defined work or of a provisional sum for Performance specified Work) and/or clause 3.27 of the Management Contract Conditions; and/or

.2   under clause 1.5 and section 8; and/or

.3   in regard to the opening up for inspection of any work covered up or the testing of any of the work, materials or goods in accordance with clause 3.10 of the Management Contract Conditions and/or clause 3.3.2 (including making good in consequence of such opening up or testing) unless the inspection or test showed that the work,

materials or goods were not in accordance with the Management
Contract or the Works Contract as the case may be;

.6   the Management Contractor, or the Works Contractor through the Management Contractor, not having received in due time necessary Instructions (including those for or in regard to the expenditure of provisional
sums), drawings, details or levels from the Professional Team for which
the Management Contractor or the Works Contractor, through the Management Contractor, specifically applied in writing provided that such
application was made on a date which having regard to the Completion
Date or the period or periods for the completion of the Works was
neither unreasonably distant from nor unreasonably close to the date
on which it was necessary for the Management Contractor or the Works
Contractor to receive the same;

.7   .1   delay on the part of other works contractors in respect of the Project
which the Works Contractor has taken all practicable steps to avoid
or reduce;

.2   delay on the part of Nominated Suppliers which the Works Contractor has taken all practicable steps to avoid or reduce;

.8   .1   the execution of work not forming part of the Project by the Em-
ployer himself or by persons employed or otherwise engaged by the
Employer as referred to in clauses 3.23 to 3.25 of the Management
Contract Conditions or the failure to execute such work;

.2   the supply by the Employer of materials and goods which the
Employer has agreed to provide for the Project or the failure so to
supply;

.9   the exercise after the Base Date by the United Kingdom Government of
any statutory power which directly affects the execution of the Project
by restricting the availability or use of labour which is essential to the
proper carrying out of the Project or preventing the Management Contractor or Works Contractor from, or delaying the Management Contractor or Works Contractor in, securing such goods or materials or such
fuel or energy as are essential to the proper carrying out of the Project;

.10   .1   the Works Contractor's inability for reasons beyond his control and
which he could not reasonably have foreseen at the Base Date to
secure such labour as is essential to the proper carrying out of the
Works;

.2   the Works Contractor's inability for reasons beyond his control and
which he could not reasonably have foreseen at the Base Date to
secure such goods or materials as are essential to the proper carrying out of the Works;

.11   the carrying out by a local authority or statutory undertaker of work in
pursuance of its statutory obligations in relation to the Project, or the
failure to carry out such work;

.12   failure of the Employer to give in due time ingress to or egress from the
site of the Project or any part thereof through or over any land, build-

ings, way or passage adjoining or connected with the site and in the possession and control of the Employer, in accordance with the Management Contract after receipt by the Architect of such notice, if any, as the Management Contractor is required to give, or failure of the Employer to give such ingress or egress as otherwise agreed between the Architect and the Management Contractor;

.13 where it is stated in Works Contract/1 Section 1, item 5 that clause 2.3.2 of the Management Contract Conditions applies to the Project, any deferment by the Employer in giving possession of the site of the Project to the Management Contractor;

.14 by reason of the execution of work for which an Approximate Quantity is included in the Numbered Documents which is not a reasonably accurate forecast of the quantity of work required;

.15 the use or threat of terrorism and/or the activity of the relevant authorities in dealing with such use or threat;

.16 delay which the Works Contractor has taken all practicable steps to avoid or reduce consequent upon a change in the Statutory Requirements after the Base Date which necessitates some alteration or modification to any Performance Specified Work;

.17 compliance or non-compliance by the Employer with clause 5.18 of the Management Contract; *(Employer's obligation – Planning Supervisor – Principal Contractor where not the management Contractor)*

.18 subject to clause 2.3.2 delay arising from a suspension by the Works Contractor of the performance of his obligations under the Works Contract to the Management Contractor pursuant to clause 4.28.

.19 save as provided for in clauses 2.10.1 to 2.10.18 any impediment, prevention or default, whether by act or omission by the Employer or any person for whom the Employer is responsible except to the extent that it was caused or contributed to by any default, whether by act or omission, of the Works Contractor or his servants, agents or sub-contractors.

17.6.2 Commentary

Other than the terminology of management contractor and works contractor instead of contractor and sub-contractor, these provisions parallel the provisions of NSC/C. There are some points to note as follows.

Notice

The works contractor's obligation to give notice in clause 2.2.1 is similar to the obligation in DSC/C, clause 2.1, in so far as the works contractor's ability to identify the cause or causes of delay is concerned.

Time period

The time period for response is 12 weeks as in NSC/C and DSC/C.

Architect's dissent

The question of the architect's dissent has been discussed in Chapter 10, section 10.6.2, when considering the extension of time provisions in MC 98 management contract. The works contract provides in clause 2.3.1 that the management contractor must notify the architect before either giving or refusing an extension of time. The terms of the architect's dissent must be passed on to the works contractor. It is difficult to see the practical effect of such passing on of the architect's dissent, because the contract is clear that it is the management contractor's duty to consider whether the works contractor should have an extension of time. It seems that the management contractor cannot rely on the architect's dissent and he must take action as though the architect had never passed any opinion. No doubt the management contractor will carefully weigh anything the architect has to say, but final responsibility remains with the management contractor. Although the architect must again be notified before the management contractor carries out his final review of extensions of time under clause 2.7, there is no reference to the architect's dissent nor to the terms of any such dissent being passed to the works contractor. In practice, no doubt the architect will express his views and the management contractor will pass them to the works contractor. In practice, the management contractor may be reluctant to give an extension of time in the face of the architect's dissent.

Relevant events

The relevant events closely follow those in NSC/C.

17.6.3 The clauses dealing with loss and/or expense

The text of clauses 4.45 and 4.46 is as follows:

> **Matters affecting regular progress – direct loss and/or expense – Management Contractor's and works contractor's rights (4.45 to 4.51)**

> 4.45   If the Works Contractor makes written application to the Management Contractor stating that he has incurred or is likely to incur direct loss and/or expense in the execution of the Works Contract for which he would not be reimbursed by a payment under any other provision in the Works Contract by reason of the regular progress of the Works or any part thereof having been or being likely to be materially affected by any deferment by the Employer in giving possession of the site of the Project to the Management Contractor where it is stated in Works Contract/1, Section 1,

item 5 that clause 2.3.2 of the Management Contract Conditions applies to the Project or by any one or more of the matters set out in clause 4.46, the Architect or if so instructed by the Architect, the Quantity Surveyor in consultation with the Management Contractor shall ascertain the amount, if any, of that direct loss and/or expense. Provided always that:

.1   the Works Contractor's application shall be made as soon as it has become, or should reasonably have become, apparent to him that the regular progress of the Works or of any part thereof has been or was likely to be affected as aforesaid, and

.2   the Works Contractor, in order reasonably to enable the direct loss and/or expense to be ascertained, shall submit to the Management Contractor such information in support of his application including details of such loss and/or expense as the Management Contractor is requested by the Architect or the Quantity Surveyor to obtain from the Works Contractor, and

.3   the Works Contractor has complied with clause 4.48.

4.46   The following are the matters referred to in clause 4.45:

.1   the Management Contractor, or the Works Contractor through the Management Contractor, not having received in due time necessary instructions (including those for or in regard to the expenditure of provisional sums), drawings, details or levels from the Professional Team for which the Management Contractor, or the Works Contractor through the Management Contractor, specifically applied in writing provided that such application was made on a date which having regard to the Completion Date or the period or periods for completion of the Works was neither unreasonably distant from nor unreasonably close to the date on which it was necessary for the Management Contractor or the Works Contractor to receive the same; or

.2   the opening up for inspection of any work covered up or the testing of any of the work, materials or goods in accordance with clause 3.10 of the Management Contract Conditions and/or clause 3.3.2 (including making good in consequence of such opening up or testing), unless the inspection or test showed that the work, materials or goods were not in accordance with the Management Contract or the Works Contract as the case may be; or

.3   any discrepancy in or divergence between the documents provided by the Employer as referred to in Article 7 of the Management Contract and/or any Instructions or Directions issued by the Management Contractor (save insofar as any such Instruction or Direction requires a Variation) and/or the Works Contract; or

.4   Instructions issued under clause 3.5 of the Management Contract Conditions in regard to the postponement of any work to be executed under the provisions of the Management Contract or the Works Contract; or

.5   the execution of work not forming part of the Project by the Employer himself or by persons employed or otherwise engaged by the Employer as referred to in clauses 3.23 to 3.25 of the Management

Contract Conditions or the failure to execute such work or the supply by the Employer of materials and goods which the Employer has agreed to provide for the Project or the failure so to supply; or

.6 failure of the Employer to give in due time ingress to or egress from the site of the Project or any part thereof through or over any land, buildings, way or passage adjoining or connected with the site and in the possession and control of the Employer, in accordance with the Management Contract, after receipt by the Architect of such notice, if any, as the Management Contractor is required to give or failure of the Employer to give such ingress or egress as otherwise agreed between the Architect and the Management Contractor; or

.7 Instructions issued under clause 3.4 (other than, where bills of quantities are included in the Numbered Documents, an instruction for the expenditure of a provisional sum for defined work) and 3.27 of the Management Contract Conditions; or

.8 the execution of work for which an Approximate Quantity is included in the Numbered Documents which is not a reasonably accurate forecast of the quantity of work required.

.9 compliance or non-compliance by the Employer with clause 5.18 of the Management Contract; *(Employer's obligation – Planning Supervisor – Principal Contractor where not the Management Contractor)*

.10 suspension by the Works Contractor of the performance of his obligations under this Works Contract to the Management Contractor pursuant to clause 4.28 provided the suspension was not frivolous or vexatious.

.11 save as provided for in clauses 4.46.1 to 4.46.10 any impediment, prevention or default, whether by act or omission by the Employer or any person for whom the Employer is responsible except to the extent that it was caused or contributed to by any default, whether by act or omission, of the Works Contractor or his servants, agents or sub-contractors.

4.47 Any amount from time to time ascertained as a result of the operation of clause 4.45 shall be added to the Works Contract Sum or included in the computation of the Ascertained Final Works Contract Sum.

4.48 The Works Contractor shall comply with all directions of the Management Contractor which are reasonably necessary to enable the ascertainment which results from the operation of clauses 4.45 to 4.47 to be carried out.

4.49 If the regular progress of the Works (including any part thereof which is sub-contracted) is materially affected by any act, omission or default of the Management Contractor (including, where the Management Contractor is the Principal Contractor, any omission or default in the discharge of his obligations as the Principal Contractor) or any person for whom the Management Contractor is responsible as defined in clause 6.1, the Works Contractor shall within a reasonable time of such material effect becoming apparent give written notice thereof to the Management Contractor and the agreed amount of any direct loss and/or expense thereby caused to the

Works Contractor shall be recoverable by the Works Contractor from the Management Contractor as a debt. Provided always that:

.1   the Works Contractor's application shall be made as soon as it has become, or should reasonably have become, apparent to him that the regular progress of the works or of any part thereof has been or was likely to be affected as aforesaid; and

.2   the Works Contractor, in order to enable the direct loss and/or expense to be ascertained, shall submit to the Management Contractor such information in support of his application including details of the loss and/or expense as the Management Contractor may reasonably require from the Works Contractor.

4.50   If the regular progress of the Project (including any part thereof which is part of a Works Contract) or the regular progress of the Works (including any part thereof which is sub-contracted) is materially affected by any act, omission or default of the Works Contractor or any person for whom the Works Contractor is responsible as defined in clause 6.1, the Management Contractor shall within reasonable time of such material effect becoming apparent give written notice thereof to the Works Contractor and the agreed amount of any direct loss and/or expense thereby caused to the Management Contractor (whether suffered or incurred by the Management Contractor or by other works Contractors employed by the Management Contractor on the Project from whom claims under similar provisions in the relevant Works Contracts have been agreed by the Management Contractor, other Works Contractors and the Works Contractor) may be deducted from any monies due or to become due to the Works Contractor or may be recoverable by the Management Contractor from the Works Contractor as a debt. Provided always that:

.1   the Management Contractor's application shall be made as soon as it has become, or should reasonably have become, apparent to him that the regular progress of the Project or of any Works Contract has been or was likely to be affected as aforesaid; and

.2   the Management Contractor, in order to enable the direct loss and/or expense to be ascertained, shall submit to the Works Contractor such information in support of his application including details of the loss and/or expense as the Works Contractor may reasonably request from the Management Contractor.

4.51   The provisions of clauses 4.45 to 4.51 are without prejudice to any other rights or remedies which the Management Contractor or the Works Contractor may possess.

17.6.4 Commentary

The commentary on NSC/C clauses 4.38 to 4.41 is relevant, as is the consideration of the position under clause 26 of JCT 98[692]. There are some points to note, as follows.

---

[692] See Chapter 12, section 12.2.4.

Application

The wording of the proviso is slightly different, but it appears to be to the same effect as the equivalent NSC/C. However, the works contractor is required to have complied with clause 4.48. That clause places an obligation on the works contractor to comply with all directions of the management contractor reasonably necessary to enable the ascertainment which results from clauses 4.45 to 4.47 to be carried out. The effect is that if the works contractor fails to comply with any directions of the management contractor, the ascertainment cannot be done.

Ascertainment

The ascertainment is to be carried out by the architect or if he so instructs, by the quantity surveyor in consultation with the management contractor. This should be contrasted with the position under MC 98[693]. There, it is quite clear that the opinion of the architect triggers ascertainment. Under the works contract, the works contractor is not specifically informed that the architect's opinion about the validity of the application is paramount. There may be rare occasions when the works contractor can take some advantage from this.

Claims between management contractor and works contractor

These provisions in clauses 4.49 and clause 4.50 cover claims against the management contractor which he cannot pass up to the employer or against the works contractor respectively. The wording refers to persons for whom the management contractor or the works contractor is responsible. This includes sub-sub-contractors. The provisions are very similar to the equivalent provisions in NSC/C.

17.7 The ACA form of sub-contract

17.7.1 Introduction

This form of sub-contract was issued unilaterally by the Association of Consultant Architects in October 1982 and is designed for use with the ACA Form of Building Agreement[694]. The form current at the time of writing was revised in 2003.

The use of the sub-contract form is not obligatory under the main contract but its drafting follows very closely the main contract provisions and it is advisable to use it in conjunction with that form. Claims under the

---

[693] See Chapter 12, section 12.7.2.
[694] See Chapter 15 for consideration of the ACA Building Contract 1998 (ACA 3).

sub-contract form (as under the ACA contract itself) are limited, and this applies to claims both for extensions of time and for money.

17.7.2 Extensions of time

The sub-contract provisions for extensions of time are contained in clause 7, which reads as follows:

7.   COMMENCEMENT, COMPLETION AND DELAYS IN THE EXECUTION OF THE SUB-CONTRACT WORKS

7.1   The Sub-Contractor shall commence the execution of the Sub-Contract Works within 10* working days of receipt of the Contractor's written instruction so to do. The Sub-Contractor shall proceed with the same regularly and diligently and in accordance with the Sub-Contract Time Schedule and shall complete the same on or before the date or dates for completion stated in the Sub-Contract Time Schedule, subject as hereinafter provided in this Clause 7.

7.2   No extension of time shall be granted to the Sub-Contractor except in the case of:

(a)   any circumstance entitling the Contractor to an extension of time for completion of the Works or (if appropriate) of any Section in which the Sub-Contract Works are comprised; or

(b)   any act, instruction, default or omission of the Contractor, whether authorised by or in breach of this Sub-Contract which the Sub-Contractor shall prove to the satisfaction of the Contractor has prevented the completion of the Sub-Contract Works by the date or dates for completion stated in the Sub-Contract Time Schedule; or

(c)   any instruction to postpone given in accordance with Clause 7.5;

In any such event the Contractor shall grant to the Sub-Contractor such extension of time to the date or dates for completion of the Sub-Contract Works as he then estimates to be fair and reasonable. Provided that no account shall be taken of any of the circumstances referred to in Clause 7.2(a) unless the Sub-Contractor shall have given written notice and full and detailed particulars to the Contractor at the time and in the manner in which notice and particulars are required to be given by the Contractor to the Architect or the Supervising Officer, as the case may be, under the Agreement.

7.3   If the Sub-Contract Works or any part of them are not completed in accordance with the provisions of Clause 7.1, the Contractor shall give to the Sub-Contractor written notice to that effect.

7.4   In considering any extension of time the Contractor shall be entitled to take into account at any time before the Taking-Over of the Works or of the last Section (if appropriate) in which the Sub-Contract Works are comprised, the effects of any omission from the Sub-Contract Works. Where the Architect reviews all or any of his previous decisions under Clause 11.7 of the Agreement and fixes a date for the Taking-Over of the Works or of any Section later than that previously stated, adjusted or fixed, the Contractor shall review the extensions of time (if any) previously granted by him under Clause 7.2 of this Sub-Contract in respect of the circumstances referred to in Clause 7.2(a) accordingly.

17.7.3 Commentary on the extension of time clause

The first matter which should be noted – for it is highly relevant to claims – is that the sub-contractor's obligation in clause 7.1 is to 'commence the execution of the sub-contract works … proceed with the same regularly and diligently and in accordance with the … Time Schedule and … complete the same *on or before* the date or dates for completion …'. The views expressed by Megarry J in *London Borough of Hounslow* v. *Twickenham Garden Developments Ltd*[695] appear to be equally applicable to sub-contractor's claims under this clause. If the sub-contractor:

> 'is well ahead with his works and is then delayed [by something giving him a claim for extension of time], the [contractor] may nevertheless reach the conclusion that completion of the works is not likely to be delayed beyond the date for completion'.[696]

Whether the sub-contractor has proceeded regularly and diligently with the works is a question of fact[697].

Extensions of time as such are dealt with by clause 7.2, the wording of which makes it clear that the sub-contractor is entitled to extensions of time only on the grounds stated in the clause. These grounds fall into three groups:

'Any circumstance entitling the Contractor to an extension of time' under the main contract

Assuming that the ACA sub-contract is used in conjunction with the ACA main form, there are two possibilities, since the main form contains alternative versions of clause 11.5 – Extensions of time.

Alternative 1 is the narrow version, since it entitles the main contractor (and hence the sub-contractor) to an extension of time in respect of delay or disturbance caused by 'any act, instruction, default, or omission of the employer, or of the architect on his behalf, whether authorised by or in breach of this Agreement', and this is apt to extend to breaches etc. of the express and implied terms of the contract as appropriate.

Alternative 2 lists five specific causes[698]. It is essential, therefore, that the sub-contractor is made aware of the alternative which is in operation under the main contract.

---

[695] (1970) 7 BLR 81.
[696] (1970) 7 BLR 81 at 113 per Megarry J.
[697] See Chapter 12, in section 12.2.4, 'Material effect on regular progress'.
[698] In both cases refer to the commentary on the ACA main contract terms: see Chapter 15, in section 15.4.2, 'Grounds for extension of time'.

'Any act, instruction, default or omission of the Contractor, whether authorised by or in breach of this Sub-Contract which the sub-contractor shall prove to the satisfaction of the contractor has prevented completion of the sub-contract works ...'

The standard of proof is the usual 'balance of probabilities' and the contractor's satisfaction would probably have to be reasonable. Otherwise, this ground is self-explanatory[699].

'Any instruction to postpone given in accordance with Clause 7.5.'

Clause 7.5 empowers the contractor, among other things, to postpone the completion of all or any part of the sub-contract works and obliges the sub-contractor immediately to take such measures as necessary to comply with such instruction. These provisions are very restrictive.

The giving of written notice to the contractor by the sub-contractor is clearly a condition precedent to the operation of the extension of time provisions, as is made clear by the final sentence in clause 7.2 beginning 'Provided that no account shall be taken ...'. This effectively incorporates the procedural provisions of the main contract. Written notice by the sub-contractor is not required in respect of the grounds listed in paragraphs (b) and (c) of clause 7.2.

The sub-contractor's entitlement is to such an extension of time as the contractor estimates to be fair and reasonable. Clause 7.4 states that, in considering any extension of time, the contractor is entitled to have regard to any omission of work from the sub-contract – the contract wording is 'take into account' and also that where, under clause 11.7 of the main contract, the architect reviews his previous decisions on extensions of time and fixes a *later* date than that previously set, the contractor must (in the first edition of the form the word was 'may') review any extensions of time which he has granted previously to the sub-contractor, *but only in respect* of the matters referred to in clause 7.2(a), i.e. grounds under the main contract.

17.7.4 Money claims under ACA sub-contract

This important matter is covered by clause 9 of the sub-contract, which parallels clause 7 of the ACA main contract[700] and effectively incorporates the procedural and allied provisions of the main contract form in the sub-contract. Clause 9 reads as follows:

9. DAMAGE, LOSS AND/OR EXPENSE

9.1 The Sub-Contractor shall give to the Contractor any notices, particulars, estimates or other information which the Contractor is required to give to

---

[699] Reference should be made to Chapter 15, section 15.2.2, which considers whether there must be a legal wrong involved.
[700] Considered in Chapter 15, in section 15.4.2, 'Grounds for extension of time'.

the Architect or to the Supervising Officer, as the case may be, under the provisions of the Agreement in sufficient time to enable the Contractor to claim any adjustment to the Contract Sum or any damage, loss and/or expense to which he is entitled under the provisions of the Agreement.

9.2   If the regular progress of the Sub-Contract Works is disrupted or delayed by any circumstance other than Architect's or Supervising Officer's instructions (to which the provisions of Clause 5 shall apply) entitling the Contractor to claim from the Employer damage, loss and/or expense under clause 7 of the Agreement, the Contractor shall at the request of the Sub-Contractor recover for the Sub-Contractor any damage, loss and/or expense suffered or incurred by the Sub-Contractor by reason of such disruption or delay to the Sub-Contract Works: Provided always that the Sub-Contractor shall comply with his obligations under clause 9.1.

9.3   If, pursuant to the provisions of Clause 9.2 of this Sub-Contract, the Contractor recovers any damage, loss and/or expense in respect of any circumstance which has affected the execution of the Sub-Contract Works, the Contractor shall in turn pay to the Sub-Contractor such proportion of it (if any) as may, in the Contractor's opinion, be fair and reasonable and such proportion shall be added to the Sub-Contract Sum.

9.4   Save in the case of Contractor's instructions (to which the provisions of clause 5 shall apply), if any act, omission, default or negligence of the Contractor or his employees, agents or sub-contractors disrupts the regular progress of the Sub-Contract Works or delays the execution of them in accordance with the dates stated in the Sub-Contract Time Schedule and, in consequence of such disruption or delay, the Sub-Contractor suffers or incurs damage, loss and/or expense, he shall be entitled to recover the same from the Contractor as a debt.

17.7.5 Commentary on the money claims clause

Clause 9.1 obliges the sub-contractor to give to the contractor any notices, estimates or other information which the contractor is obliged to give to the architect under the main form. He must do this in sufficient time to enable the contractor to claim any adjustment to the contract sum or any damage, loss and/or expense on his own or the sub-contractor's behalf. Once again, this is an example of contractual claims being passed up the contract chain.

Claims under clause 9 are of two types, dealt with separately in clauses 9.2 and 9.4.

Clause 9.2 is concerned with 'Claims against the Employer', i.e. those claims for which the employer has a responsibility to the main contractor. It is quite plain from the wording that such claims extend to both 'disturbance' and 'prolongation' (the sub-clause refers to work being 'disrupted or delayed') and all that is required is that the *regular progress* of the sub-contract works should be disrupted or delayed by 'any circumstance other than Architect's instructions … entitling the Contractor to claim from the

Employer' under the terms of clause 7 of the main contract'[701]. This clause obliges the contractor to 'recover for the sub-contractor' any 'damage, loss or expense suffered or incurred' by the latter 'by reason of such disruption or delay to the Sub-Contract Works', i.e. the main contractor must pass on the sub-contractor's claims. This is subject to the overall proviso (which is a condition precedent) that the sub-contractor shall have given the main contractor the requisite notices etc. under clause 9.1. In simple terms, the sub-contractor provides details of his claim to the main contractor who then passes them on to the employer under the main contract terms.

Clause 9.3 deals with settlement of claims, and the vague wording is to be deprecated. In short, the contractor is to pay over to the sub-contractor such proportion (if any) of any monies which he recovers under the main contract claims provisions 'as may, in the Contractor's opinion, be fair and reasonable'; the contractor's opinion would be something which could be challenged in adjudication, arbitration or litigation. It is arguable that this provision is caught under the Housing Grants, Construction and Regeneration Act 1996 as being essentially a pay-when-paid provision.

Clause 9.4 deals with claims by the sub-contractor against the main contractor and is concerned with claims which the main contractor cannot pass on to the employer. The contractor's instructions are dealt with separately in clause 5, which is in almost identical terms to the corresponding main contract provision, clause 8. Apart from the contractor's instructions, clause 9.4 is merely declaratory of what would be the position at common law in any case. It says that the sub-contractor is entitled to recover, as against the main contractor, any damage, loss and/or expense that he suffers or incurs as a result of 'any act, omission, default or negligence of the contractor or his employees, agents or sub-contractors which *disrupts* the regular progress of the sub-contract works, or delays the execution of them in accordance with the dates stated in the sub-contract time schedule'. This provision is wholly unsatisfactory since it provides no express machinery for the recovery of such claims; and it is an extraordinary provision to find in what is clearly a carefully-drafted contractual document. At the very least one would have expected to find provisions equivalent to those in the JCT sub-contract forms discussed earlier in this chapter; as it is, it would seem that sub-contractor's claims under clause 9.4 must be pursued in adjudication, arbitration or litigation, as appropriate. In practice, main contractors and sub-contractors would be well advised to try and agree such claims rather than pursue formal proceedings. Any sums agreed would be payable outside the bounds of the contract.

These provisions are unsatisfactory from the point of view of both contractors and sub-contractors and it may well be that sub-contractors operating under this form may be better advised to sue at common law.

---

[701] Architect's instructions are dealt with separately by main contract clause 8; see the commentary about that in Chapter 15, section 15.3.2.

APPENDICES

Appendix A

Example of build-up of contractor's entitlement to reimbursement of direct loss and/or expense under JCT 98, clause 26

A.1 Introduction

In the first edition of his book the authors were in some difficulty in devising a suitable example of a claim under the JCT 80 contract. Even at the time of writing the second edition there had been little experience of the use of the JCT form in practice and so there was virtually no historical data against which to measure the courts' or arbitrators' attitudes towards claims under that contract. JCT 80 turned into JCT 98. Importantly the terms of clauses 3 and 26 of the form provide that, if all parties play their parts properly, a contractor's entitlement to reimbursement under the clause should be dealt with monthly during the course of the contract. Hence, there should be no such thing as a claim submission, consisting of several lever arch files, some time after practical completion.

The terms of clause 26 have not changed significantly with the passage of time. However, as anyone engaged in the commercial and contractual aspects of the construction industry will know only too well, since the previous edition was written, there has been a wealth of judicial and arbitral decisions relating to claims under the JCT form and so there is now a considerable amount of historical data.

However, that wealth of data serves to confuse rather than clarify the situation. There are, it seems, as many approaches to claiming under clause 26 of the JCT form as there are contractors preparing them. It seems that more often than not the greatest factor constraining the constituents – and often the amount – of the claim is the imagination of its draftsman. Over recent years 'claimsmanship' has been elevated to an art whereby the draftsman searches for theoretical and hypothetical arguments and calculations by which it is hoped to convince the employer or his advisers that the loss and/or expense claimed, in all probability, must have accrued. More often than not, the theories and hypothesis are enhanced with computer-aided charts, diagrams and spreadsheets. Although often of doubtful use, nonetheless they are the means by which the architect and quantity surveyor are intended to be convinced of the accuracy of the claim presented.

To comprehensively discuss and provide examples of each such claim is certainly beyond the scope of this book. Moreover, the imaginative, artistic and hypothetical approach is likely to disguise one fundamental point. The contractor, if entitled to claim at all, is entitled to reimbursement of *his* direct loss and/or expense, not some hypothetical contractor's theoretical loss and/or expense. There can be few companies in the industry, small or large, who could operate without the use of computers. From drafting correspondence and minutes of meetings on word processing software through to simple labour and wage records, estimating, programming and progress reporting and measurement and valuation on spreadsheets, even the smallest contractors have no excuse for poor record keeping.

What follows, therefore, is not a 'sample claim', but a final build-up, warts and all, of a contractor's entitlement under a hypothetical contract as it might be done by the architect or quantity surveyor at the end of the job, based upon information given by the contractor. It is to be assumed, of course, that payments on a provisional basis have been made to the con-tractor from time to time under interim certificates during the course of the contract, and that only a small balance will be left to be paid to him after this exercise.

A.2 *Contract particulars*

The contract is for the construction of a speculative, but prestigious, office development. The essential details for purposes of this exercise are: a reinforced concrete structure (no basement) supported upon bored piles; cladding is curtain walling with tinted anti-sun glass; full air-conditioning; marble flooring and wall and column linings to the prestige entrance hall; piling, cladding, air-conditioning and marble by nominated sub-contractors.

Contract sum	£5,400,000
Date of possession	3 September 2002
Date for completion	7 September 2004
Liquidated and ascertained damages	£30,000 per week
Fluctuations	As is commonly the case the parties have not expressly deleted the fluc-tuations clauses but have simply inserted against the Contract Sum in Article 2 the words 'fixed price'

Delays occurred to the works as follows:

(1) The discovery of underground brick and concrete obstructions required substantial redesigning of foundations, necessitating reinforced concrete rafts

(2) A period of heavy snow during February and March 2003 delayed the erection of structural steel
(3) Nominated sub-contractors for the lifts caused delay over a period of 4 months
(4) Public holidays fell within the contract period due to the delayed completion resulting from (1) to (3) above.

The contractor has given the notices of delay required under clause 25 and he subsequently applied for direct loss and/or expense under clause 26 based on the same facts. The architect granted extensions of time at appropriate times during the progress of the works and he has made his final decision under clause 25.3.3. He has confirmed his previous interim extension of time awards which together total 20 weeks. The revised completion date is 25 January 2005. In accordance with clause 25.3.1 the architect has stated the relevant events which he has taken into account in awarding his extension. He has calculated the extensions of time due to the contractor by inputting delays into a computer model of the work programme. The architect made the following summary of his conclusions for the office file:

	Delay to completion (weeks)
(1) *Underground obstructions*	
(a) Little work done during weeks 4 to 7 inclusive of the project.	4
(b) Extra work in rafts	1
(2) *Heavy snow*	
Steel delivery and erection delayed	3
(3) *Nominated lift sub-contractor*	
Delays in obtaining relevant drawings to enable floors to be correctly dimensioned, late delivery of key pieces of equipment for building in at each floor level led to protracted delays over a period of 4 months	10
(4) Construction work was pushed into 2005 over the Christmas holidays	2
Total extension	20 weeks

On receipt of the contractor's application for loss and/or expense, the architect made reference to his notes on the extended period and instructed the quantity surveyor accordingly. The quantity surveyor has incorporated that opinion in his build-up notes.

A.3 The ascertainment

In summary the quantity surveyor has ascertained the contractor's total entitlement as follows:

(1) Underground obstructions

(a)	Costs of delay during redesign		£36,242.06
(b)	Costs of unproductive labour		£ 7,872.46
(c)	Costs of delay caused by additional works.		£ 9,060.52
(2)	Heavy snow		£ nil
(3)	Nominated lift sub-contractor		£ nil
(4)	Contractor's public holidays		£18,121.04
(5)	Increased costs resulting from delays		£18,750.00
(6)	Under recovery of contribution to head office overheads resulting from delays		£ 654.27
	Total		£90,700.35

For the purpose of this example it has been assumed that, in carrying out the ascertainment, the quantity surveyor made notes. What follows are imaginary figures which, the reader is asked to assume, are based upon proper evidence submitted by the contractor. In fact, of course, the figures have no basis in fact or reality and they are totally unsuitable for use with a real claim, even for comparison purposes. Figures are used simply to avoid putting £yyyy or £xxxxx. The notes also record the architect's decisions about those aspects of the claim which it is for the architect alone to decide – essentially questions of validity. Further assume that, before work commenced, the quantity surveyor and the contractor have prepared a month-by-month forecast of preliminaries expenditure on the basis of the contractor's priced preliminaries bill of quantities. The extent to which the quantity surveyor has ascertained correctly is discussed in the 'observations' which follow each part of the ascertainment.

Claim No. 1 Underground obstructions – delay 4 weeks

(a) *Management and staff*

Very little work was carried out after the underground obstructions were located until a decision was made about the foundations, and the redesign was finished. Although, obviously, it is the end date of the project that is extended, the delay caused to that end date occurs at an early stage in the project. Since the loss and/or expense to which the contractor is entitled must be directly attributable to the delay concerned, the quantity surveyor must establish what additional costs or losses accrued during the period of delay; that is to say during the time it took from finding the obstructions to the redesign of the works.

At this early stage in the project, in accordance with his programme and the predicted preliminaries expenditure, the contractor had already committed to site the following management and other staff:

1 No site agent	resident on site	£500.00/wk	£ 500.00
1 No sub agent	resident on site	£450.00/wk	£ 450.00
1 No site clerk	resident on site	£350.00/wk	£ 350.00
1 No section engineer	average 2 days/wk	£440.00/wk	£ 176.00

1 No chainman	average 2 days/wk £100/wk	£ 20.00
1 No quantity surveyor		
resident on site £430.00/wk		£ 430.00
Weekly preliminaries cost of management and staff		£1,926.00/wk

A contracts manager, bonus surveyor and senior quantity surveyor all visit site regularly from head office throughout the project. However, the contractor has priced his tender in such a way that it is intended that their costs are recovered through the contractor's overall allowance for overheads. These costs are dealt with later in this analysis.

The contractor's site management and staff are retained on site during the whole of the delay period and the quantity surveyor calculates the loss and expense on the cost of time-related management and staff to be:

Period of critical delay 4 weeks @ £1,926.00/wk = £7,704.00

to collection

(b)   Site accommodation:

By the time the delay occurred, the contractor had already brought to site much of his intended site establishment in accordance with his preliminaries forecast. On site between weeks 4 to 7 inclusive were:

3 No site offices	£160.00/wk
1 No site toilet unit	£140.00/wk
1 No site store	£130.00/wk
1 No canteen/drying room	£165.00/wk

The accommodation is retained on site during the whole period of investigation, decision making and redesign and the quantity surveyor calculates the loss and expense of the time-related site accommodation to be:

3 No site offices	£160.00/wk each ×3 × 4 weeks	= £ 1,920.00
1 No site toilet unit	£140.00/wk ×4 weeks	= £ 560.00
1 No site store	£130.00/wk ×4 weeks	= £ 520.00
1 No canteen/drying rm	£165.00/wk×4 weeks	= £ 660.00
		£ 3,660.00

to collection

(c)   Services and facilities

Against the items for power, lighting, fuels, water, telephone and administration the contractor has made an overall allowance of £1.20/m^2/wk in his tender. The total floor area of accommodation referred to above is 220 m^2 and the quantity surveyor calculates the loss and expense of the time-related services and facilities to be:

220 m^2 × £1.20 × 4 weeks =   £1,056.00

to collection

(d) Safety, health and welfare

Against this item the contractor has a time-related allowance in his tender of £200.00/wk. The quantity surveyor calculates the loss and expense associated with the safety, health and welfare costs to be:

£200.00 × 4 weeks =   £800.00

to collection

(e) Storage of materials

The contractor has made no separate allowance for either the stores and/or attendance by a resident storeman/site clerk associated with the storage of material. Their costs are catered for in the site accommodation and staff items noted above. The quantity surveyor therefore calculates the loss and expense associated with storage of materials costs to be:

£ incl.

to collection

(f) Rubbish disposal

Against an item for rubbish disposal the contractor has a time-related allowance of an average of 5 skips per week over the period of the contract. Each skip is valued at £165.00 which is deemed to be the cost of the skip hire, skip removal and a further allowance for site labour associated with the task of rubbish removal. The quantity surveyor calculates the loss and expense associated with rubbish removal to be:

£355.00 × 4 weeks =   £1,420.00

to collection

(g) Cleaning

Against an item for cleaning, the contractor has a time-related allowance of £2,750.00 for the whole of the contract period. The quantity surveyor calculates the loss and expense associated with rubbish removal to be:

£2,750.00 ÷ 106 weeks × 4 weeks =   £103.77

to collection

(h) Drying out

Against this item the contractor has entered £nil time-related allowance. The quantity surveyor has therefore made no allowance in his calculation for any costs associated with this item:

£nil

to collection

(i)   *Protection of work*

Against this item the contractor has made a time-related allowance totalling £7,200.00 for the whole period of the contract. The quantity surveyor calculates the loss and expense associated with protection during the period of delay to be:

£7,200.00 ÷ 106 weeks × 4 weeks =   £271.70
to collection

(j)   *Security*

Against this item the contractor has a time-related allowance of £15,000.00 for the whole of the contract period. The quantity surveyor calculates the loss and expense associated with security to be:

£15,000.00 ÷ 106 weeks × 4 weeks =   £566.04
to collection

(k)   *Maintenance of public and private roads*

Against this item the contractor has provided for a time-related allowance of £3,750.00. In the analysis of programmed expenditure of preliminaries agreed between the contractor and quantity surveyor before work began, that amount was shown as being expended in the following way:

Month 1   35% of the total, expended in equal proportions weekly
Month 2   25% of the total, expended in equal proportions weekly
Months 3–24   35% of the total, expended in equal proportions weekly
Month 25   5% of the total, expended in equal proportions weekly

The quantity surveyor calculates the loss and expense associated with maintenance of public and private roads to be:

£3,750.00 × 35% ÷ 4 weeks × 1 week =   £ 328.13
£3,750.00 × 25% ÷ 4 weeks × 3 weeks =   £ 703.13
                                                         £1,031.26
                                                         to collection

(l)   *Small plant and tools*

Against this item the contractor has made an overall time-related allowance of £9,500.00. In the analysis of programmed expenditure of preliminaries agreed between the contractor and quantity surveyor before work began, that amount was simply spread equally throughout the contract period. The quantity surveyor calculates the loss and expense associated with small plant and tools to be:

£9,500.00 ÷ 106 × 4 weeks =   £358.49
to collection

(m) Mechanical plant – earthmoving and concrete plant

By the time underground obstructions had been discovered the contractor was committed to sending – and had already sent to site – the following plant, in anticipation of beginning excavation and construction of pile caps and ground beams.

Tracked excavator	1 No
Dumper truck	1 No
Concrete mixing plant and cement silo	1 No

During the period of delay, the excavator and dump truck were retained on site, as were the mixing plant and silo which were both constructed in position ready for operation. Based on standard hire charges for plant of the type concerned (without operatives), the quantity surveyor has calculated the loss and expense on these items to be:

Tracked excavator	£1,500.00 × 4 weeks =	£ 6,000.00
Dumper truck	£ 500.00 × 4 weeks =	£ 2,000.00
Concrete mixing plant and silo	£2,500.00 × 4 weeks =	£10,000.00
		£18,000.00
		to collection

(n) Mechanical plant transport

A site vehicle for general use throughout the period of the project has been allowed in the contractor's tender. It has been valued by simply inserting a lump sum of £14,000.00 (including fuel, running costs and driver). The quantity surveyor has calculated that the contractor is due to loss and expense associated with this vehicle amounting to:

£14,000.00 ÷ 106 × 4 weeks =   £528.30
                                                                                                     to collection

(o) Temporary works

The contractor has allowed in his tender for ongoing maintenance necessary to repair the normal degradation of specific hardstandings which will need to be created on the site during the construction process. The amount concerned is provided as a lump sum of £10,000.00. One such hardstanding is already constructed by the time the underground obstructions are discovered and the quantity surveyor has calculated that the contractor is due loss and expense associated with additional maintenance throughout the 4 weeks delay of:

Total area of hardstanding $= 2,000\,\text{m}^2$
Area laid at time of delay $= 297\,\text{m}^2$
Programmed duration
for provision of $297\,\text{m}^2$ of
temporary hardstanding $= 8$ weeks

Cost attributed to maintenance of 297 m² for
4 week delay = £10,000.00 ÷ 2,000 × 297 ÷ 8 × 4 =   £742.50
                                                                    to collection

Collection of costs for underground obstructions

Management and staff	£ 7,704.00
Site accommodation	£ 3,660.00
Services and facilities for power, lighting, fuels, water, telephone and administration	£ 1,056.00
Services and facilities for safety, health and welfare	£ 800.00
Services and facilities for storage of materials	£ incl.
Services and facilities for rubbish disposal	£ 1,420.00
Services and facilities for cleaning	£ 103.77
Services and facilities for drying out	£ Nil
Services and facilities for protection of works	£ 271.70
Services and facilities for security	£ 566.04
Services and facilities for maintenance of public and private roads	£ 1,031.26
Services and facilities for small plant and tools	£ 358.49
Services and facilities for earthmoving and concrete plant	£ 18,000.00
Services and facilities for site vehicle	£ 528.30
Services and facilities for temporary works	£ 742.50
Total loss and expense for 4 weeks delay to summary	£ 36,242.06

(p)   Labour cost – measured work

In addition to the losses associated with time-related cost items, the contractor has suffered losses in respect of certain costs of measured works which the quantity surveyor must also take into account. For example, within the rates and prices for the measured work in sub-structures programmed to be done at this time, the contractor made allowance for:

3 No civils ganger   wage cost at £340.00/wk/each
6 No labourers   The labour element of the measured rates for sub-structure works are calculated on an average cost of £203.72/wk per operative, based on a 37 1/2 hr week, including employer's national insurance contribution, graduated pension, holidays with pay, wet

and guaranteed time, bonus payments and other like oncosts and with an allowance for head office overheads, oncosts and profit of 11%.

Both gangers are retained on site. Each ganger is almost wholly unproductive during that time except that, for some 20% of the time one was moved to another site. So far as the labourers are concerned, they too are largely unproductive throughout the period of delay. During that time the contractor was prevented from carrying out a programmed amount of work worth £16,300.00 (incl. overheads and profit). Instead, he carried out only £2,750.00 (incl. overheads and profit) worth of that work in the same 4-week period. Given that the work is labour intensive and assuming a labour/plant/material ratio of 80:12.5:7.5, the programmed labour return compared with the actual net labour return in the period will be £11,747.75 and £1,981.98 respectively. The labour has, therefore, been unproductive to the tune of £9,765.77 or 83.13% of net programmed return on labour.

Loss of revenue on measured work:

4 weeks × £340.00 × 2 ganger	= £2,720.00
4 weeks × £340.00 × 1 ganger × 80%	= £1,088.00
4 weeks × 6 labourers@ £203.72 × 83.13%	= £4,064.46
	£7,872.46
	to summary

(q) *Underground obstructions – additional work – 1 week delay*

The additional work involved the contractor in engaging another civil engineering supervisor and other staff on the project for the period of this extra work. The costs of the supervisor and the staff concerned will already have been taken into account in the valuation of the additional work in accordance with clause 13.5.3.3. of the contract. The contractor's entitlement under clause 26.1 is, therefore, confined to the costs, if any, associated with the critical delay which results in the further 1 week overrun to the completion of the contract. The quantity surveyor has calculated that loss and expense to be:

1 week at the total weekly rate from the collection of (a) to (o) above:

1 week × £9,060.52 per week =	£9,060.52
	to summary

Observations on claim No. 1

(a) *Management and staff*

The quantity surveyor has, quite correctly, ascertained the loss and expense that accrued at the time the delay arose. So far as the individual operatives are concerned for which a claim was made, it is difficult to see how each of the individuals was totally unproductive. The people concerned are all likely to have carried out productive work even though operations out on site are suspended. For example, the site clerk's day-to-day duties may well not be directly affected by a delay to progress on site, the quantity surveyor will no doubt still have much that he could be doing that was not affected by the delay and the site agent and sub-agent may well have been able to carry out preparatory work for other future planned operations on this site or perhaps might even be able to lend short-term assistance associated with work on other projects.

The contractor's duty is to minimise the loss occasioned by the delay, so far as he reasonably can do so. To that end he must make the best possible productive use of the staff involved and the quantity surveyor will want to satisfy himself that the personnel claimed for were, in fact, unavoidably unproductive as a consequence of the delay in question.

So far as the calculation of the contractor's loss is concerned, it appears from this example that the quantity surveyor has adopted the contractor's rates in the preliminaries bill of quantities as equating to the cost of the claimed staff and management. In practice, for various reasons, those rates are likely to differ from the true cost to the contractor of employing each of the individuals concerned. Since the contractor's entitlement is to be reimbursed his direct loss and/or expense, the quantity surveyor should calculate that loss and expense by reference to actual cost.

Most contractors, irrespective of size, these days will have computerised wage and other cost records and so it would not be unreasonable to presume that an accurate record of staff costs can readily be made available to the quantity surveyor. In any event he should start from the position of establishing the contractor's true cost and much of the speculation and conflict which surrounds the ascertainment of loss and expense claims could no doubt be substantially reduced and perhaps even avoided altogether if, as soon as the contractor has notified the architect that it has become apparent that the works are being delayed, the architect and/or quantity surveyor made it clear to the contractor what records they will expect to have available to them and in what format they should be produced. Indeed, that is one of the reasons why the contractor's application must be made promptly and why, if the contractor delays in making his application, he cannot be given the benefit of any doubt.

On the general question of cost versus preliminaries, there is a school of thought which says the contractor is not, in fact, precluded from being reimbursed at the rates used in the preliminaries bill. Indeed, it has been argued that the contractor is entitled to recover *either* his actual staff cost or his preliminaries allowance, whichever is the greater. The rationale behind this proposition seemingly is that, since the contractor is entitled to recover his loss, if the true cost exceeds the preliminaries rate the true cost will clearly be the measure of that loss.

However, if preliminaries rates exceed the true cost, then provided the contractor can demonstrate that those preliminaries rates are rates that he consistently and successfully recovers on similar contracts, the fact that he is retaining the relevant staff on this project longer than envisaged means that he is prevented from deploying same staff on other work where they would provide the return. Arguably, since the 'time-related charges' are generally intended to properly reflect the contractor's true preliminaries costs, the quantity surveyor will have reviewed the potential accuracy of those rates and prices when initially assessing the contractor's tender. If they are obviously seriously inflated (or 'front loaded') he will no doubt have raised and hopefully have resolved the point before the employer executes the contract.

However, on the principle that loss and/or expense is to be recovered on the same basis as damages for breach of contract, actual costs are to be used unless there are powerful reasons to the contrary.

(b) Site accommodation

Whereas it may well be relatively simple to redeploy staff or to temporarily re-allocate them to alternative productive tasks during short periods of delay so as to mitigate the cost implications of the delay, the same is not generally true of site accommodation. Even if it could be redeployed, the economics of disestablishing it and re-establishing it at a later date would be likely to militate against such redeployment and so it is reasonable to suppose, in the present circumstances, that the contractor should be reimbursed for the cost of the various facilities claimed for.

However, once again the quantity surveyor appears to have based his calculation on the contractor's rates in the preliminaries bill of quantities and not on the true cost of the accommodation concerned. What is more, with the possible exception of the toilet unit the cabins in question may well be the contractor's own property, as is commonly the case, and not hired in for this project. The quantity surveyor will therefore want to determine whether or not they do belong to the contractor. If they are his property the amount to which the contractor will be entitled by way of loss and expense will be considerably less and will amount simply to a figure calculated by reference to the depreciation on the asset value (see Chapter 6, 'Contractor's own plant' in section 6.5.6) with, perhaps, a further allowance for additional repair and maintenance costs. In the event that any service costs associated with the cabins (such as toilet facilities) are also included in the rates, the quantity surveyor will also want to be satisfied that any genuine tangible reduction in those service costs resulting from the suspension/delay will be credited against the loss and expense that would otherwise be due to the contractor.

(c) Power, lighting, fuel, water, telephone and administration

Following the principle that the contractor is entitled only to recover his actual cost in respect of each of the items referred to, the quantity surveyor should ascertain the contractor's losses by reference to the actual charges incurred for the period concerned. However, the quantity surveyor must also apply a realistic approach to the exercise of ascertainment. Clearly, the quantity surveyor is not free to disregard the contract terms in undertaking the ascertainment process. But he must apply his judgment and expertise to the task. That is to say, there can be little or no justification for spending an inordinate amount of time and cost in insisting on cost records which themselves may produce an inexact result. That is particularly so when, taking into account the possible margin of error likely to result, applying a properly considered tender rate may well prove more cost effective in the long run. The occasions when that is likely to be the case are probably few but, there is some merit in approaching the ascertainment of these particular items on the basis adopted in the example above. That is to say, by reference to the tender allowance related to the superficial floor area of the site accommodation.

Accounts for items such as fuel, telephones, lighting, etc. are often rendered a month or more in arrears. They may often be based on estimated readings and may often include fixed rental or similar charges for the forthcoming quarter or other accounting period. It is difficult, if not in fact impossible, to differentiate between costs attributable directly to the delay and those which may well have little or no bearing on the delay.

(d)   Safety, health and welfare

As a general principle the quantity surveyor should look to establishing the true cost of the items concerned and not some theoretical average weekly cost included in the tender. It is quite possible that specific relevant cost records for things such as safety officer's site visits and the like will not necessarily be readily available, and it has to be conceded that, like the electricity, fuel and other costs referred to above, there is some practical merit in approaching valuation of this item by reference to the preliminaries allowance. However, at the very least the quantity surveyor must satisfy himself that the allowance concerned realistically reflects the costs that the contractor is likely to incur in respect of such items. Although it is of course reasonable to expect the contractor to make a time-related allowance for this item in his tender, and whilst for convenience that allowance may well be calculated by reference to average weekly (or perhaps monthly) costs, many of the recurring weekly or monthly costs associated with safety, health and welfare will largely depend on the amount of work done and the level of staffing in any particular week or month of the project.

Quite simply, the time and cost dedicated to the safety, health and welfare matters will be relative to the number of operatives on site and to the activities in which they are involved. The extent to which the contractor has chosen to sub-let work and the terms on which he may have agreed to provide safety, health and welfare facilities to those sub-contractors may also have a bearing on the cost that the main contractor actually bears in this respect. It follows that if work is delayed and the labour force (or sub-contract attendance) is consequently significantly, temporarily, reduced – or is largely inactive – then it is quite conceivable that costs of these time-related preliminaries may also be significantly below the average.

(e) Storage of material

The contractor has made no separate allowance for either the stores and/or attendance by a resident storeman/site clerk associated with the storage of material. As to the quantity surveyor's calculation of this item see above.

(f) Rubbish disposal

Once again the quantity surveyor appears to have erroneously referred to the contractor's preliminaries allowance as opposed to the true cost of providing facilities for the removal of rubbish from the site when ascertaining the contractor's loss. Having quite possibly arranged for skips to be delivered to site before realising the work was to be suspended, it could reasonably be the case that the costs of temporarily removing them from site and then subsequently returning them again later would far outweigh the ongoing hire cost, if any, pending resumption of the work. Moreover, the contractor might legitimately argue that, without the aid of a crystal ball, he could have had no way of knowing from one day to the next when the delay would come to an end. Consequently, even if it transpires that it would have been more economical to remove and return the skips at a later date, that would not have been a practical option and so the contractor would reasonably expect to be paid any additional hire charges resulting from the delay.

The quantity surveyor also appears to have ignored the fact that, even if in the particular circumstances it was appropriate to value the contractor's loss by reference to the preliminaries rate, the rate includes allowance for skip removal and a further allowance for site labour associated with the task of rubbish removal. An adjustment for those allowances is appropriate, in any event.

(g) Cleaning

So far as cleaning is concerned, generally the same criticisms as those above apply

(h) Drying out

Here the quantity surveyor has valued the item at £nil on the premise that the contractor entered no value against this cost in his tender preliminaries. The fact that the contractor is entitled only to his costs and not to some theoretical preliminaries valuation in this case works to the contractor's advantage, because irrespective of the absence of a tender rate for this work, if it transpires that the contractor has incurred a cost for this work there is no reason in principle why he should not recover the additional cost which he will then incur due to the delay.

(j)   Security

It is entirely reasonable to expect the contractor to have maintained security throughout the additional period associated with the delay. Indeed, it may even be argued that during that period of inactivity increased security may have been necessary. It is not uncommon for contractors nowadays to engage the services of private security companies and even if an in-house watchman and other security arrangements were adopted, there is no obvious reason why the quantity surveyor should refer to the preliminaries rates rather than proper cost records to provide evidence of the contractor's loss and expense.

(l)   Small plant and tools

The cost records should form the basis of the quantity surveyor's calculation. Moreover, it should not be forgotten that the contractor has an underlying duty to mitigate the costs associated with the delay. Whether or not the items of small plant, tools and equipment are owned or are hired in by the contractor, they may well have been readily taken off hire or transferred for use on other projects. The quantity surveyor should, therefore, satisfy himself that the contractor is not claiming for what could have been avoidable costs. Once satisfied that this is not the case, then again the amount to which the contractor is entitled will vary greatly depending on whether or not the tools and equipment concerned are hired or owned in-house (see the commentary under (b) above). Even if the use of the preliminaries rates could be justified on practical grounds, given the relatively insignificant value of this item and the relative margin for error likely to arise from adopting those figures, the quantity surveyor has used what amounts to an average weekly cost. Since the nature, extent and quantity of small plant, tools and equipment necessary on the site depend on the nature of the work being done at any given time, it is unjustifiable to merely adopt the average weekly rate in the contractor's tender.

(m)   Mechanical plant – earthmoving and concrete plant

The ascertainment will involve the quantity surveyor in more than simply applying, as he has, tender rates to a multiplicand which represents the number of weeks of delay. He must once more consider whether the contractor could realistically have arranged for the plant and machinery concerned to be taken off hire or temporarily removed from site pending resumption of the works. Since it costs money to take plant from yard to site, back to yard and then back to site again, the financial consequences of this would have to be carefully considered. That is to say, would the removal and remobilisation costs have outweighed the likely savings?

There is also the fundamental question whether or not the plant and equipment sent to site is hired or owned by the contractor. It may of course be that the contractor's tender rate in the preliminaries properly reflects the correct hire charge and that the contractor fully intended to hire in the plant concerned, but when the time came he utilised in-house plant that was available at the time. There will be a significant difference in what the contractor is entitled to be reimbursed depending on whether the items concerned are owned or hired in. Finally, even when considering cost records, if the equipment is hired the quantity surveyor will want to consider whether the rate(s) claimed include some allowance for an operator/ driver or for fuel, maintenance or other such work-related costs that would be unlikely to arise during periods of inactivity.

(n) Transport – site vehicle

The principles that apply to heavy mechanical plant and equipment referred to above do not differ when any legitimate loss associated with site transport is ascertained.

(o) Temporary works – hardstandings and roads

Although the preliminaries items in the bill of quantities offer a good checklist of the likely heads of cost and/or loss that the contractor may incur as a consequence of delay to the works, they should not be blindly followed. Leaving to one side the obviously erroneous use of theoretical rates and calculations derived from the contractor's tender, the quantity surveyor must also consider whether or not the head of claim is likely to cause additional loss and/or expense. For example, despite convenient reference to it as a time-related cost, the need for ongoing maintenance of road(s) and/or hardstanding(s) may well have more to do with the work done than the time taken to complete the project. In short, although described as time-related, it is difficult to see how, subject to weather conditions, the road/hardstanding in question will degrade over a two or three week period so as to warrant the contractor being paid additional loss and expense when the site is effectively at a standstill during that time and the hardstanding and/or road concerned is not, therefore, in use.

(p) Labour cost – measured work

Here, the quantity surveyor has quite rightly had regard to the contractor's wage records for the costs associated with the two gangers concerned. Quite correctly, too, he has taken account of the time when the second of the two gangers was deployed elsewhere and so was productively employed. However, he appears to have disregarded the fact that some productive work was done by the labour force and so one would expect a further proportion of the gangers' costs to be offset against that, albeit reduced, productivity.

Regarding the loss associated with the alleged reduction in the productivity of labourers, the quantity surveyor will want to satisfy himself that the reduced output is not due simply to either poor management on the contractor's part or to some other inefficiencies of the operatives concerned. He will also want to be convinced that the measure of efficient working, namely the rate of output envisaged in the tender is, in fact, a reasonably realistic prospect. In essence the questions are:

(1) was the expected output reasonable in the first place? *and*
(2) could the contractor have achieved a far better rate of production than he did in fact achieve? It follows from this question that the quantity surveyor must reasonably be satisfied that there is a link between the alleged delay and the alleged financial consequences of it.

The quantity surveyor will only be in a position to properly ascertain the resulting loss if those two facts have been established.

Claim No. 2   Delays due to heavy snow

The architect has notified the quantity surveyor that this is not a matter under clause 26 and, therefore, the contractor is not entitled to any additional costs for delay associated with this item.

Claim No. 3   Delays due to nominated lift sub-contractor

The architect has notified the quantity surveyor that this is not a matter under clause 26 and, therefore, the contractor is not entitled to any additional costs for delay associated with this item.

Observations on claims Nos. 2 and 3

The architect has granted an extension of time in respect of these two issues. However, there is no direct relationship between the provisions for extension of time and those for reimbursement of loss and expense. The contract makes clear that the contract sum shall not be adjusted otherwise than in strict accordance with the express provisions of the contract. So far as losses resulting from delay and/or disruption are concerned, clause 26 of the contract alone sets out the circumstances giving rise to entitlement and nowhere in clause 26 is the contractor given the right to recover these costs. The architect has informed the quantity surveyor that in his opinion this part of the claim for loss and/or expense has no validity. The quantity surveyor, therefore, has no option but to ignore any such costs in his ascertainment of the contractor's overall entitlement under the contract.

Claim No. 4 Contractor's holidays

Delays caused by the preceding relevant events have pushed the completion beyond the Christmas/New Year holiday 2004/5. It will not be delayed beyond Easter 2005, but because the project will extend beyond the 2-week construction industry holiday period, the quantity surveyor calculates the contractor's entitlement as:

2 weeks × the total weekly rate from the collection of (a) to (o) above:
2 weeks × £9,060.52 per week =   £18,121.04
to summary

Observations on claim No. 4

On a strict reading of the contract, the contractor is not entitled either to an extension of time or to payment of loss and expense resulting from the contract period being carried over a previously irrelevant holiday. This is on the simple basis that the contractor entered into a contract which provided for extensions of time and loss and/or expense in certain circumstances and the contractor must have been aware that the contract period might be extended or prolonged into a holiday period. However, the courts have not been slow to acknowledge the fact that this type of consequential loss will generally give rise to an entitlement, despite the absence of an express term of the contract to that effect.

   The principle is quite simple. If a delay is caused by an event which gives rise to entitlement and that delay then pushes the contract over into a holiday or other disruptive or delaying period, then the consequential loss and expense will be recoverable[702]. In this example the situation for the architect is, of course, somewhat complicated since the contractor has suffered three delays. Two are not the financial responsibility of the employer, and the question therefore is: did the contract period extend beyond the holiday period as a direct result of a matter referred to in clause 26 or, alternatively, was it the result of the heavy snow or nominated lift subcontractor, neither of which carry any entitlement. Only when he is reasonably satisfied that it is the former can the architect confidently instruct the quantity surveyor to include the loss and expense concerned in his ascertainment under the contract.

   Even if it is assumed that the architect has carried out the necessary analysis, for the reasons already explained the quantity surveyor must have regard to actual costs and not to preliminaries rates and prices to

---

[702] See the consideration of the 'knock-on' effect in Chapter 6, section 6.4.

ascertain the financial consequences. In this instance, the architect has wrongly taken account of all the extension of time and not simply the prolongation resulting from clause 26.2 matters, which amount to only 5 weeks. Since the period during which the loss is incurred is the unforeseen holiday break, then it is the cost associated with that particular period in the contract on which the quantity surveyor should concentrate his attention.

He has, however, simply applied an all-in weekly rate for time-related costs derived from an analysis of an entirely unrelated period, and even then, for the reasons already outlined, that figure is itself open to criticism. The quantity surveyor should consider closely whether, and if so to what extent, the costs accruing during the holiday period might be significantly lower than those of a normal working period. For example, plant may easily be taken off hire, certain wages may be payable from holiday credits for which the contractor has already made allowance in his rates and prices in the tender, and a distinct close-down of the site will inevitably mean a tangible reduction in costs such as light, heating, telephone, administration, maintenance of plant and equipment and the like.

Claim No. 5 Increased costs

The contract is let on a fixed price basis. The quantity surveyor is aware that throughout the project and up until the actual completion date there have been increases in the cost of labour and certain materials. The contractor has claimed that he is entitled to be compensated for those increases and the quantity surveyor has calculated the contractor's entitlement adopting the National Economic Development Office (NEDO) indices (calculation not shown), and the resulting amount is:

> £18,750.00
> *to summary*

Observations on claim No. 5

Clearly, if due to delays the contractor incurs labour or material increases that he would not otherwise have encountered, then this is a real loss. However, in arriving at a proper ascertainment of that loss a number of factors must be considered. Use of a NEDO formula is undoubtedly a convenient expedient by which to mechanically ascertain the loss and so avoid what would otherwise be a difficult and time consuming exercise. It is for the contractor to supply detailed evidence supporting his claim for increases. Those difficulties are not underestimated. However, that is clearly not sufficient justification for the quantity surveyor to ignore ascertaining the contractor's true loss and cost and in doing so having to consider the particular circumstances of this project instead of applying general theoretical principles. For example, it may be that the contractor took delivery of 100% of his requirements before the expected increase. The difficulties in ascertaining the true extent of that loss may well be compounded further by the fact that when the contractor came to place his order, market forces allowed him to secure a buy down from the supplier which resulted in him paying a significantly reduced price. An increase of 5% some 8 months on might then result in a purchase price of less than the material rate on which the contractor's tender is based. Those and other such particular circumstances might well make the quantity surveyor's task of ascertaining the increased cost a difficult and onerous one and there can be little justification for rejecting a proper analysis in favour of a theoretical approach.

Claim No. 6 Loss of contribution to head office overheads

The quantity surveyor accepts the principle that, where the contractor is delayed and must retain resources on site for longer than anticipated, if those resources do not produce a significant corresponding increase in the turnover achieved on this particular project there will be a shortfall in the contribution which will be made by this site to defray head office overhead costs.

The quantity surveyor has calculated the extent of that shortfall by establishing that the margin to be recovered if based on the average margin recovered over the last 2 years will be 11% as evidenced by an auditor's certificate to that effect.

The contribution to overheads anticipated to be recovered over the contract period of 108 weeks from the tender sum of £5,400,000 is therefore:

$$\frac{11}{111} \times \frac{£5,400.000}{106 \text{ weeks}} = £5,048.44 \text{ per week}$$

The final account (without allowance for recovery of loss and expense) is £2,500,750 and so the actual contribution to overheads that will be recovered over the actual contract period from the final account figure will be:

$$\frac{11}{111} \times \frac{£5,750,000}{126 \text{ weeks}} = £4,522.38 \text{ per week}$$

The shortfall in recovery is, therefore, said to be:

$$(£5,048.44 - £4,522.38) \times 126 \text{ weeks} \qquad\qquad = £66,283.56$$

However, since only events (1) and (4) above carry any entitlement to reimbursement, the quantity surveyor has calculated the loss and expense due pursuant to clause 26 to be:

$$\frac{11}{111} \times \frac{£5,750,000}{113 \text{ weeks}} = £5,042.65 \text{ per week}$$

The shortfall in recovery is, therefore, said to be:

$$(£5,048.44 - £5,042.65) \times 113 \text{ weeks} \qquad = £654.27$$

to summary

Observations on Claim No. 6

The quantity surveyor has adopted what is commonly referred to as a formula approach to ascertaining the contractor's entitlement here. There has been much debate about the applicability and use of such formulae. The principle behind their use is conveniently set out in Chapter 6, 'Head office overheads' in section 6.5.2. The proposition put forward is essentially this: after making their best estimate of the prime cost of the works, contractors when tendering tend to add a single percentage uplift to those prime cost rates and prices. In this way they provide, in the distribution of those rates across the whole scope of the work, for a contribution to the general overheads and profit of the company. When the contractor is delayed, the contractor's loss and expense resulting from that delay will hopefully include a proportionate extension of this percentage. Put simply, if the contractor expects this project to contribute 10% of the contract sum to the company's overheads and profit throughout, say, a 106 week contract period, and if the contract period becomes 126 weeks and there is no corresponding increase in contribution, the contractor will have recovered less than expected in respect of overheads and profit. On the face of it the simplicity and logic of that proposition is attractive and difficult to deny.

Indeed, since first mooted in *Hudson's Building and Engineering Contracts*, other respected works have been cited as giving express approval to the proposition. But that conveniently ignores the warnings that all proponents of the formula approach give concerning its use.

The most fundamental of these must be that its application depends on a basic presumption, which the contractor must of course substantiate, that the percentage claimed would have been capable, in fact, of being earned elsewhere had it not been for the delay concerned. This itself depends on two further presumptions. First, that the contractor did not habitually underestimate the true cost and therefore the predicted resource level and/or duration of the works. Second, that subsequent market trends have not changed so as to reduce the potential for achieving that contribution from work elsewhere.

Moreover, the criteria which must be met before consideration can legitimately be given to using a formula approach to ascertainment have been further expanded since a formula first appeared in *Hudson's* 10th edition. Finally, it should not be forgotten that the philosophy behind this head of claim depends largely on the argument that the contractor was unable to deploy his resources on this delayed project to work elsewhere during the period of delay. With the extensive use of sub-contractors and, in particular, labour only sub-contractors over recent years, and with the flexibility that sub-contracting offers main contractors when deciding on their capacity to take on further work, the argument that the contractor was forced to turn work away because his labour force was already committed on the late project is probably now more difficult to sustain.

A.4 *Generally*

It is worth repeating that all figures used in these examples are entirely imaginary. They are merely a means to an end. It is important to read the observations with the examples in order to understand the principles of ascertainment, because the imaginary quantity surveyor has not always carried out the ascertainment in the best possible way. The loss must have followed directly from the event and evidence should be produced to support the amount claimed and/or ascertained.

It may be true that the contractor's recovery against overheads and profit during the original contract period may, week-by-week, have been less than anticipated. The shortfall may, in fact, have been made up in the period of overrun during which the contract will have been earning an overhead recovery on additional work done, but not previously anticipated. Additional site supervision costs would clearly be a recoverable head of claim but a direct causal relationship would have to be established between the alleged cause of delay and the necessity to employ supervisory staff for the additional period. The figures would also require substantiation by way of wages sheets or the like.

A claim for increased costs is in principle permissible, but the method adopted is wholly unsatisfactory. Unless the parties have agreed otherwise, it is inappropriate to use a theoretical index-related approach. Indeed, over the period concerned this could even work to the contractor's disadvantage, and insistence on a proper investigative approach should not be seen as merely an attempt to make life difficult for the contractor but more a means of arriving at what, after all, is said in the contract to be an ascertainment of the losses incurred.

Loss of productivity or uneconomic working is a possible head of claim, and is particularly difficult to assess as regards labour. Some indication has been given earlier in this book about the best way of keeping labour records, and it is certainly not permissible, as this quantity surveyor's ascertainment tends to suggest, to add what amounts to an arbitrary percentage to the allegedly anticipated labour costs.

It is evident that the loss will vary according to the circumstances of the case and even where the actual labour cost is compared with that contemplated at the time of tender, this will necessitate abstracting the labour element from the contract price and the actual labour costs from the contractor's records. The gross difference between these two figures must be further adjusted to take account of any actual labour costs expended as a result of unclaimable circumstances, such as contemporaneous delay and disruption arising from *force majeure* or from the contractor's own inefficiency. Even this approach falls short of the ideal since it relates facts in the form of the ascertained actual costs to an estimate of what the costs otherwise would have been. The difference between the two, of course, is itself an estimate and not a fact. For this kind of approach to have any reasonable prospect of success, the contractor must put forward convincing evidence that the labour costs contemplated were reasonable at the time *and* capable in fact of achieving the work output required by the original work content.

Finally, there is no reference to interest or to finance costs under the principles established in *F. G. Minter Ltd* v. *Welsh Health Technical Services Organisation* (1980). It seems that the substantial differences in wording between clause 24 of JCT 1963 and clause 26 of the 1998 edition virtually rule out any such entitlement in practice, although in theory it still exists, because, if properly administered the contract should now ensure that the contractor's loss and/or expense is largely reimbursed to him from month to month as he incurs it; finance charges therefore should be minimal (see Chapter 6, 'Financing charges and interest' in section 6.5.8). Nevertheless, it *may* be possible for a contractor to establish such an entitlement, and it should certainly not be ruled out altogether.

Appendix B:

Extension of Time and Money Claims Under DOM/1

The following consists of extracts from the third edition of this book. Although DOM/1 is no longer recommended for use, it will no doubt continue in use for some time and for this reason it is included as an appendix, but without significant amendment.

B.1 Introduction

Domestic sub-contractors are work and material sub-contractors who are not nominated by the architect under clause 35 of JCT 80. The full title of this document, which is issued jointly by BEC, FASS and CASEC, is 'The Standard Form of Sub-Contract for Domestic Sub-Contractors for use where the main contract is in one of the JCT Local Authorities/Private Editions either with or without quantities' and, unlike the former Blue sub-contract form, which it has superseded, it is intended for use only with the JCT main form. The Blue Form could be used in conjunction with other standard form contracts, such as GC/Works/1[703]. DOM/1 was published in 1980.

B.2 Extensions of time under DOM/l

Extensions of time are dealt with by clause 11, which reads as follows:

11 Sub-Contractor's obligation – carrying out and completion of Sub-Contract Works – extension of Sub-Contract time

11.1 The Sub-Contractor shall carry out and complete the Sub-Contract Works in accordance with the details in the Appendix, part 4, and reasonably in accordance with the progress of the Works but subject to receipt of the notice to commence work on site as stated in Appendix, part 4, and to the operation of clause 11.

11.2 .1 If and whenever it becomes reasonably apparent that the commencement, progress or completion of the Sub-Contract Works or any part

---

[703] Anyone interested in the detailed provisions of the Blue Form should refer to Professor V. Powell-Smith's *The Standard (Non-Nominated) Form of Building Sub-Contract*, 1980, IPC Building & Contract Journals Ltd.

thereof is being or is likely to be delayed, the Sub-Contractor shall forthwith give written notice to the Contractor of the material circumstances including, insofar as the Sub-Contractor is able, the cause or causes of the delay and identify in such notice any matter which in his opinion comes within clause 11.3.1.

.2   In respect of each and every matter which comes within clause 11.3.1, and identified in the notice given in accordance with clause 11.2.1, the Sub-Contractor shall, if practicable in such notice, or otherwise in writing as soon as possible after such notice:

.1   give particulars of the expected effects thereof; and

.2   estimate the extent, if any, of the expected delay in the completion of the Sub-Contract Works or any part thereof beyond the expiry of the period or periods stated in the Appendix, part 4, or beyond the expiry of any extended period or periods previously fixed under clause 11 which results therefrom whether or not concurrently with delay resulting from any other matter which comes within clause 11.3.1; and

.3   the Sub-Contractor shall give such further written notices to the Contractor as may be reasonably necessary or as the Contractor may reasonably require for keeping up to date the particulars and estimate referred to in clause 11.2.2.1 and .2 including any material change in such particulars or estimate.

11.3   If on receipt of any notice, particulars and estimate under clause 11.2 the Contractor properly considers that:

.1   any of the causes of the delay is an act, omission or default of the Contractor, his servants or agents or his sub-contractors, their servants or agents (other than the Sub-Contractor, his servants or agents) or is the occurrence of a Relevant Event; and

.2   the completion of the Sub-Contract Works is likely to be delayed thereby beyond the period or periods stated in the Appendix, part 4, or any revised such period or periods.

then the Contractor shall, in writing, give an extension of time to the Sub-Contractor by fixing such revised or further revised period or periods for the completion of the Sub-Contract Works as the Contractor then estimates to be reasonable.

11.4   .1   When fixing such revised period or periods, the Contractor shall, if reasonably practicable having regard to the sufficiency of the notice, particulars and estimate, fix such revised period or periods within the following time limit:

.1   not later than 16 weeks from the receipt by the Contractor of the notice and of reasonably sufficient particulars and estimates, or

.2   where the time between receipt thereof and the expiry of the period or periods for the completion of the Sub-Contract Works is less than 16 weeks, not later than the expiry of the aforesaid period or periods.

.2   The Contractor, when fixing such revised period or periods, shall state:

.1 which of the matters, including any of the Relevant Events, referred to in clause 11.3.1 he has taken into account; and

.2 the extent, if any, to which the Contractor has regard to any direction requiring as a Variation the omission of any work or obligation or restriction issued since the previous fixing of any such revised period or periods for the completion of the Sub-Contract Works.

11.5 If, upon receipt of any notice, particulars and estimate under clause 11.2, the Contractor properly considers that he unable to give, in writing, an extension of time to the Sub-Contractor, the Contractor shall, if reasonably practicable having regard to the aforesaid notice, particulars and estimate, so notify the Sub-Contractor in writing not later than 16 weeks from receipt of the notice particulars and estimate, or, where the time between such receipt and the expiry of the period or periods for the completion of the Sub-Contract Works is less than 16 weeks, not later than the expiry of the aforesaid period or periods.

11.6 After the first exercise by the Contractor of the duty under clause 11.3, the Contractor may fix a period or periods for completion of the Sub-Contract Works shorter than that previously fixed under clause 11.3 if, in the opinion of the Contractor, the fixing of such shorter period or periods is fair and reasonable having regard to any direction issued requiring as a Variation the omission of any work or obligation or restriction where such issue is after the last occasion on which the Contractor made a revision of the aforesaid period or periods.

11.7 If the expiry of the period when the Sub-Contract Works should have been completed in accordance with clause 11.1 occurs before the date of practical completion of the Sub-Contract Works established under clause 14.1 or 14.2, the Contractor may

and

not later than the expiry of 16 weeks from the date of practical completion of the Sub-Contract Works, the Contractor shall:

either

.1 fix such a period or periods for completion of the Sub-Contract Works longer than that previously fixed under clause 11 as the Contractor properly considers to be fair and reasonable having regard to any of the matters referred to in clause 11.3.1 whether upon reviewing a previous decision or otherwise and whether or not the matters referred to in clause 11.3.1 have been specifically notified by the Sub-Contractor under clause 11.2; or

.2 fix such a period or periods for completion of the Sub-Contract Works shorter than that previously fixed under clause 11 as the Contractor properly considers to be fair and reasonable having regard to any direction issued requiring as a Variation the omission of any work where such issue is after the last occasion on which the Contractor made a revision of the aforesaid period or periods; or

.3 confirm to the Sub-Contractor the period or periods for the completion of the Sub-Contract Works previously fixed.

11.8 The operation of clause 11 shall be subject to the proviso that the Sub-Contractor shall use constantly his best endeavours to prevent delay in the progress of the Sub-Contract Works or any part thereof, however caused, and to prevent any such delay resulting in the completion of the Sub-Contract Works being delayed or further delayed beyond the period or periods for completion, and the Sub-Contractor shall do all that may reasonably be required to the satisfaction of the Architect and the Contractor to proceed with the Sub-Contract Works.

11.9 No decision of the Contractor under clauses 11.2 to .7 inclusive shall fix a period or periods for completion of the Sub-Contract Works which will be shorter than the period or periods stated in the Appendix, part 4.

The following are the Relevant Events referred to in clause 11.3.1:

.1 force majeure;

.2 exceptionally adverse weather conditions;

.3 loss or damage occasioned by any one or more of the Specified Perils;

.4 civil commotion, local combination of workmen, strike or lockout affecting any of the trades employed upon the Works or any of the trades engaged in the preparation, manufacture or transportation of any of the goods or materials required for the Works;

.5 compliance by the Contractor with the Architect's instructions (which shall be deemed to include compliance by the Sub-Contractor with the Contractor's directions which pass on such instructions):

   .1 under clauses 2.3, 13.2, 13.3, 23.2, 34, 35 or 36 of the Main Contract Conditions, or
   .2 in regard to the opening up for inspection of any work covered up or the testing of any of the work, materials or goods in accordance with clause 8.3 of the Main Contract Conditions and/or clause 4.3.1 (including making good in consequence of such opening up or testing) unless the inspection or test showed that the work, materials or goods were not in accordance with the Main Contract or the Sub-Contract as the case may be:

.6 the Contractor, or the Sub-Contractor through the Contractor, not having received in due time necessary instructions, drawings, details or levels from the Architect for which the Contractor or the Sub-Contractor, through the Contractor, specifically applied in writing provided that such application was made on a date which having regard to the Completion Date or the period or periods for the completion of the Sub-Contract Works was neither unreasonably distant from nor unreasonably close to the date on which it was necessary for the Contractor or the Sub-Contractor to receive the same;

.7 delay on the part of Nominated Sub-Contractors or of Nominated Suppliers in respect of the Works which the Contractor has taken all practicable steps to avoid or reduce;

.8 .1 the execution of work not forming part of the Main Contract by the Employer himself or by persons employed or otherwise engaged by the Employer as referred to in clause 29 of the Main Contract Conditions or the failure to execute such work;

.8 .2 the supply by the Employer of materials and goods which the Employer has agreed to provide for the Works or the failure so to supply;

.9 the exercise after the Base Date by the United Kingdom Government of any statutory power which directly affects the execution of the Works by restricting the availability or use of labour which is essential to the proper carrying out of the Works, or preventing the Contractor or Sub-Contractor from, or delaying the Contractor or Sub-Contractor in, securing such goods or materials or such fuel or energy as are essential to the proper carrying out of the Works;

.10 .1 the Contractor's or Sub-Contractor's inability for reasons beyond his control and which he could not reasonably have foreseen at the Base Date for the purposes of the Main Contract or the Sub-Contract as the case may be to secure such labour as is essential to the proper carrying out of the Works;

.2 the Contractor's or Sub-Contractor's inability for reasons beyond his control and which he could not reasonably have foreseen at the Base Date for the purposes of the Main Contract or the Sub-Contract as the case may be to secure such goods or materials as are essential to the proper carrying out of the Works;

.11 the carrying out by a local authority or statutory undertaker of work in pursuance of its statutory obligations in relation to the Works, or the failure to carry out such work;

.12 failure of the Employer to give in due time ingress to or egress from the site of the Works or any part thereof through or over any land, buildings, way or passage adjoining or connected with the site and in the possession and control of the Employer, in accordance with the Contract Bills and/or the Contract Drawings, after receipt by the Architect of such notice, if any, as the Contractor is required to give, or failure of the Employer to give such ingress or egress as otherwise agreed between the Architect and the Contractor;

.13 the valid exercise by the Sub-Contractor of the right in clause 21.6 to suspend the further execution of the Sub-Contract Works;

.14 where it is stated in the Appendix part 1, Section B that clause 23.1.2 of the Main Contract Conditions applies to the Main Contract, any deferment by the Employer in giving possession of the site of the Works to the Contractor.

B.3 Commentary

To all intents and purposes this parallels the provisions of NSC/C, clauses 2.2 to 2.6, so that the commentary thereon is relevant. However, the wording has differences in places, and the following points should be noted.

B.3.1 Progress

The words of clause 11.1 have been construed very broadly by contractors over the years as obliging the sub-contractor to work in accordance with the contractor's progress on the works and that if the contractor's progress slowed or quickened, the sub-contractor was obliged to follow suit. It is now established that, under clause 11.1 of DOM/1, the sub-contractor may plan and perform the work as he pleases if there is no indication to the contrary, provided that he finishes it by the time fixed in the contract. The sub-contractor's only obligations so far as programming requirements are concerned are those requirements expressly contained in the sub-contract itself[704]. It is likely that the same principle applies in the case of other similarly worded sub-contracts such as NSC/C.

B.3.2 Notice

In clause 11.2.1 the sub-contractor's obligation to give notice of the material circumstances is a qualified one. The material circumstances are to include the cause or causes of the delay, *insofar as the sub-contractor* is able to identify them. In practice, it is thought that this qualification will make little difference.

B.3.3 Extension of time

The architect is not involved at all in granting extensions of time. The duty is the contractor's alone. This apart, the procedure is the same as under NSC/C.

B.3.4 Time period

The time period within which the contractor must respond to the sub-contractor's notice is 16 weeks, as opposed to 12 weeks in NSC/C; this time limit runs from receipt of notice and what the contractor considers to be 'reasonably sufficient' particulars and estimate.

---

[704] *Pigott Foundations* v. *Shepherd Construction* (1996) 67 BLR 48

B.3.5 Omissions

Omission directions given by the contractor can be taken into account.

B.3.6 Best endeavours

In clause 11.8, the operation of the extension of time provisions is made subject to the proviso that the sub-contractor shall 'use constantly his best endeavours to prevent delay ... and the Sub-Contractor shall do all that may reasonably be required to the satisfaction of the *Architect and the Contractor to proceed with* the Sub-Contract Works'. Why the architect should have been introduced at this point is not clear; but the dual obligation is of no practical significance because the architect has no right to require anything directly from a domestic sub-contractor.

B.3.7 Completion date

Clause 11.9 is vital; it parallels clause 25.3.6 of JCT 80 and means that, no matter how much work is omitted, the sub-contractor is always entitled to his original sub-contract completion date.

These points apart, the provision operates in the same way as the corresponding clause in NSC/C.

B.4 Direct loss and/or expense claims under DOM/1

This matter is dealt with in clause 13 and is the matching provision to NSC/C clauses 4.38 to 4.41. The text reads as follows:

13 Matters affecting regular progress – direct loss and/or expense – Contractor's and Sub-Contractor's rights

13.1 If due to deferment of giving to the Contractor possession of the site of the Works where it is stated in the Appendix part 1 Section B that clause 23.1.2 of the Main Contract Conditions applies to the Main Contract or the regular progress of the Sub-Contract Works is materially affected by any act, omission or default of the Contractor, his servants or agents, or any sub-contractor, his servants or agents (other than the Sub-Contractor, his servants or agents), or is materially affected by any one or more of the Relevant Matters referred to in clause 13.3 and if the Sub-Contractor shall within a reasonable time of such material effect becoming apparent make written application to the Contractor, the agreed amount of any direct loss and/or expense thereby caused to the Sub-Contractor shall be recoverable from the Contractor as a debt. Provided always that:

.1   the Sub-contractor's application shall be made as soon as it has become, or should reasonably have become, apparent to him that the regular progress of the Sub-Contract Works or of any part thereof has been or is likely to be affected as aforesaid; and

.2   the Sub-Contractor shall submit to the Contractor such information in support of his application as is reasonably necessary to show that the regular progress of the Sub-Contract Works or any part thereof has been or is likely to be affected as aforesaid; and

.3   the Sub-contractor shall submit to the Contractor such details of such loss and/or expense as the Contractor requests in order reasonably to enable that direct loss and/or expense as aforesaid to be agreed.

13.2   If, and to the extent that, it is necessary for the agreement of any direct loss and/or expense applied for under clause 13.1, the Contractor shall state in writing to the Sub-contractor what extension of time, if any, has been made under clause 11 in respect of the Relevant Events referred to in clause 11.10.5.1 (so far as that clause refers to clauses 2.3, 13.2, 13.3 and 23.2 of the Main Contract Conditions) and in clauses 11.10.5.2, 11.10.6, 11.10.8 and 11.10.12.

13.3   The following are the Relevant Matters referred to in clause 13.1:

.1   the Contractor, or the Sub-Contractor through the Contractor, not having received in due time necessary instructions, drawings, details or levels from the Architect for which the Contractor or the Sub-Contractor through the Contractor specifically applied in writing provided that such application was made on a date which having regard to the Completion Date or the period or periods for completion of the Sub-Contract Works was neither unreasonably distant from nor unreasonably close to the date on which it was necessary for the Contractor or the Sub-Contractor to receive the same;

.2   the opening up for inspection of any work covered up or the testing of any work, materials or goods in accordance with clause 8.3 of the Main Contract Conditions and/or clause 4.3.1 (including making good in consequence of such opening up or testing), unless the inspection or test showed that the work, materials or goods were not in accordance with the Main Contract or the Sub-Contract as the case may be:

.3   any discrepancy in or divergence between the Contract Drawings and/or the Contract Bills and/or the Numbered Documents;

.4   the execution of work not forming part of the Main Contract by the Employer himself or by persons employed or otherwise engaged by the Employer as referred to in clause 29 of the Main Contract Conditions or the failure to execute such work or the supply by the Employer of materials and goods which the Employer has agreed to provide for the Works or the failure so to supply;

.5   Architect's instructions issued in regard to the postponement of any work to be executed under the provisions of the Main Contract or the Sub-Contract, which shall be deemed to include Contractor's directions issued under clause 4 in respect of such matters;

.6   failure of the Employer to give in due time ingress to or egress from the site of the Works, or any part thereof through or over any land, buildings, way or passage adjoining or connected with the site and in the possession and control of the Employer, in accordance with the Contract Bills and/or the Contract drawings, after receipt by the Architect of such notice, if any, as the Contractor is required to give or failure of the Employer to give such ingress or egress as otherwise agreed between the Architect and the Contractor;

.7   Architect's instructions issued under clause 13.2 of the Main Contract Conditions requiring a Variation or under clause 13.3 of the Main Contract Conditions in regard to the expenditure of provisional sums (other than work to which clause 13.4.2 of the Main Contract Conditions refers), which shall be deemed to include Contractor's directions issued under clause 4 which pass on such instructions.

13.4   If the regular progress of the Works is materially affected by any act, omission or default of the Sub-Contractor, his servants or agents, and if the Contractor shall within a reasonable time of such material effect becoming apparent make written application to the Sub-Contractor, the agreed amount of any direct loss and/or expense thereby caused to the Contractor may be deducted from any monies due or to become due to the Sub-Contractor or may be recoverable from the Sub-Contractor as a debt. Provided always that:

.1   the Contractor's application shall be made as soon as it has become, or should reasonably have become, apparent to him that the regular progress of the Works or of any part thereof has been or is likely to be affected as aforesaid; and

.2   the Contractor shall submit to the Sub-Contractor such information in support of his application as is reasonably necessary to show that the regular progress of the Works or of any part thereof has been or is likely to be affected as aforesaid; and

.3   the Contractor shall submit to the Sub-Contractor such details of such loss and/or expense as the Sub-Contractor requests in order reasonably to enable the ascertainment and agreement of that direct loss and/or expense as aforesaid.

13.5   The provisions of clause 13 are without prejudice to any other rights or remedies which the Contractor or Sub-Contractor may possess.

B.5 Commentary

B.5.1 Sub-contractor's claims

Again, the commentary on NSC/C clauses 4.38 to 4.41 is relevant, as is the discussion of the position under clause 26 of JCT 80[705].

---

[705] See Chapter 12, section 12.2.

Here, however, the draftsman has not slavishly followed the wording of the similar clauses in other forms, although it seems to have the same effect.

The text should be compared and contrasted with NSC/C and the following points should be noted:

(1) Clause 13.1: The wording indicates that compliance with the provisions as to written notice is a condition precedent. The sub-contractor must make written application to the main contractor 'within a reasonable time' of the material effect on progress becoming apparent. Proviso .1 adjusts the requirement so that the sub-contractor must make application 'as soon as it has become, or should reasonably have become apparent'. The timescale in the proviso is stricter than that in the introductory part of clause 13.1 and, in our view, it would be enforceable.

(2) The sub-contractor's obligation is to 'submit...such information in support of his application as is reasonably necessary to show that...regular progress...has been or is likely to be affected ...'.

(3) He is also to submit 'such details of such loss and/or expense as the Contractor requests in order reasonably to enable that direct loss and/or expense...to be agreed', i.e. moneyed-out claims backed up by supporting evidence.

(4) Clause 13.2, like JCT 80, clause 26.3, appears to link extensions of time to direct loss and/or expense claims and the list of relevant events should be noted.

(5) Clause 13.1 refers to 'Relevant Matters', i.e. grounds giving rise to a potential claim. Apart from acts, omissions, or defaults of the main contractor or those for whom he is responsible in law, the 'Relevant Matters' are listed in clause 13.3.

B.5.2 Main contractor's claims

Clause 13.4 deals with claims by the main contractor against the sub-contractor in respect of 'disturbance of regular progress' of the main contract works. Unlike the corresponding provision in NSC/C (clause 4.40) sub-sub-contractors are not included. The contractor is required to make written application to the sub-contractor (and not merely to give written notice): the main contractor's right to claim is subject to the same three conditions as is the sub-contractor's right to claim against him.

Clause 13.5 preserves the common law rights of both contractor and sub-contractor.

The contractor's right to set-off is dealt with by clause 23, and the adjudication provision is in clause 24[706].

---

[706] Subsequently amended by the Housing Grants, Construction and Regeneration Act 1996.

Table of cases

The following abbreviations of reports are used:

AC, App Cas – Law reports, Appeal Cases
All ER – All England Law Reports
ALJR – Australian Law Journal Reports
ALR – Australian Law Reports
APCLR – Asia-Pacific Construction Law Reports
BCL – Building and Construction Law
BCLC – Butterworth's Company Law Cases
BLM – Building Law Monthly
BLR – Building Law Reports
Burr – Burrow's King's Bench Reports
Ch App – Chancery Appeal Cases
Ch, Ch D – Law Reports, Chancery Cases/Division
CILL – Construction Industry Law Letter
CLC – Commercial Law Cases
CLD – Construction Law Digest
CLR – Commercial Law Reports (Canada)
Com Cas, Comm – Commercial Cases
Con LR – Construction Law Reports
Const LJ – Construction Law Journal
CP – Common Pleas
Ct Cl – Court of Claims Reports
DLR – Dominion Law Reports
EG – Estates Gazette Cases
Ex, Exch, Ex D – Law Reports, Exchequer Cases/Division
FSR – Fleet Street Reports
H & C – Hurlstone and Coltman's Exchequer Reports
H & N – Hurlstone & Norman's Exchequer Reports
JP – Justice of the Peace & Local Government Review
KB – Law Reports, King's Bench
LJCP – Law Journal Reports, Common Pleas
LJQB – Law Journal Queen's Bench
Lloyds Rep – Lloyd's Reports
LT – Law Times Reports
M & W – Meeson and Welsby's Exchequer Reports
NE – North Eastern Reports
NSWLR – New South Wales Law Reports
NY – New York Court of Appeal
NZLR – New Zealand Law Reports
P & CR – Planning and Compensation Reports
PD – Law Reports, Probate Division
QB – Law Reports, Queen's Bench

Table of standard form contract clauses

Index